METHODS IN MOLECULAR BIOLOGY

For further volumes:
http://www.springer.com/series/7651

SH2 Domains

Methods and Protocols

Edited by

Kazuya Machida

Department of Genetics and Genome Sciences, University of Connecticut School of Medicine, Farmington, CT, USA

Bernard A. Liu

Broad Institute of MIT and Harvard, Cambridge, MA, USA

Humana Press

Editors
Kazuya Machida
Department of Genetics and Genome Sciences
University of Connecticut School of Medicine
Farmington, CT, USA

Bernard A. Liu
Broad Institute of MIT and Harvard
Cambridge, MA, USA

ISSN 1064-3745 ISSN 1940-6029 (electronic)
Methods in Molecular Biology
ISBN 978-1-4939-6760-5 ISBN 978-1-4939-6762-9 (eBook)
DOI 10.1007/978-1-4939-6762-9

Library of Congress Control Number: 2016961345

Printed on acid-free paper

This Humana Press imprint is published by Springer Nature
The registered company is Springer Science+Business Media LLC
The registered company address is: 233 Spring Street, New York, NY 10013, U.S.A.

Preface

Since the serendipitous discovery of the Src homology 2 (SH2) domain by Tony Pawson and colleagues, our understanding of the molecular recognition and posttranslational modifications has greatly advanced. The SH2 domain lies at the heart of phosphotyrosine signaling, coordinating signaling events downstream of receptor tyrosine kinases (RTKs), adaptors, and scaffolds. Over a hundred SH2 domains are present in mammals, each having a unique specificity which determines its interactions with multiple binding partners. Over the past two decades, there has been a tremendous advancement in technologies and an array of techniques available for studying SH2 domains and phosphotyrosine signaling. As many of these techniques and protocols are becoming standard laboratory tools, there was a need for a single source that would provide an in-depth introduction and analysis of the technologies being used—hence this handbook. This book is primarily intended as a methodological reference for computational biologists, biochemists, structural biologists, cell biologists, pathologists, and people who are just embarking on SH2 domains or phosphotyrosine signaling. More broadly, the people who are looking to investigate how protein interaction domain mediate specificity in signaling systems may also find these methodologies informative.

Organization of the book: Part I (Chaps. 1–2) outlines the history of the SH2 domain, technology development, and key concepts of cell signaling; Part II (Chaps. 3–7) focuses on computational approaches and tools for identification, classification, and prediction of SH2 domains and their interactions; Part III (Chaps. 8–13) details various approaches for preparation of the SH2 domains as experimental tools; Part IV (Chaps. 14–18) presents methods for structural analysis and conventional binding assays using SH2 domains; Part V (Chaps. 19–27) describes high-throughput and proteomics approaches analyzing SH2-mediated interactions; and finally, Part VI (Chaps. 28–31) covers applications of SH2 domain to functional and imaging analyses. Together, these protocols include some useful adaptations and applications that may have significance for those in the wider research community who are already familiar with protein interaction domains. See Chap. 1 for an overview of SH2 domains and details regarding chapters in this book.

Dedication to Tony Pawson (1952–2013)

We wish to dedicate this volume to the memory of Tony Pawson, who was a scientific pioneer and a truly outstanding biologist. In his lab at the Lunenfeld-Tanenbaum Research Institute at Mt. Sinai Hospital in Toronto, Canada, Tony made groundbreaking contributions in biology where he pioneered the detailed mechanisms of protein-protein interactions and phosphotyrosine signaling. Tony and colleagues first demonstrated that a region outside the catalytic kinase region could regulate the kinase activity and transformation in cells. His pioneering work eventually led to the identification of the SH2 domain and furthermore its fundamental role in phosphotyrosine signaling. His work was characterized by his remarkable scientific insight by applying new technologies and innovative approaches toward tackling complex biological questions.

Beyond science, Tony was always a kind and charming individual with diverse interests from fly-fishing to running. Interacting with him both scientifically and personally was always inspiring and a great pleasure. It was a privilege to have him as both a mentor and a leader. Tony's loss has left a hole in the field, but his memories live on in his research, his students, postdocs, and his many amazing colleagues.

Farmington, CT, USA *Kazuya Machida*
Cambridge, MA, USA *Bernard A. Liu*

Contents

Contributors

AKIRA ASAI • *Center for Drug Discovery, Graduate School of Pharmaceutical Sciences, University of Shizuoka, Shizuoka, Japan*

YUMI ASANO • *Tamano Technology Center, Research and Development Headquarters, Mitsui Engineering & Shipbuilding Co., Ltd., Tamano, Japan*

JEFFREY J. BABON • *The Walter and Eliza Hall Institute of Medical Research, Melbourne, VIC, Australia; Department of Medical Biology, University of Melbourne, Melbourne, VIC, Australia*

ROLF BACKOFEN • *Bioinformatics Group, Department of Computer Science, University of Freiburg, Freiburg, Germany; Centre for Biological Signaling Studies (BIOSS), University of Freiburg, Freiburg, Germany*

SOPHIA BUHS • *Research Institute Children's Cancer Center and Department of Pediatric Hematology and Oncology, University Medical Center Hamburg-Eppendorf, Hamburg, Germany*

LIEVEN BUTS • *Jean Jeener NMR Center, Structural Biology Brussels, Vrije Universiteit Brussel, Brussel, Belgium; Structural Biology Research Center, VIB, Brussels, Belgium*

BIZHU CHU • *Department of Chemistry, South University of Science and Technology of China, Shenzhen, People's Republic of China*

MARK F. CIACCIO • *Independent Biotechnology Professional, Chicago, IL, USA; The Ben May Department of Cancer Research, The Institute for Genomics and Systems Biology, University of Chicago, Chicago, IL, USA*

BRETT W. ENGELMANN • *Department of Human Genetics, The University of Chicago, Chicago, IL, USA*

MARI FUJIOKA, • *Department of Cell Physiology, Hokkaido University Graduate School of Medicine, Kita-ku, Sapporo, Japan*

ABEL GARCIA-PINO • *Jean Jeener NMR Center, Structural Biology Brussels, Vrije Universiteit Brussel, Brussel, Belgium; Structural Biology Research Center, VIB, Brussels, Belgium*

HELWE GERULL • *Research Institute Children's Cancer Center and Department of Pediatric Hematology and Oncology, University Medical Center Hamburg-Eppendorf, Hamburg, Germany*

HAIMING HUANG • *Banting and Best Department of Medical Research and Department of Molecular Genetics, University of Toronto, Toronto, ON, Canada*

RADU HUCULECI • *Jean Jeener NMR Center, Structural Biology Brussels, Vrije Universiteit Brussel, Brussel, Belgium; Structural Biology Research Center, VIB, Brussels, Belgium*

KARL JABLONOWSKI • *Division of Emergency Medicine, Department of Medicine, University of Washington, Seattle, WA, USA; Division of Biomedical and Health Informatics, Department of Biomedical Informatics and Medical Education, University of Washington, Seattle, WA, USA*

JOSHUA A. JADWIN • *Raymond and Beverly Sackler Laboratory of Genetics and Molecular Medicine, Department of Genetics and Genome Sciences, University of Connecticut School of Medicine, Farmington, CT, USA*
RICHARD B. JONES • *Independent Biotechnology Professional, Moscow, ID, USA; The Ben May Department of Cancer Research and the Institute for Genomics and Systems Biology, University of Chicago, Chicago, IL, USA*
TOMONORI KANEKO • *Department of Biochemistry and Siebens-Drake Medical Research Institute, Schulich School of Medicine and Dentistry, University of Western Ontario, London, ON, Canada*
MI KE • *Department of Chemistry, South University of Science and Technology of China, Shenzhen, People's Republic of China*
FABIEN KIEKEN • *Jean Jeener NMR Center, Structural Biology Brussels, Vrije Universiteit Brussel, Brussel, Belgium; Structural Biology Research Center, VIB, Brussels, Belgium*
KOUSIK KUNDU • *Department of Human Genetics, The Wellcome Trust Sanger Institute, Hinxton, Cambridge, UK; Department of Haematology, University of Cambridge, Cambridge, UK; Bioinformatics Group, Department of Computer Science, University of Freiburg, Freiburg, Germany*
ARTEM LAKTYUSHIN • *The Walter and Eliza Hall Institute of Medical Research, Melbourne, VIC, Australia; Department of Medical Biology, University of Melbourne, Melbourne, VIC, Australia*
TOM LENAERTS • *MLG, Département d'Informatique, Université Libre de Bruxelles, Brussels, Belgium; AI-lab, Vakgroep Computerwetenschappen, Vrije Universiteit Brussel, Brussels, Belgium; Interuniversity Institute of Bioinformatics Brussels (IB^2), ULB-VUB, Brussels, Belgium*
LEI LI • *Department of Biochemistry, Schulich School of Medicine and Dentistry, Western University, London, ON, Canada; Medical College of Qingdao University, Qingdao, China*
SHAWN S.C. LI • *Department of Biochemistry and Siebens-Drake Medical Research Institute, Schulich School of Medicine and Dentistry, University of Western Ontario, London, ON, Canada*
NICHOLAS P.D. LIAU • *The Walter and Eliza Hall Institute of Medical Research, Melbourne, VIC, Australia; Department of Medical Biology, University of Melbourne, Melbourne, VIC, Australia*
LIN LIN • *Department of Chemistry, South University of Science and Technology of China, Shenzhen, People's Republic of China; Materials Characterization and Preparation Center, South University of Science and Technology of China, Shenzhen, People's Republic of China*
BERNARD A. LIU • *Broad Institute of MIT and Harvard, Cambridge, MA, USA*
HUADONG LIU • *Department of Biochemistry, Siebens-Drake Medical Research Institute, Schulich School of Medicine and Dentistry, University of Western Ontario, London, ON, Canada*
KAZUYA MACHIDA • *Raymond and Beverly Sackler Laboratory of Genetics and Molecular Medicine, Department of Genetics and Genome Sciences, University of Connecticut School of Medicine, Farmington, CT, USA*

BRUCE J. MAYER • *Raymond and Beverly Sackler Laboratory of Genetics and Molecular Medicine, Department of Genetics and Genome Sciences, University of Connecticut School of Medicine, Farmington, CT, USA*

MARISSA A. MCKERCHER • *Department of Chemistry and Biochemistry, University of Colorado, Boulder, CO, USA*

SHIGEYUKI NAKADA • *Tamano Technology Center, Research and Development Headquarters, Mitsui Engineering & Shipbuilding Co., Ltd., Tamano, Japan*

KHONG Y. NG • *Raymond and Beverly Sackler Laboratory of Genetics and Molecular Medicine, Department of Genetics and Genome Sciences, University of Connecticut School of Medicine, Farmington, CT, USA*

PETER NOLLAU • *Research Institute Children's Cancer Center and Department of Pediatric Hematology and Oncology, University Medical Center Hamburg-Eppendorf, Hamburg, Germany*

NICO VAN NULAND • *Jean Jeener NMR Center, Structural Biology Brussels, Vrije Universiteit Brussel, Brussel, Belgium; Structural Biology Research Center, VIB, Brussel, Belgium*

MARI OGIUE-IKEDA • *Raymond and Beverly Sackler Laboratory of Genetics and Molecular Medicine, Department of Genetics and Genome Sciences, University of Connecticut School of Medicine, Farmington, CT, USA*

DONGMYUNG OH • *Cell Analysis and Modeling, University of Connecticut School of Medicine, Farmington, CT, USA; Mechanobiology Institute, National University of Singapore, Singapore, Singapore*

YUSUKE OHBA • *Department of Cell Physiology, Hokkaido University Graduate School of Medicine, Kita-ku, Sapporo, Japan*

SACHDEV S. SIDHU • *Banting and Best Department of Medical Research and Department of Molecular Genetics, University of Toronto, Toronto, ON, Canada*

RADOSLAW M. SOBOTA • *Systems Structural Biology Group, Institute of Molecular and Cell Biology (IMCB), Agency of Science, Technology and Research (A*STAR), Singapore, Singapore*

KAZUYUKI TAKAKUMA • *Center for Drug Discovery, Graduate School of Pharmaceutical Sciences, University of Shizuoka, Shizuoka, Japan; Innovative Research Division, Mitsubishi Tanabe Pharma Corporation, Saitama, Japan*

RUIJUN TIAN • *Department of Chemistry, South University of Science and Technology of China, Shenzhen, People's Republic of China; Shenzhen Key Laboratory of Cell Microenvironment, South University of Science and Technology of China, Shenzhen, People's Republic of China*

COURTNEY VOSS • *Department of Biochemistry, Siebens-Drake Medical Research Institute, Schulich School of Medicine and Dentistry, University of Western Ontario, London, ON, Canada*

DEBORAH S. WUTTKE • *Department of Chemistry and Biochemistry, University of Colorado, Boulder, CO, USA*

Part I

Overview

Chapter 1

Introduction: History of SH2 Domains and Their Applications

Bernard A. Liu and Kazuya Machida

Abstract

The Src Homology 2 (SH2) domain is the prototypical protein interaction module that lies at the heart of phosphotyrosine signaling. Since its serendipitous discovery, there has been a tremendous advancement in technologies and an array of techniques available for studying SH2 domains and phosphotyrosine signaling. In this chapter, we provide a glimpse of the history of SH2 domains and describe many of the tools and techniques that have been developed along the way and discuss future directions for SH2 domain studies. We highlight the gist of each chapter in this volume in the context of: the structural biology and phosphotyrosine binding; characterizing SH2 specificity and generating prediction models; systems biology and proteomics; SH2 domains in signal transduction; and SH2 domains in disease, diagnostics, and therapeutics. Many of the individual chapters provide an in-depth approach that will allow scientists to interrogate the function and role of SH2 domains.

Key words SH2 domain, SH2 domain discovery, High-throughput screening, Phosphoproteomics, pTyr peptides, SPOT, SH2 specificity, SH2 binding assays

1 Introduction

The Src Homology 2 (SH2) domain is the major protein interaction module that is central to phosphotyrosine signaling. Since its original discovery as a protein interaction domain involved in regulating the catalytic activity of the viral oncogene v-Fps/Fes, many other protein interaction domains have been discovered and shown to play an important role in assembling protein complexes mediated through protein–protein interactions, and thus an essential component in signal transduction. Over the last 25 years, particularly in the last 10 years, there has been tremendous advancement in technologies and an array of techniques available for studying SH2 domains and phosphotyrosine signaling. With over 120 SH2 domains in the human genome, approaches can range from single SH2 domain studies (for example X-ray crystallography) to large high-throughput approaches (protein arrays, peptide array, phage

Kazuya Machida and Bernard A. Liu (eds.), *SH2 Domains: Methods and Protocols*, Methods in Molecular Biology, vol. 1555,
DOI 10.1007/978-1-4939-6762-9_1,

display). As many of these techniques and protocols are well on their way to becoming standard laboratory tools, there was a need for a single source that would provide an in-depth introduction and analysis of the technologies being used. And who best to describe these algorithms and their nuances than the people who developed these approaches—hence this handbook. These protocols include some useful adaptations and applications that may have significance for those in the wider research community who are already familiar with protein interaction domains. Many of these chapters will provide an in-depth approach that will allow scientists to interrogate the function and role of SH2 domains. Each individual chapter in this volume contains complete experimental detail for the protocols described so that even a newcomer should be able to perform the techniques. Here we will provide a glimpse of the history of SH2 domains and describe many of the tools and techniques that have been developed along the way and discuss future directions for SH2 domain studies.

1.1 Discovery and Identification of SH2 Domains

The SH2 domain was initially discovered by Tony Pawson and colleagues while studying the v-Fps/Fes oncoprotein from the Fujinami sarcoma virus and its role during cellular transformation [1]. In this elegant experiment, Tony and colleagues introduced in-frame dipeptide motif insertional mutagenesis to disrupt the folded structure of the cytoplasmic oncoprotein v-Fps. The inserts that were found impaired kinase activity and resulted in the loss of the transforming ability of v-Fps. This region lay amino-terminal to the start of the kinase domain [1–3]. Analysis of this amino-terminal noncatalytic region consisting of ~100 amino acids was observed to share protein sequence homology to Src-related kinases in numerous eukaryotic organisms and subsequently named the SH2 domain [1, 4]. This led other researchers to the identification of numerous other catalytic and non-catalytic proteins that contained SH2 domains particularly among adaptors and signaling scaffolds [5, 6]. Shortly after this initial discovery, SH2 domains were later shown to share the capacity to bind activated RTKs and determined to be the element that coupled phosphorylated RTKs to their downstream targets [7–11]. Bruce Mayer outlines three key concepts of cell signaling that can be traced back to the early seminal studies of SH2 domains (*see* Chapter 2).

After the discovery of SH2 domains, numerous other protein interactions domains were later discovered including the SH3, PTB, PDZ, and many others. With more and more proteins containing protein domains, methods such as protein sequence alignments were applied for identifying these domains. Following this overview, two chapters in this book (Part II: Computational Approaches) deal with the various computational approaches developed that can be used to identify SH2 domains. The chapter by Karl Jablonowski lies central to modular protein interactions where he describes the use of hidden Markov models (HMMs) and protein sequence alignments to search and identify SH2 domains throughout proteome databases (*see* Chapter 3). The application of this approach identified 122 SH2 domains present in 112 proteins in the human genome [12, 13] (a complete list of human SH2 domains can be found in Table 1). Today many web resources utilize these models for predicting protein domains including a few such as SMART, Pfam, and Conserved Domains Database (CDD) (*see* Chapter 4). The compendium of SH2 domains in mouse and human has led to further explorations and understanding of phosphotyrosine signaling at a systems-level.

The discovery of SH2 domains in early single celled eukaryotes including *Monosiga brevicollis* suggested phosphotyrosine signaling through the advent of SH2 domains might have paved the path towards multicellularity and metazoan development [14–16]. Not only does searching for SH2 domains in lower eukaryotes help us understand the evolution of metazoans but also may help gauge our understanding of the evolutionary trajectory of SH2 domains throughout the tree of life. This book includes a chapter that examines the classification of SH2 domain families and describes computational methods for tracing SH2 domain family back through lower eukaryotes to understand how specificity and SH2 domain proteins may have evolved for diversification and expansion of phosphotyrosine signaling (*see* Chapter 4). Thus, the discovery of the SH2 domain by Tony Pawson has led to a greater understanding of protein domains and how critical they are in nearly all biological processes. This book is dedicated to his immeasurable scientific contribution to this field and his everlasting impact on our research.

1.2 Structural Biology and Phosphotyrosine Binding

The first structures solved by John Kuriyan and colleagues of the SH2 domain from v-Src revealed a central antiparallel β-sheet structure flanked by two α-helices [17, 18]. The β-sheet, intervening loops, and one of the helices in the SH2 domain mediate phosphopeptide binding. The SH2 domain structure shows a high degree of structural homology consisting of a core fold of three or four β strands, which make up an antiparallel β-sheet, flanked by two α-helices. Today over 300 structures have been deposited into the protein databank covering about 50% of all human SH2 domains [12].

An essential technique for solving structures and performing any biochemical assays with SH2 domains involves recombinant protein expression. Multiple chapters in this book discuss the technique of cloning and expression of SH2 domains using various systems and tags (*see* Chapter 8). While close to half of all the human SH2 domains structures have been solved by either X-ray crystallography or by Nuclear Magnetic Resonance (NMR), solving structures for the other half remain a challenge due to lack

Table 1
List of human SH2 domains

Name	Gene ID	Domains	Alternate names	Genomic location	Mouse homolog	Gene ID	Alternate symbols	Cytogenetic location	Genetic location
ABL1	25	1	ABL, JTK7, p150, c-ABL	9q34.1	Abl1	11350	Abl, c-Abl, E430008G22Rik	2 B	2 21.0cM
ABL2	27	1	ARG, ABLL	1q24-q25	Abl2	11352	Arg, Abll	1 G3	1 82.1 cM
APS	10603	1		7q22	Aps	23921	Aps	5 G1	
BCAR3	8412	1	SH2D3B, NSP2, KIAA0554	1p22.1	Bcar3	29815	AND-34	3 G3	
BKS	55620	1	STAP2, FLJ20234	19p13.3	Stap2	106766	AW049765	17 C	
BLK	640	1	MGC10442	8q23-p22	Blk	12143		14 C3	14 28.0 cM
BLNK	29760	1	Ly57, SLP65, BLNK-s, SLP-65, B-cell linker	10q23.2-q23.33	Blnk	17060	BCA, Bca, BASH, BLNK, Ly57, Ly-57, Lyw-57, SLP-65	19 C3	19 31.0 cM
BMX	660	1	ETK, PSCTK2, PSCTK3	Xp22.2	Bmx	12169	Etk, Etk/Bmx	X F5	
BRDG1	26228	1	STAP-1	4q13.2	Brdg1	56792	STAP-1, AI586015	5 E1	
BRK	5753	1	PTK6	20q13.3	Ptk6	20459	BRK, Sik, tks, Tksk	2 H4	2 110.0 cM
BTK	695	1	AT, ATK, BPK, XLA, IMD1, AGMX1, PSCTK1	Xq21.33-q22	Btk	12229	xid, AI528679	X E3	X 51.0 cM
CBL	867	1	CBL2, c-CBL, RNF55, Cas-Br-M	11q23.3	Cbl	12402	cbl, Cbl-2, c-Cbl, 4732447J05Rik	9 A5.2	9 26.0 cM
CBLB	868	1	RNF56	3q13.11	Cblb	208650		16 A1	
CBLC	23624	1	CBL-3, CBL-SL, RNF57	19q13.2	Cblc	80794	Cbl3, 2310076I21Rik, 2310079L19Rik	7 A2	
CHN1	1123	1	CHN, ARHGAP2, RHOGAP2, n-chimerin, chimerin	2q31-q32.1	Chn1	108699	1700026N20Rik, 1700112L09Rik, 2900046J01Rik, ARHGAP2	2 C3	
CHN2	1124	1	ARHGAP3, RHOGAP3, beta chimerin, chimerin 2	7p15.3	Chn2	69993	1700026N20Rik, Chimerin (chimaerin) 2, Bch, ARHGAP3	6 B3	

CISH	1154	1	CIS, G18, SOCS, CIS-1	3p21.3	Cish	12700	Cis, F17, F23, CIS1	9 F1	9 59.0 cM
CRK	1398	1	CRKII	17p13.3	Crk	12928	Crko, Crk-I, Crk-II, Crk3, Crkol, CrkIII, Crk-III	11 A-D	11 44.15 cM
CRKL	1399	1	Crk-like	22q11.21	Crkl	12929	Crkol, 1110025F07Rik	16 A3	
CSK	1445	1	CYL	15q23-q25	Csk	12988		9 B	9 32.0 cM
DAPP1	27071	1	BAM32, DKFZp667E0716	4q25-q27	Dapp1	26377	Bam32	3 G3	
FER	2241	1	TYK3, c-FER, p94-FER	5q21	Fert2	14158	Fer, Fert, C330004K01Rik	17 E1.1	11 46.0 cM
FES	2242	1	FPS, c-FES	15q26.1	Fes	14159	FPS, c-fes	7 D2	7 39.0 cM
FGR	2268	1	SRC2, c-fgr, p55c-fgr	1p36.2-p36.1	Fgr	14191		4 D2.3	4 64.6 cM
FRK	2444	1	GTK, RAK, PTK5	6q21-q22.3	Frk	14302	BSK, GTK, RAK, BSK/IYK	10 B1	
FYN	2534	1	SLK, SYN, MGC45350	6q21	Fyn	14360		10 B1	10 25.0 cM
GRAP	10750	1	MGC64880	17p11.2	Grap	71520	8430435N19Rik	11 B2	
GRAPL	400581	1		17p11.2					
GADS	9402	1	P38, GRAP2, GRID, GRPL, GrbX, Mona, GRB2L, GRBLG, Grf40, GRAP-2	22q13.2	Mona	17444	GRID, Gads	15 E2	
GRB2	2885	1	ASH, Grb3-3, EGFRBP-GRB2	17q24-q25	Grb2	14784	Ash	11 E2	11 75.0 cM
GRB7	2886	1		17q12	Grb7	14786		11 D	11 57.0 cM
GRB10	2887	1	RSS, IRBP, MEG1, GRB-IR, KIAA0207	7p12-p11.2	Grb10	14783	Meg1, 5730571D09Rik, 5730571D09Rik	11 A1	11 8.0 cM
GRB14	2888	1		2q22-q24	Grb14	50915		2 C1.3	
HCK	3055	1	JTK9	20q11-q12	Hck	15162	Bmk, Hck-1, MGC18625	2 H1	2 86.0 cM
HSH2D	84941	1	FLJ14886, HSH2, ALX	19p13.11	Hsh2d	209488	Alx, Hsh2	8 B3.3	
ITK	3702	1	EMT, LYK, PSCTK2	5q31-q32	Itk	16428	Emt, Tsk, Tcsk	11 B1.1	11 22.0 cM

(continued)

Table 1
(continued)

Name	Gene ID	Domains	Alternate names	Genomic location	Mouse homolog	Gene ID	Alternate symbols	Cytogenetic location	Genetic location
JAK1	3716	1	JAK1A, JAK1B, JAK-1	1p32.3-p31.3	Jak1	16451	MGC37919, C130039L05Rik	4 C6	4 46.3 cM
JAK2	3717	1	JAK-2	9p24	Jak2	16452	Fd17, AI504024	19 C1	19 24.0 cM
JAK3	3718	1	JAK-3, JAKL, LJAK, L-JAK	19p13.1	Jak3	16453		8 B3.3	8 33.0 cM
LCK	3932	1	LSK	1p34.3	Lck	16818	Hck-3, p56	4 D2.2	4 59.0 cM
LNK	10019	1		12q24	Lnk	16923		5 F	5 65.0 cM
LYN	4067	1	JTK8	8q13	Lyn	17096	Hck-2	4 A1	4 0.0 cM
MATK	4145	1	LSK, CHK, CTK, HYL, HHYLTK, MGC1708, MGC2101, DKFZp434N1212	19p13.3	Matk	17179	CHK, HYL, Ntk	10 C1	10 43.0 cM
MIST	116449	1	CLNK	4p16.1	Clnk	27278	MIST	5 B3	
NCK1	4690	1	NCK, NCKalpha, MGC12668	3q21	Nck1	17973	Nck	9 E4	
NCK2	8440	1	GRB4, NCKbeta, NCK-2	2q12	Nck2	17974	Grb4, NCKbeta	1 C1.1	
PIK3R1	5295	2	P85A, PIK3R1, GRB1	5q13.1	Pik3r1	18708	PI3K, p50alpha, p55alpha, p85alpha, C530050K14	13 D1	13 50.0 cM
PIK3R2	5296	2	P85B, PIK3R2	19q13.2-q13.4	Pik3r2	18709		8 B3.3	8 33.5 cM
PIK3R3	8503	2	P85G, PIK3R3, p55-GAMMA	1p34.1	Pik3r3	18710	p55pik	4 D1	
PLCG1	5335	2	PLC1, PLC148, PLC-II	20q12-q13.1	Plcg1	18803	Plc-1, Plcg-1, Plc-gamma1, Cded	2 H2	2 92.0 cM
PLCG2	5336	2	PLC-IV	16q24.1	Plcg2	234779	Plcg-2, MGC38590	8 E1	

PTPN11	5781	2	CFC, NS1, SHP2, SHP-2, BPTP3, PTP2C, PTP-1D, SH-PTP2, SH-PTP3, MGC14433, PRO1847	12q24	Ptpn11	19247	Syp, Shp2, PTP1D, PTP2C, SAP-2, SHP-2, SH-PTP2, SH-PTP3, 2700084A17Rik	5 F	
PTPN6	5777	2	HCP, HCPH, SHP-1, HPTP1C, PTP-1C, SHP-1 L, SH-PTP1	12p13	Ptpn6	15170	me, hcp, Hcph, Ptp1C, SHP-1, motheaten	6 F3	6 60.22 cM
RASA1	5921	2	RASA, GAP, RASA, RASGAP, p120GAP, PKWS, CMAVM	5q13.3	Rasa1	218397	Gap, Rasa, RasGAP, MGC7759	13 C3	
RIN1	9610	1		11q13.2	Rin1	225870		19 A	
RIN2	54453	1	RASSF4	20p11.22	Rin2	74030	RASSF4, 2010003K16Rik, 4632403N06Rik	2 G1	
RIN3	79890	1	FLJ11700, FLJ22439, DKFZp762H1613	14q32.12	Rin3	217835	6430500 K07	12 E	
SH2B	25970	1	SH2-B, DKFZP547G1110	16p11.2	Sh2bpsm1	20399	Irip, SH2-B, SH2-Bb	7 F3	7 61.0 cM
SH2D1A	4068	1	DSHP, LYP, SAP, XLP, EBVS, IMD5, XLPD, MTCP1	Xq25-q26	Sh2d1a	20400	SAP	X A5	
SH2D1B	117157	1	EAT2, EAT-2	1q23.3	Eat2	26904	EAT-2	1 H2	
SH2D1C[a]	N/A	1	N/A	N/A	Sh2d1c	545378	EAT-2B	1 H3	
SH2D2A	9047	1	VRAP, TSAD, TSAd, F2771	1q21	Sh2d2a	27371	Lad, RIBP, Ribp, TSAd	3 F1	3 42.6 cM
SH2D3A	10045	1	NSP1	19p13.3	N/A	N/A	N/A	N/A	N/A
SH2D3C	10044	1	NSP3, CHAT, PRO34088	9q34.11	Sh2d3c	27387	Chat, Nsp3, Shep1	2 B	
SH2D4A	63898	1	SH2A, FLJ20967	8p21.2	Sh2d4a	72281	2210402M20Rik, SH2A	8 C1	
SH2D4B	387694	1	FLJ41984, CAI14998	10q21.1	Sh2d4b	328381	A430109M18Rik	14 B	

(continued)

Table 1
(continued)

Name	Gene ID	Domains	Alternate names	Genomic location	Mouse homolog	Gene ID	Alternate symbols	Cytogenetic location	Genetic location
SH2D5	400745	1	LOC400745	1p36.12	SH2D5	230863	BC036961	4 D3	
SH2D7	646892	1		15q25.1	Sh2d7	244885	4933412E14Rik	9 A5.3	
SH3BP2	6452	1	CRPM, RES4-23, CRBM	4p16.3	Sh3bp2	24055	3BP2	5 B1	
SHB	6461	1	RP11-3J10.8	9p12-p11	Shb	230126		4 B1	
SHC1	6464	1	SHC, SHCA, SHC-1, p52SHC, p66SHC, FLJ26504	1q21	Shc1	20416	Shc, p66, ShcA	3 F1	3 45.0 cM
SHC2	25759	1	SLI, SCK, SHCB	19p13.3	Shc2	216148	SCK, Sli, ShcB, 6720466E06	10 C1	10 42.0 cM
SHC3	53358	1	NSHC, SHCC, N-Shc, Rai	9q22.1-q22.2	Shc3	20418	Rai, ShcC, N-Shc	13 A5	13 29.0 cM
SHC4	399694	1	MGC34023, RaLP, SHCD	15q21.1-q21.2	Shc4	271849	RaLP, Gm685, 9930029B02Rik, 6230417E10Rik	2 E5	
SHD	56961	1	LOC56961	19p13.3	Shd	20420		17 C	
SHE	126669	1	LOC126669	1q21.3	She	214547	LOC214547, 9430022A14	3 F1	
SHF	90525	1	LOC90525, hypothetical protein BC007586	15q21.1	Shf	435684	Shf, SHB like adaptor, LOC435684	2 F1	
SHIP1	3635	1	INPP5D, HCK, MGC104855, SHIP, SIP-145, hp51CN	2q36-q37	Inpp5d	16331	SHIP, 145 kDa, s-SHIP	1 C5	1 57.0 cM
SHIP2	3636	1	INPPL1	11q23	Inppl1	16332	51C, SHIP2	7 E1	
SLAP	6503	1	SLA1, SLAP1, SLA	8q22.3-qter	Sla	20491	Slap, Slap-1	15 D2	15 37.5 cM

SLAP2	84174	1	SLAP-2, SLA-2, FLJ21992, MGC49845, C20orf156	20q11.23	Sla2	77799	SLAP2, SLAP-2, A930009E21Rik	2 H1	
SLNK	284948	1	SH2D6, LOC284948, B cell linker protein, SH2 linker protein related to BLNK	2p11.2	Slnk	71130	4933424C13Rik, MGC118336	6 C3	
SLP76	3937	1	SLP-76, LCP2	5q33.1-qter	Lcp2	16822	SLP-76	11 A4	
SOCS1	8651	1	JAB, CIS1, SSI1, TIP3, CISH1, Cish1, SSI-1, SOCS-1	16p13.13	Socs1	12703	JAB, Cish1, Cish7, SOCS1, SSI-1	16 A1	
SOCS2	8835	1	CIS2, SSI2, Cish2, SSI-2, SOCS-2, STATI2	12q	Socs2	216233	hg, JAB, CIS2, Cish2, SSI-2, SOCS-2, 8030460 M17	10 C2	10 52.0 cM
SOCS3	9021	1	CIS3, Cish3, SSI-3, SOCS-3, MGC71791	17q25.3	Socs3	12702	CIS3, Cish3, EF-10, SSI-3, SOCS-3	11 E2	
SOCS4	122809	1	SOCS7, CIS4, STAT4, SSI4, DKFZp686J1568	14q22.2	Socs4	67296	SOCS7, 3110032M18Rik	14 B	
SOCS5	9655	1	CIS6, CISH6, Cish5, SOCS-5, KIAA0671	2p21	Socs5	56468	Cish5, SOCS-5, 1810018L08Rik	17 E4	
SOCS6	9306	1	CIS4, SSI4, Cish4, SOCS4, STAI4, STAT4, STATI4, HSPC060	18q22.2	Socs6	54607	CIS4, Cis4, SSI4, Cish4, STAI4, STAT4, SOCS-4, SOCS-4, STATI4, HSPC060, 5830401B18Rik	18 E4	
SOCS7	30837	1	NAP4, NAP-4, SOCS4, SOCS-7	17q12	Socs7	192157	Nap4, 2310063P06Rik	11 D	
SRC	6714	1	ASV, SRC1, c-src, p60-Src	20q12-q13	Src	20779		2 H1	2 89.0 cM
SRMS	6725	1	SRM, C20orf148, dJ697K14.1	20q13.33	Srms	20811	srm, A230069J08Rik	2 H4	2 102.0 cM
STAT1	6772	1	ISGF-3, STAT91	2q32.2	Stat1	20846	2010005J02Rik	1 C1.1	1 25.9 cM
STAT2	6773	1	P113, ISGF-3, STAT113, MGC59816	12q13.13	Stat2	20847	1600010G07Rik	10 D3	10 70.0 cM

(continued)

Table 1
(continued)

Name	Gene ID	Domains	Alternate names	Genomic location	Mouse homolog	Gene ID	Alternate symbols	Cytogenetic location	Genetic location
STAT3	6774	1	APRF, FLJ20882, MGC16063	17q21.31	Stat3	20848	Aprf, 1110034C02Rik	11 D	11 60.5 cM
STAT4	6775	1		2q32.2-q32.3	Stat4	20849		1 C1.1	1 25.9 cM
STAT5A	6776	1	MGF, STAT5	17q11.2	Stat5a	20850	Stat5	11 D	11 60.5 cM
STAT5B	6777	1	STAT5	17q11.2	Stat5b	20851		11 D	11 60.5 cM
STAT6	6778	1	STAT6B, STAT6C, D12S1644, IL-4-STAT	12q13	Stat6	20852		10 D3	10 70.0 cM
SUPT6H	6830	1	SPT6, SPT6H, emb-5, KIAA0162	17q11.2	Supt6h	20926	SPT6, mKIAA0162	11 B5	11 44.91 cM
SYK	6850	2		9q22	Syk	20963		13 B-C2	13 37.0 cM
TEC	7006	1	PSCTK4	4p12	Tec	21682		5 C3.2	5 41.0 cM
TENC1	23371	1	C1-TEN, KIAA1075, C1TEN, TNS2	12q13.13	Tenc1	209039	C1-ten	15 F2	
TNS1	7145	1	TNS, TENSIN, PRO0929, DKFZP434G162	2q35-q36	Tns1	21961	Tensin1, Tns1200014E20Rik, 1110018I21rik	1 C3	1 44.5 cM
TNS3	64759	1	TEM6, FLJ13732, TENS1, H_NH04I23.2	7p13-12.3	Tns3	319939	Tem6, Tensin3, Tens1, F830010I22Rik	11 A1	
TNS4	84951	1	CTEN , FLJ14950	17q21.2	Cten	217169	AU016405, 9930017A07Rik	11 D	
TXK	7294	1	RLK, TKL, BTKL, PSCTK5, PTK4	4p12	Txk	22165	Rlk, Btkl, PTK4	5 C3.2	5 40.0 cM
TYK2	7297	1	JTK1	19p13.2	Tyk2	54721	JTK1	9 A3	

VAV1	7409	1	VAV	19p13.2	Vav1	22324	Vav	17 D	17 32.7 cM
VAV2	7410	1	VAV-2	9q34.1	Vav2	22325	2810040F13Rik	2 A3	2 15.3 cM
VAV3	10451	1	VAV-3	1p13.3	Vav3	57257	MGC27838, A530094I06Rik	3 G1	
YES	7525	1	Yes, C-YES, c-yes, P61-YES, HST441, YES1	18p11.31-p11.21	Yes	22612		5 B1	5 18.2 cM
ZAP70	7535	2	SRK, STD, ZAP-70, TZK	2q12	Zap70	22637	Srk, 70 kDa, TZK, ZAP-70	1 A4-C1	1 20.5 cM

[a]Pseudogene

of soluble and stable protein expression for many SH2 domains including the STAT and SOCS families. In this book, several chapters will discuss new approaches to tackling the expression systems for some challenging SH2 domains in order to generate soluble protein for crystallography or NMR structure determination. Akira Asai and Kazuyuki Takakuma describe a method of generating a deletion mutant of STAT5B and STAT3 that is soluble and retains phosphotyrosine binding (*see* Chapter 9). Jeffrey Babon's group describes a method for generating functional SOCS family of SH2 domains by refolding unfolded protein back into its native state (*see* Chapter 10). For those SH2 domains that are insoluble in bacteria, baculovirus-insect cell expression systems can be an alternative to producing soluble and functionally active proteins (*see* Chapter 11). In addition, dimerization can pose as an issue during protein expression and purification as well as provide biological functions for particular SH2 domains like Fyn. Tom Lenaerts' group discusses expression and purification steps necessary for separating the monomeric Fyn SH2 from the dimeric forms (see Chapter 14). This section of the book provides multiple flavors for producing recombinant SH2 domain protein necessary not just for solving the structure but also for other applications discussed in the later sections of this book.

The fundamental element of SH2 domains is their ability to distinguish binding between phosphorylated and non-phosphorylated peptides. The specific recognition of phosphotyrosine involves amino–aromatic interactions between lysine and arginine side chains and the ring system in addition to hydrogen-bonding interactions with the negatively charged phosphate. A positively charged binding pocket utilizes a critical Arg residue in the β2 strand (within a very highly conserved FLVR motif; [19, 20] to bind the pTyr of target ligands. Mari Ikeda et al. describe approaches creating loss of function mutations and domain concatenation (*see* Chapter 12). Huang et al., describes protein engineering of the SH2 domain by mutating amino acid residues to generate SH2 domain superbinders that bind phosphotyrosine with a super high affinity (*see* Chapter 13).

The canonical mode of binding depends upon the residues surrounding the critical Arg, which directly determine the specificity of amino acids from position +1 to +6 (C-terminal of the pTyr) [21–23]. For example, binding of the preferred pYEEI motif involves regions on either side of the SH2 domain central β-sheet; between the β-sheet and the N-terminal αA helix is a deep binding pocket that accommodates the pTyr, while on the opposite side of the β-sheet, adjacent to the C-terminal αB helix, residues form a hydrophobic pocket into which the isoleucine chain extends [18]. Therefore the pTyr binding pocket of an SH2 domain is adjacent to the region that dictates binding specificity and determines whether or not the signaling effector is recruited to a given RTK or scaffold protein. This, in turn, determines which pathway(s) is subsequently activated downstream of the receptor (reviewed in [21–23]). For example, the SH2 domain of Grb2 preferentially binds pYXNX motifs (where X represents any of the natural 20 amino acids) present among its interaction partners including the Shc proteins. Grb2's preference for Asn at the +2 position is mediated by a Trp residue (W121) in the EF loop of the SH2 domain whereby mutation of this position (W121T) displays weak binding to pYXNX motifs [24]. The canonical mode of binding for SH2 domains has evolved to increase ligand selectivity and signal complexity (a list of SH2 domain specificities are provided in Table 2).

Table 2
Binding specificities of SH2 domains

SH2 domain	Binding motif	Reference
Abl1	[pY] [E/T/M] [N/E/D] [P/V/L] [P] [P] [pY] [E/D/M/V] [M/V/N] [V/L/P/I/A/T] [M/G] [pY] [D/E/S] [D/E] [P/N/D/E]	[22] [30] [58]
Abl2	[P/L/V/H] [P/M/L] [pY] [E/M/D/S/A] [M/V/E/I] [P/T/V/M] [M/L/V] [pY] [V] [N/Q]	[30] [58]
APS	[V/L] [I/P] [pY] [I/V/L] [N/L] [L/I/F] [L/I]	[30]
BCAR3	[L] [X] [pY] [S/G/D] [D] [X] [W/F]	[30]
BLK	[D/E] [X] [pY] [E/D] [E/N] [V/L] [D] [pY] [D/E/ϕ] [D/E/L] [P/I]	[30] [58]
Blnk	[pY] [D] [D] [pY] [D/Q] [D] [V]	[30] [85]
BMX	[P/E] [pY] [D/E] [N] [E/D]	[30]
Brdg1	[E/P] [pY] [S/E/D] [Y/T/L/F] [X] [I/L/F]	[30]
Brk	[N] [P] [pY] [D/E/T] [T/E/N/M] [V/T/P/A/Y] [D/M/E] [pY] [D/E] [X] [D/E/ϕ]	[30] [58]
Btk	[Q/V/E] [E] [pY] [D/E] [E] [V/T/E]	[30]
Cbl	[D/N] [X] [pY] [S/T] [X] [X] [P] [X] [D] [pY] [R] [X] [R] [A] [ϕ] [X] [N] [Q] [pY] [S/T]	[86] [87] [87]
Crk	[pY] [D/K/N] [H/F/R] [P/V/L] [N] [X] [pY] [M/D/N/K/S] [T/M/S] [[P/L/V] [R/M/A/S] [D] [X] [pY] [D] [V/L] [P] [P]	[22] [30] [58]

(continued)

Table 2
(continued)

SH2 domain	Binding motif	Reference
CrkL	[N/P/T] [P] [pY] [M/K/N/D/S] [T/K/M/R/S] [P/L/V] [R/A/K/M] [D] [X] [pY] [D] [ϕ] [P] [P/R]	[30] [58]
Csk	[pY] [T/A/S] [K/R/Q/N] [M/I/V/R] [pY] [A/S/T] [N] [V/P]	[28] [30]
Dapp1	[pY] [X] [X] [D/E/ϕ] [E]	[58]
Fer	[D/E] [E/P] [pY] [E] [N/E] [Y/E] [D] [D/E] [D/E] [pY] [D/G] [D/E] [ϕ]	[30] [58]
Fes	[pY] [E/Y/I] [N/M/E] [V/Y/L/E] [S/G/D/T] [pY] [E] [X] [V/I]	[28] [30]
Fgr	[N/P/I/Q] [E] [pY] [E/D] [N/E] [L/E] [E/D] [pY] [E/Y/D] [E/N/D] [I/V] [E] [P/D/E] [X] [pY] [D/E/G] [X] [D/E/ϕ] [Y]	[22, 30] [58]
Frk	[F/L] [pY] [E/Y/T/S] [E/D] [I/L/F/V] [D/E]	[30]
Fyn	[pY] [E/T] [E/D/Q] [I/V/M] [N/P] [Y/F] [pY] [E/D/Y] [N/E/T/M] [I/L/V/P] [D/E] [pY] [X] [D/[illegible]] [ϕ]	[22] [30] [58]
GADS	[P/V/L] [P/L/I/T] [pY] [V/E/S/A/I] [N] [E/V/M/L] [L/M/I/V] [pY] [V] [N]	[30] [58]
GRAP	[pY] [E/V/I/A] [N] [V/L/E/I] [M/E] [pY] [V/E] [N] [[illegible]]	[30] [58]
Grb2	[pY] [Q/Y/V] [N] [Y/Q/F] [V] [pY] [Q] [N] [W/F] [pY] [I/V] [N] [I/L/V] [V/P] [X] [pY] [V/I/M] [N] [M/V] [pY] [V/E] [N]	[88] [34] [31] [30] [58]
(SEM-5)	[pY] [L/V/I] [N] [V/P]	[22]
Grb7	[F/Y] [pY] [E/T/Y/S] [N] [I/L/V/P/T/Y/S] [pY] [D/Y/Q] [N/Y/T/M] [X] [N] [pY] [E] [N/Y]	[31] [30] [58]
Grb10	[F/Y] [pY] [E/T/Y/S] [N] [I/L/V/P/T/Y/S] [N/P/Q/T] [E/P/Y/D] [pY] [E/D] [N/T] [V/P] [D/E]	[31] [30]
Grb14	[pY] [E/D/S] [N/T/M] [V]	[30]
Hck	[N] [E/P] [pY] [E/D] [N] [P/V/I/L] [D/E] [D/P] [X] [pY] [D/E/G] [D/E/ϕ] [P/I/L]	[30] [58]
Hsh2d	[M] [pY] [E/N/D] [N] [X] [M]	[30]

(continued)

Table 2
(continued)

SH2 domain	Binding motif	Reference
Itk	[pY] [A/E/V] [Y/F/E/S/N/V] [P/F/I/H]	[89]
	[E] [pY] [E/D] [D/N] [M/V] [M]	[30]
	[pY] [ϕ] [X] [D/ϕ]	[58]
Lck	[pY] [E/T/Q] [E/D] [I/V/M]	[22]
	[V/F] [V/I] [pY] [E/V/Y/S/T] [E] [I/L/V/E/F] [E/Y/W/I/D]	[30]
	[pY] [D/E/G] [D/E] [P/L] [P]	[58]
Lyn	[T] [X] [pY] [E/D/T/S] [N/F] [I/L/A] [F/W/Y/L]	[30]
	[pY] [D/E/G] [D/E] [P] [P]	[58]
MATK	[I/V/M/L] [M/E] [pY] [Y/A/E/T] [M] [A/V/M/I] [M]	[30]
MIST	[pY] [ϕ] [ζ] [ϕ] [D/E] [ϕ]	[58]
Nck1	[pY] [D] [E] [P/D/V]	[22]
	[pY] [D/Y] [D/E/L/Y] [V]	[90]
	[P/V] [V/Y/I/L] [pY] [D/E/S] [E/N/M] [P/V/A] [S]	[30]
	[pY] [D/E] [E/L] [P/V]	[58]
Nck2	[P/D/E/T] [V/I/E] [pY] [E/D/S] [N/E/M/T] [V/P/A/T] [D]	[30]
PIK3R1_N	[pY] [M/I/V/E] [X] [M]	[22]
	[X] [H/N/M] [pY] [V/E/I/T] [N/M/E] [M/E]	[30]
	[pY] [V/D] [X] [I/M/V]	[58]
PIK3R1_C	[pY] [M/L/I] [X] [M]	[22]
	[pY] [E/V] [N] [M/L]	[30]
	[pY] [V/M/E] [N/T/M] [M]	[58]
PIK3R2_N	[N] [E/G] [pY] [V/E/I] [N/M/T/L] [M/Y/P] [S/D/G/M]	[30]
PIK3R2_C	[pY] [V/I/L/E] [N/I/M/L] [M/V/L/T]	[30]
PIK3R3_N	[D] [E] [pY] [V/I/E] [N/E/L/M] [M]	[30]
PIK3R3_C	[pY] [V/I/E/A] [N/I/M] [M/L/V/T] [M/G/A/T]	[30]
PLCG1_N	[pY] [V/I/L] [E/D] [L/I/V]	[22]
PLCG1_C	[pY] [V/I/L] [E/D] [P/V/I]	[22]
	[pY] [L/V/I/T] [L/M] [P/I/L]	[30]
	[pY] [ϕ] [X] [D/E]	[58]
PLCG2_C	[Y/F] [Y] [pY] [E/V/I] [Y/E] [E/D/F/Y] [Y/F/E/W]	[30]
PTPN6_N	[pY] [F] [X] [F/P/L/Y]	[28]
	[L] [Y/H] [pY] [M/F] [X] [F/M]	[91]
PTPN6_C	[V/I/L] [X] [pY] [A] [X] [L/V]	[91]
	[X] [X] [pY] [Y] [M] [K/R]	[91]
	[pY] [I/V/L] [E] [E/L/M] [E/F]	[30]

(continued)

Table 2
(continued)

SH2 domain	Binding motif	Reference
PTPN11_N	[pY] [I/V] [X] [V/I/L/P] [I/L/V/M] [X] [pY] [T/V/A] [X] [I/V/L/F] [L/V/H] [X] [pY] [V/L/I] [X] [L/M] [L/M] [pY] [ϕ] [X] [ϕ] [D/E] [ϕ]	[22] [92] [30] [58]
PTPN11_C	[I/L/V] [I/L/V] [I/F/V] [pY] [T/V] [I/L] [I/L/V/P] [T/V/I/Y] [X] [pY] [A/S/T/V] [X] [I/V/L) [M] [pY] [V/I/T/L/S/E] [D/M] [I/M/V/A] [M]	[31, 30, 92]
Rasal_N	[pY] [I/L/V] [X] [)] [P] [pY] [T/S/D/V] [E/D/M] [M/P/V/D] [M/V/I] [pY] [ϕ] [X] [D/ϕ]	[93] [30] [58]
Rasal_C	[pY] [X] [X] [P] [pY] [S] [X] [P/V] [pY] [[X] [X] [D/E/ϕ]	[93] [30] [58]
SH2B	[P/W] [I/L] [pY] [Y/F/E] [F/Y/W] [F/Y/L] [D] [pY] [X] [X] [D/E/ϕ]	[30] [58]
SH2D1A	[T] [I] [pY] [X] [X] [V/I] [T/S] [X] [pY] [X] [X] [V/I] [S/T] [I/V] [pY] [Y/F/S] [L/F/Y/M] [V/I/L] [G]	[94] [95] [30]
SH2D1B	[T] [I] [pY] [X] [X] [V/I] [T/I/V] [I/V] [pY] [S/L/A/V] [F/V/T/M] [V/A] [S/G/M/D] [pY] [[X] [X] [ϕ]	[94] [30] [58]
SH2D2A	[H/E/P] [pY] [D/E/S] [N] [pY] [E] [N/T] [D/ϕ]	[30] [58]
SH3BP2	[pY] [E/M/V] [N/V/I] [X] [V/I] [pY] [E] [N/V] [pY] [D/E] [N] [V]	[28] [30] [58]
SHB	[pY] [T/V/I] [X] [L] [E/P/L] [pY] [D/E] [E] [L] [E/D] [pY] [ϕ] [X] [ϕ] [D/E] [ϕ]	[96] [30] [58]
SHD	[N/E/D] [pY] [E/D] [N] [L/Y] [pY] [ϕ] [X] [ϕ] [D/E] [ϕ]	[30] [58]
SHE	[E/M] [pY] [E] [N/D] [X] [M] [pY] [ϕ] [X] [ϕ] [D/E] [ϕ]	[30] [58]
SHF	[pY] [Y/F] [E] [L/Y] [D] [pY] [ϕ] [X] [ϕ] [D/E] [ϕ]	[30] [58]
Shc1	[pY] [I/E/Y/L] [X] [I/L/M] [pY] [I/E/T] [X] [I/L/M] [N] [I/V] [pY] [E/G] [T] [I/V/L] [W/F] [H/P] [pY] [E/S] [N/T/M/A] [L/I/V/A] [pY] [D/E/G] [D/E/ϕ] [ϕ]	[28] [97] [34] [30] [58]

(continued)

Table 2
(continued)

SH2 domain	Binding motif	Reference
Shc2	[pY] [I/M/T/D/L] [M/I] [M/F/I/Y/V] [pY] [I/E/L/S] [M/L/I/T/A/V] [L/V]	[97] [30]
Shc3	[pY] [L/M/I/Q] [M/Y] [I/L/M/V] [pY] [E/L] [N/M/T] [L/I]	[30, 97]
Shc4	[R/E] [pY] [E/Y/L/I/M] [M/L/T/F/I] [L/I/Y/V/M] [W/F/Y]	[30]
SHIP1	[pY] [Y/S/T/V] [L/Y/N/L/E/F] [L/N/E/I/V] [pY] [S/Y] [L] [L/V]	[92] [30]
SHIP2	[N/T/P] [P] [pY] [D/E/S] [T/M] [L/I/V/A] [M/R/G]	[30]
SLNK	[E] [pY] [D/Y/E] [F/D] [X] [E/D/W/M] [pY] [G/D/V] [D/T] [D/ϕ]	[30] [58]
SOCS2	[D] [pY] [I/L/V] [L/I] [I/V]	[98]
SOCS3	[pY] [S/A/V/Y/F] [ɿ] [V/I/L] [ɿ] [H/V/I/Y]	[99]
SOCS4	[pY] [I/L/V] [X] [I/V]	[98]
SOCS4	[A] [pY] [V/I/L] [D/E] [V/I/S/T/G] [H/D/E/W]	[98]
SOCS6	[pY] [V/I] [ɿ] [I/F/M/V]	[100]
SOCS7	[pY] [V/W/L] [X] [W/F/L]	[100]
Src	[pY] [E/D/T] [E/N/Y] [I/M/L] [P/H] [I/P] [pY] [E/D/V/I] [L/I/E] [I/L/V] [D] [D/E] [X] [X] [pY] [D] [D/E/ϕ] [P/I]	[22] [30] [58]
Srms	[P] [P] [pY] [E] [N] [V/L]	[30]
STAT1	[pY] [D/E] [P/R] [R/P/Q]	[101]
STAT3	[pY] [X] [X] [Q]	[102]
SUPT6H	[K/P/M] [K/M] [pY] [P/I] [P] [K/R] [M]	[30]
Syk_N	[S/M] [M/S] [pY] [S/Y] [F] [L/I] [W/K]	[30]
Syk_C	[pY] [Q/T/E] [E/Q] [L/I] [G] [G/A] [pY] [T/A/Q/S/E] [T/D/S/A/E] [L] [S/T/Q/A/D] [ϕ] [pY] [V] [X] [D/E/ϕ] [D/E]	[28] [30] [58]
Tec	[D/I/V] [pY] [D/S/A/N] [E/N/D] [V/P/A/I/L] [D]	[30]
TenC1	[pY] [E] [N] [F/I/V] [N/Q] [E/D/P] [pY] [D/E] N/Y] [V] [pY] [E]	[84] [30] [58]
TNS1	[pY] [E] [N] [F/I/V] [pY] [D/E] [N] [I/F/V] [V/L] [P/L] [pY] [E/D/S] [N/E/D/M] [L/M/V/E]	[84] [88] [30]

(continued)

Table 2 (continued)

SH2 domain	Binding motif	Reference
TNS4	[G] [pY] [D/E] [N] [L/V/F/P/I] [F/M/A]	[30]
Txk	[M] [M] [pY] [D/E] [F/N] [M/V] [W/M/D/E]	[30]
Vav1	[pY] [M/L/E] [E] [P] [L] [pY] [L/E/V/I/M] [N/M/Q] [P/E/A/D] [D/S/T/A] [pY] [E/L/V] [X] [P]	[28] [30] [58]
Vav2	[N/Q/P] [P/E] [pY] [V/E/L] [T/M/E/N] [D/E/M]	[30]
Vav3	[E/H/D] [pY] [E/L/V/M] [N/M/E/F] [D/E/P] [D]	[30]
Yes	[pY] [E/D/S] [N/T] [L/V/I] [pY] [D/E/G] [D/E/ϕ] [ϕ]	[30] [58]
Zap70_N	[pY] [E] [N] [V/L] [D] [P] [X] [pY] [X] [X] [ψ/ϕ]	[30] [58]

The binding specificity of SH2 domains can be assessed using a number of different approaches. Of these, pull down and far-Western are traditional yet still popular SH2 binding assays in the signaling research field. Both assays are suitable to visualize/identify the presence of SH2 binding sites/proteins in complex protein samples, e.g., lysates and tissues. These established methods optimized for SH2 domains are collectively described in the fourth section of this book (*see* Chapter 17). In addition, a miniaturized in-solution SH2 binding assay based on proximity ligation has been recently developed (*see* Chapter 18). The assay termed SH2-PLA, has low sample requirement, high sensitivity with PCR-based detection, and the potential to be utilized in diagnostics and high-throughput settings.

SH2 domains mediate binding to phosphotyrosine peptides with affinities ranging from double-digit nM to low μM dissociation constants (K_D). To measure the binding affinities between the SH2 binding and phosphotyrosine ligand multiple techniques are available including isothermal calorimetry (ITC), NMR, surface plasmon resonance (SPR), time-resolved fluorescence energy transfer (TR-FRET), Alpha-Screen, and fluorescence polarization (FP). A rapid and scalable method for measuring the K_D between a fluorescently labeled peptide and a purified SH2 domain in solution is by FP (*see* Chapter 17). Marissa McKercher and Deborah Wuttke describe two biophysical approaches, NMR for gaining conformational insights into SH2-mediated interactions, and isothermal calorimetry (ITC) for calculating the SH2 binding affinities with unlabeled peptides (*see* Chapters 15 and 16). The advantage of utilizing NMR is the ability to identify regions within the SH2 domain that mediate contact with the phosphopeptide while other approaches such as ITC, SPR, and FP lack this capability. Other biophysical approaches, while not discussed in full detail in this book, include SPR for interrogating the biochemical properties of SH2-phosphotyrosine peptides interactions (see other published chapters in Methods in Molecular Biology [25, 26]).

1.3 Characterizing SH2 Specificity and Generating Prediction Models

SH2 domains differ in their binding preferences for specific phosphorylated ligands, resulting in specificity in signal transduction [27]. Pioneering studies by Songyang and colleagues developed degenerate chemically synthesized peptide libraries that were passed over a column displaying the target SH2 domain protein. Peptides captured by the immobilized SH2 protein are subsequently eluted and subjected to sequencing, either by Edman degradation or by mass spectrometry, thereby revealing the intrinsic binding preferences or specificity profiles of a subset of SH2 domains [22, 28]. These studies revealed that, in addition to phosphotyrosine, amino acids in positions from −2 to +4 relative to phosphotyrosine contribute to high-affinity binding in most cases; more extended contacts (−6 to +6) have been observed in some structural studies. Later with the advancement of SPOT peptide arrays [29], peptides could be synthesized onto an acid-hardened nitrocellulose membrane (*see* Chapters 20 and 21). This approach is powerful in its ability to panel SH2 domains again hundreds up to thousands of phosphotyrosine peptides, where each SPOT represents a peptide of a known sequence, without the need for peptide sequencing. Peptide arrays approaches such as oriented peptide array libraries (OPAL) are an alternative method, similar to degenerate peptide arrays, for determining the specificity profiles for a large set of soluble SH2 domains [30, 31]. However, these approaches capture general binding motifs through paneling individual positions (for example at +1 or +2 residues C-terminal from the phosphotyrosine) independently of neighboring positions. Secondary effects such as neighboring residue effects are likely to be missed, and hence contextual peptide sequence information is overlooked. Further studies of the FGF and insulin/IGF-1 receptor systems using physiological peptides revealed the importance of contextual peptide sequence specificity for SH2 domains [32]. Not only are favorable residues (permissive factors) important for achieving specificity for SH2 domains, but nonpermissive factors (anti-motifs) are disfavorable by abolishing interactions. Utilizing peptides arrays with natural physiological peptides has revealed that binding can be contextual as nonpermissive factors (negative factors) can dramatically affect SH2 domain binding. In this book, SPOT arrays will be described in further detail for determining the specificity of SH2 domains by identifying positive and negative (nonpermissive) residues critical for phosphotyrosine binding (*see*

Chapter 20). However one of the limitations to SPOT arrays is its semiquantitative ability to measure binding. To advance the SPOT approach, Brett Engelmann describes a quantitative method for measuring SH2 and peptides using the CPCMA approach (*see* Chapter 21) [33]. By solubilizing the peptides, synthesized using the SPOT approach on a dissolvable nitrocellulose membrane, they can be array printed on multiple glass slides which is much more scalable than a single SPOT membrane. With multiple arrays of peptides printed on a single slide, each array can be incubated with an SH2 domain at varying concentrations. From this data, each SH2 domain and peptide interaction may have a calculated dissociation constant (K_D). This information is extremely powerful for modeling protein interactions as it relates to competitive binding events between various SH2 domain proteins within a cell. Other approaches including but not limited to phage display [34], high-throughput fluorescence polarization [35, 36], and protein microarrays [37] have been developed to study the binding specificity of SH2 domains. Moreover, new technologies including mass-spectrometry are now being implemented to screen and identify SH2 binding peptides.

While degenerate peptide libraries and SPOT peptide arrays are powerful approaches for analyzing SH2 domain binding specificity limited to the binding pocket some information may not be captured using these approaches. For example, specificity outside of the phosphotyrosine binding pocket or understanding the context of the binding partner outside of the peptide may not be informative from this data. In some cases, specificity can be achieved through a secondary binding site, which cannot be identified using peptide binding studies but may be achieved through X-ray crystallography and other biophysical methods [38]. Structural studies by Schlessinger and colleagues determined the amino-terminal SH2 domain of PLCgamma1 binds to the pTyr containing tail of the kinase domain and makes additional contacts with the kinase domain via a secondary binding site on the SH2 domain. The secondary site involves the BC and DE loops of the SH2 domain and contributes to an added level of specificity unforeseen by peptide library approaches. Hence, SH2 domains can achieve high selectivity with kinases and substrates through mediating interactions outside the binding regions of the classical pTyr binding pocket. It has been recently reported that plasma membrane lipids directly bind SH2 domains with high affinity and specificity and modulate SH2 domain-mediated protein–protein interactions [39].

Many of these specificity-profiling experiments provide specific matrix information that has led to predictive algorithms capable of predicting potential SH2 binding partners. For example Songyang's degenerate peptide library data gave rise to the development of Scansite, where users can search among 13 SH2 domains for potential binding partners across the proteome [40]. Furthermore, Shawn Li and colleagues utilizing their SPOT OPAL library were able to generate SMALI, a database capable of predicting and mapping potential SH2-pTyr interactions (*see* Chapter 5) [41]. However, there are limitations to utilizing these position weighted matrix (PWM)-based predictions as they do not take into account cellular context, expression, and peptide-sequence context. With more advanced programming and a growing number of SH2-peptide dataset, other algorithms such as a support vector machine (SVM) or machine learning can enhance the predictive power for certain SH2 domains [42]. The SVM predictor, DomPep, was developed using a combination of proteomic data (e.g., peptide and protein arrays) and sequence information (e.g., domain sequence identity and PWM distance calculated using domain-binding peptides) to achieve high accuracy and greater coverage in predicting protein–protein interactions by SH2 domains (*see* Chapter 5). In the chapter by Kousik Kundu and Rolf Backofen, the authors describe a supervised learning approach for predicting SH2 domain interactions utilizing positive and negative datasets (*see* Chapter 6). With more data arriving each year, algorithms for predicting SH2 domain interactions will dramatically improve.

1.4 Systems Biology and Proteomics

In the era of -omics biology, ongoing research has led to an ever-increasing integrative web of exploratory science that includes bioinformatics, evolutionary biology, genomics, proteomics, and systems biology. Bioinformatic tools and computational algorithms have unveiled remarkable insights into phosphotyrosine signaling and SH2 domains. For example, these tools aid in the identification of human encoded SH2 domain proteins [12, 13] and provide insight into the evolution of phosphotyrosine signaling (*see* Chapter 4). From these results, they offer a glimpse into the emergence of multicellularity and the use of phosphotyrosine regulatory mechanisms [43]. As another example of the power of bioinformatics, Karl Jablonowski applied clustering analysis to previous collections of SH2 domains datasets to bring insight into new binding or specificity patterns between the different SH2 domains (*see* Chapter 7). In addition to bioinformatics tools, data can be easily extracted from databases regarding SH2 domain related diseases, structures and mouse knockout information. With more and more resources available, scientists have ready access to a wealth of information for investigating SH2 domains.

In the arena of proteomics mass spectrometry, technology has radically changed the field of biology. In the last 10 years, the field of phosphoproteomics has grown significantly revealing over 200,000 phosphorylation sites (40,000 pTyr sites) on more than 20,000 proteins [44]. With the growing number of identified phosphopeptides, software programs have been designed to utilize this data to develop hypotheses for protein interactions and infer protein interaction networks or pathways. Programs such as Phosida [44, 45], Phosphosite [46], PhosphoELM [47, 48], ProteomeScout [49], and PhosphoBLAST [50] contain a wealth of data on phosphorylated proteins and peptides derived from biochemical and proteomic based studies. Other sophisticated algorithms such as NetworkIn and NetPhorest compile data from the specificity of phospho-binding domains such as FHA, 14-3-3, PTB, BRCT, WW, SH2, and kinases, together with phosphoproteomics to generate an in vivo network of interactions taking into account context information [51, 52]. These computational approaches demonstrate the power of bioinformatics in studying not only protein interaction domains but also in gaining a systems-level understanding of the cell. By mining these wealthy datasets, computational scientists are able to address some important biological questions from dynamic changes in phosphorylation in cancer cells to evaluating how protein interaction domains may be affected by mutations in the network during cancer progression. A recent study was able to examine the effect of mutations on a signaling network using the Cancer Genome Atlas (TCGA) together with specificity analysis of SH2 domains [53]. Furthermore, network rewiring in cancer has become a critical area of study in phosphotyrosine signaling. The recent analysis using computational modeling of cancer genomes revealed network-attacking mutations (NAMs) within SH2 domain modulation [54]. These systems-level studies can bring insight into the role of SH2 domains in diseases and will prove fruitful in the future in leading to new diagnostic tools and may eventually lead to new therapeutic interventions.

A key step towards understanding the function of SH2 domains is to determine potential substrate binding partners. With the use of physiological ligands, large-scale systems-analysis of receptor tyrosine kinase systems such as the EGFR family of kinases reveals a dynamic network of potential SH2 domain interactions. Extensive proteomic studies not only determine the specificity but also are important in defining the protein interaction networks for SH2 domains. Protein microarray studies using human SH2 and PTB domains were first to perform a systems-level approach to quantitatively measure the affinities between SH2 domains and phosphosites on the EGFR family of RTKs [37, 55, 56]. An approach of this nature captures the network surrounding RTKs. As above mentioned, the SPOT peptide array is also an extremely useful approach to test binding of SH2 domains to physiological peptides from various receptor systems or even large sets of phosphotyrosine peptides to reveal the extensive network of potential SH2 interactions and to understand the specificity of certain SH2 domains [57, 58].

Mass spectrometry (MS) is an extremely powerful tool in phosphoproteomics for identifying and quantifying tyrosine phosphorylated peptides in cells and tissues [59, 60] Because of the low abundance of pTyr peptides in cell, phosphopeptide enrichment, often by various combinations of anti-pTyr immunoprecipitation and immobilized metal ion affinity chromatography (IMAC) is required. As a signaling reader domain for phosphotyrosine sequences in cells, SH2 domains are biologically relevant affinity reagents for enriching phosphotyrosine peptides prior to MS analyses. Three contributors to this book describe techniques combining SH2 domain-based enrichment and LC-MS. Ruijun Tian's group employs pull down with GST-SH2 to enrich pTyr proteins from pervanadate-treated cell samples to maximize identification of the SH2 binding proteins (*see* Chapter 22). Peter Nollau's group utilizes a dual tag strategy by which a GST-tag is used for purification of SH2 domain proteins and a biotin tag is used for pull-down with streptavidin agarose (*see* Chapter 23). In this approach, the GST portion, a potential cause of background, can be removed from the SH2 bait prior to pull down. Radoslaw Sobota introduces a method of combining Stable Isotope Labeling by Amino Acids in Cell Culture (SILAC) and SH2 pull-down (*see* Chapter 24). An advantage of this approach is that different SH2 binding profiles can be quantified and compared using SILAC. As nonspecific binding is the major issue in pull down-based proteomics, the authors of these chapters repeatedly stress the importance of using assay controls such as mutated SH2, GST protein, and phosphopeptide competitors. The inverse approach using pTyr peptides from the ErbB family for pull down identified unique combinations of SH2 and PTB domain-containing proteins for each receptor [50].

With any high-throughput approach, the method used may lead to false-positive or false—negative data, which could interfere with downstream computational and biological studies. With the diminishing costs of these technologies, applying reciprocal or complementary approaches can be used to generate high quality datasets. In this book, Shawn Li and colleagues describe a comprehensive phosphotyrosine-based signaling using reciprocal protein-arrays and peptide arrays in their screen to identify SH2 domain interactions with high confidence [61] (*see* Chapter 25).

As we move forward, the main fundamental role of SH2 domains is recognizing phosphotyrosine residues. One of the major challenges still faced within the SH2 is identifying or predicting an SH2 domain ligand. With a range of proteomic based approaches, predicting SH2 domain ligands has become feasible using various computational and web-based approaches such as Scansite, SMALI, NetworkIn, and others. Moreover, new computational algorithms are arising where large datasets can be analyzed in greater detail specificity using semi-supervised learning approach to accurately predict SH2 domain interactions. These computational approaches are fundamental for establishing a basis for developing new hypotheses regarding SH2 domains. All together this brings multiple in silico, in vitro, and in vivo assays that future scientists may use to study SH2 domains singularly or at a system-wide level.

1.5 SH2 Domains in Signal Transduction

SH2 domain containing proteins play a critical role in phosphotyrosine signaling. Loss of these proteins has been associated with severe phenotypes in studies using genetically modified mice [12]. Conceptually, the phosphotyrosine signaling pathways are regulated by a three-part toolkit, namely tyrosine kinase (TK), tyrosine phosphatase (PTP), and pTyr-reader domains, e.g., SH2 and PTB [16]. The phosphorylation state of each site is controlled by a dynamic balance between the rate of kinase-mediated phosphorylation and phosphatase-mediated dephosphorylation, while signal specificity is largely conferred by the sequence selectivity of the reader domains [62, 63]. In addition, ligand selection of SH2 domains in vivo depends on cellular factors such as local concentration, phosphotyrosine site stoichiometry, and the presence of competitors. As a result, SH2-ligand interactions identified by in vitro affinity-based binding assays using homogenous conditions could significantly differ from those occurring in vivo.

One way to capture the spectrum of SH2 binding sequences in vivo is by monitoring the change in phosphorylation of individual sites after exogenous expression of an SH2 domain, with the assumption that engagement of an SH2 domain with its physiological binding sites protects them from phosphatases, and results in increased net phosphorylation. Joshua Jadwin describes the utilization of such an approach, the SH2 protection assay, to capture the sequence specificity of SH2 domains in vivo (*see* Chapter 28). On the other hand, interrogation of localized SH2 binding dynamics requires direct visualization of individual molecules at the site of phosphorylation. Donmyung Oh describes the application of total internal reflection (TIR) microscopy and single-molecule imaging to quantify the kinetic parameters of fluorescent-labeled SH2 domains interacting with pTyr sites near the cell membrane (*see* Chapter 29). One interesting finding obtained using this approach is that the dwell time of SH2 modules within the TIR illumination field (i.e., plasma membrane) is significantly longer than predicted by chemical dissociation rate constants, suggesting that SH2 modules rebind nearby pTyr sites multiple times before dissociating from membrane [64]. These findings exemplify the importance of techniques applying SH2 domains as tools for studying phosphotyrosine signaling in vivo.

As mentioned above, generating superbinders of SH2 domains using phage display can be a great tool to examine signaling networks (*see* Chapter 13). For example, the expression of superbinders in cells can antagonize the recruitment of SH2 domains to actively phosphorylated signals thus suppressing signals from RTKs

[65]. These extremely high affinity superbinders can also be engineered into existing SH2 domain proteins to synthetically rewire the cell or to interrogate activation of these proteins for understanding towards phosphotyrosine ligands. In this case adding a superbinder SH2 of Grb2 can trigger precocious stem cell differentiation, presumably because of an artificially enhanced affinity to transmembrane receptors [66]. Other tools using phage-display selection system include small antibody mimic or monobodies generated against SH2 domains. For example directed evolution of monobodies using the fibronectin scaffold that blocks the phosphopeptide binding pocket of the SH2 domains of the Abl tyrosine kinase and the N-SH2 of Shp2 tyrosine phosphatase can be used to inhibit the SH2 domain in downstream signaling functions [67, 68]. Lastly, engineered tools such as clamps are highly applicable for interrogating SH2 domain function. For example, directing Grb2 to a single phosphotyrosine site of the Ptpn11/Shp2 phosphatase, while excluding other sites tested, is sufficient for differentiation of the essential primitive endoderm lineage from embryonic stem cells [69]. With these directed evolution-based methods available reagents can be generated to reveal complex SH2 domain functions at the cellular level.

1.6 SH2 Domains in Disease, Diagnostics and Therapeutics

Mutations have been identified among many SH2 domain-containing proteins and within the SH2 domain itself [12, 70]. These mutations are implicated in many human diseases ranging from cancers, diabetes, to immunodeficiencies. With advanced next-generation sequencing technologies rapidly sequencing cancer genomes, mutations within SH2 domains can be readily cataloged and explored [71, 72]. As these mutations are identified, tools are necessary to explore the consequences arising from these mutations. Standard biophysical approaches comparing mutant and wild-type domains may reveal changes in binding affinity to phosphotyrosine peptides. Other approaches such as SPOT peptide arrays may be applicable for determining whether mutations alter specificity rather than affinity. While these approaches often rely on a priori knowledge of the binding partner, a different approach is needed to observe global changes in SH2 binding. For example, the aforementioned SILAC-MS has been used to capture SH2 mutation-dependent changes in the global profile of phosphotyrosine ligand selectivity [73](*see* Chapter 24). Together, these techniques have the potential to reveal the role of SH2 domain mutations in disease and mechanisms by which they drive cancer progression.

Aberrant tyrosine phosphorylation is particularly associated with cancer. Mutationally activated tyrosine kinases such as Abl and members of the EGF receptor (EGFR) family are known drivers of tumorigenesis, and have been successfully targeted by small molecule tyrosine kinase inhibitors (TKI) [74, 75]. Given the central role of tyrosine kinases in many aspects of tumor biology, and their ability to be targeted by new therapies, studying the global state of tyrosine phosphorylation of tumors is likely to provide predictive insights difficult or impossible to achieve using other existing approaches. SH2 profiling is one such method, which utilizes a battery of SH2 domains to profile the global state of tyrosine phosphorylation state in cells expressing oncogenic tyrosine kinases [76, 77]. Reverse phase protein array (RPPA) allows for simultaneous measurement of protein expression and modification in a large number of samples making the platform suitable for detection of aberrant signaling pathways in a group of cell lines or tumor specimens. Rosette assay is a high-throughput application of SH2 profiling that uses a dot blot-based RPPA and is applicable for samples such as peptides, recombinant proteins, and cell lysates (*see* Chapter 26). Microwestern, another RPPA-based method developed by the Richard Jones group, has the dual features of resolution and throughput allowing for micro scale western blotting suitable for molecular profiling (*see* Chapter 27). These assays may assist in the identification of biomarkers and drug targets within pathogenic tyrosine kinase signaling networks. For assessment of aberrant signaling in individual tumor cells, direct staining of fixed tumor specimens with SH2 domains holds promise. Sophia Buhs and Peter Nollau describe a histochemical method that uses SH2 domains as probes (*see* Chapter 31). This approach may allow not only for an assessment of the global change in SH2 binding sites but also for detection of a small number of pathway-activated cells within a larger tumor mass.

Other advanced systems for studying SH2 domains in human diseases include FRET biosensors and Alpha-based inhibitor screen. For example, Ohba and colleagues have designed a CRKL SH2-based FRET biosensor that can be used to measure the level of BCR-ABL activity in CML cells (*see* Chapter 30). This technique can be applied to study the effect of small molecule inhibitors, e.g., BCR-ABL TKI, in patient cells. Alpha-Screen, a high-throughput fluorescent bead-based assay, is another promising technique [78]. Akira Asai and Kazuyuki Takakuma describe their application of Alpha-Screen for identification of novel inhibitors of STAT SH2-ligand interactions (*see* Chapter 19).

1.7 Conclusions and Future Directions

Targeting the SH2 has been thought as a promising target for drug design and cancer therapy [79]. However, the success at inhibiting SH2 domains has been fairly limiting as most progress has been made in targeting tyrosine kinases. Novel chemical approaches such as cyclic peptides and phosphomimetics and novel therapeutic options for targeting SH2 domains have shown to be promising [80, 81]. Recent technologies such as phage display and the use of fibronectin scaffolds have been able to disrupt SH2 domain signaling in cells [67, 82, 83]. While this book does not capture this topic, scientists interested in this area of research and exploring new therapeutic strategies may find the methods and technologies mentioned in this book useful in the development of therapeutics.

1.8 Who Might Find This Handbook of Use

This volume is intended as both a textbook on SH2 domains and as a reference book for researchers looking to investigate or study SH2 domains but also to utilize SH2 domains to understand phosphotyrosine signaling and more broadly how protein interaction domain mediate specificity in signaling systems. The people who might find this handbook of use are computational biologists, biochemists, structural biologists, cell biologists, pathologists, and people who are just embarking on SH2 domains or phosphotyrosine signaling.

Acknowledgments

We are incredibly grateful to all the authors who contributed to this volume. We thank the series editor, John Walker, for his continuous encouragement and providing constructive comments. We also thank Joshua Jadwin, Piers Nash, and Bruce Mayer for their support. Finally, we would like to extend our thanks to David Casey and Anna Rakovsky of Springer USA, and L. Sujitha of Spi Global for their help in the compilation of this book. This study was partly supported by grant CA1154966 from the National Institutes of Health and Quest for CURES (QFC) grant from the Leukemia and Lymphoma Society (to K.M.).

References

1. Sadowski I, Stone JC, Pawson T (1986) A noncatalytic domain conserved among cytoplasmic protein-tyrosine kinases modifies the kinase function and transforming activity of Fujinami sarcoma virus P130gag-fps. Mol Cell Biol 6(12):4396–4408
2. DeClue JE, Sadowski I, Martin GS, Pawson T (1987) A conserved domain regulates interactions of the v-fps protein-tyrosine kinase with the host cell. Proc Natl Acad Sci U S A 84(24):9064–9068
3. Sadowski I, Pawson T (1987) Catalytic and non-catalytic domains of the Fujinami sarcoma virus P130gag-fps protein-tyrosine kinase distinguished by the expression of v-fps polypeptides in Escherichia coli. Oncogene 1(2):181–191
4. Stone JC, Atkinson T, Smith M, Pawson T (1984) Identification of functional regions in the transforming protein of Fujinami sarcoma virus by in-phase insertion mutagenesis. Cell 37(2):549–558
5. Mayer BJ, Hamaguchi M, Hanafusa H (1988) A novel viral oncogene with structural similarity to phospholipase C. Nature 332(6161):272–275. doi:10.1038/332272a0
6. Matsuda M, Mayer BJ, Fukui Y, Hanafusa H (1990) Binding of transforming protein, P47gag-crk, to a broad range of phosphotyrosine-containing proteins. Science 248(4962):1537–1539
7. Olivier JP, Raabe T, Henkemeyer M, Dickson B, Mbamalu G, Margolis B, Schlessinger J, Hafen E, Pawson T (1993) A Drosophila SH2-SH3 adaptor protein implicated in coupling the sevenless tyrosine kinase to an activator of Ras guanine nucleotide exchange, Sos. Cell 73(1):179–191
8. Rozakis-Adcock M, Fernley R, Wade J, Pawson T, Bowtell D (1993) The SH2 and SH3

domains of mammalian Grb2 couple the EGF receptor to the Ras activator mSos1. Nature 363(6424):83–85. doi:10.1038/363083a0

9. Anderson D, Koch CA, Grey L, Ellis C, Moran MF, Pawson T (1990) Binding of SH2 domains of phospholipase C gamma 1, GAP, and Src to activated growth factor receptors. Science 250(4983):979–982

10. Moran MF, Koch CA, Anderson D, Ellis C, England L, Martin GS, Pawson T (1990) Src homology region 2 domains direct protein-protein interactions in signal transduction. Proc Natl Acad Sci U S A 87(21):8622–8626

11. Mohammadi M, Honegger AM, Rotin D, Fischer R, Bellot F, Li W, Dionne CA, Jaye M, Rubinstein M, Schlessinger J (1991) A tyrosine-phosphorylated carboxy-terminal peptide of the fibroblast growth factor receptor (Flg) is a binding site for the SH2 domain of phospholipase C-gamma 1. Mol Cell Biol 11(10):5068–5078

12. Liu BA, Jablonowski K, Raina M, Arce M, Pawson T, Nash PD (2006) The human and mouse complement of SH2 domain proteins-establishing the boundaries of phosphotyrosine signaling. Mol Cell 22(6):851–868. doi:10.1016/j.molcel.2006.06.001

13. Liu BA, Shah E, Jablonowski K, Stergachis A, Engelmann B, Nash PD (2011) The SH2 domain-containing proteins in 21 species establish the provenance and scope of phosphotyrosine signaling in eukaryotes. Sci Signal 4(202):83. doi:10.1126/scisignal.2002105

14. King N, Westbrook MJ, Young SL, Kuo A, Abedin M, Chapman J, Fairclough S, Hellsten U, Isogai Y, Letunic I, Marr M, Pincus D, Putnam N, Rokas A, Wright KJ, Zuzow R, Dirks W, Good M, Goodstein D, Lemons D, Li W, Lyons JB, Morris A, Nichols S, Richter DJ, Salamov A, Sequencing JG, Bork P, Lim WA, Manning G, Miller WT, McGinnis W, Shapiro H, Tjian R, Grigoriev IV, Rokhsar D (2008) The genome of the choanoflagellate Monosiga brevicollis and the origin of metazoans. Nature 451(7180):783–788. doi:10.1038/nature06617

15. Manning G, Young SL, Miller WT, Zhai Y (2008) The protist, Monosiga brevicollis, has a tyrosine kinase signaling network more elaborate and diverse than found in any known metazoan. Proc Natl Acad Sci U S A 105(28):9674–9679. doi:10.1073/pnas.0801314105

16. Lim WA, Pawson T (2010) Phosphotyrosine signaling: evolving a new cellular communication system. Cell 142(5):661–667. doi:10.1016/j.cell.2010.08.023

17. Waksman G, Kominos D, Robertson SC, Pant N, Baltimore D, Birge RB, Cowburn D, Hanafusa H, Mayer BJ, Overduin M et al (1992) Crystal structure of the phosphotyrosine recognition domain SH2 of v-src complexed with tyrosine-phosphorylated peptides. Nature 358(6388):646–653. doi:10.1038/358646a0

18. Waksman G, Shoelson SE, Pant N, Cowburn D, Kuriyan J (1993) Binding of a high affinity phosphotyrosyl peptide to the Src SH2 domain: crystal structures of the complexed and peptide-free forms. Cell 72(5):779–790, doi:0092-8674(93)90405-F [pii]

19. Johnson LN, Noble ME, Owen DJ (1996) Active and inactive protein kinases: structural basis for regulation. Cell 85(2):149–158

20. Hidaka M, Homma Y, Takenawa T (1991) Highly conserved eight amino acid sequence in SH2 is important for recognition of phosphotyrosine site. Biochem Biophys Res Commun 180(3):1490–1497

21. Pawson T, Gish GD, Nash P (2001) SH2 domains, interaction modules and cellular wiring. Trends Cell Biol 11(12):504–511, doi:S0962-8924(01)02154-7 [pii]

22. Songyang Z, Shoelson SE, Chaudhuri M, Gish G, Pawson T, Haser WG, King F, Roberts T, Ratnofsky S, Lechleider RJ et al (1993) SH2 domains recognize specific phosphopeptide sequences. Cell 72(5):767–778, doi:0092-8674(93)90404-E [pii]

23. Pawson T (1995) Protein modules and signalling networks. Nature 373(6515):573–580. doi:10.1038/373573a0

24. Marengere LE, Songyang Z, Gish GD, Schaller MD, Parsons JT, Stern MJ, Cantley LC, Pawson T (1994) SH2 domain specificity and activity modified by a single residue. Nature 369(6480):502–505. doi:10.1038/369502a0

25. Panayotou G, Ladbury J (2001) Analysis of SH2 domain—phosphopeptide interactions by isothermal titration calorimetry and surface plasmon resonance. Methods Mol Biol 124:295–311

26. Vely F, Trautmann A, Vivier E (2000) BIAcore analysis to test phosphopeptide-SH2 domain interactions. Methods Mol Biol 121:313–321. doi:10.1385/1-59259-044-6:313

27. Pawson T (2004) Specificity in signal transduction: from phosphotyrosine-SH2 domain interactions to complex cellular systems. Cell 116(2):191–203, doi:S0092867403010778 [pii]

28. Songyang Z, Shoelson SE, McGlade J, Olivier P, Pawson T, Bustelo XR, Barbacid M, Sabe

H, Hanafusa H, Yi T et al (1994) Specific motifs recognized by the SH2 domains of Csk, 3BP2, fps/fes, GRB-2, HCP, SHC, Syk, and Vav. Mol Cell Biol 14(4):2777–2785

29. Frank R (2002) The SPOT-synthesis technique. Synthetic peptide arrays on membrane supports—principles and applications. J Immunol Methods 267(1):13–26, doi:S0022175902001370 [pii]
30. Huang H, Li L, Wu C, Schibli D, Colwill K, Ma S, Li C, Roy P, Ho K, Songyang Z, Pawson T, Gao Y, Li SS (2008) Defining the specificity space of the human SRC homology 2 domain. Mol Cell Proteomics 7(4):768–784. doi:10.1074/mcp.M700312-MCP200
31. Rodriguez M, Li SS, Harper JW, Songyang Z (2004) An oriented peptide array library (OPAL) strategy to study protein-protein interactions. J Biol Chem 279(10):8802–8807. doi:10.1074/jbc.M311886200, M311886200 [pii]
32. Liu BA, Jablonowski K, Shah EE, Engelmann BW, Jones RB, Nash PD (2010) SH2 domains recognize contextual peptide sequence information to determine selectivity. Mol Cell Proteomics 9(11):2391–2404. doi:10.1074/mcp.M110.001586
33. Engelmann BW, Kim Y, Wang M, Peters B, Rock RS, Nash PD (2014) The development and application of a quantitative peptide microarray based approach to protein interaction domain specificity space. Mol Cell Proteomics 13(12):3647–3662. doi:10.1074/mcp.O114.038695
34. Dente L, Vetriani C, Zucconi A, Pelicci G, Lanfrancone L, Pelicci PG, Cesareni G (1997) Modified phage peptide libraries as a tool to study specificity of phosphorylation and recognition of tyrosine containing peptides. J Mol Biol 269(5):694–703
35. Leung KK, Hause RJ Jr, Barkinge JL, Ciaccio MF, Chuu CP, Jones RB (2014) Enhanced prediction of Src homology 2 (SH2) domain binding potentials using a fluorescence polarization-derived c-Met, c-Kit, ErbB, and androgen receptor interactome. Mol Cell Proteomics 13(7):1705–1723. doi:10.1074/mcp.M113.034876
36. Hause RJ Jr, Leung KK, Barkinge JL, Ciaccio MF, Chuu CP, Jones RB (2012) Comprehensive binary interaction mapping of SH2 domains via fluorescence polarization reveals novel functional diversification of ErbB receptors. PLoS One 7(9), e44471. doi:10.1371/journal.pone.0044471
37. Jones RB, Gordus A, Krall JA, MacBeath G (2006) A quantitative protein interaction network for the ErbB receptors using protein microarrays. Nature 439(7073):168–174. doi:10.1038/nature04177, nature04177 [pii]
38. Bae JH, Lew ED, Yuzawa S, Tome F, Lax I, Schlessinger J (2009) The selectivity of receptor tyrosine kinase signaling is controlled by a secondary SH2 domain binding site. Cell 138(3):514–524. doi:10.1016/j.cell.2009.05.028, S0092-8674(09)00631-X [pii]
39. Park MJ, Sheng R, Silkov A, Jung DJ, Wang ZG, Xin Y, Kim H, Thiagarajan-Rosenkranz P, Song S, Yoon Y, Nam W, Kim I, Kim E, Lee DG, Chen Y, Singaram I, Wang L, Jang MH, Hwang CS, Honig B, Ryu S, Lorieau J, Kim YM, Cho W (2016) SH2 domains serve as lipid-binding modules for pTyr-signaling proteins. Mol Cell 62(1):7–20. doi:10.1016/j.molcel.2016.01.027
40. Obenauer JC, Cantley LC, Yaffe MB (2003) Scansite 2.0: proteome-wide prediction of cell signaling interactions using short sequence motifs. Nucleic Acids Res 31(13):3635–3641
41. Li L, Wu C, Huang H, Zhang K, Gan J, Li SS (2008) Prediction of phosphotyrosine signaling networks using a scoring matrix-assisted ligand identification approach. Nucleic Acids Res 36(10):3263–3273. doi:10.1093/nar/gkn161, gkn161 [pii]
42. Kundu K, Costa F, Huber M, Reth M, Backofen R (2013) Semi-supervised prediction of SH2-peptide interactions from imbalanced high-throughput data. PLoS One 8(5), e62732. doi:10.1371/journal.pone.0062732
43. Liu BA, Nash PD (2012) Evolution of SH2 domains and phosphotyrosine signalling networks. Philos Trans R Soc Lond B Biol Sci 367(1602):2556–2573. doi:10.1098/rstb.2012.0107
44. Olsen JV, Blagoev B, Gnad F, Macek B, Kumar C, Mortensen P, Mann M (2006) Global, in vivo, and site-specific phosphorylation dynamics in signaling networks. Cell 127(3):635–648. doi:10.1016/j.cell.2006.09.026, S0092-8674(06)01274-8 [pii]
45. Gnad F, Ren S, Cox J, Olsen JV, Macek B, Oroshi M, Mann M (2007) PHOSIDA (phosphorylation site database): management, structural and evolutionary investigation, and prediction of phosphosites. Genome Biol 8(11):R250. doi:10.1186/gb-2007-8-11-r250, gb-2007-8-11-r250 [pii]
46. Hornbeck PV, Chabra I, Kornhauser JM, Skrzypek E, Zhang B (2004) PhosphoSite: a bioinformatics resource dedicated to physiological protein phosphorylation. Proteomics 4(6):1551–1561. doi:10.1002/pmic.200300772

47. Diella F, Cameron S, Gemund C, Linding R, Via A, Kuster B, Sicheritz-Ponten T, Blom N, Gibson TJ (2004) Phospho.ELM: a database of experimentally verified phosphorylation sites in eukaryotic proteins. BMC Bioinformatics 5:79. doi:10.1186/1471-2105-5-79 1471-2105-5-79 [pii]
48. Diella F, Gould CM, Chica C, Via A, Gibson TJ (2008) Phospho.ELM a database of phosphorylation sites—update. Nucleic Acids Res 36(Database issue):D240–D244. doi:10.1093/nar/gkm772, gkm772 [pii]
49. Matlock MK, Holehouse AS, Naegle KM (2015) ProteomeScout: a repository and analysis resource for post-translational modifications and proteins. Nucleic Acids Res 43(Database issue):D521–D530. doi:10.1093/nar/gku1154
50. Wang Y, Klemke RL (2008) PhosphoBlast, a computational tool for comparing phosphoprotein signatures among large datasets. Mol Cell Proteomics 7(1):145–162. doi:10.1074/mcp.M700207-MCP200, M700207-MCP200 [pii]
51. Linding R, Jensen LJ, Pasculescu A, Olhovsky M, Colwill K, Bork P, Yaffe MB, Pawson T (2008) NetworKIN: a resource for exploring cellular phosphorylation networks. Nucleic Acids Res 36(Database issue):D695–D699. doi:10.1093/nar/gkm902, gkm902 [pii]
52. Miller ML, Jensen LJ, Diella F, Jorgensen C, Tinti M, Li L, Hsiung M, Parker SA, Bordeaux J, Sicheritz-Ponten T, Olhovsky M, Pasculescu A, Alexander J, Knapp S, Blom N, Bork P, Li S, Cesareni G, Pawson T, Turk BE, Yaffe MB, Brunak S, Linding R (2008) Linear motif atlas for phosphorylation-dependent signaling. Sci Signal 1(35):ra2. doi:10.1126/scisignal.1159433, 1/35/ra2 [pii]
53. AlQuraishi M, Koytiger G, Jenney A, MacBeath G, Sorger PK (2014) A multiscale statistical mechanical framework integrates biophysical and genomic data to assemble cancer networks. Nat Genet 46(12):1363–1371. doi:10.1038/ng.3138
54. Creixell P, Schoof EM, Simpson CD, Longden J, Miller CJ, Lou HJ, Perryman L, Cox TR, Zivanovic N, Palmeri A, Wesolowska-Andersen A, Helmer-Citterich M, Ferkinghoff-Borg J, Itamochi H, Bodenmiller B, Erler JT, Turk BE, Linding R (2015) Kinome-wide decoding of network-attacking mutations rewiring cancer signaling. Cell 163(1):202–217. doi:10.1016/j.cell.2015.08.056
55. Kaushansky A, Allen JE, Gordus A, Stiffler MA, Karp ES, Chang BH, MacBeath G (2010) Quantifying protein-protein interactions in high throughput using protein domain microarrays. Nat Protoc 5(4):773–790. doi:10.1038/nprot.2010.36
56. Kaushansky A, Gordus A, Chang B, Rush J, MacBeath G (2008) A quantitative study of the recruitment potential of all intracellular tyrosine residues on EGFR, FGFR1 and IGF1R. Mol Biosyst 4(6):643–653. doi:10.1039/b801018h
57. Tinti M, Kiemer L, Costa S, Miller ML, Sacco F, Olsen JV, Carducci M, Paoluzi S, Langone F, Workman CT, Blom N, Machida K, Thompson CM, Schutkowski M, Brunak S, Mann M, Mayer BJ, Castagnoli L, Cesareni G (2013) The SH2 domain interaction landscape. Cell Rep 3(4):1293–1305. doi:10.1016/j.celrep.2013.03.001
58. Liu BA, Engelmann BW, Jablonowski K, Higginbotham K, Stergachis AB, Nash PD (2012) SRC homology 2 domain binding sites in insulin, IGF-1 and FGF receptor mediated signaling networks reveal an extensive potential interactome. Cell Commun Signal 10(1):27. doi:10.1186/1478-811X-10-27
59. Humphrey SJ, Azimifar SB, Mann M (2015) High-throughput phosphoproteomics reveals in vivo insulin signaling dynamics. Nat Biotechnol 33(9):990–995. doi:10.1038/nbt.3327
60. Curran TG, Zhang Y, Ma DJ, Sarkaria JN, White FM (2015) MARQUIS: a multiplex method for absolute quantification of peptides and posttranslational modifications. Nat Commun 6:5924. doi:10.1038/ncomms6924
61. Liu H, Li L, Voss C, Wang F, Liu J, Li SS (2015) A comprehensive immunoreceptor phosphotyrosine-based signaling network revealed by reciprocal protein-peptide array screening. Mol Cell Proteomics 14(7):1846–1858. doi:10.1074/mcp.M115.047951
62. Mayer BJ (2015) The discovery of modular binding domains: building blocks of cell signalling. Nat Rev Mol Cell Biol 16(11):691–698. doi:10.1038/nrm4068
63. Mayer BJ (2012) Perspective: dynamics of receptor tyrosine kinase signaling complexes. FEBS Lett 586(17):2575–2579. doi:10.1016/j.febslet.2012.05.002
64. Jadwin JA, Oh D, Curran TG, Ogiue-Ikeda M, Jia L, White FM, Machida K, Yu J, Mayer BJ (2016) Time-resolved multimodal analysis of Src Homology 2 (SH2) domain binding in signaling by receptor tyrosine kinases. Elife 5, e11835. doi:10.7554/eLife.11835
65. Kaneko T, Huang H, Cao X, Li X, Li C, Voss C, Sidhu SS, Li SS (2012) Superbinder

SH2 domains act as antagonists of cell signaling. Sci Signal 5243:68. doi:10.1126/scisignal.2003021

66. Findlay GM, Smith MJ, Lanner F, Hsiung MS, Gish GD, Petsalaki E, Cockburn K, Kaneko T, Huang H, Bagshaw RD, Ketela T, Tucholska M, Taylor L, Bowtell DD, Moffat J, Ikura M, Li SS, Sidhu SS, Rossant J, Pawson T (2013) Interaction domains of Sos1/Grb2 are finely tuned for cooperative control of embryonic stem cell fate. Cell 152(5):1008–1020. doi:10.1016/j.cell.2013.01.056
67. Wojcik J, Hantschel O, Grebien F, Kaupe I, Bennett KL, Barkinge J, Jones RB, Koide A, Superti-Furga G, Koide S (2010) A potent and highly specific FN3 monobody inhibitor of the Abl SH2 domain. Nat Struct Mol Biol 17(4):519–527. doi:10.1038/nsmb.1793
68. Sha F, Gencer EB, Georgeon S, Koide A, Yasui N, Koide S, Hantschel O (2013) Dissection of the BCR-ABL signaling network using highly specific monobody inhibitors to the SHP2 SH2 domains. Proc Natl Acad Sci U S A 110(37):14924–14929. doi:10.1073/pnas.1303640110
69. Yasui N, Findlay GM, Gish GD, Hsiung MS, Huang J, Tucholska M, Taylor L, Smith L, Boldridge WC, Koide A, Pawson T, Koide S (2014) Directed network wiring identifies a key protein interaction in embryonic stem cell differentiation. Mol Cell 54(6):1034–1041. doi:10.1016/j.molcel.2014.05.002
70. Lappalainen I, Thusberg J, Shen B, Vihinen M (2008) Genome wide analysis of pathogenic SH2 domain mutations. Proteins 72(2):779–792. doi:10.1002/prot.21970
71. Forbes SA, Bindal N, Bamford S, Cole C, Kok CY, Beare D, Jia M, Shepherd R, Leung K, Menzies A, Teague JW, Campbell PJ, Stratton MR, Futreal PA (2011) COSMIC: mining complete cancer genomes in the catalogue of somatic mutations in cancer. Nucleic Acids Res 39(Database issue):D945–D950. doi:10.1093/nar/gkq929
72. Cancer Genome Atlas Research N, Weinstein JN, Collisson EA, Mills GB, Shaw KR, Ozenberger BA, Ellrott K, Shmulevich I, Sander C, Stuart JM (2013) The cancer genome atlas pan-cancer analysis project. Nat Genet 45(10):1113–1120. doi:10.1038/ng.2764
73. Muller PJ, Rigbolt KT, Paterok D, Piehler J, Vanselow J, Lasonder E, Andersen JS, Schaper F, Sobota RM (2013) Protein tyrosine phosphatase SHP2/PTPN11 mistargeting as a consequence of SH2-domain point mutations associated with Noonan Syndrome and leukemia. J Proteomics 84:132–147. doi:10.1016/j.jprot.2013.04.005
74. Druker BJ, Guilhot F, O'Brien SG, Gathmann I, Kantarjian H, Gattermann N, Deininger MW, Silver RT, Goldman JM, Stone RM, Cervantes F, Hochhaus A, Powell BL, Gabrilove JL, Rousselot P, Reiffers J, Cornelissen JJ, Hughes T, Agis H, Fischer T, Verhoef G, Shepherd J, Saglio G, Gratwohl A, Nielsen JL, Radich JP, Simonsson B, Taylor K, Baccarani M, So C, Letvak L, Larson RA, Investigators I (2006) Five-year follow-up of patients receiving imatinib for chronic myeloid leukemia. N Engl J Med 355(23):2408–2417. doi:10.1056/NEJMoa062867
75. Byrd JC, Furman RR, Coutre SE, Flinn IW, Burger JA, Blum KA, Grant B, Sharman JP, Coleman M, Wierda WG, Jones JA, Zhao W, Heerema NA, Johnson AJ, Sukbuntherng J, Chang BY, Clow F, Hedrick E, Buggy JJ, James DF, O'Brien S (2013) Targeting BTK with ibrutinib in relapsed chronic lymphocytic leukemia. N Engl J Med 369(1):32–42. doi:10.1056/NEJMoa1215637
76. Machida K, Khenkhar M, Nollau P (2012) Deciphering phosphotyrosine-dependent signaling networks in cancer by SH2 profiling. Genes Cancer 3(5-6):353–361. doi:10.1177/1947601912459048
77. Machida K, Thompson CM, Dierck K, Jablonowski K, Karkkainen S, Liu B, Zhang H, Nash PD, Newman DK, Nollau P, Pawson T, Renkema GH, Saksela K, Schiller MR, Shin DG, Mayer BJ (2007) High-throughput phosphotyrosine profiling using SH2 domains. Mol Cell 26(6):899–915. doi:10.1016/j.molcel.2007.05.031
78. Takakuma K, Ogo N, Uehara Y, Takahashi S, Miyoshi N, Asai A (2013) Novel multiplexed assay for identifying SH2 domain antagonists of STAT family proteins. PLoS One 8(8), e71646. doi:10.1371/journal.pone.0071646
79. Machida K, Mayer BJ (2005) The SH2 domain: versatile signaling module and pharmaceutical target. Biochim Biophys Acta 1747(1):1–25. doi:10.1016/j.bbapap.2004.10.005
80. Watson GM, Gunzburg MJ, Ambaye ND, Lucas WA, Traore DA, Kulkarni K, Cergol KM, Payne RJ, Panjikar S, Pero SC, Perlmutter P, Wilce MC, Wilce JA (2015) Cyclic peptides incorporating phosphotyrosine mimetics as potent and specific inhibitors of the Grb7 breast cancer target. J Med Chem 58(19):7707–7718. doi:10.1021/acs.jmedchem.5b00609
81. Iwata T, Tanaka K, Tahara T, Nozaki S, Onoe H, Watanabe Y, Fukase K (2013) A conformationally fixed analog of the peptide mimic Grb2-SH2 domain: synthesis and evaluation against the A431 cancer cell. Mol Biosyst 9(5):1019–1025. doi:10.1039/c3mb25462c

82. Wojcik J, Lamontanara AJ, Grabe G, Koide A, Akin L, Gerig B, Hantschel O, Koide S (2016) Allosteric inhibition of Bcr-Abl kinase by high-affinity monobody inhibitors directed to the SH2-kinase interface. J Biol Chem. doi:10.1074/jbc.M115.707901
83. Grebien F, Hantschel O, Wojcik J, Kaupe I, Kovacic B, Wyrzucki AM, Gish GD, Cerny-Reiterer S, Koide A, Beug H, Pawson T, Valent P, Koide S, Superti-Furga G (2011) Targeting the SH2-kinase interface in Bcr-Abl inhibits leukemogenesis. Cell 147(2):306–319. doi:10.1016/j.cell.2011.08.046
84. Songyang Z, Cantley LC (1995) Recognition and specificity in protein tyrosine kinase-mediated signalling. Trends Biochem Sci 20(11):470–475
85. Kabak S, Skaggs BJ, Gold MR, Affolter M, West KL, Foster MS, Siemasko K, Chan AC, Aebersold R, Clark MR (2002) The direct recruitment of BLNK to immunoglobulin alpha couples the B-cell antigen receptor to distal signaling pathways. Mol Cell Biol 22(8):2524–2535
86. Lupher ML Jr, Songyang Z, Shoelson SE, Cantley LC, Band H (1997) The Cbl phosphotyrosine-binding domain selects a D(N/D)XpY motif and binds to the Tyr292 negative regulatory phosphorylation site of ZAP-70. J Biol Chem 272(52):33140–33144
87. Ng C, Jackson RA, Buschdorf JP, Sun Q, Guy GR, Sivaraman J (2008) Structural basis for a novel intrapeptidyl H-bond and reverse binding of c-Cbl-TKB domain substrates. EMBO J 27(5):804–816. doi:10.1038/emboj.2008.18, emboj200818 [pii]
88. Auger KR, Songyang Z, Lo SH, Roberts TM, Chen LB (1996) Platelet-derived growth factor-induced formation of tensin and phosphoinositide 3-kinase complexes. J Biol Chem 271(38):23452–23457
89. Bunnell SC, Diehn M, Yaffe MB, Findell PR, Cantley LC, Berg LJ (2000) Biochemical interactions integrating Itk with the T cell receptor-initiated signaling cascade. J Biol Chem 275(3):2219–2230
90. Jones N, Blasutig IM, Eremina V, Ruston JM, Bladt F, Li H, Huang H, Larose L, Li SS, Takano T, Quaggin SE, Pawson T (2006) Nck adaptor proteins link nephrin to the actin cytoskeleton of kidney podocytes. Nature 440(7085):818–823
91. Beebe KD, Wang P, Arabaci G, Pei D (2000) Determination of the binding specificity of the SH2 domains of protein tyrosine phosphatase SHP-1 through the screening of a combinatorial phosphotyrosyl peptide library. Biochemistry 39(43):13251–13260
92. Sweeney MC, Wavreille AS, Park J, Butchar JP, Tridandapani S, Pei D (2005) Decoding protein-protein interactions through combinatorial chemistry: sequence specificity of SHP-1, SHP-2, and SHIP SH2 domains. Biochemistry 44(45):14932–14947
93. Holland SJ, Gale NW, Gish GD, Roth RA, Songyang Z, Cantley LC, Henkemeyer M, Yancopoulos GD, Pawson T (1997) Juxtamembrane tyrosine residues couple the Eph family receptor EphB2/Nuk to specific SH2 domain proteins in neuronal cells. EMBO J 16(13):3877–3888
94. Poy F, Yaffe MB, Sayos J, Saxena K, Morra M, Sumegi J, Cantley LC, Terhorst C, Eck MJ (1999) Crystal structures of the XLP protein SAP reveal a class of SH2 domains with extended, phosphotyrosine-independent sequence recognition. Mol Cell 4(4):555–561
95. Hwang PM, Li C, Morra M, Lillywhite J, Muhandiram DR, Gertler F, Terhorst C, Kay LE, Pawson T, Forman-Kay JD, Li SC (2002) A "three-pronged" binding mechanism for the SAP/SH2D1A SH2 domain: structural basis and relevance to the XLP syndrome. EMBO J 21(3):314–323
96. Karlsson T, Songyang Z, Landgren E, Lavergne C, Di Fiore PP, Anafi M, Pawson T, Cantley LC, Claesson-Welsh L, Welsh M (1995) Molecular interactions of the Src homology 2 domain protein Shb with phosphotyrosine residues, tyrosine kinase receptors and Src homology 3 domain proteins. Oncogene 10(8):1475–1483
97. O'Bryan JP, Martin CB, Songyang Z, Cantley LC, Der CJ (1996) Binding specificity and mutational analysis of the phosphotyrosine binding domain of the brain-specific adaptor protein ShcC. J Biol Chem 271(20):11787–11791
98. Bullock AN, Rodriguez MC, Debreczeni JE, Songyang Z, Knapp S (2007) Structure of the SOCS4-ElonginB/C complex reveals a distinct SOCS box interface and the molecular basis for SOCS-dependent EGFR degradation. Structure 15(11):1493–1504
99. De Souza D, Fabri LJ, Nash A, Hilton DJ, Nicola NA, Baca M (2002) SH2 domains from suppressor of cytokine signaling-3 and protein tyrosine phosphatase SHP-2 have similar binding specificities. Biochemistry 41(29):9229–9236
100. Krebs DL, Uren RT, Metcalf D, Rakar S, Zhang JG, Starr R, De Souza DP, Hanzinikolas K, Eyles J, Connolly LM,

Simpson RJ, Nicola NA, Nicholson SE, Baca M, Hilton DJ, Alexander WS (2002) SOCS-6 binds to insulin receptor substrate 4, and mice lacking the SOCS-6 gene exhibit mild growth retardation. Mol Cell Biol 22(13):4567–4578

101. Wiederkehr-Adam M, Ernst P, Muller K, Bieck E, Gombert FO, Ottl J, Graff P, Grossmuller F, Heim MH (2003) Characterization of phosphopeptide motifs specific for the Src homology 2 domains of signal transducer and activator of transcription 1 (STAT1) and STAT3. J Biol Chem 278(18):16117–16128

102. Stahl N, Farruggella TJ, Boulton TG, Zhong Z, Darnell JE Jr, Yancopoulos GD (1995) Choice of STATs and other substrates specified by modular tyrosine-based motifs in cytokine receptors. Science 267(5202):1349–1353

Chapter 2

What Have We Learned from SH2 Domains?

Bruce J. Mayer

Abstract

SH2 domains first shed light on the key role of modular binding domains in cell signaling. Much of what we now know about the logic and design principles underlying cell signaling mechanisms, and how such mechanisms might have evolved, can be traced back to early work on SH2 domains. Here we briefly outline several key concepts that emerged from such studies.

Key words Modular binding domains, Tyrosine phosphorylation, Cell signaling

1 Introduction

More than 30 years ago, the first description of the SH2 domain was published by Tony Pawson's lab [1]. At that time, the SH2 domain was known only as a region of sequence similarity among diverse nonreceptor tyrosine kinases; mutations in this region affected the activity of the adjacent tyrosine kinase domain, but the region itself was not required for catalytic activity. From these modest beginnings, breathtaking advances in our understanding of cell signaling soon emerged. Over the ensuing half decade, the SH2 domain went from a mysterious "modulatory domain" to the archetype of a novel kind of protein structure: the modular binding domain [2]. In the current post-genomic world, we know that the SH2 domain is just one of the many hundreds of conserved domains that serve as the building blocks for the diverse proteins that make up a cell [3–5]. But of these, the SH2 domain takes pride of place as the one that first and most clearly illuminated many key principles of cell signaling. I would like to focus on three such key concepts that can be traced back to SH2 domains: reader domains, the modularity of signaling proteins, and the importance of changes in subcellular localization and protein–protein binding in signaling. Each is briefly discussed below.

Kazuya Machida and Bernard A. Liu (eds.), *SH2 Domains: Methods and Protocols*, Methods in Molecular Biology, vol. 1555,
DOI 10.1007/978-1-4939-6762-9_2, © Springer Science+Business Media LLC 2017

2 Reader Domains and the Writer/Eraser/Reader Paradigm

The defining feature of the SH2 domain is that it binds specifically to short peptides only when they are phosphorylated on tyrosine residues [6, 7]. Thus they serve to "read" the posttranslational marks on proteins generated by tyrosine kinases, thereby converting changes in enzymatic activity (of tyrosine kinases, and of tyrosine phosphatases which remove the marks) into changes in protein–protein binding, which in turn may cause other changes such as in subcellular localization (Fig. 1). Many signaling systems are now known to involve similar "writer/eraser/reader" systems, in which modular reader domains, by binding to posttranslationally modified sites, read out the activity of enzymes that generate and remove posttranslation marks ("writers" and "erasers" respectively) [8].

Although the SH2 domain was the first modular domain shown to depend on a specific posttranslational modification, the idea of a signaling system in which information is encoded by reversible posttranslational modification and read out by the binding of modular domains in a modification-dependent fashion is probably most fully appreciated in the context of chromatin modifications [9, 10]. The histones that package genomic DNA into chromatin can be enzymatically modified in a variety of ways, including by phosphorylation, lysine acetylation, lysine and arginine methylation, and ubiquitinylation; each of these modifications can in turn be recognized by one or more modular binding domains [11]. Many of the modifying or demodifying enzymes

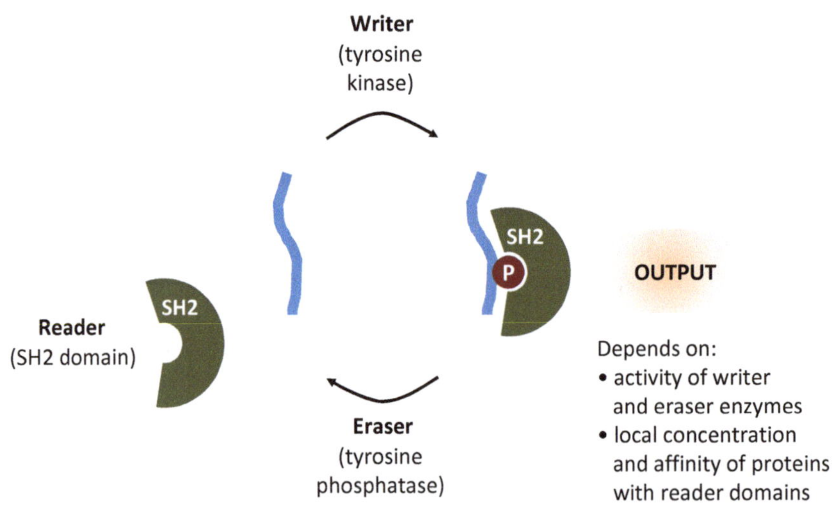

Fig. 1 SH2 domains function as readers in a Writer/Eraser/Reader signaling system. SH2 domains serve to "read" the level of their tyrosine-phosphorylated binding sites in the cell, which in turn is determined by the relative activity of tyrosine kinases and phosphatases. Signal output ultimately depends on the amount (and location) of complexes between the SH2 domain-containing protein and its phosphorylated binding sites

themselves contain reader domains that bind specifically to modified sites, and thus can mediate positive and negative feedback. Thus it is easy to envision how such writer/eraser/reader systems could serve as the basis for the extraordinarily complex regulation of chromatin structure that underlies the control of transcription, replication, and DNA repair.

3 Modularity of Signaling Proteins

The SH2 domain first highlighted one of the key properties of signaling proteins (and indeed most proteins)—that they are built by stringing together multiple modular domains, each conferring particular functions. While the SH2 domain was first identified in nonreceptor tyrosine kinases, it soon was also found in combination with other domains in a variety of other signaling proteins, including enzymes, scaffold/adaptor proteins, transcription factors, and others [12]. In most cases the isolated SH2 domain could be expressed in the absence of other domains, retaining its ability to bind specific tyrosine phosphorylated peptides [13–15]. So, much like the standardized electronic or mechanical components found in complex human-made devices, SH2 domains can be plugged into many different specific contexts to provide its unique function. The ability of such modular domains to confer their function to a variety of different proteins in combinatorial fashion is a common property of signaling proteins [16].

The idea of protein modularity that began with the SH2 domain has profound implications for how signaling proteins are regulated and how signaling circuits could evolve. The existence of a modular "parts list" of functional domains that can be genetically shuffled and recombined provides a ready mechanism for the evolution of new signaling proteins and new regulatory relationships in signaling networks [17]. Modularity ensures that important functions can be imported into new contexts—that is, a recombination event that results in insertion of an SH2 domain into a protein where it previously did not exist, has a good chance of producing a novel protein that is both functional and has acquired the ability to interact with (and potentially to regulate and be regulated by) specific tyrosine phosphorylated proteins. Modularity also provides a ready mechanism to regulate enzymes: intramolecular autoinhibition of a catalytic domain by a modular binding domain can be alleviated by binding of the domain to its ligand in trans [18] (Fig. 2). In the absence of modularity (that is, if each protein must incrementally evolve on its own, as an integrated unit), it is much more difficult to imagine how complex regulatory circuits, or the regulation of signaling proteins by multiple inputs, could easily evolve.

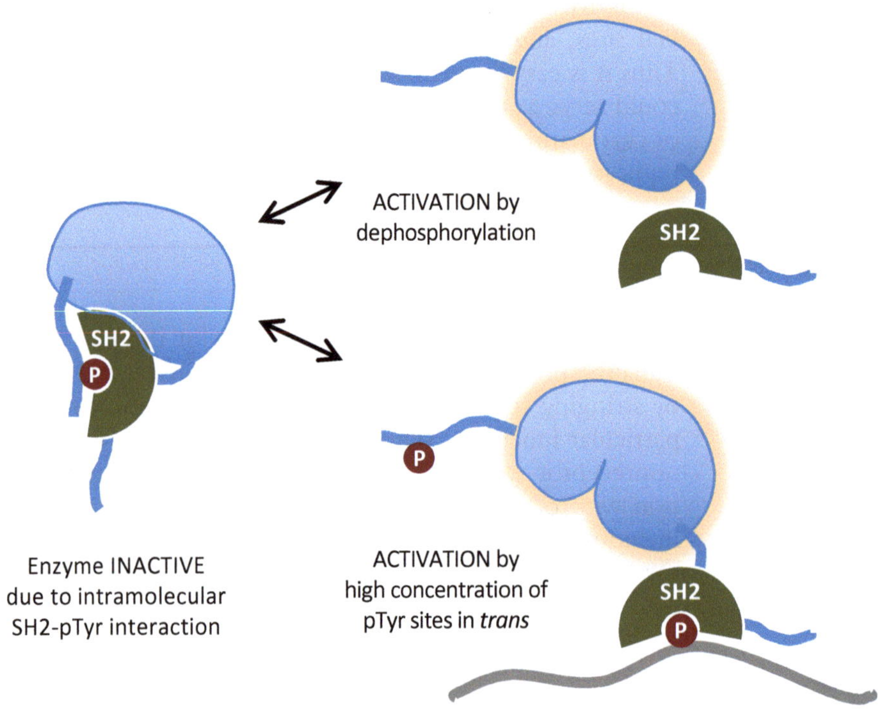

Fig. 2 Modular binding domains provide a mechanism to regulate enzyme activity. In this example, a catalytic domain is held in an inactive conformation by intramolecular interactions between an SH2 domain and a tyrosine-phosphorylated site on the same protein (*left*). This arrangement allows several straightforward mechanisms to relieve inhibition of the enzyme: either dephosphorylation of the intramolecular regulatory site (*top*), or binding of the SH2 domain to exogenous tyrosine phosphorylated sites in *trans* (*bottom*). While this example is generic, such a scheme is actually used to regulate the activity of nonreceptor tyrosine kinases such as Src

4 Subcellular Localization and Protein–Protein Interaction as Fundamental Currencies of Signaling

While the essential biochemical activity of the SH2 domain is to bind tyrosine-phosphorylated peptides, the most common effect of binding of an SH2 domain to its target is to relocalize the protein containing the domain. For example, relocalization of cytosolic SH2-containing proteins to the plasma membrane is the critical step for signaling by receptor tyrosine kinases, as well as by receptors coupled to nonreceptor tyrosine kinases, such as B-cell and T-cell receptors, cytokine receptors, and integrins [19]. This simple act of changing the localization of SH2-containing proteins in the cell has important consequences, such as increasing the effective activity of enzymes whose substrates are localized to membranes. This was first demonstrated in the case of the Ras activator Sos, which is recruited to phosphorylated receptors on the membrane via the SH2 domain containing adaptor protein Grb2; recruitment to the membrane was critical because its target, the small G protein Ras, is confined to the membrane by covalent lipid modifications [20]. The substrates of other SH2 domain-containing enzymes, such as phospholipase C and phosphatidylinositol 3-kinase, are themselves membrane lipids, and thus recruitment to the membrane increases their access to these substrates (Fig. 3).

We now appreciate that changing the subcellular distribution of proteins, especially from cytosol to plasma membrane or nucleus, is a very common mechanism for transmitting signals in the cell. Indeed, optogenetic approaches that use light to precisely control protein–protein interactions and subcellular localization, mimicking the normal signal-induced relocalization of proteins, are now powerful tools for dissecting downstream effects of signaling [21, 22].

In a broader sense, the discovery that most of the effects of tyrosine phosphorylation are exerted indirectly, by creating binding sites for SH2 domains on downstream effector proteins, was fundamentally important in focusing attention on the specific types of changes that lead to signal transmission. The transmission or processing of information depends on change in some property of the system (for example, changes in the frequency or amplitude of sound or radio waves). It turns out that changes in a limited number of biophysical properties are used over and over in cell signaling systems; these can be considered the "currencies" of signaling. These include changes

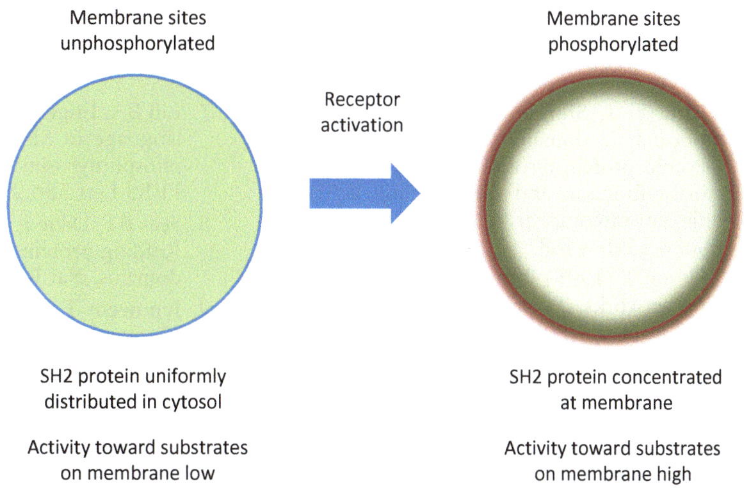

Fig. 3 SH2-pTyr interactions drive relocalization to the membrane. In an unstimulated cell (*left*), phosphorylation of receptors and associated proteins on the plasma membrane is low, and SH2-containing effectors (*green color*) are uniformly distributed. Upon receptor activation, tyrosine phosphorylation increases on the membrane (*red*), recruiting SH2-containing effector proteins and greatly increasing their concentration near the membrane. When effector substrates are also localized on the membrane, total activity increases due to high local concentrations of both substrate and effector, even though the total amount of the effector in the cell is unchanged

in enzymatic activity, posttranslational modification, protein–protein interactions, subcellular distribution, and a few others. Often these changes are linked together—as seen when change in the activity of tyrosine kinases leads to increased phosphorylation, which in turn leads to increased SH2-mediated binding and relocalization to the membrane. It is fair to say that the insights first appreciated in SH2 domains were important in moving the field beyond the "black box" toward a much fuller understanding of the simple logic that underlies all signaling mechanisms.

5 Prospects

The discovery of the SH2 domain occurred during a golden age for cell signaling, when the availability of new and improved techniques for DNA sequencing, PCR-based molecular cloning, methods to express and purify recombinant proteins efficiently, advances in structural biology, and many others innovations led to rapid progress and startling new discoveries. However even a quarter-century later, the SH2 domain still provides many new insights into biology and signaling, and serves as a test-bed for ever more powerful technologies and experimental approaches (*see* Chapter 1). It is a lot to expect from a mere hundred or so amino acids, but as shown in the following volume, the future promises to be just as brilliant.

References

1. Sadowski I, Stone JC, Pawson T (1986) A noncatalytic domain conserved among cytoplasmic protein-tyrosine kinases modifies the kinase function and transforming activity of fujinami sarcoma virus P130*gag-fps*. Mol Cell Biol 6:4396–4408
2. Pawson T (1995) Protein modules and signalling networks. Nature 373:573–579
3. Orengo CA, Thornton JM (2005) Protein families and their evolution-a structural perspective. Annu Rev Biochem 74:867–900
4. Müller A, MacCallum RM, Sternberg MJ (2002) Structural characterization of the human proteome. Genome Res 12:1625–41
5. Vogel C, Bashton M, Kerrison ND, Chothia C, Teichmann SA (2004) Structure, function and evolution of multidomain proteins. Curr Opin Struct Biol 14:208–216
6. Machida K, Mayer BJ (2005) The SH2 domain: versatile signaling module and pharmaceutical target. Biochim Biophys Acta 1747:1–25
7. Liu BA, Engelmann BW, Nash PD (2012) The language of SH2 domain interactions defines phosphotyrosine-mediated signal transduction. FEBS Lett 586:2597–2605
8. Seet BT, Dikic I, Zhou MM, Pawson T (2006) Reading protein modifications with interaction domains. Nat Rev Mol Cell Biol 7:473–483
9. Jenuwein T, Allis CD (2001) Translating the histone code. Science 293:1074–1080
10. Strahl BD, Allis CD (2000) The language of covalent histone modifications. Nature 403: 41–45
11. Zentner GE, Henikoff S (2013) Regulation of nucleosome dynamics by histone modifications. Nat Struct Mol Biol 20:259–266
12. Liu BA, Jablonowski K, Raina M, Arce M, Pawson T, Nash P (2006) The human and mouse complement of SH2 domain proteins–establishing the boundaries of phosphotyrosine signaling. Mol Cell 22:851–868
13. Huang H, Li L, Wu C, Schibli D, Colwill K, Ma S, Li C, Roy P, Ho K, Songyang Z, Pawson

T, Gao Y, Li SS (2008) Defining the specificity space of the human SRC homology 2 domain. Mol Cell Proteomics 7:768–784

14. Jones RB, Gordus A, Krall JA, MacBeath G (2006) A quantitative protein interaction network for the ErbB receptors using protein microarrays. Nature 439:168–174
15. Machida K, Thompson CM, Dierck K, Jablonowski K, Kärkkäinen S, Liu B, Zhang H, Nash PD, Newman DK, Nollau P, Pawson T, Renkema GH, Saksela K, Schiller MR, Shin DG, Mayer BJ (2007) High-throughput phosphotyrosine profiling using SH2 domains. Mol Cell 26:899–915
16. Basu MK, Carmel L, Rogozin IB, Koonin EV (2008) Evolution of protein domain promiscuity in eukaryotes. Genome Res 18:449–461
17. Bhattacharyya RP, Reményi A, Yeh BJ, Lim WA (2006) Domains, motifs, and scaffolds: the role of modular interactions in the evolution and wiring of cell signaling circuits. Annu Rev Biochem 75:655–680
18. Lim WA (2002) The modular logic of signaling proteins: building allosteric switches from simple binding domains. Curr Opin Struct Biol 12:61–68
19. Schlessinger J, Lemmon MA (2003) SH2 and PTB domains in tyrosine kinase signaling. Sci STKE 2003:RE12
20. Margolis B, Skolnik EY (1994) Activation of Ras by receptor tyrosine kinases. J Am Soc Nephrol 5:1288–1299
21. Toettcher JE, Voigt CA, Weiner OD, Lim WA (2011) The promise of optogenetics in cell biology: interrogating molecular circuits in space and time. Nat Methods 8:35–38
22. Zhang K, Cui B (2015) Optogenetic control of intracellular signaling pathways. Trends Biotechnol 33:92–100

Part II

Computational Approaches

Chapter 3

Hidden Markov Models for Protein Domain Homology Identification and Analysis

Karl Jablonowski

Abstract

Protein domain identification and analysis are cornerstones of modern proteomics. The tools available to protein domain researchers avail a variety of approaches to understanding large protein domain families. Hidden Markov Models (HMM) form the basis for identifying and categorizing evolutionarily linked protein domains. Here I describe the use of HMM models for predicting and identifying Src Homology 2 (SH2) domains within the proteome.

Key words Hidden Markov model, Sequence alignment, Neighbor-joining phylogenetic tree, SH2 domains

1 Introduction

SH2 domains partake in some of the canonical Darwinian evolution paradigms such as duplication, divergence, and convergence. All members of the SH2 domain family are linked through evolution. It is this link that forms the basis of similarity and allows for identification and analysis.

This chapter exhibits an example of protein homology identification and analysis via Hidden Markov Models (HMM) and protein sequence alignment algorithms. Both of these methods are meaningfully usable because protein domains from the same family are evolutionarily linked and thus share amino acid transitions utilized by HMMs and amino acid sequences utilized by sequence alignment algorithms. The conservation of each of these two features between proteins is a function of evolutionary distance.

Proteomic evolution gives rise to protein families [1]. A protein, once created from its genetic coding, is selected for through a biological function and thus retained in the genome. When nature imposes a survival pressure for the organism to select a new protein based biological function, the currents of evolution have three primary choices: *convergent evolution*—a different protein

Kazuya Machida and Bernard A. Liu (eds.), *SH2 Domains: Methods and Protocols*, Methods in Molecular Biology, vol. 1555, DOI 10.1007/978-1-4939-6762-9_3, © Springer Science+Business Media LLC 2017

mutating to fulfill this function; *de novo evolution*—creating a protein from proverbial scratch with unrelated genetic material; or through *gene duplication and mutation*—copying the existing protein and mutating it to fulfill the additional biological function.

Gene duplication and mutational events are employed widely throughout evolution and is the basis that gives rise to protein families [1, 2]. All SH2 domains likely evolved from a singular predecessor. When the gene that codes for a protein domain duplicates to create a second protein-domain coding gene, the relationship between the two protein domains is called a *paralog*. When the organism's evolution diverges into two distinct species, the protein domain from one species is an *ortholog* to the matching protein domain in the other. Paralogs and orthologs together comprise the protein domain evolutionary relationships known as *homologs*.

Related protein domains share features (for example the amino acid sequences that determine the secondary structure and protein fold). Tools were developed to identify these features such as Multiple Sequence Alignment and HMM. Databases are dedicated to storing the information amassed while identifying these domain-specific and family-specific evolutionary relationships. Pfam [3] is a database of sequence alignments and HMMs of protein families. PROSITE [4] is a database of protein families, domains and functional sites as well as patterns to identify them. SUPERFAMILY [5] is a database of annotated families and superfamilies in completely sequenced genomes. InterPro [6] is a database of identifiable features for characterizing families, domains, and functional sites. PROFESS [7] is a database of protein function, evolution, structure, and sequence. PANDIT [8] is a database of protein domains that assesses phylogentic relationships based on coding-DNA of the proteins. Many more databases contribute to the body of knowledge of proteomic evolutionary relationships, too numerous to enumerate here. Those that require no human annotation may even be automatically updated as new species' proteomes or additions to existing proteomes are published, making for nearly real-time proteomic analysis. The aforementioned databases and many others are quintessential to proteome-wide analyses.

After establishing the evolutionary relatedness of all SH2 domains, one may use it to further analyze other facets of the protein domains such as peptide ligand binding patterns and inferred function (*see* Chapter 7).

1.1 Hidden Markov Models

HMMs are a means of modeling transitions from one state to another. For proteomic sequence analyses the states are amino acids and the transitions are the subsequent reads for position 1 and position 2. For example, consider the phosphotyrosine peptide sequence Y-D-D-P, that binds to many SH2 domains [9]. The states are the amino acids tyrosine (Y), aspartic acid (D), and proline (P) and the transitions are Y→D, D→D, and D→P.

Before creating an HMM the model must be trained on a sizable seed-set that is broad enough to identify disparate protein domains. The Pfam seed-set for the SH2 domain, for example, is made up of 58 sequences from 15 different species and over many protein families. A seed-set is the initial set of examples that is used to build a model, in this case it is the SH2 domain sequences and sequence transitions that will inform the HMM. There is a tradeoff between sensitivity and specificity. Too small a seed-set and you will miss domains distantly related to the seed-set proteins. Too large a seed-set and you will introduce much noise and duplicative proteins that can compromise the accuracy of the model. For example, if an HMM was trained on STAT1, the HMM would only be able to identify STAT1 state transitions. This model would effectively be able to identify only STAT1 and lose other SH2 domains in the false positive and false negative noise of the proteome. Therefore, the seed-set must also be varied and broad. Moreover, if an HMM was trained only on the STAT family SH2 domains (STAT1 and STAT2), the ability to identify the family members of SLAP, SHIP, and CHN SH2 domains would be severely limited as they lie at the opposite end of the SH2 phylogenetic tree [10]. Thus, a reasonably sized and varied seed-set will often derive a more versatile HMM (Fig. 1).

1.2 Multiple Sequence Alignment

Multiple sequence alignments are created by an algorithm like the Smith–Waterman algorithm. The Smith–Waterman [11] is an implementation of dynamic programming (an instance of the Dijkstra algorithm [12]). Dynamic programming is a method for finding an optimal solution of a complex problem by partitioning it into many simple sub-problems. To illustrate dynamic programming one may perform a global alignment of the first alpha-helices of SH2D4B and SH2D5. Create a matrix with a column for every amino acid in the SH2D4B sequence, and a row for every amino acid in the SH2D5 sequence (Fig. 2a). Pad the matrix with 1s in the first row and first column. Every matrix element is a conditional maximization function. Start at the upper left corner. If the amino acid of the column and of the row match, the matrix element is set equal to the upper-left diagonal value plus one. Otherwise, it is set equal to the maximum value of either the matrix element above or to the left. Repeat this protocol to populate the whole of the matrix (Fig. 2b). To find the path through this matrix, start at the bottom right corner. Follow the matrix elements of the same number to the left, and then up. A number change can only occur diagonally up and to the left, and always decreases by one. This path represents a global alignment of matching, inserting, or deleting amino acids.

Multiple sequence alignment algorithms can compute the "shortest path" between two sequences. That is, it counts the minimum number of insertions, deletions, or substitution of amino acids one must make for sequence A to match sequence B.

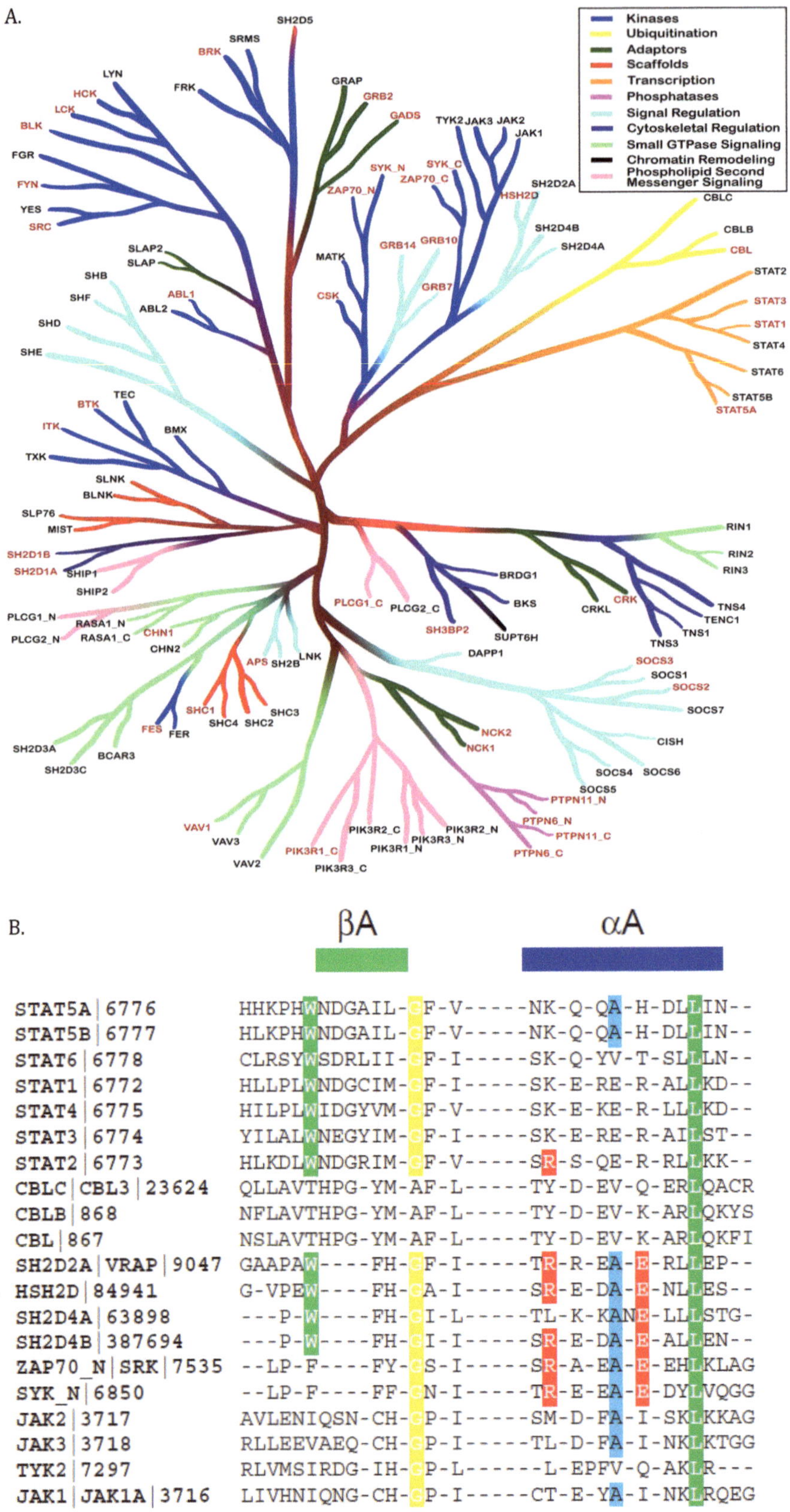

Fig. 1 SH2 Domain phylogenetic tree (**a**) and partial sequence alignment (**b**). The full alignment may be found at https://sites.google.com/site/sh2domain/alignment. Reprinted with permission from Elsevier and Rightslink by Copyright Clearance Center [10]

A.

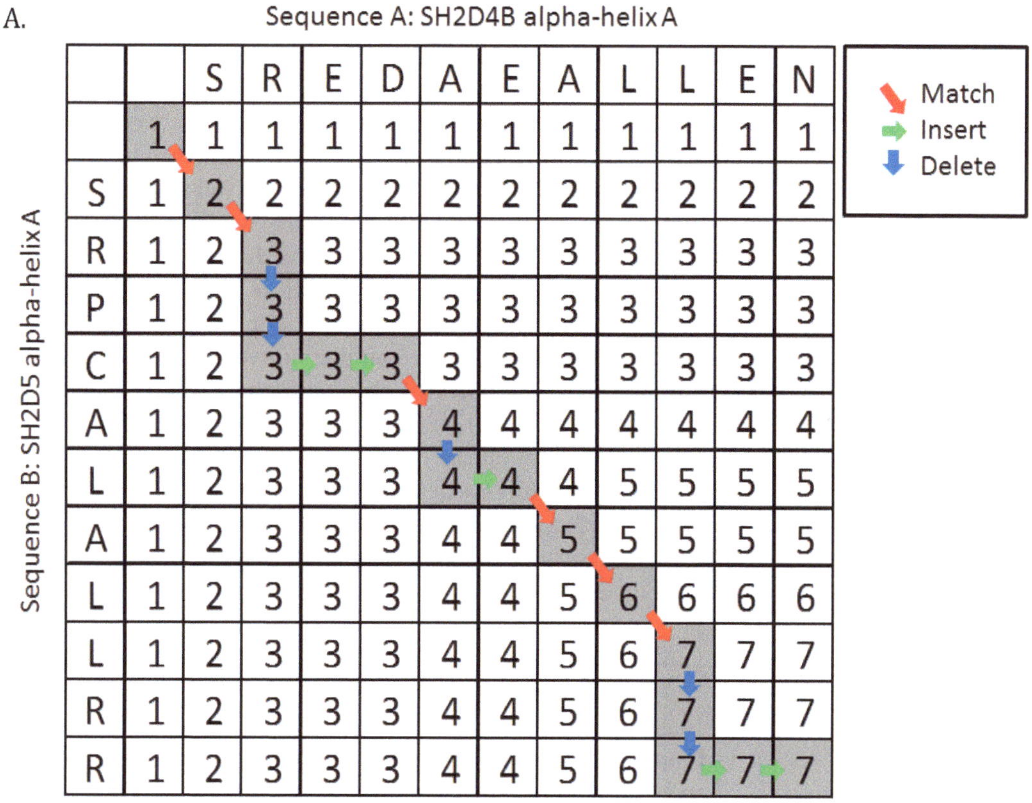

B.

A, B are sequences

A_i is the ith element of the sequence A

$M(i,j)$ is a position in the Matrix, row i and column j

If $(A_i=B_j)$: $M(i,j) = M(i-1,j-1) + 1$

else: $M(i,j) = Max(M(i-1,j), M(i,j-1))$

C.

$S(A_i,B_j)$ is the weight of mutation from the ith element of sequence A to the jth element of sequence B, derived from the substitution matrix

If $(A_i=B_j)$: $M(i,j) = M(i-1,j-1) + S(A_i,B_j)$

else: $M(i,j) = Max(M(i-1,j), M(i,j-1), M(i-1,j) + S(A_{i-1},B_j), M(i,j-1) + S(A_i,B_{j-1}))$

Fig. 2 Multiple sequence alignment based on dynamic programming. (**a**) Sequence matrix for SH2D4B and SH2 D5. (**b**) Conditional maximization function. (**c**) Weighting scheme of a substitution matrix

Weightings may also be defined to treat insertions, deletions, or substitution of specific amino acids unequally. These weightings take the form of a gap-scoring scheme and a substitution matrix, which will affect the shortest path calculation.

One may want to use a substitution matrix tailored to random genetic drift. For example, mutating a methionine to a threonine would require one codon nucleotide mutation, methionine to a serine would require two codon nucleotides mutations, and methionine to a tyrosine would require all three codon nucleotides to be mutated. Hence, random genetic drift would not act equally on all amino acid substitutions. For an alignment algorithm to reflect this inequality of substitutions, the inclusion of a weighting matrix as a parameter is necessary. A weighting matrix would take the form of a 20 by 20 matrix. Rows would represent the original amino acids, columns would represent the substitution amino acids, and the intersection of a row and a column in the matrix would be the weighting associated with the original converting to the substitution amino acid. Examples of these are the PAM [13] (Percent Accepted Mutation) and the BLOSUM [14] (Block Substitution Matrix) series of mutation matrices. The dynamic programming example above requires a deletion and insertion if amino acids do not match. With a slight adjustment it may be modified to consider mutations, incorporating the weighting scheme of a substitution matrix (Fig. 2c).

2 Materials

2.1 Databases

1. Human Proteome databases. Uniprot [15], Ensembl [16], and NCBI [17].
2. Hidden Markov model web-based algorithms (e.g., SMART [18], Pfam [3], Interpro [6], PROSITE [19]).

2.2 HMMER (HMMBuild and HMMSearch)

1. HMMER [20] is a suite of software widely used to identify homologous protein (and nucleotide) sequences. HMMER can be used to both generate HMMs and to use HMMs to identify matches in a single sequence or a database of sequences. This section focuses on two of HMMER's utilities, HMMBuild and HMMSearch (*see* **Notes 1–3**).

2.3 ClustalW

1. ClustalW [21] is a command line interface program. It is one of the most popular multiple sequence alignment series of programs utilized in bioinformatics. ClustalW also has the additional functionality of creating neighbor-joining phylogenetic trees based on the multiple sequence alignments it performs. ClustalX is a widely used Graphical User Interface (GUI) for ClustalW (*see* **Notes 4** and **5**).

3 Methods

3.1 Identification of SH2 Domains Using HMMs

1. Download a human proteome database (Uniprot, Ensembl, or NCBI) in the FASTA file format (such as human.fasta). The example we will use, the reviewed Swiss-Prot FASTA file, may

be downloaded from UniProt Knowledge Base download website (http://www.uniprot.org/downloads).

2. Download and install HMMER. The installation should contain a minimum of the HMMBuild and HMMSearch utilities.
3. Download PFam's SH2 seed set from the PFam website (http://pfam.xfam.org/family/PF00017/alignment/seed) and save it locally to a file on your computer (e.g., "sh2.aln").
4. Build an HMM. The Linux command "hmmbuild model.hmm sh2.aln" will create an HMM from the seed set and save it into the file "model.hmm". *See* Fig. 3a for the command line input and output, and Fig. 3b for the model.
5. Search the human proteome using the SH2 HMM. The Linux command "hmmsearch –A result.txt --incE 0.001 model.hmm uniprot_sport.fasta" will provide the search results as a multiple sequence alignment into the file "result.txt". The incE is an E-value threshold cutoff, and as set to 0.001 would allow one false positive for every 1000 true positives. The results (*see* Fig. 4) will have an inclusion threshold line, where sequences above the line meet the threshold cutoff and below the line do not.
6. This set of SH2 domains will contain duplicates depending on the source database. Since the Swiss-Prot data source we used was reviewed, the step is not necessary for this example. If the data source has not been reviewed, it may be necessary to manually filter the set of domains based on identical gene loci, plurality due to sequence errors or polymorphisms, gene duplication events, splice variants, and pseudogenes. Entrez Gene [22] is a great resource for tracking genetic provenance of protein sequences.
7. The file "result.txt" now contains the statistically significant SH2 domain HMM matches in the proteome. The N and C terminal sequences extend beyond the conserved central core of SH2 domains and are important for protein folding and formation of the phosphotyrosine peptide binding pocket (*see* **Note 6**).
8. As a validation step, it is good practice to confirm the prediction of the SH2 domain in the sequences from the "result.txt" file. The web interfaces for SMART, InterPro, and Pfam (listed in author's preferred order) offer protein sequence matching and analyzing tools for such a purpose. Confirm predictions using SMART or InterPro.

3.2 Sequence Alignments Using ClustalW

1. Download and install ClustalW.
2. Convert the "result.txt" HMMSearch file from the Stockholm format to the Clustal format. Execute the Linux command line "grep '#' –v result.txt > result.clu" to remove the Stockholm mark-up lines. Execute "sed –e 's/\./-/g' result.clu > result.

A.

```
Command Line Executed:
        hmmbuild model.hmm sh2.aln

Command Line Output:
# hmmbuild :: profile HMM construction from multiple sequence alignments
# HMMER 3.0 (March 2010); http://hmmer.org/
# Copyright (C) 2010 Howard Hughes Medical Institute.
# Freely distributed under the GNU General Public License (GPLv3).
# - - - - - - - - - - - - - - - - - - - - - - - - - - - - - - - - - - - -
# input alignment file:            sh2.aln
# output HMM file:                 model.hmm
# - - - - - - - - - - - - - - - - - - - - - - - - - - - - - - - - - - - -

# idx name                 nseq  alen  mlen eff_nseq re/pos description
#---- -------------------- ----- ----- ----- -------- ------ -----------
1    SH2                     53   109    77     3.73  0.735 SH2 domain

# CPU time: 0.16u 0.02s 00:00:00.18 Elapsed: 00:00:00.22
```

B.

```
HMMER3/b [3.0 | March 2010]
NAME  SH2
ACC   PF00017.21
DESC  SH2 domain
LENG  77
ALPH  amino
RF    no
CS    yes
MAP   yes
DATE  Fri Nov 11 00:32:10 2016
NSEQ  53
EFFN  3.729797
CKSUM 3656216282
GA    21.00 21.00
TC    21.00 21.00
NC    20.90 20.90
STATS LOCAL MSV       -9.2540  0.71860
STATS LOCAL VITERBI   -9.7040  0.71860
STATS LOCAL FORWARD   -4.1295  0.71860
HMM          A        C        D        E        F        G        H        I        K
L        M        N        P        Q        R        S        T        V        W        Y
            m->m     m->i     m->d     i->m     i->i     d->m     d->d
  COMPO   2.76986  4.57526  3.04297  2.64778  3.06323  2.84906  3.46020  2.89586  2.66087
2.40067  3.83003  3.10186  3.68847  3.04927  2.76141  2.57406  2.88732  2.76035  4.23622
3.10764
          2.68618  4.42225  2.77519  2.73123  3.46354  2.40513  3.72494  3.29354  2.67741
2.69355  4.24690  2.90346  2.73739  3.18146  2.89801  2.37887  2.77519  2.98518  4.58477
3.61503
          0.00909  5.10114  5.82348  0.61958  0.77255  0.00000        *
      1   5.13582  6.03777  5.97324  5.92810  2.30743  5.60694  4.21808  4.68400  5.68932
3.15181  5.19619  5.13025  5.85106  5.30870  5.46303  5.01350  5.32285  4.64530  0.54865
1.65640      1 - T
          2.68618  4.42225  2.77519  2.73123  3.46354  2.40513  3.72494  3.29354  2.67741
2.69355  4.24690  2.90346  2.73739  3.18146  2.89801  2.37887  2.77519  2.98518  4.58477
3.61503
```

Fig. 3 Hidden Markov model (HMM). (**a**) Command line input and output. (**b**) Model HMM

clu" to replace unpaired region annotation with the Clustal compatible annotation. Insert "CLUSTAL X (1.81) multiple sequence alignment" at the top of the file, and delete the "//" at the bottom.

3. Command ClustalW to generate a neirghbor-joining tree with the line "clustalw –infile=result.clu -tree". The resulting Newick/PHYLIP formatted file "result.ph" will automatically

```
Command Line Output:
# hmmsearch :: search profile(s) against a sequence database
# HMMER 3.0 (March 2010); http://hmmer.org/
# Copyright (C) 2010 Howard Hughes Medical Institute.
# Freely distributed under the GNU General Public License (GPLv3).
# - - - - - - - - - - - - - - - - - - - - - - - - - - - - - - - - - - - -
# query HMM file:                    model.hmm
# target sequence database:          uniprot_sprot.fasta
# MSA of all hits saved to file:     result.txt
# sequence inclusion threshold:      E-value <= 0.001
# - - - - - - - - - - - - - - - - - - - - - - - - - - - - - - - - - - - -

Query:        SH2  [M=77]
Accession:    PF00017.21
Description:  SH2 domain
Scores for complete sequences (score includes all domains):
   --- full sequence ---    --- best 1 domain ---    -#dom-
    E-value  score  bias    E-value  score  bias    exp  N  Sequence
    ------- ------ -----    ------- ------ -----    ---- --  --------
    5.1e-50  172.1   0.8    2.2e-24   90.0   0.0    3.6  2  sp|P29349|CSW_DROME
    6.6e-50  171.8   0.0    4.9e-23   85.7   0.0    2.8  2  sp|P41499|PTN11_RAT
    7.1e-50  171.7   0.0    4.8e-23   85.7   0.0    2.7  2  sp|P35235|PTN11_MOUSE
...
    0.00041   25.0   0.3     0.0019   22.9   0.0    2.4  1  sp|Q54G43|SHKE_DICDI
    0.00079   24.1   0.0     0.0028   22.3   0.0    2.0  1  sp|Q24151|STAT_DROME
    0.00084   24.0   0.0     0.0023   22.6   0.0    1.8  1  sp|Q7KZ85|SPT6H_HUMAN
  ------ inclusion threshold ------
      0.001   23.8   0.0      0.003   22.3   0.0    1.9  1  sp|Q4HYQ4|SPT6_GIBZE
     0.0011   23.7   0.0     0.0026   22.5   0.0    1.7  1  sp|Q5KLR1|SPT6_CRYNE
     0.0017   23.0   0.0     0.0042   21.8   0.0    1.8  1  sp|Q9NAD6|STAT1_CAEEL
...
```

Fig. 4 HMM SH2 search output

be generated in the same directory. This tree represents the neighbor-joining tree as HMMSearch aligns protein sequences.

4. For a complete ClustalW sequence alignment, and a necessary step if the sequences have been augmented as suggested in Subheading 3.1, **step** 7, the full multiple sequence alignment command is "clustalw –infile=result.clu -align". This will generate an alignment file "result.aln" and a neighbor-joining tree "result.dnd".

5. Create a dendrogram as an unrooted neighbor-joining tree (*see* **Note** 7). The resulting dendrogram is the basis for Fig. 1.

4 Notes

1. The HMMBuild utility is used to create an HMM profile. The utility takes as an input a multiple sequence alignment file and produces a file containing the HMM profile based on a few parameters such as weighting scheme, alphabet selection, and format. An HMM profile file is a file that contains the transitions and their associated probabilities that make up the essence of the model.

2. The user may then search a database or data file using that particular HMM profile via the utility HMMSearch. The result of this search is a score (or e-value—Expectation Value: a measure

of statistical significance of the sequence score) for each sequence against which the profile was compared. The output of this search utility is customizable and may include threshold cutoffs based on score or e-value.

3. HMMER is the primary Hidden Markov Model suite of software used by many protein family and protein domain databases for identifying homologous protein sequences. The software suite streamlined the process of creating HMM profiles and searching vast databases for matches to those profiles. The software suite also embodies many more utilities for the customization of that process.
4. Choosing a quantitative comparison algorithm (like the Smith–Waterman sequence alignment algorithm) and selecting the appropriate parameters (like the substitution matrix and the gap-scoring scheme) may have a drastic impact on the outcome of the neighbor-joining phylogenetic tree. The researcher must take care to understand how the algorithm chosen optimizes the relevant parameters.
5. There exist a plethora of alternative multiple sequence alignment algorithms to choose from, each with their own method for optimizing a form of alignment. BLAST [23] (Basic Local Alignment Search Tool) segments sequences into many smaller subsequence and matches then further computes on those smaller subsequences. Clustal Omega [24] uses HMM and a guide tree technique. MAFFT [25] (Multiple Alignment using Fast Fourier Transform) uses, as the name implies, a fast Fourier transform technique to assist alignments. PRANK [26] (Probabilistic Alignment Kit) uses phylogentic information and rebalances the cost-function associated with insertions and deletions based on evolutionary distances between sequences. The available multiple sequence alignment algorithms (too numerous to enumerate here) continues to expand and continues to improve.
6. These regions are typically not as well conserved and therefore are not included in the matched domain regions. It is good practice to extend the SH2 domain sequence 20 amino acids on both the N and C termini. These sequences may be found in the original "uniprot_sprot.fasta" file.
7. Neighbor-joining tree can be visualized using programs such as Phylip's Drawtree [27], TreeView [28], or NJPlot [29].

Acknowledgments

The knowledge amassed to write this chapter was based on work supported by the University of Chicago Cancer Research Foundation Women's Board and Piers Nash's laboratory at the University of Chicago Ben May Department for Cancer Research.

References

1. Liu BA, Shah E, Jablonowski K, Stergachis A, Engelmann B, Nash PD (2011) The SH2 domain-containing proteins in 21 species establish the provenance and scope of phosphotyrosine signaling in eukaryotes. Sci Signal 4(202):ra83. doi:10.1126/scisignal.2002105
2. Liu BA, Nash PD (2012) Evolution of SH2 domains and phosphotyrosine signalling networks. Philos Trans R Soc Lond B Biol Sci 367(1602):2556–2573. doi:10.1098/rstb.2012.0107
3. Finn RD, Coggill P, Eberhardt RY et al (2016) The Pfam protein families database: towards a more sustainable future. Nucleic Acids Res 44(Database issue):D279–D285. doi:10.1093/nar/gkv1344
4. Sigrist CJA, de Castro E, Cerutti L, Cuche BA, Hulo N, Bridge A, Bougueleret L, Xenarios I (2012) New and continuing developments at PROSITE. Nucleic Acids Res. doi:10.1093/nar/gks1067
5. Gough J, Karplus K, Hughey R, Chothia C (2001) Assignment of homology to genome sequences using a Library of Hidden Markov Models that represent all proteins of known structure. J Mol Biol 313(4):903–919
6. Hunter S, Jones P, Mitchell A, Apweiler R, Attwood TK, Bateman A, Bernard T, Binns D, Bork P, Burge S, de Castro E, Coggill P, Corbett M, Das U, Daugherty L, Duquenne L, Finn RD, Fraser M, Gough J, Haft D, Hulo N, Kahn D, Kelly E, Letunic I, Lonsdale D, Lopez R, Madera M, Maslen J, McAnulla C, McDowall J, McMenamin C, Mi H, Mutowo-Muellenet P, Mulder N, Natale D, Orengo C, Pesseat S, Punta M, Quinn AF, Rivoire C, Sangrador-Vegas A, Selengut JD, Sigrist CJ, Scheremetjew M, Tate J, Thimmajanarthanan M, Thomas PD, Wu CH, Yeats C, Yong SY (2012) InterPro in 2011: new developments in the family and domain prediction database. Nucleic Acids Res 40(Database issue):D306–D312. doi:10.1093/nar/gkr948
7. Triplet T, Shortridge M, Griep M, Stark J, Powers R, Revesz P (2010) PROFESS: a PROtein Function, Evolution, Structure and Sequence database. Database (Oxford) 2010:baq011
8. Whelan S, de Bakker PIW, Quevillon E, Rodriguez N, Goldman N (2006) PANDIT: an evolution-centric database of protein and associated nucleotide domains with inferred trees. Nucleic Acids Res 34(Database issue):D327–D331. doi:10.1093/nar/gkj087
9. Liu BA, Engelmann BW, Jablonowski K, Higginbotham K, Stergachis AB, Nash PD (2012) SRC Homology 2 Domain Binding Sites in Insulin, IGF-1 and FGF receptor mediated signaling networks reveal an extensive potential interactome. Cell Commun Signal 10(1):27. doi:10.1186/1478-811X-10-27
10. Liu BA, Jablonowski K, Raina M, Arce M, Pawson T, Nash PD (2006) The human and mouse complement of SH2 domain proteins-establishing the boundaries of phosphotyrosine signaling. Mol Cell 22(6):851–868. doi:10.1016/j.molcel.2006.06.001
11. Smith TF, Waterman MS (1981) Identification of common molecular subsequences. J Mol Biol 147(1):195–197
12. Dijkstra EW (1959) A note on two problems in connexion with graphs. Numer Math 1:260–271
13. Dayhoff MO, Schwartz RM, Orcutt BC (1978) A model of evolutionary change in proteins. In: Dayhoff MO (ed) Atlas of protein sequece and structure, supplement 3. National Biomedical Research Foundation, Washington, DC, pp 345–352
14. Henikoff S, Henikoff JG (1992) Amino acid substitution matrices from protein blocks. Proc Natl Acad Sci U S A 89(22):10915–10919
15. Bairoch A, Apweiler R, Wu CH, Barker WC, Boeckmann B, Ferro S, Gasteiger E, Huang H, Lopez R, Magrane M, Martin MJ, Natale DA, O'Donovan C, Redaschi N, Yeh LS (2005) The Universal Protein Resource (UniProt). Nucleic Acids Res 33(Database issue):D154–D159. doi:10.1093/nar/gki070
16. Hubbard T, Andrews D, Caccamo M, Cameron G, Chen Y, Clamp M, Clarke L, Coates G, Cox T, Cunningham F, Curwen V, Cutts T, Down T, Durbin R, Fernandez-Suarez XM, Gilbert J, Hammond M, Herrero J, Hotz H, Howe K, Iyer V, Jekosch K, Kahari A, Kasprzyk A, Keefe D, Keenan S, Kokocinsci F, London D, Longden I, McVicker G, Melsopp C, Meidl P, Potter S, Proctor G, Rae M, Rios D, Schuster M, Searle S, Severin J, Slater G, Smedley D, Smith J, Spooner W, Stabenau A, Stalker J, Storey R, Trevanion S, Ureta-Vidal A, Vogel J, White S, Woodwark C, Birney E (2005) Ensembl 2005. Nucleic Acids Res 33(Database issue):D447–D453. doi:10.1093/nar/gki138
17. Pruitt KD, Brown GR, Hiatt SM, Thibaud-Nissen F, Astashyn A, Ermolaeva O, Farrell CM, Hart J, Landrum MJ, McGarvey KM, Murphy MR, O'Leary NA, Pujar S, Rajput B, Rangwala SH, Riddick LD, Shkeda A, Sun H, Tamez P, Tully RE, Wallin C, Webb D, Weber J, Wu W, DiCuccio M, Kitts P, Maglott DR, Murphy TD, Ostell JM (2014) RefSeq: an update on mammalian reference sequences.

Nucleic Acids Res 42(Database issue):D756–D763. doi:10.1093/nar/gkt1114

18. Letunic I, Doerks T, Bork P (2009) SMART 6: recent updates and new developments. Nucleic Acids Res 37(Database issue):D229–D232. doi:10.1093/nar/gkn808
19. Sigrist CJ, Cerutti L, de Castro E, Langendijk-Genevaux PS, Bulliard V, Bairoch A, Hulo N (2010) PROSITE, a protein domain database for functional characterization and annotation. Nucleic Acids Res 38(Database issue):D161–D166. doi:10.1093/nar/gkp885
20. Finn RD, Clements J, Eddy SR (2011) HMMER web server: interactive sequence similarity searching. Nucleic Acids Res 39(Web Server issue):W29–W37. doi:10.1093/nar/gkr367
21. Larkin MA, Blackshields G, Brown NP, Chenna R, McGettigan PA, McWilliam H, Valentin F, Wallace IM, Wilm A, Lopez R, Thompson JD, Gibson TJ, Higgins DG (2007) Clustal W and Clustal X version 2.0. Bioinformatics 23(21):2947–2948. doi:10.1093/bioinformatics/btm404
22. Maglott D, Ostell J, Pruitt KD, Tatusova T (2005) Entrez Gene: gene-centered information at NCBI. Nucleic Acids Res 33(Database Issue):D54–D58. doi:10.1093/nar/gki031
23. Altschul SF, Gish W, Miller W, Myers EW, Lipman DJ (1990) Basic local alignment search tool. J Mol Biol 215:403–410
24. Sievers F, Wilm A, Dineen DG, Gibson TJ, Karplus K, Li W, Lopez R, McWilliam H, Remmert M, Söding J, Thompson JD, Higgins D (2011) Fast, scalable generation of high-quality protein multiple sequence alignments using Clustal Omega. Mol Syst Biol 7:539. doi:10.1038/msb.2011.75
25. Katoh K, Misawa K, Kuma K, Miyata T (2002) MAFFT: a novel method for rapid multiple sequence alignment based on fast Fourier transform. Nucleic Acids Res 30(14):3059–3066
26. Loytynoja A, Goldman N (2005) An algorithm for progressive multiple alignment of sequences with insertions. Proc Natl Acad Sci U S A 102:10557–10562
27. Felsenstein J (1989) PHYLIP - Phylogeny Inference Package (Version 3.2). Cladistics 5:164–166
28. Page RD (2002) Visualizing phylogenetic trees using TreeView. Curr Protoc Bioinformatics 00:6.2:6.2.1–6.2.15
29. Perrière G, Gouy M (1996) WWW-Query: an on-line retrieval system for biological sequence banks. Biochimie 78:364–369

Chapter 4

Classification and Lineage Tracing of SH2 Domains Throughout Eukaryotes

Bernard A. Liu

Abstract

Today there exists a rapidly expanding number of sequenced genomes. Cataloging protein interaction domains such as the Src Homology 2 (SH2) domain across these various genomes can be accomplished with ease due to existing algorithms and predictions models. An evolutionary analysis of SH2 domains provides a step towards understanding how SH2 proteins integrated with existing signaling networks to position phosphotyrosine signaling as a crucial driver of robust cellular communication networks in metazoans. However organizing and tracing SH2 domain across organisms and understanding their evolutionary trajectory remains a challenge. This chapter describes several methodologies towards analyzing the evolutionary trajectory of SH2 domains including a global SH2 domain classification system, which facilitates annotation of new SH2 sequences essential for tracing the lineage of SH2 domains throughout eukaryote evolution. This classification utilizes a combination of sequence homology, protein domain architecture and the boundary positions between introns and exons within the SH2 domain or genes encoding these domains. Discrete SH2 families can then be traced across various genomes to provide insight into its origins. Furthermore, additional methods for examining potential mechanisms for divergence of SH2 domains from structural changes to alterations in the protein domain content and genome duplication will be discussed. Therefore a better understanding of SH2 domain evolution may enhance our insight into the emergence of phosphotyrosine signaling and the expansion of protein interaction domains.

Key words Evolution, SH2 domain, Protein tyrosine kinase, Phosphotyrosine, Sequence alignment, Gene duplication, Domain shuffling, Structure alignments

1 Introduction

Posttranslational modification (PTM) by protein phosphorylation of tyrosine residues is a mechanism for relaying signals from external stimuli to internal signals found in metazoan cells [1, 2]. The three major components of phosphotyrosine (pTyr) signaling are protein tyrosine kinases (PTKs), protein tyrosine phosphatases (PTPs), and protein interaction domains (PIDs) that recognize and bind tyrosine phosphorylated residues [3]. Several PIDs that bind pTyr include the phosphotyrosine binding domain (PTB),

Kazuya Machida and Bernard A. Liu (eds.), *SH2 Domains: Methods and Protocols*, Methods in Molecular Biology, vol. 1555, DOI 10.1007/978-1-4939-6762-9_4, © Springer Science+Business Media LLC 2017

Src homology 2 (SH2) domains, C2 and the Hakai phosphotyrosine binding (HYB) domain [4]. SH2 domains represent the largest and most predominant protein domain dedicated to pTyr signaling.

Determining the origins of pTyr signaling can provide insight into the emergence and expansion of multicellular organisms [5, 6]. The complete set of pTyr signaling components emerged approximately 900 million years ago (mya) at the pre-metazoan boundary between single-celled and multicellular organisms. This divergence point suggests that pTyr signaling may have facilitated the evolution of metazoans [3, 7]. Additional evolutionary events, such as the apparent global loss of tyrosine residues [8], together with the expansion in the number of genes dedicated to this mode of signaling, further suggests that the acquisition of tyrosine phosphorylation and SH2 domain-mediated signaling promoted metazoan development [3, 9]. As the primary pTyr recognition modules in metazoans, SH2 domains appear alongside, and coevolve with, PTKs and PTPs [5, 10]. In parallel, SH2 domains expanded, promoting coordinated emergence and increased utilization of pTyr signaling during eukaryotic evolution. Within the human genome, 121 SH2 domains are found present within 111 proteins [11, 12]. Prior studies have identified SH2 domains within representative organisms [5, 12] or SH2 families across a small subset of organisms [13–15]. However, tracing the lineage and origin of SH2 domains across organisms throughout the phylogenetic tree remains a challenge. By understanding the origins of SH2 domains and how they became integrated into genomes to expand the utilization of pTyr may help us better understand metazoan evolution and how signaling networks evolved.

In this chapter, I describe a methodology for classifying SH2 domain into discrete families. By utilizing multiple parameters such as sequence alignments, protein domain architecture and the boundary positions between introns and exons, this provides a robust approach for tracing SH2 families across multiple genomes of living organisms. In addition, I present insights and methods for analyzing the divergence within SH2 domain families.

2 Materials

2.1 Identification of SH2 Domains in Eukaryotic Genomes

1. Generate a list of organisms and confirm the completion of their genomes. Sequenced genomes can be identified at the NCBI Genome website (http://www.ncbi.nlm.nih.gov/genome/browse/). Additional sequenced genomes can be found at the Joint Genome Institute (http://genome.jgi.doe.gov/).

2. Several algorithms that utilize Hidden Markov models (HMM) aid in the identification of protein domains (*see* Chapter 3). Such HMM search tools include Pfam (http://www.sanger.ac.uk/Software/Pfam/) [16] and SMART (http://smart.embl-heidelberg.de/) [17, 18]. These online tools can predict and identify proteins that contain SH2 domains.
3. Conserved Domain Database (CDD) (http://www.ncbi.nlm.nih.gov/cdd/) predicts protein domains.

2.2 Classification of SH2 Domain Proteins by Sequence, Domain Organization, and Splice Patterns

1. Sequence alignments can be performed using ClustalW or ClustalX, which can be downloaded at www.clustal.org. Other options include using running alignments using a web browser at http://www.ebi.ac.uk/Tools/msa/clustalw2/.
2. For displaying phylogenies, download TreeView from http://taxonomy.zoology.gla.ac.uk/rod/treeview.html or Dendroscope at http://ab.inf.uni-tuebingen.de/software/dendroscope/.
3. Ensembl database (http://www.ensembl.org/) provides information intron and exon boundaries at the amino acid sequence level.
4. NCBI Blast (http://blast.ncbi.nlm.nih.gov/Blast.cgi).
5. InParanoid (http://inparanoid.sbc.su.se/).

2.3 Diversification of SH2 Families

1. Sequence alignment of SH2 families members using ClustalW or ClustalX.
2. Identified phosphorylation sites can be collected from various databases including PhosphoSite (http://www.phosphosite.org) or PhosphoELM (http://phospho.elm.eu.org/).
3. SMART and PFam.

2.4 Structural Conservation in SH2 Domain Structure and Specificity

1. The protein data bank (PDB) deposits 3D structures of SH2 domains at http://www.rcsb.org/. For a list of all known SH2 domain structures and their PBD IDs visit https://sites.google.com/site/sh2domain/sh2-structures.
2. For structure visualization PyMOL is recommended and can be downloaded at https://www.pymol.org (*see* **Note 1**).
3. UCSF-Chimera is recommended for integrating sequence alignment and displaying sequence homology onto structures. Chimera can be downloaded at https://www.cgl.ucsf.edu/chimera/.
4. Sequence alignments of SH2 domains from specific families using ClustalX/W.

3 Methods

3.1 Identification of SH2 Domain in Eukaryotic Genomes

There are multiple methods for predicting and identifying SH2 domains. This section will describe several popular methods for predicting an SH2 domain from sequence or extracting proteins containing SH2 domains within an entire organism. For example, an analysis of 16 eukaryotes revealed varying numbers of SH2 domain proteins among these Unikonts (Fig. 1). A table containing the prefix, common names, genus species, and taxonomy ID of several commonly studies eukaryotes can be found in Table 1 (*see* **Notes 2** and **3**). One method is to generate your own HMMer using a multiple sequence alignment, which was used to identify SH2 domains in human (*Homo sapiens)* and mouse *(Mus musculus)* [12] (*see* Chapter 3). Below are example methods for using the HMM search tools SMART (**steps 1** and **2**) or Pfam (**steps 3** and **4**) to predict and identify proteins that contain SH2 domains. An alternative method is using the Conserved Domains Database (CDD) to identify proteins within the NCBI database that contain an SH2 domain (**step 5**).

1. *Searching for SH2 domains using SMART.* If you have a single protein or a list of protein sequences, input the amino acid sequence (single letter annotation) into the Sequence Analysis box (left side of the website) and click 'Sequence SMART' (Fig. 2a). SMART will then determine whether the protein sequence contains an SH2 domain. Click on the SH2 domain

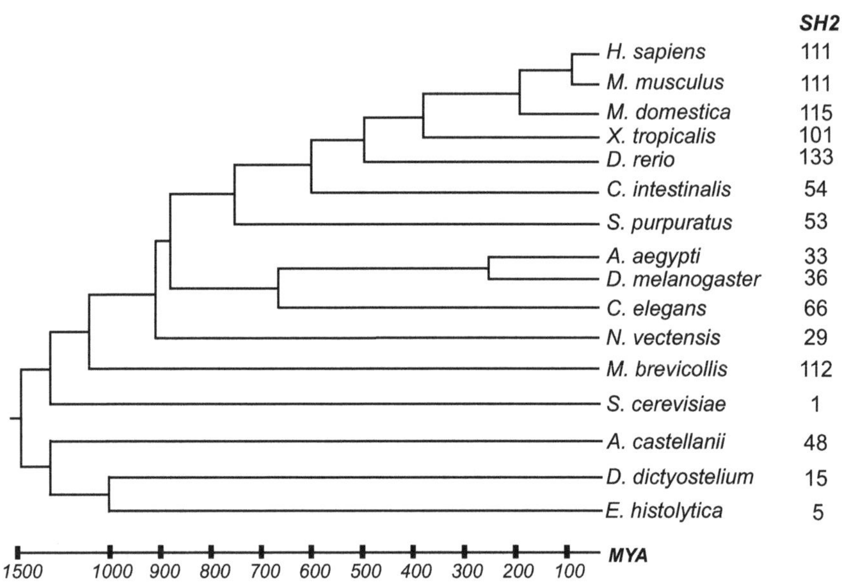

Fig. 1 SH2 domain across eukaryotes. The total number of SH2 encoded genes across 16 eukaryote genomes. The branched divergence times and lengths in millions of years (mya) are indicated. Figure modified from original article [19]

Table 1
List of eukaryotic organisms

Prefix	Common name	Genus species	Taxonomy ID
Hs	Human	*Homo sapiens*	9606
Mm	House mouse	*Mus musculus*	10090
Md	Gray short-tailed opossum	*Monodelphis domestica*	13616
Xt	Western clawed frog	*Xenopus tropicalis*	8364
Xl	African clawed frog	*Xenopus laevis*	8355
Dr	Zebrafish	*Danio rerio*	7955
Bf	Florida lancelet	*Branchiostoma floridae*	7739
Ci	Sea squirt	*Ciona intestinalis*	7719
Sp	Sea urchin	*Strongylocentrotus purpuratus*	7668
Dm	Fruit fly	*Drosophila melanogaster*	7227
Aa	Yellow fever mosquito	*Aedes aegypti*	7159
Ce	Roundworm	*Caenorhabditis elegans*	6239
Hm	Hydra	*Hydra magnipapillata*	6085
Nv	Sea anemone	*Nematostella vectensis*	45351
Ta	Trichoplax	*Trichoplax adhaerens*	10228
Sd	Sponge	*Suberites domuncula*	55567
Ef	Freshwater sponge	*Ephydatia fluviatilis*	31330
Mb	Choanoflagellate	*Monosiga brevicollis MX1*	431895
Mo	Choanoflagellate	*Monosiga ovata*	81526
Sc	Budding yeast	*Saccharomyces cerevisiae*	4932
Dd	Slime mold	*Dictyostelium discoideum AX4*	352472
Dp	Slime mold	*Dictyostelium purpureum*	5786
Eh	Amoeba	*Entamoeba histolytica*	5759
Ng	Amoeba-flagellate	*Naegleria gruberi*	5762
Pc	Oomycetes plant pathogen	*Phytophthora capsici*	4784
At	Thale cress	*Arabidopsis thaliana*	3702
Tt	Ciliated protozoan	*Tetrahymena thermophila*	5911
Tv	Parasitic protozoan	*Trichomonas vaginalis*	5722

logo to visualize the amino acid sequence. Copy and paste this sequence if desired.

2. *Identifying all SH2 domains within organisms using SMART.* Using the Architecture analysis section input 'SH2' into the

Fig. 2 SMART and Pfam domain searches. (**a**) Screenshot of the SMART domain webpage. On the left half of the site includes a Protein sequence box where one can search, by pasting the single letter amino acid sequence, whether the protein includes an SH2 domain. On the right is a search for SH2 domains across Taxonomy. (**b**) Screenshot of the Pfam website. The sequence search box is where one can search, by pasting the single letter amino acid sequence, whether the protein includes an SH2 domain. (**c**) The Pfam website can search for SH2 domains using the PF00017 identifier to browse all proteins containing a Pfam predicted SH2 domain

Domain Selection and press 'Architecture query'. If you know the organism you are interested in, input the genus and species into the Taxonomy selection box before clicking 'Architecture query' (Fig. 2a). If no selection is chosen, during the search you can select the organisms from a broad list across many Kingdoms in order to identify SH2 containing proteins and sequences. Displayed will be a page of proteins with various domain architectures (*see* **Note 4**).

3. *Searching for SH2 domains using Pfam.* If you have a single protein sequence, click on Sequence Search on from front page. Input your amino acid sequence (single letter annotation) into the search box and click Go (Fig. 2b). Pfam will display the list of protein domains. By clicking Show on the right side of the table, one can view the alignment and sequence based on the Pfam HMM model.
4. *Identifying all SH2 domains within organisms using Pfam.* Click on View a Pfam Family and type in either "SH2" or the accession "PF00017" and click Go (Fig. 2c). You will be taken to the SH2 family webpage. On the top of the page there were be a list of the total number architectures, sequences, interactions, species, and structures. Click on species and it will take you to a sunburst tree covering thousands of Eukaryotic organisms. Click on the organism or tree of interest. When selected on the right panel will show the number of sequences and species. Click on Generate a FASTA-format file and you will be able to download all the sequences containing an SH2 domain from that organism(s) (*see* **Note 4**).
5. *Identifying SH2 domains using the Conserved Domains Database (CDD).* Using NCBI identify the organism you are interested in using the taxonomy browser. Click on Gene and search for SH2 containing genes using the Accession: cl15255 or/and cd00173. Check your list of SH2 domains and eliminate any redundancies using an identifier such as GeneID. Export the protein sequence into a FASTA file or document for later use (*see* **Note 5**). Repeat the above steps to identify all SH2 containing genes in your list of organisms.
6. *Cross reference results across databases.* Confirm whether the SH2 sequence prediction from one method predicts an SH2 domain in another. Note any SH2 domain sequences that are found in one method and not another. These sequences can be checked using multiple sequence alignments to confirm whether they contain elements of an SH2 domain fold (*see* Subheading 3.2).

3.2 Classification of SH2 Domain Proteins by Sequence, Domain Organization, and Splice Patterns

A combination of sequence homology comparison, analysis of protein domain architecture, and the boundary positions between introns and exons within the SH2 domain or genes encoding these domains was utilized to assign SH2 domains into discrete SH2 families. With these three comparative analysis approaches, one can separate individual SH2 domains into discrete families (Fig. 3). In

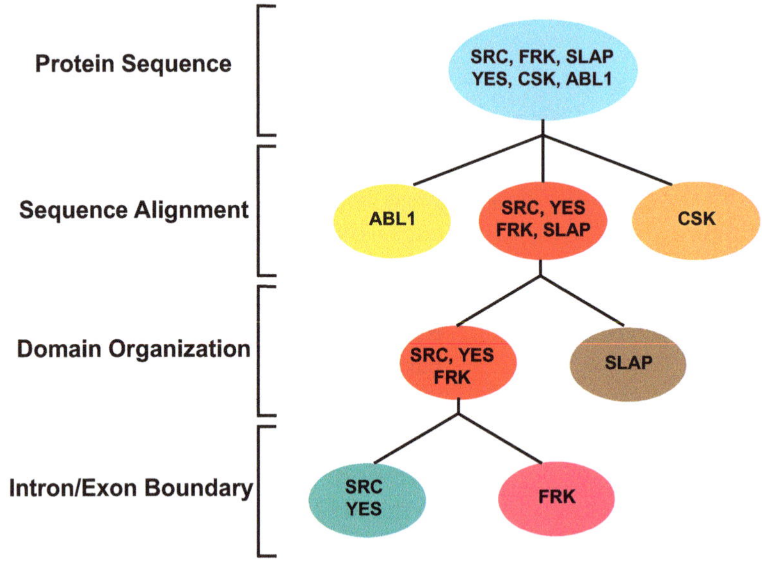

Fig. 3 Cataloging SH2 proteins into families. A hierarchical method for defining SH2 families by protein sequence alignment (ClustalW), domain organization (Ensembl and SMART) and intron/exon splice patterns (Ensembl and SMART). The SH2 domain proteins in the *blue circle* are divided into separate families according to the different methods. Figure modified from original article [11]

this section, a hierarchical approach begins with multiple sequence alignments to generate groups of SH2 domains, followed by domain organization and subsequently using intron–exon boundaries to further separate SH2 domains into discrete families. Defining an SH2 family will make it more feasible to trace SH2 families across organisms. An alternative approach is to use InParanoid to search for paralogs and orthologs across multiple organisms (**step 8**) [20].

1. *Compile SH2 domain sequences*

   With the comprehensive list of SH2 proteins across various species, isolate the full-length sequences for each individual protein that contains an SH2 domain. Utilizing SMART, Pfam, or CDD isolate and extract the sequence for just the SH2 domain and save these sequences into a FASTA file (*see* **Note 6**).

2. *Clustal alignment of SH2 domains*

   With the FASTA file of SH2 domain sequences, input your file into the ClustalW or ClustalX software. You should see your sequences loaded on the screen. Click Alignment on the menu and then click Do Complete Alignment (*see* **Note 7**).

3. *Generating a Phylogenetic tree*

   Assemble a phylogenetic tree of the SH2 domains using the Neighbor-Joining method with bootstrap replicates [21] (*see* **Note 8**). Open the neighbor-joining file in TreeView or

Dendroscope to view the subtrees and roots for specific SH2 domains.

4. *Identification of splice sites within SH2 domains.* Splicing of the nascent pre-mRNA by removing the intron while joining exons gives rise to an mRNA transcript that is translated into a protein. Each splice site position (where exons are joined) and phase can be presented at the mRNA codon including the amino acid position within the protein. The splice site junctions within the SH2 domain were visualized at the protein sequence level within the Ensembl database. The Ensembl sequence identifier for each SH2 domain protein can be used and entered into SMART, which provides a visual of the domain organization with splice site positions and phase indication (0, 1, 2) of the splice junction (Fig. 4). Splice sites with phase 0 are located between codons, phase 1 introns are located between the first and second nucleotides of a codon, and phase 2 introns are located between the second and third nucleotides. Overlay the data gathered by SMART with the Ensembl splice site position with the ClustalW alignment of SH2 domains (from **item 1** of Subheading 2.3) to generate a comprehensive set of splice patterns for your list of SH2 domains. Map the splice sites onto the SH2 domain protein secondary structure using published 3D structures. When two SH2 domains contain a splice site with the same phase that falls at identical splice junctions (within a window of 1 to 2 positions N- or C-terminal in respect to the splice site) in the sequence alignment, we consider this splice site position to be conserved. Perform standard hierarchical clustering on the compiled splice sites using the sequence alignment as a background template. The phases (0, 1, 2) and the positions within the secondary structure were used for weighting the conservation of the splice sites.
5. Determine the complete domain organization (also referred as domain architecture) of each SH2 domain protein. With the full-length sequences of SH2 domain containing proteins, compile the list of protein interaction domains also found within the protein and note the order (N-terminus to C-terminus of the protein). The HMM definitions from SMART, PFAM, and CDD are recommended to identify the domain descriptions within the full-length protein sequences (*see* **Note 9**).
6. *Determining members of an SH2 Family.* The branches from the neighbor-joining tree (from **step 3** in Subheading 3.2) should be used as a primary method for distinguishing SH2 domains into unique families. SH2 domains with shorter branches are more likely related than those separated as long branches. Secondary disambiguation of SH2 families is assisted

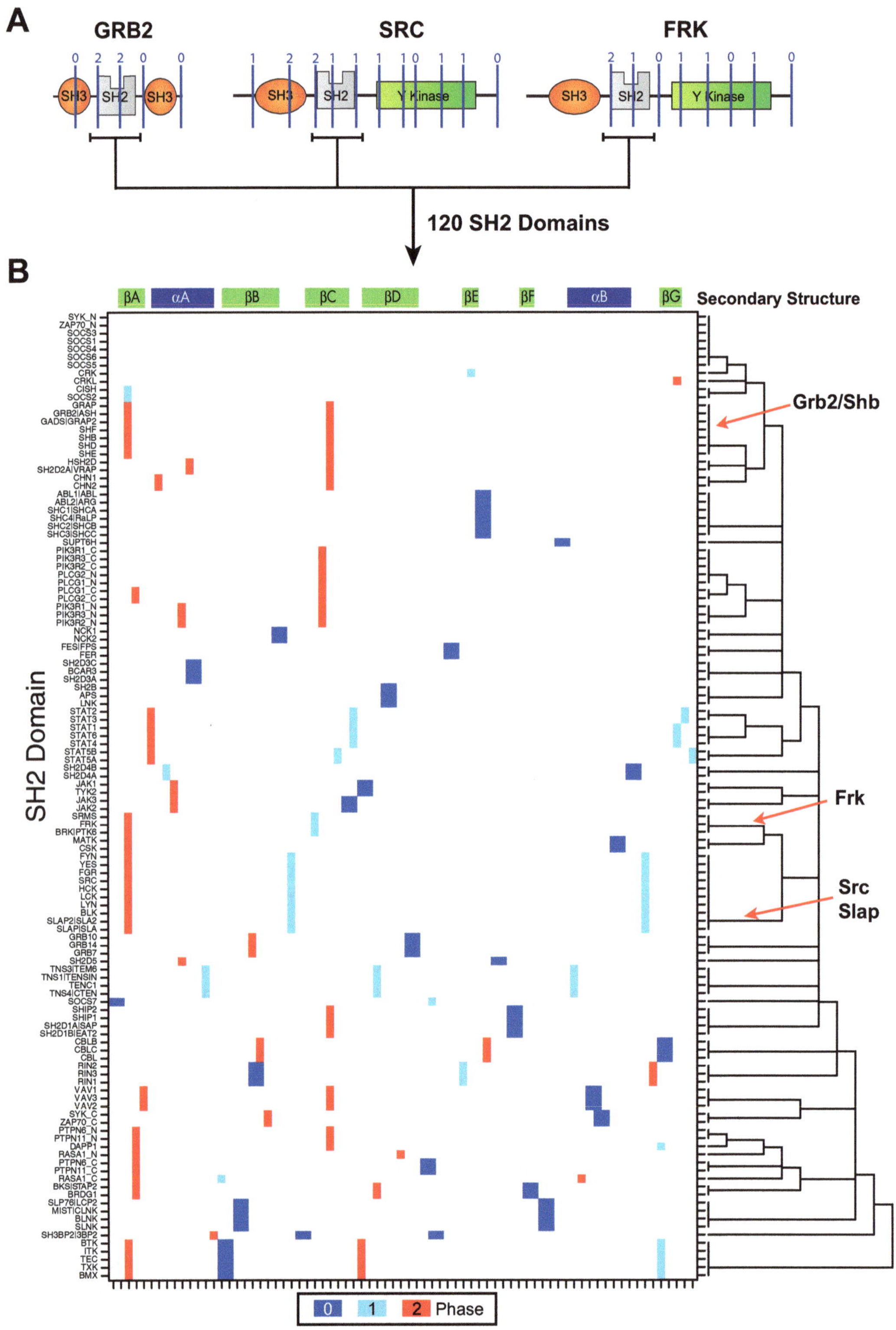

Fig. 4 The splice site patterns within SH2 domains. (**a**) Three examples intron and exon splice sites on SH2 proteins. The splice sites indicated with lines and numbers above were identified with SMART and Ensembl. (**b**) A hierarchical clustering of human SH2 domains splice sites displayed across the secondary structure of the SH2 domain (top of the graph) [11]. The splice sites are colored based on the three different phases

by analysis of protein domain organization (from **step 5** in Subheading 3.2). Tertiary disambiguation of SH2 families utilized splice patterns within the SH2 domain to assign SH2 proteins into families with identical splice-junction patterns when protein domain organization and near-neighbor sequence phylogeny was not sufficient to make the distinction (from **step 4** in Subheading 3.2).

7. *Tracing SH2 domains across Eukaryotes.* The protein domain organization (also referred as domain architecture) provided another parameter for identifying orthologs of SH2 families. The approaches described above were used to trace orthologs in order to identify precursors of genes and domains involved. Human orthologs were identified through sequence alignment using NCBI's BlastP and by matching conserved domain organization. The top alignment hit from BlastP or the lowest E-value was used to best identify orthologs and paralogs. (*see* **Note 10**). Use the Ensembl ortholog prediction to confirm BLAST results (*see* **Note 11**). When neither BLAST nor domain organization is able to confidently place an SH2 domain protein into one of the 38 families consider these proteins as unique to a specific lineage or organism.
8. *Identifying and confirming orthologs using InParanoid.* Open a web browser and visit the InParanoid website. Search by using GeneID/proteinID or through a Blast search. If using GeneID/proteinID, input either the GeneID/proteinID name, select the identifier type and the list of species you wish to search and click submit. For using a Blast search, input your SH2 domain sequence and click submit. Confirm results from **step 7** with the search using the InParanoid database.

3.3 Diversification of SH2 Families

Genes encoding catalytic PTKs and SH2 domains, like many gene families involved in cellular communication, evolved from a single or small number of ancestral genes by gene duplication and domain shuffling [22]. Duplicate copies of genes can diverge through means of evolving novel protein domain specificity for ligands, gain or loss of a protein domain, insertion or deletion of short linear motifs and lastly an alteration in the promoter of the gene allowing for specialized tissue expression and transcription control (Fig. 5a). These events may help explain why 111 SH2 domain-containing proteins are found in humans; whereas the unicellular eukaryotic yeast, *Saccharomyces cerevisiae*, contains a single protein [12] Domain shuffling of existing genes to generate proteins with novel domain organizations was likely a driving force in the transition from unicellular eukaryotes to differentiated multicellular animals [23] (Fig. 5b). This process placed the modular SH2 domain in the context of other domains that have varied functions, allowing SH2 domains to participate in diverse cellular processes.

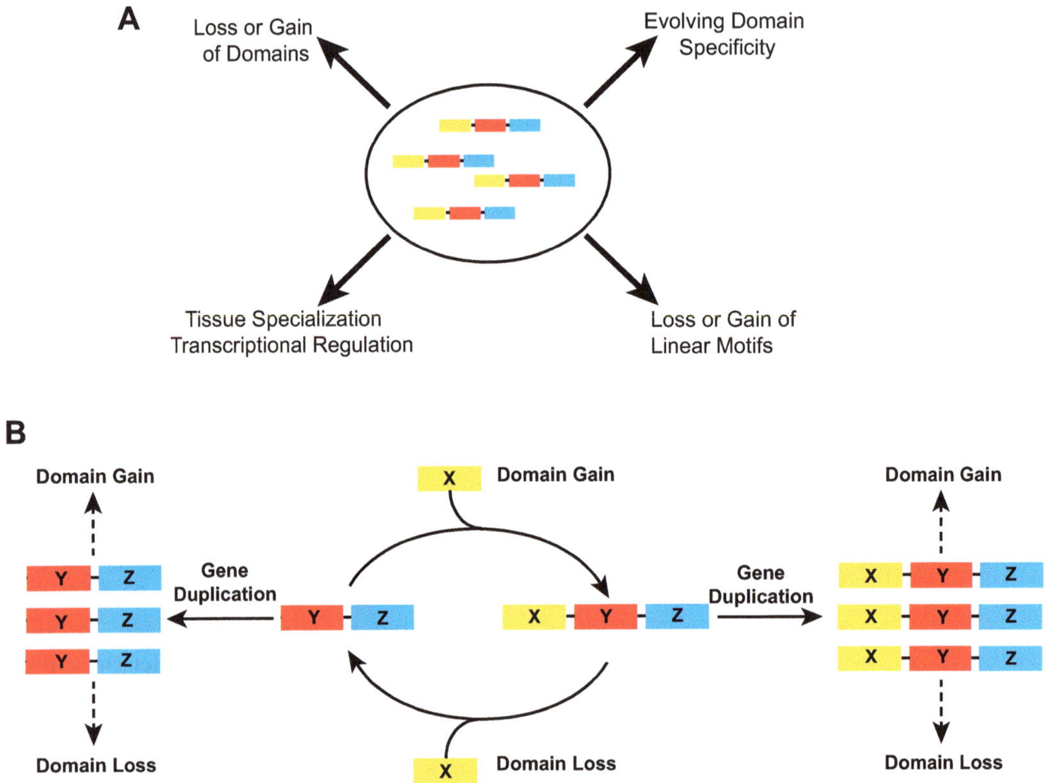

Fig. 5 Genetic events that lead to diversification of SH2 families. (**a**) Duplicate copies of genes can diverge through means of evolving novel protein domain specificity for ligands, gain or loss of a protein domain, insertion or deletion of short linear motifs, and lastly an alteration in the promoter of the gene allowing for specialized tissue expression and transcription control. (**b**) Diagram representing the events of domain gain or loss either before or after events of gene duplication. Figure modified from original article [11, 19]

1. Compile a list and table of all members of an SH2 family across multiple species.
2. Determine if multiple copies of a family exists within a species. Gene duplication events are likely to give rise to duplicate copies of a gene.
3. Determine whether the number of family members increase or decrease from species to species. Note events such as gene loss throughout evolution.
4. Neofunctionalization and subfunctionalization are mechanisms for retaining SH2 domains. Examine the domain organization of each members of the family by using the SMART of Pfam algorithms (*see* **Note 12**). List or draw the domain organization and note all the domains within a family Determine whether members of the family display gains or loss of domains. Classify families into events of gene duplication, gene loss, domain gain, and domain loss.

5. *Identifying diversity between family members.*
6. Identify the full-length sequences of SH2 domain proteins across families.
7. Load and run a sequence alignment using ClustalW.
8. Check your alignment and confirm that the domains properly align.
9. Examine the alignment for regions of diversity. For instance some family members may include regions of protein insertions or deletions. Identify other features such as post-translational modification sites including phosphorylation and ubiquitination.

3.4 Structural Conservation in SH2 Domain Structure and Specificity

The specificity of an SH2 domain can either change or remain conserved over time (Fig. 6). Conservation of peptide ligand specificity pocket allows interactions to be conserved throughout evolution. Insertion of linear motifs and evolving novel binding sites can allow for an SH2 domain to recognize multiple binding partners. Specificity can also evolve for a domain, yet this can result in acquiring a completely new set of binding partners while the ancient partners are lost. Lastly, a domain can coevolve over time to maintain ancient binding partners but also evolve novel binding partners by evolving its peptide-binding pocket. A protein domain can become inactivated through mutations of its binding pocket such that it no longer can recognize any of its original ligands.

1. *Obtain the 3D structure of an SH2 domain.*

   Downloaded SH2 domain structures as PDB files from the RCSB protein databank (http://www.rcsb.org) (*see* **Note 13**). Perform a sequence alignment of SH2 domain sequences (*see* Subheading 3.2) by obtaining all SH2 domain sequences of that particular domain family across species. Compile SH2 domain sequences into a Fasta file format. Input Fasta file into ClustalX or ClustalW and run the alignment. Save the alignment file (ending in .aln).
2. *Mapping conservation on SH2 domain structures.*
3. Sequence conservation mapped onto the SH2 domain structure was achieved using the program UCSF Chimera (*see* **Note 14**).
4. Load sequence alignments from ClustalX and the structure files manually into the Chimera software.
5. Under the alignment window, select under the Structure heading and click Render by Conservation. A tool-box should appear termed "Render/Select by Attribute".
6. To set the sequence conservation settings, under Render at the Attribute pull down, select mavPercentConserved.

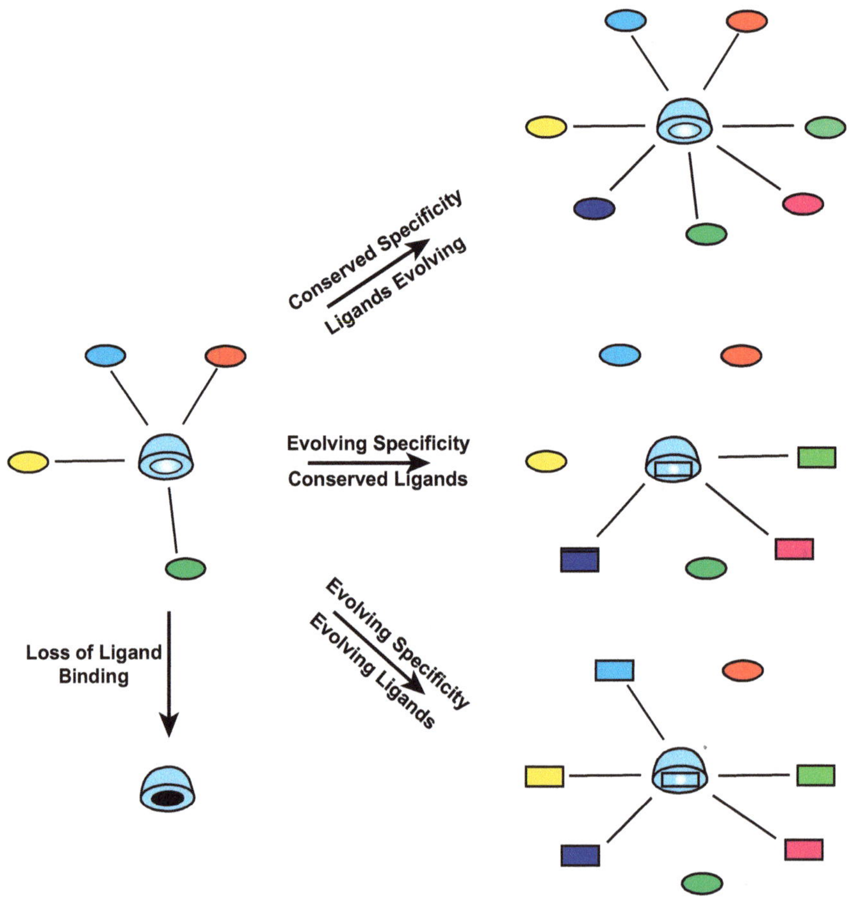

Fig. 6 Evolution of SH2 domain specificity. The specificity of a protein interaction domain, such as an SH2 domain, can either change or remain conserved over time. Conservation of peptide ligand specificity pocket allows interactions to be conserved throughout evolution. Insertion of linear motifs and evolving novel binding sites can allow for an SH2 domain to recognize multiple binding partners. Specificity can also evolve for a domain, yet this can result in acquiring a completely new set of binding partners while the ancient partners are lost. Lastly, a domain can coevolve over time to maintain ancient binding partners but also evolve novel binding partners by evolving its peptide-binding pocket. A protein domain can become inactivated through mutations of its binding pocket such that it no longer can recognize any of its original ligands. Figure modified from original article [19]

7. A histogram is shown below. Select a bar to adjust the colors for the represented percentage conservation tool. The value can be adjusted according to the percent conservation. A new bar can be introduced by clicking "Ctrl + Right Mouse Click".
8. After adjusting the colors and percent conservation, click OK or Apply.
9. Visualization of the structure surface or ribbon can be adjusted and saved.

4 Notes

1. UCSF-Chimera is recommended for visualizing conservation in the structure. PyMOL can be used to visualize amino-acid conservation. Use ConSurf Serve (http://consurf.tau.ac.il/2016/) to generate a PDB file with conservation scores written to the b-factors and then for example use PyMOL to apply a color gradient according to them.
2. For example, the annotated genomes of *Aedes aegypti*, *Caenorhabditis elegans*, *Entamoeba histolytica*, *Monodelphis domestica*, *Saccharomyces cerevisiae*, and *Strongylocentrotus purpuratus* can be found in the NCBI (http://www.ncbi.nlm.nih.gov) database. Other genomes such as *Ciona intestinalis*, *Dictyostelium purpureum*, *Monosiga brevicollis*, *Naegleria gruberi*, *Nemotostella vectensis*, *Phytophthora capsici*, *Trichomonas vaginalis*, *Trichoplax adhaerens*, and *Xenopus tropicalis* are available at the DOE Joint Genome Institute (http://genome.jgi-psf.org/). Meanwhile, *Dictyostelium discoideum* and *Dictyostelium purpureum* genomes are found in NCBI and UniProt. The *Tetrahymena thermophila* genome sequences were obtained from the Tetrahymena Genome Database (TGD) (http://www.ciliate.org/). The *Arabidopsis thaliana* genomic sequences were obtained from both NCBI and the Arabidopsis Information Resource (http://www.arabidopsis.org/).
3. Due to the draft nature of several genomes, not all genes may be found or properly annotated.
4. SMART and PFam use UniProt as an identifier and may include many redundancies. Manually remove redundancies requires tracing the UniProt to a GeneID to eliminate redundant predictions.
5. Be aware that there are several proteins that contain tandem SH2 domains.
6. The algorithms listed above tend to have shorter sequences. We recommend that you try and capture extended sequences on both ends. This will use useful when examining splice sites within SH2 domains.
7. Run the alignments using the preset settings. For details on changing the alignment settings visit the help site on the Clustal website.
8. Having the correct parameters in your alignment setting can affect your tree. In order to verify whether you alignment is as accurate as possible, we recommend using structural information to ensure specific α-helices and β-strands align correctly. For a list of solved SH2 domain structures visit http://www.sh2domain.org or visit www.rscb.org to obtain structure data.

9. Be sure to note orientation or the order of these domains as not to confuse proteins containing the same protein domains but in different orientations.
10. This analysis found BLAST to be the most reliable ortholog prediction method compared to Ensembl and InParanoid alone.
11. In certain cases where the sequence has diverged to a point where BLAST recognition is difficult to determine, examine the domain organization to categorize proteins into families.
12. Be sure to use the full-length protein sequence.
13. If no solved structures of the SH2 domain are available, structure homology based models can be used to generate a predicted structure of that SH2 domain. Submit the SH2 domain sequence into I-Tasser (http://zhanglab.ccmb.med.umich.edu/I-TASSER/) or SWISS-MODEL (http://swissmodel.expasy.org/). Download predicted structures as PDB files.
14. Chimera was chosen, as its software package is available to academic users. Other structure analysis software packages such as MOE (Molecular Operating Environment) have the capability of mapping the conservation on the protein structure.

Acknowledgments

I would like to thank Piers Nash, Tony Pawson, Chris Tan, and the members of the Nash and Pawson laboratory for helpful discussions. This work was supported by the Canadian Institutes of Health Postdoctoral Fellowship, Bernice Goldblatt Fellowship, Abbott Laboratories Graduate Fellowship, and the Cancer Research Foundation.

References

1. Hunter T (2009) Tyrosine phosphorylation: thirty years and counting. Curr Opin Cell Biol 21(2):140–146. doi:10.1016/j.ceb.2009.01.028
2. Hunter T (2014) The genesis of tyrosine phosphorylation. Cold Spring Harb Perspect Biol 6(5):a020644. doi:10.1101/cshperspect.a020644
3. Lim WA, Pawson T (2010) Phosphotyrosine signaling: evolving a new cellular communication system. Cell 142(5):661–667. doi:10.1016/j.cell.2010.08.023
4. Liu BA, Engelmann BW, Nash PD (2012) The language of SH2 domain interactions defines phosphotyrosine-mediated signal transduction. FEBS Lett 586(17):2597–2605. doi:10.1016/j.febslet.2012.04.054
5. Manning G, Young SL, Miller WT, Zhai Y (2008) The protist, Monosiga brevicollis, has a tyrosine kinase signaling network more elaborate and diverse than found in any known metazoan. Proc Natl Acad Sci U S A 105(28):9674–9679. doi:10.1073/pnas.0801314105
6. Suga H, Torruella G, Burger G, Brown MW, Ruiz-Trillo I (2014) Earliest Holozoan expansion of phosphotyrosine signaling. Mol Biol Evol 31(3):517–528. doi:10.1093/molbev/mst241
7. King N, Carroll SB (2001) A receptor tyrosine kinase from choanoflagellates: molecular

insights into early animal evolution. Proc Natl Acad Sci U S A 98(26):15032–15037. doi:10.1073/pnas.261477698,98/26/15032 [pii]

8. Tan CS, Bodenmiller B, Pasculescu A, Jovanovic M, Hengartner MO, Jorgensen C, Bader GD, Aebersold R, Pawson T, Linding R (2009) Comparative analysis reveals conserved protein phosphorylation networks implicated in multiple diseases. Sci Signal 2(81):ra39. doi:10.1126/scisignal.2000316, 2/81/ra39 [pii]
9. Kawata T, Shevchenko A, Fukuzawa M, Jermyn KA, Totty NF, Zhukovskaya NV, Sterling AE, Mann M, Williams JG (1997) SH2 signaling in a lower eukaryote: a STAT protein that regulates stalk cell differentiation in dictyostelium. Cell 89(6):909–916
10. Pincus D, Letunic I, Bork P, Lim WA (2008) Evolution of the phospho-tyrosine signaling machinery in premetazoan lineages. Proc Natl Acad Sci U S A 105(28):9680–9684. doi:10.1073/pnas.0803161105, 0803161105 [pii]
11. Liu BA, Shah E, Jablonowski K, Stergachis A, Engelmann B, Nash PD (2011) The SH2 domain-containing proteins in 21 species establish the provenance and scope of phosphotyrosine signaling in eukaryotes. Science signaling 4(202):ra83. doi:10.1126/scisignal.2002105
12. Liu BA, Jablonowski K, Raina M, Arce M, Pawson T, Nash PD (2006) The human and mouse complement of SH2 domain proteins-establishing the boundaries of phosphotyrosine signaling. Mol Cell 22(6):851–868. doi:10.1016/j.molcel.2006.06.001
13. Serfas MS, Tyner AL (2003) Brk, Srm, Frk, and Src42A form a distinct family of intracellular Src-like tyrosine kinases. Oncol Res 13(6–10):409–419
14. Steele RE, Stover NA, Sakaguchi M (1999) Appearance and disappearance of Syk family protein-tyrosine kinase genes during metazoan evolution. Gene 239(1):91–97, doi:S0378-1119(99)00373-X [pii]
15. Colicelli J (2010) ABL tyrosine kinases: evolution of function, regulation, and specificity. Sci Signal 3(139):re6. doi:10.1126/scisignal.3139re6, scisignal.3139re6 [pii]
16. Finn RD, Tate J, Mistry J, Coggill PC, Sammut SJ, Hotz HR, Ceric G, Forslund K, Eddy SR, Sonnhammer EL, Bateman A (2008) The Pfam protein families database. Nucleic Acids Res 36(Database issue):D281–D288. doi:10.1093/nar/gkm960, gkm960 [pii]
17. Schultz J, Milpetz F, Bork P, Ponting CP (1998) SMART, a simple modular architecture research tool: identification of signaling domains. Proc Natl Acad Sci U S A 95(11):5857–5864
18. Letunic I, Copley RR, Pils B, Pinkert S, Schultz J, Bork P (2006) SMART 5: domains in the context of genomes and networks. Nucleic Acids Res 34(Database issue):D257–D260. doi:10.1093/nar/gkj079, 34/suppl_1/D257 [pii]
19. Liu BA, Nash PD (2012) Evolution of SH2 domains and phosphotyrosine signalling networks. Philos Trans R Soc Lond B Biol Sci 367(1602):2556–2573. doi:10.1098/rstb.2012.0107
20. O'Brien KP, Remm M, Sonnhammer EL (2005) Inparanoid: a comprehensive database of eukaryotic orthologs. Nucleic Acids Res 33(Database issue):D476–D480. doi:10.1093/nar/gki107
21. Tamura K, Dudley J, Nei M, Kumar S (2007) MEGA4: Molecular Evolutionary Genetics Analysis (MEGA) software version 4.0. Mol Biol Evol 24(8):1596–1599. doi:10.1093/molbev/msm092, msm092 [pii]
22. Shiu SH, Li WH (2004) Origins, lineage-specific expansions, and multiple losses of tyrosine kinases in eukaryotes. Mol Biol Evol 21(5):828–840. doi:10.1093/molbev/msh077, msh077 [pii]
23. Jin J, Xie X, Chen C, Park JG, Stark C, James DA, Olhovsky M, Linding R, Mao Y, Pawson T (2009) Eukaryotic protein domains as functional units of cellular evolution. Sci Signal 2(98):ra76. doi:10.1126/scisignal.2000546, 2/98/ra76 [pii]

Chapter 5

SH2 Ligand Prediction–Guidance for In-Silico Screening

Shawn S.C. Li and Lei Li

Abstract

Systematic identification of binding partners for SH2 domains is important for understanding the biological function of the corresponding SH2 domain-containing proteins. Here, we describe two different web-accessible computer programs, SMALI and DomPep, for predicting binding ligands for SH2 domains. The former was developed using a Scoring Matrix method and the latter based on the Support Vector Machine model.

Key words SH2, SH2-peptide interactions, Ligand prediction, SMALI, DomPep

1 Introduction

The human genome encodes ~120 SH2 domains distributed in 110 distinct proteins, which constitutes the largest family of modular domains capable of recognizing phosphotyrosine (pTyr) [1]. Identifying the specific phosphotyrosyl peptide motif recognized by an SH2 domain is the key to understanding the function of the corresponding SH2-containing protein. The substrate specificities for the majority of human SH2 domains have been determined by the oriented peptide library approach [2]. The results formed the basis of Scoring Matrix-Assisted Ligand Identification or SMALI [3], which uses position-specific scoring matrices (PSSMs) to predict ligands for 76 different SH2 domains. The PSSM method was previously used for the determination of protein kinase substrates [4] and for the development of Scansite [5], a method for predicting domain-peptide and kinase-substrate interactions. Moreover, A Support Vector machine (SVM) predictor named DomPep was developed through the combination of proteomic data (e.g., peptide and protein arrays) and sequence information (e.g., domain sequence identity and Position-Weighted-Matrix distance calculated using domain-binding peptides) to achieve high accuracy and greater coverage in predicting protein-protein interactions (PPI) mediated by modular domains [6]. The main difference between

Kazuya Machida and Bernard A. Liu (eds.), *SH2 Domains: Methods and Protocols*, Methods in Molecular Biology, vol. 1555, DOI 10.1007/978-1-4939-6762-9_5, © Springer Science+Business Media LLC 2017

PSSMs and SVMs is that the former is based on OPAL (i.e., Oriented Peptide Array Library) data [7], whereas the latter is based on peptide-array or domain-array data. Here, we describe how to use SMALI and DomPep to predict SH2-peptide interactions at the proteome scale.

2 Materials

2.1 SMALI

The program is accessible via http://ext0641.biochem.fmd.uwo.ca/cgi-bin/SMALI_domain_search.pl.

2.2 DomPep

The program is accessible via http://ext0641.biochem.fmd.uwo.ca/dompep.html.

3 Methods

3.1 SMALI (See Note 1)

SMALI is used to identify SH2 domains preferred by a query protein or proteins that harbors one or more tyrosine-phosphorylation sites (Fig. 1). The PSSM matrices used for the prediction are available upon request. The specific steps involved in SMALI are:

Fig. 1 Sample output for SMALI. (**a**) A query protein may be entered by typing or pasting the sequence in the space provided. Partial sequence can also be accepted. One or more SH2 domains in the pull-down menu may be selected for the prediction. (**b**) Tabulated results showing the query protein name, sequence, locations of Tyr residues, and SH2 domains predicted to bind a particular Tyr site (assuming the site is phosphorylated). A relative SMALI score is given in parenthesis beside a selected SH2 domain. Only those SH2 domains with a relative score of >1.0 are listed

1. Query proteins may be specified by the complete or partial sequence entered in FASTA format in the space provided (Fig. 1a).
2. Prior to activating a search, the user has the option of selecting a given SH2 domain, a subgroup, or all SH2 domains (default) (Fig. 1a).
3. The output file lists the query protein sequence with all tyrosine residues highlighted (Fig. 1b). In a separate panel, the Tyr-containing peptides are listed along with a group of SH2 domains preferred by the peptides. The numbers in the parenthesis besides an SH2 domain denote its relative SMALI score for a given Tyr site. The relative SMALI score is defined as the ratio of SMALI score over a cut-off value, corresponding to the score that separates the top 4.5% of peptides from the remaining Tyr-containing peptides taken from all human proteins in the Swiss-Prot database (*see* Ref. 3 for details). An SH2 domain with a larger relative score has a greater tendency to bind to a Tyr (pTyr) site. The output file lists only those SH2 domains that have a relative SMALI score >1.0, suggests a strong potential for binding.

3.2 DomPep

DomPep was developed to predict protein-protein interactions mediated by any modular domains, although the current version of Dompep contains only predictors for the SH2 and PDZ domains. The main steps of DomPep are:

1. A query protein with complete or partial sequence is entered in FASTA format in the space provided (as in Fig. 2a).
2. The domain type for prediction is selected (as in Fig. 2a).
3. The domains of interest may be selected if listed; otherwise, input the query domain sequence (as in Fig. 2b). The user has the option to select a level of stringency for the prediction (*see* Ref. 6 for details).
4. The output file of a DomPep prediction consists of a list of peptides with prediction scores, where positive scores are strong predictors. Their locations in the corresponding proteins and the protein names are shown in a separate column (as in Fig. 2c).

4 Note

1. SMALI was developed specifically from OPAL data [7]. SMALI has a few limitations, including that the assumed independence of positions within the binding regions and SMALI predictions are usually not very specific to known functional sites.

Fig. 2 Sample output for DomPep. (**a**) A query protein can be entered by typing or pasting the sequence in the space provided, followed by the selection of domain type. (**b**) The domains of interest can be selected if listed, otherwise input the query domain sequence. (**c**) The output of a DomPep prediction consists of a list of peptides with prediction scores arranged in a descending order from *top* to *bottom*

Acknowledgment

This work was supported, in part, by funds from Taishan Scholar Construction Project (to L.L.), the Canadian Cancer Society (to S.S.L.) and the Ontario Research Fund (to S.S.L.). S.S.L. holds a Canadian Research Chair in Functional Genomics and Cellular Proteomics. L.L. holds a Taishan Scholarship.

References

1. Liu BA et al (2006) The human and mouse complement of SH2 domain proteins-establishing the boundaries of phosphotyrosine signaling. Mol Cell 22(6):851–868
2. Huang H et al (2008) Defining the specificity space of the human SRC homology 2 domain. Mol Cell Proteomics 7(4):768–784
3. Li L et al (2008) Prediction of phosphotyrosine signaling networks using a scoring matrix-assisted ligand identification approach. Nucleic Acids Res 36(10):3263–3273
4. Songyang Z, Cantley LC (1998) The use of peptide library for the determination of kinase peptide substrates. Methods Mol Biol 87:87–98

5. Yaffe MB et al (2001) A motif-based profile scanning approach for genome-wide prediction of signaling pathways. Nat Biotechnol 19(4): 348–353
6. Li L et al (2011) DomPep—a general method for predicting modular domain-mediated protein-protein interactions. PLoS One 6(10): e25528
7. Rodriguez M et al (2004) An oriented peptide array library (OPAL) strategy to study protein-protein interactions. J Biol Chem 279(10): 8802–8807

Chapter 6

An Efficient Semi-supervised Learning Approach to Predict SH2 Domain Mediated Interactions

Kousik Kundu and Rolf Backofen

Abstract

Src homology 2 (SH2) domain is an important subclass of modular protein domains that plays an indispensable role in several biological processes in eukaryotes. SH2 domains specifically bind to the phosphotyrosine residue of their binding peptides to facilitate various molecular functions. For determining the subtle binding specificities of SH2 domains, it is very important to understand the intriguing mechanisms by which these domains recognize their target peptides in a complex cellular environment. There are several attempts have been made to predict SH2–peptide interactions using high-throughput data. However, these high-throughput data are often affected by a low signal to noise ratio. Furthermore, the prediction methods have several additional shortcomings, such as linearity problem, high computational complexity, etc. Thus, computational identification of SH2–peptide interactions using high-throughput data remains challenging. Here, we propose a machine learning approach based on an efficient semi-supervised learning technique for the prediction of 51 SH2 domain mediated interactions in the human proteome. In our study, we have successfully employed several strategies to tackle the major problems in computational identification of SH2–peptide interactions.

Key words Src homology 2 domain, Signal transduction, Protein–protein interaction, Phosphotyrosine peptides, Support vector machine, Semi-supervised learning

1 Introduction

In 1986, Tony Pawson and co-workers first discovered the Src homology 2 (SH2) domain from the oncogenic v-FPS/FES cytoplasmic tyrosine kinase encoded in the Fujinami sarcoma virus [1]. Since then a number of SH2 domains have been identified in several eukaryotic species [2, 3]. Although SH2 domains are found across the eukaryotes, they are more abundant in metazoans [4, 5]. Currently, 122 SH2 domains from 112 unique human proteins have been reported in the UniProtKB/Swiss-Prot database, release 2015-06 [6]. SH2 domains are approximately 100 amino acids in length and are structurally conserved domains that contain a central β sheet and two α helices [7]. These domains are known to mediate intracellular signaling pathways by specifically recog-

Kazuya Machida and Bernard A. Liu (eds.), *SH2 Domains: Methods and Protocols*, Methods in Molecular Biology, vol. 1555, DOI 10.1007/978-1-4939-6762-9_6, © Springer Science+Business Media LLC 2017

nizing short linear phosphotyrosine (pY) containing peptides [8]. Although SH2 domains mainly target phosphotyrosine (pY) residue of the binding peptides, their binding specificity is determined by the neighbor residues of the pY, particularly from −2 to +4 (pY at 0th position) [8, 9]. For example, a Leu or Pro residue at position +3 (xx-pY-xx[L/P]x) is strongly preferred by CRK SH2 domains, where x represents any naturally occurring amino acid. Alternatively, a hydrophobic residue (Φ) at position −2 (Φ x-pY-xxxx) is preferred by PTPN11 SH2 domains [10]. Previous studies showed that the mutations in some SH2 domains can cause several human diseases, such as XLP syndrome [11], X-linked α-gammaglobulinemia [12], Noonan syndrome [13], etc.

In recent years, various high-throughput techniques, such as peptide array, microarray, etc., have been introduced to define the binding specificities of SH2 domains. The enormous amount of data generated by these techniques are invaluable to build efficient computational methods. However, these data are often affected by false positive and false negative interactions. Most of the popular computational methods, that use high-throughput data as their training sets, are based on the position weight matrices (PWMs), which do not consider the dependencies among the amino acids in the peptide sequences [14, 15]. Here, we present a machine learning algorithm to build non-linear models that can exploit the dependencies between the amino acids in the binding peptides. In addition, these PWM-based models are essentially generative models, as they rely only on the information on positive interaction data and completely ignore the information on negative interaction data, whereas machine learning methods rely on both positive and negative interaction data and produce discriminative models, which have advantages over generative models [16, 17].

One of the major problems of using high-throughput data is, in general, the available information on positive interactions is much higher than the negative interactions, which leads to a severe data imbalance problem. For example, the positive interaction data can be up to 15 times more abundant than negative interaction data for an SH2 domain [18]. In the machine learning literature, it is known that the severely imbalanced class distribution in a training set negatively affects the performance of the predictive model; generally, these models are biased towards the majority class. To mitigate this problem, we employed an efficient semi-supervised technique where the *self-training* strategy was used to balance the training sets. Therefore, as a consequence, we achieved powerful discriminative models. Finally, we performed a genome-wide prediction of the SH2 domain mediated interactions in human proteome to uncover the biologically relevant interactions. The prediction tool (SH2PepInt) has been implemented in a newly developed web server, namely MoDPepInt [19].

2 Materials

2.1 Dataset

In our study, all the high-throughput interaction data were obtained mainly from three sources: one high-density peptide array [20] and two protein microarray experiments [21, 22]. Additionally, we extracted interaction data from a manually curated high-quality PhosphoELM database [23] for evaluating our models. To unveil novel interactions by genome-wide prediction, we extracted all the tyrosine containing proteins from UniProtKB/Swiss-Prot database [6].

1. **Peptide array data:** The observed binding interaction data in the peptide array experiment was deposited in the NetPhorest database [20]. From NetPhorest database, a total 14,678 positive interactions that involved 61 SH2 domains and 920 peptides were retrieved. After removing all the redundancy, we finally obtained 7544 positive interactions (Dataset I).
2. **Microarray data:** We incorporated the interaction data from two protein microarray experiments [21, 22]. From the microarray experiment in [21], we retrieved 2100 interactions with 160 positive and 1940 (2100 − 160) negative interactions, which involved 115 SH2 domains, and 20 peptides from ErbB2 and ErbB3 receptor proteins (Dataset II). Note that in Dataset II, we did not consider the interactions related to the ErbB1 receptor protein (*see* **Note 1**).

   From other microarray experiment in [22], we considered 3485 interactions with 314 positive and 3171 (3485 − 314) negative interactions, involving 85 SH2 domains, and 41 singly phosphorylated peptides from EGFR, FGFR, and IGIFR receptor proteins (Dataset III).
3. **Manually curated data:** For evaluating our models, we retrieved the binding information of SH2 domains from a manually curated database, called PhosphoELM database. We have extracted a total 878 binding interactions that involved 63 SH2 domains and 359 peptides (Dataset IV).
4. **Genome-wide prediction data:** All the human protein sequences were derived from UniProtKB/Swiss-Prot database [6]. A total 20,225 proteins, which contain 298,637 tyrosine residues, were considered. Finally, a total 298,637 tyrosine containing peptides were generated as our test set for genome-wide predictions.

2.2 Data Compilation

We have combined all the high-throughput data derived from peptide array and microarray experiments together, but surprisingly, we found there were several disagreements on the binding information between different experiments. Furthermore, these high-throughput data are often affected by a high rate of false positive and false negative interactions. The refinement of these noisy data from our training set is explained below.

1. In the two microarray experiments, i.e., Dataset II and Dataset III, there were 10 proteins that contained two SH2 domains (N and C terminal) each. Since these datasets do not report the assignment of which peptide specifically binds with which of the two SH2 domains of a protein, we discarded all the interactions related to these double-SH2 domain containing proteins.
2. We combined these two microarray data and collected a total 474 (160 + 314) positive and 5111 (1940 + 3171) negative interactions. The apparent equilibrium dissociation constant (K_D value) or affinity constant was applied to determine the positive and negative interaction classes. We used the same K_D cutoff as mentioned in [21, 22], and thus the SH2–peptide interactions with K_D values less than 2 μM were considered as binding (positive data) interactions, while remaining interactions were considered as non-binding (negative data) interactions. Nevertheless, various inconsistencies in the SH2–ErbB1 (ErbB1 or EGFR protein was common in both experiments) interactions were observed between Dataset II and Dataset III (*see* **Note 1**).
3. We could only consider 7544 positive data from the peptide array experiment [20], since there was evidence only for positive interactions (*see* **Note 2**). Surprisingly, we also observed 149 interactions for which there is a conflict between the peptide array experiment [20] and the microarray array experiments [21, 22], i.e., these interactions are positive in Dataset I, but negative in Dataset II and Dataset III. To reduce the noisy and conflicting information from our training sets, we discarded these 149 interactions. Therefore, as a consequence, the positive data in Dataset I was reduced to 7395 (7544 – 149), and the negative data in Dataset II and Dataset III was reduced to 4962 (5111 – 149).
4. Among the 474 positive interactions collected from microarray experiments, 247 interactions were already present in Dataset I. After removing the redundancy, we comprised 227 (474 – 247) positive interactions from Dataset II and Dataset III. These non-redundant positive interactions were kept for the validation.
5. We only considered those SH2 domains that have at least 40 positive interaction data, otherwise no complex model can be reliably fit. We used positive interaction data from Dataset I and the negative interaction data from Dataset II and Dataset III,

and finally, we composed our training sets with 6742 positive interactions and 2523 negative interactions for 51 human SH2 domains.

3 Methods

Here, we present a machine learning method to produce non-linear models, which can exploit the inter-dependencies between the amino acids in the binding peptides. Additionally, we present a semi-supervised technique that can efficiently refine the high-quality negative interactions from a noisy dataset.

3.1 Feature Encoding

1. Although SH2 domains specifically bind to the phosphotyrosine (pY) residue of their binding peptides, the neighbor residues of the pY are known to be highly predictive for domain–peptide interactions [14, 15].
2. We restricted the peptide sequences to 7 amino acids in length, namely we extracted the amino acids in position from −2 to +4 where the pY holds the 0th position.
3. In the feature encoding, we did not include the central residue (pY), since it was always same in the peptides from both classes (positive and negative) and thus not holding any discriminative information. Therefore, a peptide sequence was mapped into a binary vector x living in a $120 \times 6 = 120$ dimensional space, i.e., for each position, we reserved 20 dimensions for each amino acid, and encoded the amino acid type with a 1 in the corresponding dimension and 0 elsewhere.
4. A data set for each domain Dj was compiled, which was encoded as a set of pairs $(x_1, c_1), \ldots, (xn, cn)$ where xi is the binary feature vector for peptide Pi with the class label $ci \in \{-1, 1\}$. If the domain Dj interacts with the peptide Pi, then the correspondence class label is 1, otherwise, in case of non-interaction, the class label is −1.

3.2 Data Modeling

Previous research showed that the contextual dependencies between the amino acids in the cognate peptide sequences are highly important to describe the binding specificities of SH2 domains (*see* **Note 3**) [24]. Any methods that ignore these kind of dependencies often produce sub-optimal models (*see* **Note 4**). Therefore, in order to build the predictive models, which allow the dependencies between the amino acids in the binding peptides, a polynomial kernel support vector machine (SVM) has been employed. We have used the SVM implementation in C language provided in SVM*light* [25].

1. A polynomial kernel is a kernel function that computes the similarity between training samples (vectors) in the polynomial feature space to learn a non-linear model. The polynomial kernel function for two vectors: X and X' with degree d is defined in [26] as:

$$K(X,X') = (1 + \langle X,X' \rangle)^d, \tag{1}$$

where "1" is a constant, which is required to consider the effects of all degrees that are less than d. A feature space with two inputs: $X1$ and $X2$, and $d=2$ (*see* **Note 5**) is, therefore, defined as:

$$\begin{aligned} K(X,X') &= (1 + \langle X,X' \rangle)^2 \\ &= (1 + X_1X_1' + X_2X_2')^2 \\ &= 1 + 2X_1X_1' + 2X_2X_2' + (X_1X_1')^2 + (X_2X_2')^2 \\ &\quad +2X_1X_1'X_2X_2'. \end{aligned} \tag{2}$$

2. One of the main hyper-parameters in SVM is the cost parameter or C, which is used to trade off generalization of data fitting. Basically, it provides some flexibility in an enlarged feature space for data separation.
3. All the model parameters, i.e., $d \in \{1,2,3\}$ and $C \in \{0.01, 0.1, 1, 10\}$, have been optimized on training sets under a *stratified* cross-validation setting (*see* **Note 6**).
4. The optimization of these hyper-parameters is important to counter balance the overfitting phenomena (*see* **Note 7**). More specifically, for each model, the best parameter combination was chosen on a held out data set (validation set). However, the model performance was evaluated on a separate test set, which was never seen in the validation or training phase.

3.3 Semi-supervised Negative Data

Data imbalance is one of the major problems in high-throughput experiments where availability of the data from one class is much higher than the other class (*see* **Note 8**). To deal with this problem, we employed a semi-supervised learning (SSL) approach (*see* **Note 9**).

1. We resorted to the *self-training* strategy (*see* **Note 10**), although there are several strategies available to tackle the SSL problem.
2. For each domain, the initial high-throughput data was extracted from high density peptide array and microarray experiments (see dataset compilation) to train the base classifiers.
3. As an unlabeled data set, we considered the SH2–peptide pairs that did not show any positive signals in the peptide array

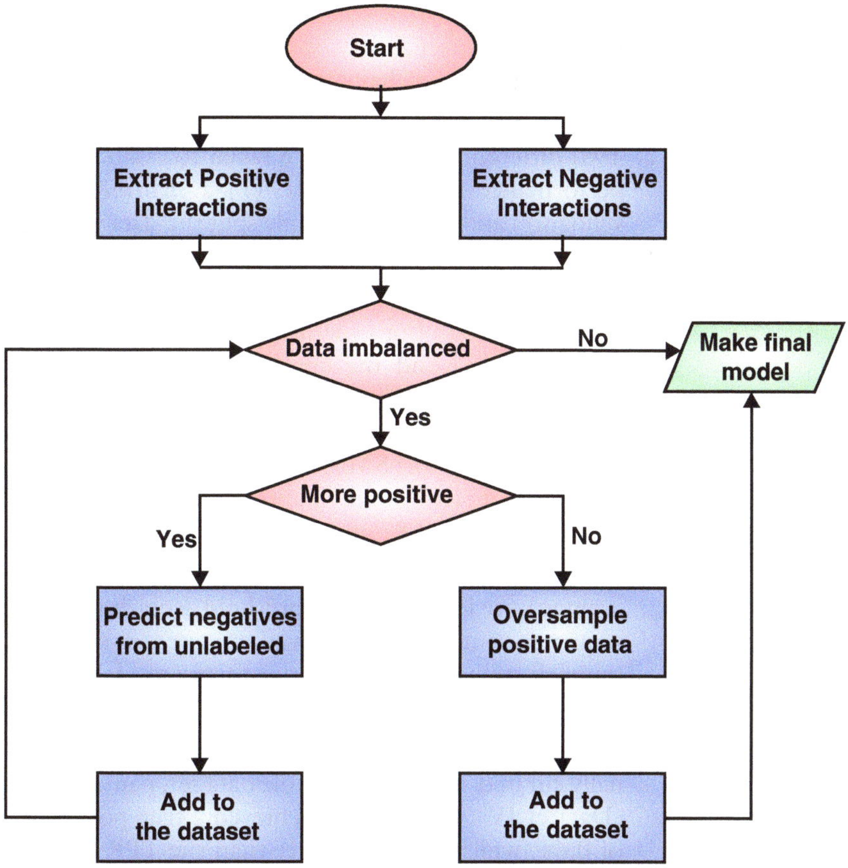

Fig. 1 For the imbalanced data sets, we encountered two types of problems: (1) for most of the domains, the positive data was much higher than the negative data and (2) for some domains, a different scenario occurred when the negative data was higher than the positive ones. To solve the first problem, we used a *self-training* strategy to predict confidence negative interaction data. The process was iteratively done until a balanced data set was reached (*left branch*). To solve the second problem, we applied a *rebalancing* technique and oversampled the positive class (*right branch*). This figure is adapted from [18]

experiment (*see* **Note 11**). Note that randomly generated negative peptides were not considered (*see* **Note 12**).

4. For each domain, a polynomial SVM was used to predict confidence negative instances from the unlabeled data set, and iteratively added them to the main training set until the data set was balanced (*see* **Note 13**). Here, the confidence of negative data was scored as the distance from the hyperplane.
5. Finally, the model selection process was performed to select the best model complexity for each specific SH2 domain. Fifty one models were built for 51 domains.
6. The flowchart for iterative balancing technique for positive and negative data is depicted in Fig. 1.

3.4 Predictive Performance

1. For measuring the predictive performance, we computed five measures, i.e., sensitivity, specificity, precision, area under the receiver operating characteristics curve (AUC ROC), and area under the precision recall curve (AUR PR) (*see* **Note 14**).
2. Two different strategies were taken to evaluate the predictive performance of our models: (1) a *stratified* fivefold cross-validation and (2) we randomly split the data where we considered 75% as training set and 25% as test set; we repeat the process 10 times to create 10 train/test data sets.
3. We compared our methods with PWM-based SMALI approach [15] and the energy model [27], and in fivefold cross-validation setting, we achieved an average AUC ROC of 0.83 and an average AUC PR of 0.93, which outperformed other two methods; SMALI and the energy model achieved an average AUC ROC of 0.71 and 0.62, respectively, and an average AUC PR of 0.87 and 0.81, respectively [18].
4. In order to achieve the same specificity as SMALI (0.95 on average), we identified our threshold accordingly, and as a consequence, our models achieved an average sensitivity of 0.45, outperforming SMALI and the energy model, which achieved an average sensitivity of 0.26 and 0.17, respectively [18].
5. For the validation purpose, we evaluated our models on a manually curated and highly reliable data set, namely PhosphoELM (Dataset IV). Our models achieved a true positive rate (TPR) of 0.64, which is much better than the TPR of 0.33 achieved by SMALI [18]. Note that the comparison with energy model could not be possible, as the class determination threshold was not specified in [27].
6. Our method efficiently predicts the binding partners of most of the SH2 domains, however, it might get challenged for some SH2 domains whose training data are very small or have *within-class imbalance* problem (*see* **Note 15**).

3.5 Genome-Wide Prediction

It is always interesting to see the interactions that are novel and biologically relevant. In order to uncover such interactions, we performed a genome-wide prediction. Subsequently, a term-centric enrichment analysis was performed to unveil novel functionalities of the predicted interactions.

1. All the generated peptides were restricted to seven amino acids in length, i.e., −2 to +4 amino acids with Tyr at 0th position.
2. We used our models to predict the binding partners of all 51 human SH2 domains.
3. All the predicted interactions were filtered based on some criteria to achieve more confidence interactions. We have used two filters: (1) phosphotyrosine (pY) (*see* **Note 16**) and (2) co-cellular localization (*see* **Note 17**).

4. For each SH2 domain, we considered top 50 predictions and performed a *term-centric* enrichment analysis using DAVID tool [28] to unveil the novel and biologically relevant interactions (*see* **Note 18**). By doing this, several biologically meaningful interactions were observed [18].
5. All the top predictions and their *term-centric* analysis for all 51 human SH2 domains are available under the URL: http://www.bioinf.uni-freiburg.de/Software/SH2PepInt/Genome-wide-predictions.tar.gz

3.6 MoDPepInt web-server

We implemented our prediction tool (SH2PepInt) for SH2-peptide interactions into a newly developed web server, called MoDPepInt (Modular Domain Peptide Interaction) [18, 19]. Currently, the MoDPepInt web server offers three different tools: (1) SH2PepInt, (2) SH3PepInt, and (3) PDZPepInt for predicting the binding interactions of three different modular domains, SH2, SH3, and PDZ, respectively [17, 18, 29].

1. The web server has two different modes: (1) basic mode and (2) expert mode. We designed a meta-web server for the basic mode, where only the input is required. The input is submitted simultaneously to all tools, and a summary table is produced. The expert mode is more flexible, where user can choose the SH2 domains of interest and use desired filters to obtain high confident interactions.
2. For SH2–peptide interactions, two filters have been used in order to increase the prediction accuracy. The filters are: (1) phosphotyrosine and (2) cellular localization (*see* **Notes 16** and **17**).
3. The MoDPepInt server is available under the URL: http://modpepint.informatik.uni-freiburg.de/SH2PepInt/Input.jsp

4 Notes

1. Eleven peptides from ErbB1 proteins were used in both microarray experiments [21, 22]. We retrieved the interaction data, which involved those 11 peptides and 85 SH2 domains (also common in both microarray experiments). Interestingly, we observed there were severe inconsistencies in the interaction data produced by these two microarray experiments, as in similar settings, one microarray experiments [21] showed positive signals ($KD < 2\,\mu M$) for 32 interactions, whereas other microarray experiment [22] showed positive signals for 120 interactions (*see* Fig. 2). All the 32 positive interactions observed in [21] were also observed in [22]. The possible reason for the lack of interactions in [21] could be caused due to

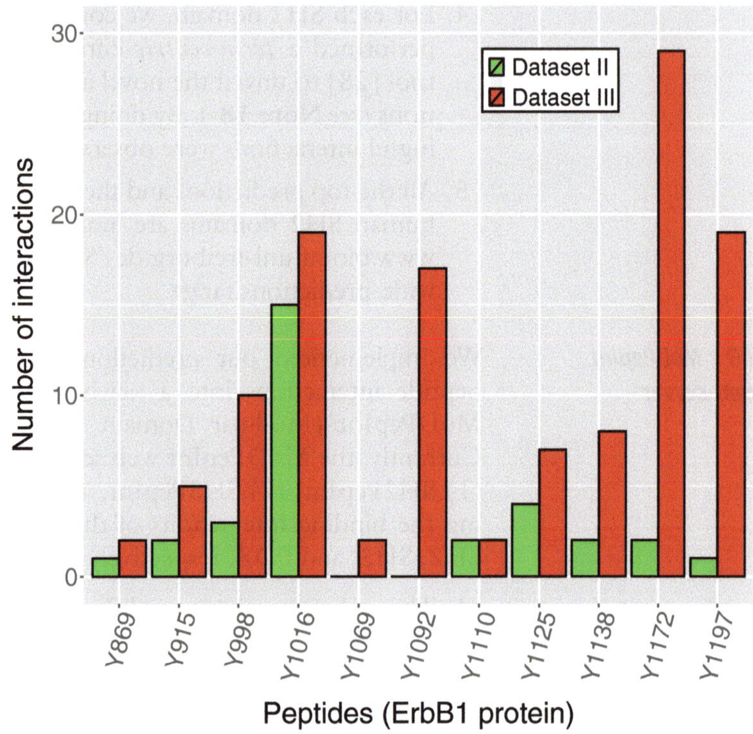

Fig. 2 Comparison of the outcome of two different microarray experiments. Eleven peptide sequences from ErbB1 protein and 85 SH2 domains were considered. The *green bars* indicate the number of SH2–peptide interactions observed in [21], and the *red bars* indicate the number of SH2–peptide interactions observed in [22]. This figure clearly shows that almost 4 times more interactions were observed in Dataset III (120 interactions observed) in comparison to Dataset II (32 interactions observed)

low concentration of proteins on the surface of the slides, which can happen when the protein printing tip is slightly misaligned. In this case, there is just not enough protein present on the surface of the slide that can cross the background threshold, even though there is, however, a tight interaction.

2. From peptide array experiment, we retrieved the interaction information for 61 domains and 920 peptides. Thus, the possible domain–peptide interactions should be $920 \times 61 = 56,120$. Among them, a total 7544 interactions showed the binding signals in the experiment, and remaining 48,576 (56120 − 7544) interactions did not show any signals. However, one should not assume all these 48,576 interactions as non-binding type, since the binding signal could not be observed may be due to some experimental stringencies. For example, an interaction might be considered as non-binding, if the binding

threshold value is less than the detection limit, even though there is, however, a weak interaction. Poor domain folding and/or peptide synthesis problem could also be the reason for these kind of error (false negative) in the data. Thus, refinement of these false negative interactions is vital to build a good predictive model.

3. In 2010, Liu et al. published an interesting study where they explained the importance of the contextual dependencies between the amino acids in the binding peptides [24]. More precisely, the binding peptides are composed of permissive and non-permissive amino acids where permissive amino acids allow the interaction and non-permissive amino acids inhibit the interactions [9, 24]. It is known that CRK SH2 domain interacts peptides where Leu or Pro amino acid is present at +3 (pY at 0th position), but surprisingly, presence of other amino acids in other positions can also influence the binding specificity. For example, basic residues (Arg and His) are disfavored at position +1 and +2, whereas Pro is prohibited at position +1. More interestingly, the acceptance of Ala at +1 completely depends on the amino acid at +3. Ala is accepted only if there is a Pro at +3, where as it is rejected, in case, there is a Leu at +3 [9].
4. All the approaches that based on position weight matrices (PWMs) [14, 15] and linear machine learning models [30] do not consider the positional dependencies between the amino acids in the binding peptides, and therefore cannot accurately determine the binding specificities of SH2 domains and eventually, produce sub-optimal results.
5. As $d = 1$ only provides linear model, we used the $d > 1$ for building non-linear models. However, the degree of the polynomial kernel is optimized via cross-validation, and therefore a simpler linear model ($d = 1$) can still be chosen for some SH2 domains when it offers better performance.
6. In a cross-validation setting, a stratification procedure is used to maintain approximately the same proportion of the two types of class labels, i.e., positive and negative, in each fold. Cross-validation with stratification procedure is known as *stratified* cross-validation.
7. Overfitting is a common problem in machine learning methods. It normally occurs when the machine learning algorithms capture the noise of the data. If the model fits too well to the data, it causes overfitting, and eventually, produces sub-optimal predictive model. Unfortunately, this important aspect is often ignored in the bioinformatics prediction methods. To overcome this overfitting problem, we used an appropriate technique, called *regularization*. The

regularized predictor is more robust to noise, and guarantees better prediction quality on unseen data. Although there are several ways to counter balance the overfitting issue, we adopted an efficient strategy where we minimized the model complexity by tuning the degree of the polynomial (d) and the cost parameter (C).

8. It is known that machine learning algorithms work poorly on highly imbalanced data, and negatively affect the performance of adaptive predictors [31]. These algorithms are generally biased towards the majority class, and hence often produce poor discriminative models.

9. In semi-supervised learning, a small amount of labeled data and a large amount of unlabeled data are trained. Note that for using the small amount of labeled data, a strong model assumptions need to be made. It is very important step as if the model assumptions do not match the nature of the problem, then it would be critical for the predictive performance. There are several techniques, such as *expectation maximization (EM)*, *co-training*, *self-training*, and *graph-based* methods, have been developed to handle the SSL problem. Each technique is used based on the requirements of the problems.

10. The *self-training* strategy relies only on the good discriminative properties of the base classifier. This is a simple wrapper method, which iteratively uses the initial labeled data to train the classifier, which then assigns a label to the remaining previously unlabeled data. In our application, this is the most suited strategy that can efficiently tackle the semi-supervised problem. Note that this approach is only applicable when at least a few confidence positive and negative data available to train the base classifier.

11. In theory, if an SH2 domain does not show a positive signal for a peptide in a peptide array experiment, the SH2 domain is considered to be a non-binder to that particular peptide, and the SH2–peptide pair is believed as a negative interaction pair. However, it is known that high-throughput data are highly affected by false negative interactions. Therefore, in order to filter high-confidence negative, we applied *self-training* strategy.

12. In common practice, random peptides are used to generate artificial negative instances. However, previous research showed that the randomly generated instances significantly decrease the prediction quality of a model [32]. Hence, instead of taking random peptides, we used experimental data.

13. For some domains, negative data was already much higher than the positive data in the base classifier. In these cases, we used a *rebalancing* technique where we over-sampled the positive class to balance the base classifier. Note that we did not under-sample of the positive class in order not to throw away the valuable information.

14. One major problem in machine learning is that the mainstream algorithms are not designed to efficiently deal with the skewed class distribution; these algorithms are more accurate only on the majority class. For example, if a data set is imbalanced, containing a few positive and many negative data, a rational choice based on maximizing the predicted accuracy (in an equal cost scenario) would most certainly be biased towards the majority class, and therefore the predictive model will almost always predict a negative response. Hence, in a binary classification, a single standard statistical measurement (e.g., accuracy, AUC ROC) will not be appropriate and can mislead the predictive performance. In this case, the model will achieve high specificity, precision, and AUC ROC but very poor sensitivity and AUC PR. Thus, it is always important to show multiple statistical measures to describe the performance of a predictive model.
15. The *within-class imbalance* and the *small-disjuncts* problem typically occur when the class concept is composed by many sub-clusters/sub-concepts, and each of the sub-cluster represented by a very small number of examples. If a small sub-cluster in training set is over-represented in the test set, our models sometimes might fail to identify those interactions. However, we could tackle this problem for the negative data, as we could select many peptides from different array experiments for which no definitive interaction information was available. Unfortunately, the information for the positive interactions was very limited for some SH2 domains, and therefore the total number of positive interactions were very few in the training set for those SH2 domains. Since this is a standard problem in machine learning, some oversampling techniques (e.g., SMOTE) have been proposed in the literature to tackle this problem, however, they have several drawbacks, e.g., requiring an explicit instance representation [18, 33].
16. This filter was used to get all the tyrosine containing peptides whose phosphorylation evidence was experimentally verified. At the time of analysis, the phosphorylation evidence of a total 30,228 peptides from 10,688 proteins was available in the PhosphoSitePlus database [34]. Note that we ignored some phosphopeptide containing proteins that were not present in the UniProtKB/Swiss-Prot database. Finally, a total number of 27,481 phosphopeptide from 9621 human proteins were used. Since SH2 domains are known to interact with phosphopeptides, this filter will provide more probable interactions.
17. It is highly unlikely to see an SH2-peptide interaction where the SH2 domain and the peptide containing protein reside different compartments of the cell. To filter out all these kind of unlikely interactions, we implemented a co-cellular localization filter. In this setting, we only considered an SH2–peptide

interaction, if the SH2 domain containing protein and the peptide containing protein share at least one GO-term that is annotated in the Gene Ontology (GO) database [35].

18. The DAVID [28] tool allows to perform a *term-centric* enrichment analysis on more than 40 different annotation categories, and reports enriched annotation terms associated with the predicted proteins. *p*-value is used to determine the enrichment; the smaller *p*-values indicate higher enrichment.

Acknowledgements

This chapter is based on our previous publication [18]. This work was funded by Bundesministerium für Bildung und Forschung (e-bio; FKZ 0316174A to Rolf Backofen), and the Centre for Biological Signalling Studies (BIOSS), University of Freiburg.

References

1. Sadowski I, Stone JC, Pawson T (1986) A noncatalytic domain conserved among cytoplasmic protein-tyrosine kinases modifies the kinase function and transforming activity of Fujinami sarcoma virus P130gag-fps. Mol Cell Biol 6(12):4396–4408
2. Mayer BJ, Hamaguchi M, Hanafusa H (1988) A novel viral oncogene with structural similarity to phospholipase C. Nature 332(6161):272–275
3. Anderson D, Koch CA, Grey L, Ellis C, Moran MF, Pawson T (1990) Binding of SH2 domains of phospholipase C gamma 1, GAP, and Src to activated growth factor receptors. Science 250(4983):979–982
4. Lim WA, Pawson T (2010) Phosphotyrosine signaling: evolving a new cellular communication system. Cell 142(5):661–667
5. Liu BA, Shah E, Jablonowski K, Stergachis A, Engelmann B, Nash PD (2011) The SH2 domain-containing proteins in 21 species establish the provenance and scope of phosphotyrosine signaling in eukaryotes. Sci Signal 4(202):ra83
6. Magrane M, UniProt Consortium (2011) UniProt Knowledgebase: a hub of integrated protein data. Database (Oxford) 2011:bar009
7. Waksman G, Kominos D, Robertson SC, Pant N, Baltimore D, Birge RB, Cowburn D, Hanafusa H, Mayer BJ, Overduin M, Resh MD, Rios CB, Silverman L, Kuriyan J (1992) Crystal structure of the phosphotyrosine recognition domain SH2 of v-src complexed with tyrosine-phosphorylated peptides. Nature 358(6388):646–653
8. Pawson T (2004) Specificity in signal transduction: from phosphotyrosine-SH2 domain interactions to complex cellular systems. Cell 116(2):191–203
9. Liu BA, Engelmann BW, Nash PD (2012) The language of SH2 domain interactions defines phosphotyrosine-mediated signal transduction. FEBS Lett 586(17):2597–2605
10. Imhof D, Wavreille A-S, May A, Zacharias M, Tridandapani S, Pei D (2006) Sequence specificity of SHP-1 and SHP-2 Src homology 2 domains. Critical roles of residues beyond the pY+3 position. J Biol Chem 281(29):20271–20282
11. Sayos J, Wu C, Morra M, Wang N, Zhang X, Allen D, van Schaik S, Notarangelo L, Geha R, Roncarolo MG, Oettgen H, De Vries JE, Aversa G, Terhorst C, (1998) The X-linked lymphoproliferative-disease gene product SAP regulates signals induced through the co-receptor SLAM. Nature 395(6701):462–469
12. Tzeng SR, Pai MT, Lung FD, Wu CW, Roller PP, Lei B, Wei CJ, Tu SC, Chen SH, Soong WJ, Cheng JW (2000) Stability and peptide binding specificity of Btk SH2 domain: molecular basis for X-linked agammaglobulinemia. Protein Sci 9(12):2377–2385
13. Tartaglia M, Mehler EL, Goldberg R, Zampino G, Brunner HG, Kremer H, van der Burgt I, Crosby AH, Ion A, Jeffery S, Kalidas K, Patton MA, Kucherlapati RS, Gelb BD (2001) Mutations in PTPN11, encoding the protein tyrosine phosphatase SHP-2, cause Noonan syndrome. Nat Genet 29(4):465–468

14. Obenauer JC, Cantley LC, Yaffe MB (2003) Scansite 2.0: Proteome-wide prediction of cell signaling interactions using short sequence motifs. Nucleic Acids Res 31(13):3635–3641
15. Li L, Wu C, Huang H, Zhang K, Gan J, Li SS-C (2008) Prediction of phosphotyrosine signaling networks using a scoring matrix-assisted ligand identification approach. Nucleic Acids Res 36(10):3263–3273
16. Ng AY, Jordan MI (2001) On discriminative vs. generative classifiers: a comparison of logistic regression and naive bayes. In: NIPS, pp 841–848
17. Kundu K, Costa F, Backofen R (2013) A graph kernel approach for alignment-free domain-peptide interaction prediction with an application to human SH3 domains. Bioinformatics 29(13):i335–i343
18. Kundu K, Costa F, Huber M, Reth M, Backofen R (2013) Semi-supervised prediction of SH2-peptide interactions from imbalanced high-throughput data. PLoS One 8(5):e62732
19. Kundu K, Mann M, Costa F, Backofen R (2014) MoDPepInt: an interactive web server for prediction of modular domain-peptide interactions. Bioinformatics 30(18):2668–2669
20. Miller ML, Jensen LJ, Diella F, Jorgensen C, Tinti M, Li L, Hsiung M, Parker SA, Bordeaux J, Sicheritz-Ponten T, Olhovsky M, Pasculescu A, Alexander J, Knapp S, Blom N, Bork P, Li S, Cesareni G, Pawson T, Turk BE, Yaffe MB, Brunak S, Linding R (2008) Linear motif atlas for phosphorylation-dependent signaling. Sci Signal 1(35):ra2
21. Jones RB, Gordus A, Krall JA, MacBeath G (2006) A quantitative protein interaction network for the ErbB receptors using protein microarrays. Nature 439(7073):168–174
22. Kaushansky A, Gordus A, Chang B, Rush J, MacBeath G (2008) A quantitative study of the recruitment potential of all intracellular tyrosine residues on EGFR, FGFR1 and IGF1R. Mol Biosyst 4(6):643–653
23. Diella F, Gould CM, Chica C, Via A, Gibson TJ (2008) Phospho.ELM: a database of phosphorylation sites–update 2008. Nucleic Acids Res 36(Database issue):D240–D244
24. Liu BA, Jablonowski K, Shah EE, Engelmann BW, Jones RB, Nash PD (2010) SH2 domains recognize contextual peptide sequence information to determine selectivity. Mol Cell Proteomics 9(11):2391–2404
25. Joachims T (1999) Making large-scale SVM learning practical. In: Schölkopf B, Burges C, Smola A (eds) Advanced in Kernel methods-support vector learning. MIT Press, Cambridge, MA, pp 169–184
26. Hastie T, Tibshirani R, Friedman J (2008) The elements of statistical learning: data mining, inference, and prediction, 2nd edn. Springer, Berlin
27. Wunderlich Z, Mirny LA (2009) Using genome-wide measurements for computational prediction of SH2-peptide interactions. Nucleic Acids Res 37(14):4629–4641
28. Huang DW, Sherman BT, Lempicki RA (2009) Bioinformatics enrichment tools: paths toward the comprehensive functional analysis of large gene lists. Nucleic Acids Res 37(1): 1–13
29. Kundu K, Backofen R (2014) Cluster based prediction of PDZ-peptide interactions. BMC Genomics 15(Suppl 1):S5
30. Li L, Zhao B, Du J, Zhang K, Ling CX, Li SS-C (2011) DomPep–a general method for predicting modular domain-mediated protein-protein interactions. PLoS One 6(10):e25528
31. He H, Garcia EA (2009) Learning from imbalanced data. IEEE Trans Knowl Data Eng 21:1263–1284
32. Ben-Hur A, Noble WS (2006) Choosing negative examples for the prediction of protein-protein interactions. BMC Bioinf 7(Suppl 1):S2
33. Chawla N, Bowyer K, Hall L, Kegelmeyer WP (2002) Smote: synthetic minority over-sampling technique. J Artif Intell Res 16: 321–357
34. Hornbeck PV, Kornhauser JM, Tkachev S, Zhang B, Skrzypek E, Murray B, Latham V, Sullivan M (2012) PhosphoSitePlus: a comprehensive resource for investigating the structure and function of experimentally determined post-translational modifications in man and mouse. Nucleic Acids Res 40(Database issue):D261–D270
35. Ashburner M, Ball CA, Blake JA, Botstein D, Butler H, Cherry JM, Davis AP, Dolinski K, Dwight SS, Eppig JT, Harris MA, Hill DP, Issel-Tarver L, Kasarskis A, Lewis S, Matese JC, Richardson JE, Ringwald M, Rubin GM, Sherlock G (2000) Gene ontology: tool for the unification of biology. The Gene Ontology Consortium. Nat Genet 25(1):25–29

Chapter 7

Proteomic Clustering Analysis of SH2 Domain Datasets

Karl Jablonowski

Abstract

Proteomic clustering analysis provides a means of identifying relationships and visualizing those relationships in an extremely complex field of study with many interacting parts. With recent high-throughput studies of Src Homology 2 (SH2) domains, many and varied datasets are being amassed. A strategy for analyzing patterns between these large datasets is required to transform the information into knowledge. The methods for creating neighbor-joining phylogenetic trees, pairs scatter plots, and two-dimensional hierarchical clustering heatmaps are just a few of the diverse methods available to a proteomic researcher. This chapter examines selecting objects to be analyzed, selecting comparison functions to apply to those objects, and pseudo-code for processing data and preparing it for various types of analyses. Here I apply clustering analysis to previous collections of SH2 domains datasets to bring insight into new binding or specificity patterns between the different SH2 domains.

Key words Neighbor-joining phylogenetic tree, Euclidean distance, Evolutionary analysis, Peptide ligand binding pattern

1 Introduction

Proteomics, as a field, employs an ever-expanding set of tools with which to study a large number of proteins, that is, many ways to inspect many objects. Proteins do not exist in isolation, and a researcher often needs many perspectives (or methods of inspection) of the complicated interacting system in molecular biology. Clustering and cladistic techniques have been developed to analyze the increasingly complex set of objects and the relationships between those objects. Clustering is a method for exploring relationships between objects. The clustering analysis produces a global view of these relationships, visualized through neighbor-joining trees and hierarchical clustering with heatmaps. The relation between multiple clustering analyses may also be investigated. Take, for example, two clusterings, one based on a set of proteins' binding patterns, and one based on a set of proteins' sequence alignment. Through this analysis, it is possible to assess the degree of evolutionary conservation or divergence by sequence alignment

Kazuya Machida and Bernard A. Liu (eds.), *SH2 Domains: Methods and Protocols*, Methods in Molecular Biology, vol. 1555, DOI 10.1007/978-1-4939-6762-9_7, © Springer Science+Business Media LLC 2017

versus binding patterns. This chapter examines the "How to compare?" and the "What to compare?" in proteomic datasets.

1.1 Foundational Nomenclature

To properly introduce the methods of clustering analysis, it is necessary to grasp a few key concepts. At the root of any form of analysis is a comparison of objects using one or more methods of inspection.

1. An object is something that is measured (SH2 domain, peptide, tissue, etc.).
2. A method of inspection is how objects are measured (difference in: pairwise sequence alignment, peptide binding specificity, gel lane signal intensity, tissue expression, etc.).
3. A 1-by-1 comparison is a comparison (using a method of inspection) between two objects that yields a numeric value. Often denoted as a function (*F*) of the two objects. For example *F*(*a*,*b*) is the comparison between *a* and *b* using method *F*.
4. The comparison function F is symmetric if $F(a,b) = F(b,a)$.
5. An N-by-N matrix is the arrangement of numeric values from a comparison of like objects vs. like objects. An N-by-N matrix can be produced using a 1-by-1 comparison iteratively over a set of objects (Fig. 1).
6. An N-by-M matrix is the arrangement of numeric values of comparison of different sets of objects (Fig. 2).
7. Euclidean distance is the distance between any two points in space. Because we live in three spatial dimensions, our spatial thinking is often confined to depth, width, and height. Figure 3

Set of 'n' objects S_n = (SHB, SHD, SHE, SHF, ...)
Method of inspection is 'pairwise alignment' PA()
such that $PA(S_1,S_2) = A_{1,2}$
Where $A_{1,2}$ is a value that represents the alignment
Matrix (B) has element $B_{i,j}$ that holds the value of column 'i' and row 'j'

For i=1 to n
For j=1 to n
$B_{i,j} = PA(S_i,S_j)$

	SHB	SHD	SHE	SHF
SHB	0	0.382	0.447	0.197
SHD	0.382	0	0.532	0.351
SHE	0.447	0.532	0	0.429
SHF	0.197	0.351	0.429	0

The matrix is symmetric about the diagonal because $PA(S_i,S_j) = PA(S_j,S_i)$

Fig. 1 N-by-N Matrix (comparing like objects)

Set of 'n' objects S_n = (SHB, SHD, SHE, SHF, ...)
Set of 'm' objects T_m = (IGSIYLFLRKR, DLKSYLRSLRP, DGMAYLNAKKF, TRDIYETDYYR, YETDYYRKGGK, ...)
Method of inspection is 'peptide binding' PB()

such that $PB(S_1,T_1) = A_{1,1}$

Where $A_{1,1}$ is a value that represents the binding between SH2 domain S_1 and peptide T_1

Matrix (C) has element $C_{i,j}$ that holds the value of column 'i' and row 'j'

For i=1 to n

For j=1 to n

$C_{i,j} = PB(S_i,T_j)$

	SHB	SHD	SHE	SHF
IGSIYLFLRKR	21.6	25.4	5.4	48.2
DLKSYLRSLRP	22.5	19.3	3.9	29.8
DGMAYLNAKKF	27.8	29.6	9.8	61.4
TRDIYETDYYR	23.5	0	0	0
YETDYYRKGGK	21.7	22.5	7.1	23.7

Fig. 2 N-by-M Matrix (comparing different objects)

The formula for computing the Euclidean Distance (ED) of any two objects in n-dimensional space is:

$ED_{a,b} = sqrt((a_1-b_1)^2+(a_2-b_2)^2+...(a_n-b_n)^2)$

We may then apply this formula to the peptide binding example: Consider a 3-dimensional space. Each dimension is a peptide from the matrix below. The coordinates for the SH2 domains in this SH2-peptide binding space are the binding specificities.

	SHB	SHD	SHE	SHF
IGSIYLFLRKR	21.6	25.4	5.4	48.2
DLKSYLRSLRP	22.5	19.3	3.9	29.8
DGMAYLNAKKF	27.8	29.6	9.8	61.4

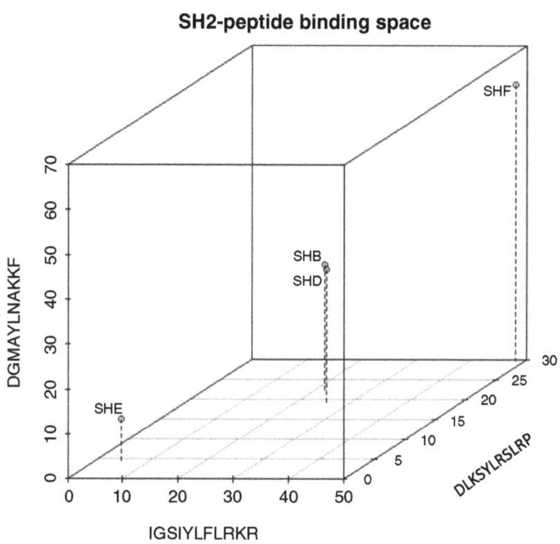

Fig. 3 Computing the Euclidean distance

outlines how to compute the Euclidean distance in peptide-binding space, where each dimension represents a peptide, and the location of each point (SH2 domain) in this space represents the binding affinity to the peptide. Though the example in Fig. 3 is three-dimensional (or three peptides), the method may be expanded to any number (n) of peptides.

8. Hierarchical clustering is a method that builds a hierarchy of clusters. The example in Fig. 4 takes a distance matrix (or an N-by-N matrix) and builds a dendrogram depicting the hierarchy of clusters. That is, SHB and SHD are clustered together as the closest domains in the peptide binding space. SHE then clusters with the SHB-SHD cluster, and finally SHF.

Compute the ED for every pair of SH2 domains in this SH2-peptide binding space, and arrange the values in a matrix (a distance matrix).
Eg. $ED_{SHB,SHD} = sqrt((21.6-25.4)^2+(22.5-19.3)^2+(27.8-29.6)^2) = 5.3$

	SHB	SHD	SHE	SHF
SHB	0	5.3	30.5	43.5
SHD	5.3	0	32.1	40.5
SHE	30.5	32.1	0	71.9
SHF	43.5	40.5	71.9	0

Distance matrices supply hierarchical clustering algorithms the necessary data to create dendrograms of the SH2 domains. The dendrogram below was created using the following R code:

```
sh2_domains<-c('SHB','SHD','SHE','SHF')
plot(hclust(distance_matrix),labels=sh2_domains, main="Clustering in peptide binding space")
```

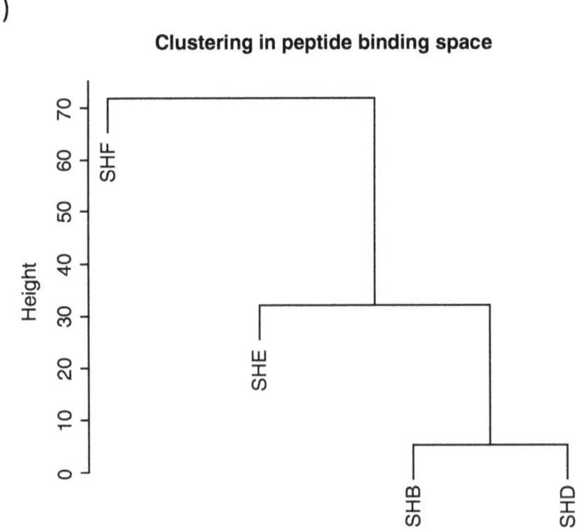

Fig. 4 Hierarchical clustering from a distance matrix

1.2 Neighbor-Joining Phylogenetic Trees

There are many software packages for generating and analyzing neighbor-joining phylogenetic trees (PHYLIP [1] for example). Any tool that allows for the creation of a quantitative comparison between two items (called a "pairwise comparison" or a "1-by-1 comparison"), will allow for the creation of a neighbor-joining phylogenetic tree (*see* **Note 1**). A neighbor-joining phylogenetic tree can be constructed from an N-by-N matrix, where each element in the matrix is a 1-by-1 comparison between two items. The N-by-N matrix is created with all items representing the rows and the columns. The intersection of a row and column is the value of the quantitative comparison between the row and column items. This is a 1-by-1 comparison repeated iteratively over each matrix element forming an N-by-N matrix of comparison. Neighbor-joining phylogenetic trees are the most common form of hierarchical clustering application in bioinformatics.

Suppose the task is to create a neighbor-joining phylogenetic tree of SH2 domains. The N-by-N matrix of comparison is constructed where the rows represent every SH2 domain to be analyzed, and every column represents a duplicate list of SH2 domains. The intersection of a row and column is the value of the quantitative comparison between the SH2 domain of the row and the SH2 domain of the column.

Choosing a quantitative comparison algorithm (like the Smith-Waterman [2] sequence alignment algorithm) and selecting the appropriate parameters (like the substitution matrix and the gap-scoring scheme) may have a drastic impact on the outcome of the neighbor-joining phylogenetic tree. The researcher must take care to understand how the algorithm chosen optimizes the relevant parameters.

Regardless of the quantitative comparison algorithm chosen, identical SH2 domains will have no difference (or total similarity) when compared. If the row and column SH2 domains are in the same order, the diagonal of the filled matrix will be observed having the value representing total similarity (100% similar or 0% difference). Most comparison algorithms of like objects are symmetric (for example compare_function (ABL1, ABL2) = compare_function (ABL2, ABL1)). This symmetry will create duplicate values as the elements of the matrix are filled. If the row and column SH2 domains are in the same order, the effect of this symmetry will result in an identical mirror image of the matrix about the diagonal. To cut down on computation time one need only to compute half of the matrix either above or below the diagonal.

The resulting matrix will contain the information required to create a neighbor-joining phylogenetic tree. The information contained in the N-by-N comparison matrix represents every SH2 domain and how it compares to every other SH2 domain. The matrix element in the row for ABL1 and the column for ABL1 will have total similarity. For ABL1 and ABL2 it will have near similarity, and for ABL1 and RIN1 it will have disparate similarity. These similarities can be visually interpreted from the SH2 domains position in the tree (*see* Fig. 1 in Chapter 3).

Any one of a variety of neighbor-joining phylogenetic tree plotting and manipulation programs (like PHYLIP) may take as input the N-by-N comparison matrix formulated above. Tree plotting programs have many parameters, but the output is generally a visual representation of similarity between all the SH2 domains in the comparison matrix.

1.3 Multiple Methodology Clustering Analysis

Some expert applications of data visualization and data analysis come from combining multiple methodologies and comparing their product. The above sections demonstrate how to use one form of comparison to create a comparison matrix for use in the neighbor-joining phylogenetic tree. From that one form of comparison, a researcher may create a comparison matrix and use software to generate and analyze a neighbor-joining tree. The challenge for combining multiple methodologies is to re-implement the above method using a different function.

In Machida et al., a binding profile was created for each SH2 domain collected from far-Western blot analysis of GST-tagged SH2 domains exposed to cell lysates [3]. A Euclidean distance function to compare binding patterns between two SH2 domains was created from the cell lysate data. Because this function is a 1-by-1 comparison (or pairwise comparison), one may utilize it to create an N-by-N similarity matrix and generate a neighbor-joining phylogenetic tree. The result of this clustering is displayed in Fig. 5a.

In this instance, there exist two comparison matrices based on different initial functions: multiple sequence alignment; and Euclidean distance of binding patterns. It is helpful for analytic purposes to create a data visualization abstraction of how SH2 domains relate to each other under the inspection of these two methods. The combination of these matrices creates a scatter plot of every pair of elements. The *Y*-axis will be the difference in multiple sequence alignment. The *X*-axis will be the Euclidean distance of binding patterns. Each point on the scatter plot is every possible pair of SH2 domains. The resulting plot will show the combined analysis based on the two methods (Fig. 5b). The analysis of the plot reveals evolutionary conservation of sequence with respect to binding pattern similarity. For example: the Src family kinases are tightly conserved both in sequence and in binding pattern; the Tensins family is conserved in sequence but not in binding pattern; and Nap4 has similar binding patterns to the Src family kinases but relatively dissimilar sequence comparisons. This clustering analysis is significant not only because it allows for the inspection of convergent and divergent evolution of SH2 domain binding patterns, but also highlights the point that the binding similarity may not be directly inferred based on sequence similarity.

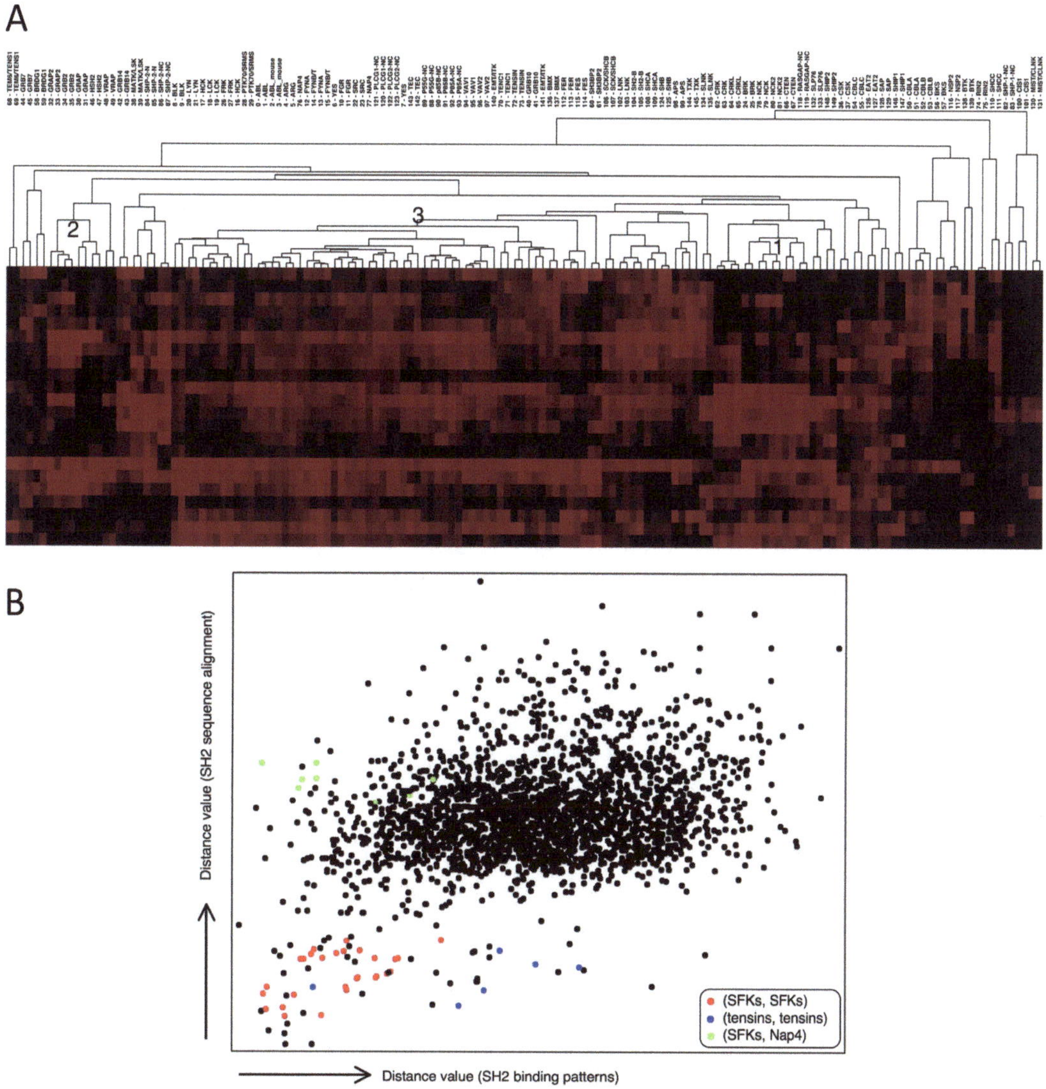

Fig. 5 (**a**) Unsupervised hierarchical clustering analysis of SH2 domain binding preferences. Each lane of SH2 far-Western blot was divided into 48 bins, the signal intensity of each bin was quantified by densitometry, and values were used as the basis for clustering. (**b**) Pairwise distance values based on sequence alignment [8] (*Y* axis) plotted against pairwise (Euclidean) distance values based on binding patterns (*X* axis) for each pair of SH2 domains. Data for Src family kinases (SFKs) versus Src family kinases (*red*), Src family kinases versus Nap4 (*green*), and tensin family versus tensin family (*blue*) are highlighted. Reprinted with permission from Elsevier and Copyright Clearance Center [3]

1.4 Two-Dimensional Hierarchical Clustering Heatmap

Hierarchical two-dimensional clustering is useful when the comparisons are not of the same object. Until now all of our comparison matrices have been N-by-N because the items of inspection have been the same set of objects: comparing SH2 domains against SH2 domains using one or more methods of inspection. A powerful bioinformatics tool is the hierarchical two-dimensional clustering. The inspection of one dimension is SH2 domains, and in the other dimension is some other object or performance measure such as peptide ligands to which the SH2 domains bind (as is the case in Fig. 6) or tissue expression profile for each SH2 domain (as is the case in Fig. 7).

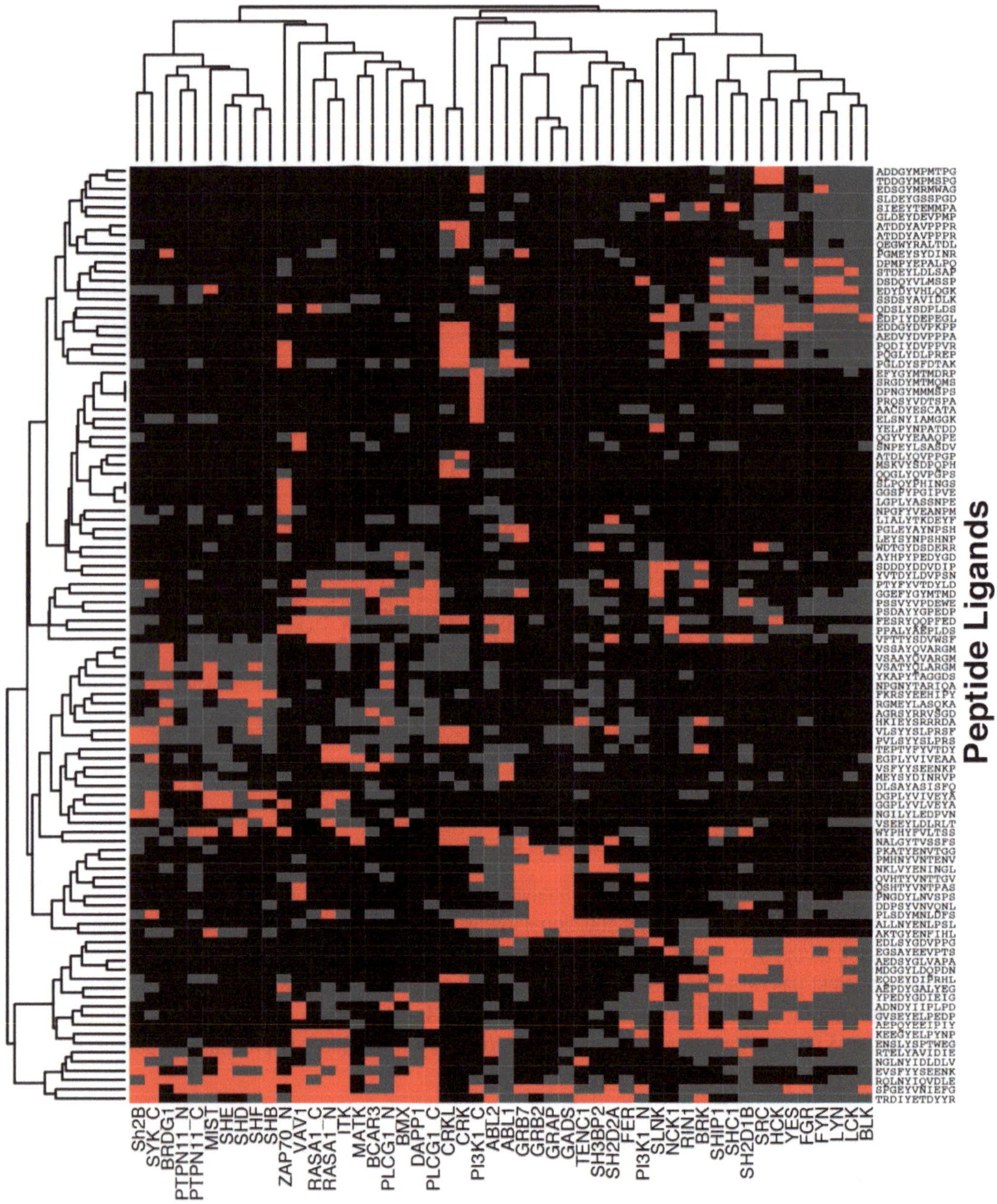

Fig. 6 SH2 domains cluster into families based on peptide binding. Hierarchical two-dimensional clustering of physiological ligands and 50 SH2 domains reveals clusters of SH2 domains motifs. The relative intensities of the SH2-peptide interaction are indicated by the color: intensities greater than 3× the mean, *red*; intensities between 1× and 3× mean, *gray*; intensities less than <1× mean, *black*. Figure modified from original article [5]

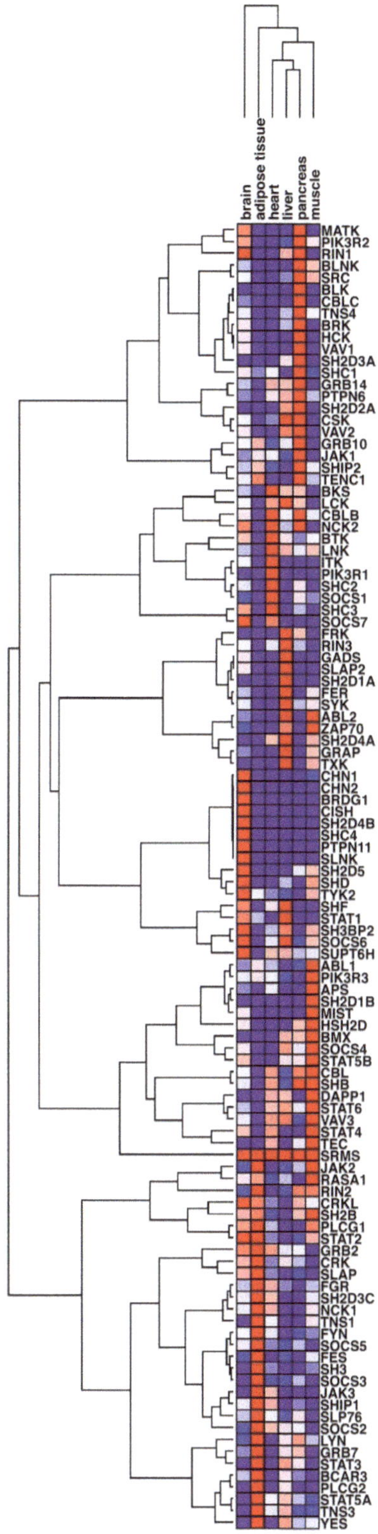

Fig. 7 Hierarchical clustering of the insulin responsive tissue expression levels for human SH2 domain-containing genes. Reprinted with permission from © Liu et al.; BioMed Central Ltd [6]

When comparing SH2 domains to a set of peptide ligands to which they bind, the comparison function is the binding affinity of the two objects. The matrix will be N-by-M where N is the set of SH2 domains and M is the set of peptide ligands. This is not necessarily a square matrix (in fact, only if N = M is this matrix square), and there will be neither mirror image nor trivial comparison along the diagonal as was the case above. The two-dimensional method calls for evaluating each dimension of clustering in a square comparison matrix of its own, one for SH2 domains and one for peptide ligands.

Each SH2 domain in the list N has a peptide ligand profile of binding affinities of length *M*. The values of binding affinities may be thought of as a projection in M-dimensional space. Suppose the SH2 domain is CRK and the peptide ligands were pYDLP, pYDVP, and pYMTM. Now suppose those normalized binding affinities are 0.9, 0.8, and 0.5 (1 being high and 0 being low). The results may be projected as a vector in three dimensional space, with each dimension being a peptide ligand affinity. Additionally, suppose CRKL has a binding affinity of 0.8, 0.9, and 0.5 of the same peptide ligands. The C-terminus SH2 domain of PIK3 could give quite a different vector of (0.1, 0.1, 0.8). Mapping the CRK, CRKL, and PIK3 (C-terminus) vectors into the same space will manifest two vectors (CRK and CRKL) close in proximity and a third (PIK3) divergent vector. The distance between any two of these vectors is calculable by Euclidean distance, which is an n-dimensional version of the Pythagorean formula (in this example, three dimensional). Euclidean distance constitutes the pairwise comparison function for this example, and, under examination, CRK and CRKL will be separated by a short Euclidean distance and PIK3 will be separated by a relatively greater distance from CRK and CRKL in this peptide ligand Euclidean space. The dimensional space can be expanded to M dimensions with M peptide ligands and still use Euclidean distance with an M dimensional version of the Pythagorean formula, which has M terms instead of the three terms in our previous example. The Euclidean distance can then be calculated for every pair of SH2 domains to create an N-by-N comparison matrix. The matrix may then allow for the creation of a neighbor-joining phylogenetic tree consistent with Subheading 3.2 to display a hierarchical clustering of SH2 domains based on their binding affinity to peptide ligands.

The second dimension of this hierarchical two-dimensional clustering may be generated by the interchange of perspective. The above example clustered SH2 domains based on binding affinities to peptide ligands, the second dimension is elucidated by clustering peptide ligands based on their binding affinities to SH2 domains. Each peptide ligand in the list M has a SH2 domain profile of binding affinities of length N. The SH2 domain profile may project into N dimensional space for each peptide ligand, and compute the M-by-M Euclidean distance comparison matrix based on the N dimensional Pythagorean formula. A neighbor-joining phylogenetic tree may then be created consistent with Subheading 3.2 to display a hierarchical clustering of peptide ligands based on their binding affinity to SH2 domains.

The hierarchical clustering in the SH2 dimension and the peptide ligand dimension may finally be spliced together for data visualization and data analysis in two dimensions, consistent with Subheading 3.4. The result of a similar process is the essence of Fig. 6. The tools for performing this hierarchical two-dimensional clustering may be found in a variety of software packages. For purposes of extensible development the Bioconductor [4] software project based in the R programming language was the suitable choice selection.

2 Materials

2.1 Datasets and Software

Select the relations between objects to be inspected. Select one or more comparison functions (sequence alignment, peptide binding pattern, quantitative western blot, SPOTS screening method, etc.) by which the objects will be compared.

1. Sequence alignment of human SH2 domains. As described in Subheading 1.2, this dataset represents phylogenetic relationship of SH2 domains and includes a single set of objects (human SH2 domains), and 120 objects (the full complement of human SH2 domains) (*see* **Note 2**).
2. Quantitative far-Western blotting data. Binding specificity of SH2 domains are determined as the band pattern in 1D gel-far-Western blotting using a phosphatase inhibited lysate mixture [3]. This data is a 130-by-49 (65 doubly measured domains, and 49 measured intensities per domain including controls) matrix based on the raw experimental data.
3. SPOT peptide array binding data. In this dataset are two sets of objects, SH2 domains and physiological phosphotyrosine ligands, with 50 and 192 objects respectively (*see* Subheading 1.4 and **Notes 3** and **4**) [5, 6].
4. Plotting software. Use Excel, Matlab, R or any other software package to create a scatter plot of the object pairs in two-dimensional space.
5. ClustalW [7] is a suite of software that provides sequence alignment and is compatible with exporting sequence relations in a distance matrix format.
6. Phylip allows users to create dendrograms from distance matrices.
7. Bioconductor (an R programming language project) allows a user to make easy utilization of a wide collection of programs and algorithms.

3 Methods

3.1 Create A Comparison Matrix

1. For every comparison method selected in Subheading 2.1 **item 1, 2,** and **3** create a comparison matrix using the following pseudo-code.
2. Input: Set of objects *N*, set of objects *M*, a quantitative comparison_function that takes as input two objects.
3. Output: A comparison matrix based on the objects and the comparison function

   for every *i* between 0 and *N*

   for every *j* between 0 and *M*

   $$\text{Matrix}(i,j) = \text{comparison\_function}(N_i, M_j)$$

3.2 Neighbor-Joining Phylogenetic Tree (Appropriate if the Comparison Matrix Is N-by-N, or Like Objects)

1. Use the sequence from Subheading 2.1 **item 1** to create a comparison matrix. This may be accomplished using the pseudo-code in Subheading 3.1, or the use of sequence alignment software such as ClustalW. ClustalW has a distance matrix output format option.
2. Save the distance matrix N-by-N into a Phylip distance matrix formatted file.
3. Use Phylip's Neighbor program to create an unrooted neighbor-joining tree file from the distance matrix. The Neighbor program will save the tree in a Newick file format.
4. Phylip's Drawtree program can create an image based on the Newick tree file.

3.3 Create an N-by-N Euclidean Distance Matrix from an N-by-M Matrix

1. An N-by-M matrix can be thought of as a collection of N objects. Each object N_i is described by a vector of length M. The Euclidean distance examines the distance between every pair of vectors. To create a Euclidean distance matrix, use the following pseudo-code.
2. Input: An N-by-M matrix, $G_{n,m}$
3. Output: An N-by-N Euclidean distance matrix

   for every *i* between 0 and *N*

   for every *j* between 0 and *N*

$$\text{Matrix}(i,j) = \text{sqrt}\left(\left(G_{i,0} - G_{j,0}\right)^2 + \left(G_{i,1} - G_{j,1}\right)^2 + \ldots\left(G_{i,m} - G_{j,m}\right)^2\right)$$

4. This distance matrix may be visualized using method in Subheading 3.2, or the one-dimensional hierarchical clustering R-code in Fig. 4.

3.4 Two-Dimensional Hierarchical Clustering Heatmap (Appropriate if the Comparison Matrix Is N-by-M, or Dissimilar Objects)

1. Use the N-by-M matrices of either Subheading 2.1 **item 2** or **3** to create a comparison matrix, as in Subheading 3.1.
2. Save the comparison matrix in a tab-delimited file, data.txt.
3. Save a list of N objects (SH2 domains for example) to a file nnames.txt.
4. Save a list of M objects (peptide ligands for example) to a file mnames.txt.
5. Execute R code that is similar to the following:

```
#Required Library
require(gplots)
#Importing Data from File
DATA <- as.matrix(read.table("data.txt",
   sep="\t")) #importing comparison matrix
Nnames <- scan("nnames.txt", what="list",
   sep="\t") #Names of 50 SH2 domains
Mnames <- scan("mnames.txt", what="list",
   sep="\t") #Names of 192 peptide ligands
#Assembling labeled matrix
Data_NxM <- matrix(c(DATA),
   nrow=50, ncol=192, byrow=TRUE,
   dimnames=list(Nnames, Mnames))
#Create and Print Two-Dimensional
   Hierarchical Clustering Heatmap
heatmap.2(Data_NxM, trace="none",
   cexRow=0.2, cexCol=0.2)
```

This R code will output a heatmap with dendrograms clustering both dimensions, both N objects and M objects.

3.5 Pairs Scatter Plot

1. Select two distinct quantitative comparison functions (comparison_functionA, comparison_functionB) that takes as input two objects each.
2. Input: Set of objects *N*.
3. Output: Three columns, Label: of each pair of objects in *N*, *X*-Axis: a value for comparison_functionA applied to the pair, *Y*-Axis: a value for comparison_functionB applied to the pair.

   for every *i* between 0 and *N*

   for every *j* between 0 and *N*

   Label = Ni, Nj

   X-Axis = comparison_functionA(Ni, Nj)

   Y-Axis = comparison_functionB(Ni, Nj)

   (*see* **Note 5**)
4. Use your preferred plotting software (Excel, Matlab, R, etc.) to create a scatter plot of the object pairs in two-dimensional space.

The *X* dimension is derived from comparison_functionA, and the *Y* dimension is derived from comparison_functionB.

4 Notes

1. For more details visit Chapter 3 (Fig. 1).
2. In Chapter 3 of this book we demonstrated clustering of SH2 domains to produce Fig. 1 of that chapter.
3. Whatever your objects (SH2 domains, peptide ligands, splice sites, alpha-helices, etc.) they should be chosen carefully, deliberately, and with controls.
4. The raw SH2-SPOT peptide binding data are normalized and organized as 50×192 matrix as described in ref [5, 6].
5. A distance matrix (like the ones created in Subheadings 3.1 and 3.3) is a composite of comparison functions iterated over every object. A comparison function of objects N_i and N_j may therefore be merely a lookup function of the distance matrix for those two corresponding objects (row *i* and column *j*).

Acknowledgments

The knowledge amassed to write this chapter was based on work supported by the University of Chicago Cancer Research Foundation Women's Board and Piers Nash's laboratory at the University of Chicago Ben May Department for Cancer Research.

References

1. Retief JD (2000) Phylogenetic analysis using PHYLIP. Methods Mol Biol 132:243–258
2. Smith TF, Waterman MS (1981) Identification of common molecular subsequences. J Mol Biol 147(1):195–197
3. Machida K, Thompson CM, Dierck K, Jablonowski K, Karkkainen S, Liu B, Zhang H, Nash PD, Newman DK, Nollau P, Pawson T, Renkema GH, Saksela K, Schiller MR, Shin DG, Mayer BJ (2007) High-throughput phosphotyrosine profiling using SH2 domains. Mol Cell 26(6):899–915. doi:10.1016/j.molcel.2007.05.031
4. Gentleman RC, Carey VJ, Bates DM, Bolstad B, Dettling M, Dudoit S, Ellis B, Gautier L, Ge Y, Gentry J, Hornik K, Hothorn T, Huber W, Iacus S, Irizarry R, Leisch F, Li C, Maechler M, Rossini AJ, Sawitzki G, Smith C, Smyth G, Tierney L, Yang JY, Zhang J (2004) Bioconductor: open software development for computational biology and bioinformatics. Genome Biol 5(10):R80. doi:10.1186/gb-2004-5-10-r80
5. Liu BA, Jablonowski K, Shah EE, Engelmann BW, Jones RB, Nash PD (2010) SH2 domains recognize contextual peptide sequence information to determine selectivity. Mol Cell Proteomics 9(11):2391–2404. doi:10.1074/mcp.M110.001586
6. Liu BA, Engelmann BW, Jablonowski K, Higginbotham K, Stergachis AB, Nash PD (2012) SRC Homology 2 Domain Binding Sites in Insulin, IGF-1 and FGF receptor mediated signaling networks reveal an extensive potential interactome. Cell Commun Signal 10(1):27. doi:10.1186/1478-811X-10-27

7. Larkin MA, Blackshields G, Brown NP, Chenna R, McGettigan PA, McWilliam H, Valentin F, Wallace IM, Wilm A, Lopez R, Thompson JD, Gibson TJ, Higgins DG (2007) Clustal W and Clustal X version 2.0. Bioinformatics 23 (21):2947–2948. doi:10.1093/bioinformatics/btm404

8. Liu BA, Jablonowski K, Raina M, Arce M, Pawson T, Nash PD (2006) The human and mouse complement of SH2 domain proteins-establishing the boundaries of phosphotyrosine signaling. Molecular cell 22(6):851–868. doi:10.1016/j.molcel.2006.06.001

Part III

Experimental Tools

Chapter 8

Expression and Production of SH2 Domain Proteins

Bernard A. Liu, Mari Ogiue-Ikeda, and Kazuya Machida

Abstract

The Src Homology 2 (SH2) domain lies at the heart of phosphotyrosine signaling, coordinating signaling events downstream of receptor tyrosine kinases (RTKs), adaptors, and scaffolds. Over a hundred SH2 domains are present in mammals, each having a unique specificity which determines its interactions with multiple binding partners. One of the essential tools necessary for studying and determining the role of SH2 domains in phosphotyrosine signaling is a set of soluble recombinant SH2 proteins. Here we describe methods, based on a broad experience with purification of all SH2 domains, for the production of SH2 domain proteins needed for proteomic and biochemical-based studies such as peptide arrays, mass-spectrometry, protein microarrays, reverse-phase microarrays, and high-throughput fluorescence polarization (HTP-FP). We describe stepwise protocols for expression and purification of SH2 domains using GST or poly His-tags, two widely adopted affinity tags. In addition, we address alternative approaches, challenges, and validation studies for assessing protein quality and provide general characteristics of purified human SH2 domains.

Key words SH2 domains, Bacterial expression, Protein purification, GST tag, His tag

1 Introduction

Eukaryotic proteins are modular in nature, built on independently folding globular domains such as kinase and SH2 domains. On the basis that SH2 and other protein interaction domains can be expressed and purified as isolated protein modules without loss of function, recombinant modular domain proteins have been widely used in both analytical and high-throughput studies. Due to their small and simple structures, modular domains are easily tagged, chemically modified, or immobilized on the solid support, allowing for their broad application in proteomics studies [1]. In forward phase arrays, SH2 domains are immobilized on a glass plate or nitrocellulose membrane (domain array) and probed with various concentrations of labeled-peptides to assess affinities [2]. In reverse phase arrays, lysates or peptides are immobilized on a solid support and probed with diluted domains to determine the presence or absence of high affinity SH2 binding sites in the samples

Kazuya Machida and Bernard A. Liu (eds.), *SH2 Domains: Methods and Protocols*, Methods in Molecular Biology, vol. 1555,
DOI 10.1007/978-1-4939-6762-9_8,

[3, 4]. In-solution and bead-based high-throughput SH2 binding assays have also been developed for pathway-wide interaction mapping, and screening of compound libraries [5, 6].

In this chapter, we focus on production of SH2 domain proteins that are suited for biochemical and proteomic studies such as pull down, far-Western, peptide arrays, mass-spectrometry, protein microarrays, reverse-phase microarrays, and high-throughput fluorescent polarization (HTP-FP). Ideally, all SH2 domain proteins can be purified in parallel in a reasonable amount of time and they are soluble and functional for those downstream assays. Empirically, however, there is no "one-size-fits-all" protocol for all SH2 domains; customization of the protocol is often required on a case-by-case basis. For optimal purification, one should consider several factors such as the length of domain boundaries, selection of fusion tags, vectors, bacterial hosts, and cell growth, induction, and lysis conditions. Therefore, it is worth revisiting the literature on genome-wide SH2 domain studies and learning from accumulated knowledge about SH2 domain proteins.

Characteristics of purified SH2 domains from Jones, Machida, Huang, Tinti, and Liu are summarized in Table 1a. Approximately 60–80% of SH2 domains expressed in bacterial hosts as soluble proteins using the standard protocols including those described in this chapter [2, 4, 7–9]. Of note, multiple groups have reported the same SH2 domains, often within the STAT, Jak, and SOCS protein families, are insoluble, have low expression, and aggregate, regardless of the type of construct or purification methods used. In such cases, one should consider modifications of the purification protocol; including expression of proteins at low temperatures, denaturing and refolding conditions (*see* Chapter 10), or the use of baculovirus (*see* Chapter 11) or a cell free expression systems [10, 11]. In Table 1b, we provide sequences of 122 human SH2 domains and additional multi-domain constructs (e.g., SH2–SH2). Together, Table 1 provides a convenient reference for researchers designing SH2 domain expression clones, with the caveat that for each clone the characteristics (tag, vector, domain boundaries, bacterial strains, etc.) and the purification method can affect the outcome.

Recombinant protein expression from *E. coli* typically requires the use of a tag that allows for protein isolation, and removal of contaminants or impurities. Multiple expression tags have been developed ranging in size, solubility, and the ability to assist in protein folding or prevent proteolysis [12]. Some tags such as glutathione s-transferase (GST) and thioredoxin aid in protein solubility, while others such as 6×His or Flag are convenient for protein isolation and detection, and some can possess both features. However, tag systems may have negative effects such as altering protein conformation, diminishing the yield, toxicity, and altering biological activity. Yet tags can be extremely useful in certain assays. For example, in peptide microarrays, an antibody against your tag can pro-

Table 1
Human complement of SH2 domain proteins

(a)

No.	Protein	Gene symbol	Gene ID	Solubility (Jones [2])	% monomer (Jones [2])	Solubility (Machida [4])	Solubility (N-GST) (Huang [9])	Solubility (N-GST) (Tinti [8])	Solubility (N-GST) (Liu [7])	Expression (mg/L) (Liu [7])	Aggregation score (TANGO) (Andersen [19])	Other references
1	Abl	ABL1	25	Good	100	Good	Good	Good	Good	11	37	
2	Arg	ABL2	27	Good	99	Good	Good	Good	Good	1	339	
3	Nsp2/SH2D3B	BCAR3	8412	Good	100	Poor	Good	Good	n/a	n/a	126	
4	Blk	BLK	640	Good	99	Good	Good	Good	Good	6	33	
5	Blnk	BLNK	29760	Good	81	Good	Good	Poor	Good	7	626	
6	Bmx	BMX	660	Good	99	Good	Good	Good	Good	6	190	
7	Btk	BTK	695	Good	97	Good	Good	Good	n/a	n/a	117	
8	CblA (SH2)	CBL	867	n/a	n/a	n/a	Poor	Good	n/a	n/a	476	
9	CblA (TKB)	CBL	867	n/a	n/a	Fair	n/a	n/a	n/a	n/a	n/a	
10	CblB (SH2)	CBLB	868	n/a	n/a	n/a	Poor	Good	n/a	n/a	590	[22]
11	CblB (TKB)	CBLB	868	n/a	n/a	Poor	n/a	n/a	n/a	n/a	n/a	
12	CblC (SH2)	CBLC	23624	n/a	n/a	n/a	Poor	Poor	n/a	n/a	393	[22]
13	CblC (TKB)	CBLC	23624	n/a	n/a	Fair	n/a	n/a	n/a	n/a	n/a	
14	Chimerin	CHN1	1123	Poor	11	Poor	Good	Poor	n/a	n/a	435	
15	Chimerin2	CHN2	1124	Good	0	Poor	Good	n/a	n/a	n/a	327	
16	Cis1/Cish	CISH	1154	Poor	0	Poor	Good	Poor	Good	n/a	504	
17	Mist/Clnk	CLNK	11644 9	Poor	0	Good	Good	Good	n/a	n/a	682	
18	Crk	CRK	1398	Good	74	Good	Good	Good	Good	5	314	

(continued)

Table 1
(continued)

(a)												
No.	*Protein*	*Gene symbol*	*Gene ID*	*Solubility (Jones [2])*	*% monomer (Jones [2])*	*Solubility (Machida [4])*	*Solubility (N-GST) (Huang [7])*	*Solubility (N-GST) (Tinti [8])*	*Solubility (N-GST) (Liu [9])*	*Expression (mg/L) (Liu [9])*	*Aggregation score (TANGO) (Andersen [10])*	*Other references*
19	CrkL	CRKL	1399	Good	97	Fair	Good	Good	Good	7	123	[23]
20	Csk	CSK	1445	n/a	n/a	Good	Good	Good	Good	18	200	
21	Dapp1	DAPP1	27071	Poor	97	Good	Good	Good	n/a	n/a	92	
22	Fer	FER	2241	Good	89	Good	Good	Good	Good	6	655	
23	Fes/Fps	FES	2242	Good	68	Good	Good	Good	Good	7	494	
24	Fgr/Src2	FGR	2268	Good	100	Good	Good	Good	Good	10	88	
25	Frk	FRK	2444	Poor	7	Good	Good	Good	Good	6	432	
26	FynA	FYN	2534	Low exp	n/a	Good	Good	Good	Good	13	34	
27	Grap	GRAP	10750	n/a	n/a	Fair	Good	n/a	Good	7	281	
28	GrapL	GRAPL	40058 1	n/a	n/a	n/a	n/a	n/a	n/a	n/a	n/a	
29	Grap2/Gads	GRAP2	9402	Good	99	Fair	Good	Good	Good	13	995	
30	Grb10	GRB10	2887	Good	100	Poor	Good	Good	Good	7	454	[24]
31	Grb14	GRB14	2888	Good	100	Fair	Good	Good	Good	10	271	
32	Grb2/Ash	GRB2	2885	Good	85	Good	Good	Good	Good	16	719	
33	Grb7	GRB7	2886	Good	100	Poor	Good	Degraded	Good	35	351	[25–27]
34	Hck	HCK	3055	Good	100	Good	Good	Good	Good	5	48	
35	Hsh2	HSH2D	84941	Good	99	Good	Good	Good	Good	7	63	
36	Ship1/Inpp5d	INPP5D	3635	n/a	n/a	Good	Good	Good	Good	5	224	
37	Ship2/Inppl1	INPPL1	3636	Good	93	Poor	Good	Good	Good	9	1192	[28, 29]

38	Emt/Itk	ITK	3702	Good	62	Good	Good	Good	Good	14	560	
39	Jak1	JAK1	3716	Poor	0	Poor	Good	Degraded	n/a	n/a	499	[30]
40	Jak2	JAK2	3717	Good	20	Poor	Good	n/a	n/a	n/a	1078	[31]
41	Jak3	JAK3	3718	Good	0	Poor	Poor	Degraded	n/a	n/a	1116	[32]
42	Lck	LCK	3932	Good	82	Good	Good	Degraded	Good	12	31	
43	Slp76/Lcp2	LCP2	3937	Good	97	Fair	Low exp	Good	n/a	n/a	1325	
44	Lyn	LYN	4067	Good	100	Good	Good	Good	Good	5	195	
45	Matk/Lsk	MATK	4145	Good	100	Good	Good	Good	Good	10	644	
46	Nck1	NCK1	4690	Good	100	Good	Good	Degraded	Good	7	114	
47	Nck2	NCK2	8440	Good	100	Good	Good	Good	Good	10	41	
48	p85a (C)	PIK3R1	5295	Good	76	Good	Good	n/a	Good	4	201	
49	p85a(N)	PIK3R1	5295	Good	99	Good	Good	n/a	Good	15	41	
50	p85a (NC)	PIK3R1	5295	Good	100	Good	n/a	Good	n/a	n/a	n/a	
51	p85b (C)	PIK3R2	5296	Good	95	Good	Good	n/a	Good	7	341	
52	p85b(N)	PIK3R2	5296	Good	88	Good	Good	n/a	Good	9	56	
53	p85b (NC)	PIK3R2	5296	Good	44	Good	n/a	n/a	n/a	n/a	n/a	
54	p55g (C)	PIK3R3	8503	Good	54	n/a	Good	n/a	Good	10	60	
55	p55g (N)	PIK3R3	8503	Good	90	n/a	Good	n/a	Good	5	33	
56	p55g (NC)	PIK3R3	8503	Good	71	Fair	n/a	Good	n/a	n/a	n/a	
57	Plcg1 (C)	PLCG1	5335	Good	99	Good	Good	n/a	n/a	n/a	52	
58	Plcg1 (N)	PLCG1	5335	Good	89	Good	Good	n/a	Good	10	196	
59	Plcg1 (NC)	PLCG1	5335	Good	100	Good	n/a	Good	n/a	n/a	n/a	
60	Plcg2 (C)	PLCG2	5336	Good	0	Fair	Good	n/a	n/a	n/a	536	

(continued)

Table 1
(continued)

(a)												
No.	Protein	Gene symbol	Gene ID	Solubility (Jones [2])	% monomer (Jones [2])	Solubility (Machida [4])	Solubility (N-GST) (Huang [7])	Solubility (N-GST) (Tinti [8])	Solubility (N-GST) (Liu [9])	Expression (mg/L) (Liu [9])	Aggregation score (TANGO) (Andersen [10])	Other references
61	Plcg2 (N)	PLCG2	5336	Good	51	Fair	Good	n/a	n/a	n/a	331	
62	Plcg2 (NC)	PLCG2	5336	Good	15	Fair	Good	Good	n/a	n/a	n/a	
63	Brk/Ptk6	PTK6	5753	Good	100	Good	Good	Good	Good	6	556	
64	SHP-2 (C)	PTPN11	5781	Good	91	Good	Good	Good	Good	11	349	
65	SHP-2 (N)	PTPN11	5781	Good	96	Good	Good	Good	Good	12	40	
66	SHP-2 (NC)	PTPN11	5781	Good	91	Good	n/a	n/a	n/a	n/a	n/a	
67	SHP-1 (C)	PTPN6	5777	Good	71	Poor	Good	n/a	Good	22	829	
68	SHP-1 (N)	PTPN6	5777	Good	12	Poor	Poor	n/a	Insoluble	n/a	23	
69	SHP-1 (NC)	PTPN6	5777	Good	34	Poor	n/a	Good	n/a	n/a	n/a	[33]
70	RasGAP (C)	RASA1	5921	Good	91	Good	Good	n/a	Good	15	563	
71	RasGAP (N)	RASA1	5921	Good	66	Good	Good	n/a	Good	12	639	
72	RasGAP (NC)	RASA1	5921	Good	93	Good	n/a	Good	n/a	n/a	n/a	
73	Rin1	RIN1	9610	Good	18	Poor	Good	Degraded	n/a	n/a	125	[34, 35]
74	Rin2	RIN2	54453	Good	0	Fair	Good	Degraded	n/a	n/a	1287	[36]
75	Rin3	RIN3	79890	Good	0	Poor	Good	Degraded	n/a	n/a	1150	[36]
76	Sh2-B	SH2B1	25970	Good	98	Good	Good	Good	Good	21	433	
77	Aps	SH2B2	10603	Good	76	Good	Good	Good	Good	3	408	
78	Lnk	SH2B3	10019	Good	84	Good	Poor	n/a	n/a	n/a	402	
79	Sap/SH2D1A	SH2D1A	4068	Good	99	Good	Good	Good	Good	8	1389	

80	Eat2/SH2D1B	SH2D1B	11715 7	Good	98	Fair	Good	Good	Good	5	290	
81	Vrap/SH2D2A	SH2D2A	9047	Good	50	Poor	Good	Good	Good	2	772	[19]
82	Nsp1/SH2D3A	SH2D3A	10045	Good	89	Good	Good	Good	n/a	n/a	39	
83	Nsp3/SH2D3C	SH2D3C	10044	Good	59	Poor	Poor	Good	n/a	n/a	81	
84	Sh2A/SH2D4A	SH2D4A	63898	n/a	n/a	Fair	Poor	Poor	n/a	n/a	73	
85	SH2D4B	SH2D4B	38769 4	n/a	n/a	Poor	Poor	n/a	n/a	n/a	310	
86	SH2D5	SH2D5	40074 5	n/a	n/a	Poor	Poor	n/a	n/a	n/a	977	
87	SH2D7	SH2D7	64689 2	n/a	n/a	n/a	n/a	n/a	n/a	n/a	n/a	
88	Slnk	SH2D6	28494 8	n/a	n/a	Good	Good	n/a	Good	26	595	
89	SH3BP2	SH3BP2	6452	Good	99	Good	Good	Good	Good	4	1019	
90	Shb	SHB	6461	Good	74	Good	Good	n/a	Good	7	23	
91	ShcA/Shc1	SHC1	6464	Good	100	Fair	Good	Good	Good	3	51	
92	Sck/Shc2/ShcB	SHC2	25759	n/a	n/a	Fair	Good	n/a	n/a	n/a	22	
93	ShcC/Shc3	SHC3	53358	Good	99	Good	Good	Good	Good	8	70	
94	ShcD/Shc4/ RaL P	SHC4	39969 4	n/a	n/a	Good	Good	Good	Good	38	12	
95	SHD	SHD	56961	n/a	n/a	Good	Good	Good	Good	4	10	
96	SHE	SHE	12666 9	n/a	n/a	Good	Good	n/a	Good	9	155	
97	SHF	SHF	90525	n/a	n/a	Good	Good	Poor	Good	15	15	
98	Slap	SLA	6503	Poor	0	Poor	Good	Poor	n/a	n/a	77	[37]
99	Slap2	SLA2	84174	Good	97	Poor	Good	n/a	n/a	n/a	93	
100	SOCS1	SOCS1	8651	Poor	100	Fair	Poor	Poor	n/a	n/a	390	

(continued)

Table 1
(continued)

(a)

No.	Protein	Gene symbol	Gene ID	Solubility (Jones [2])	% monomer (Jones [2])	Solubility (Machida [4])	Solubility (N-GST) (Huang [7])	Solubility (N-GST) (Tinti [8])	Solubility (N-GST) (Liu [9])	Expression (mg/L) (Liu [9])	Aggregation score (TANGO) (Andersen [10])	Other references
101	SOCS2	SOCS2	8835	Poor	100	Poor	Poor	Poor	Good	n/a	629	[38, 39]
102	SOCS3	SOCS3	9021	Poor	76	Poor	Poor	Poor	n/a	n/a	433	[40, 41]
103	SOCS4	SOCS4	12280 9	Poor	0	Poor	Poor	Poor	Good	n/a	654	
104	SOCS5	SOCS5	9655	Poor	49	Fair	Poor	Poor	n/a	n/a	814	[42]
105	SOCS6	SOCS6	9306	Poor	63	Poor	Poor	Poor	Good	n/a	145	
106	SOCS7	SOCS7	30837	Poor	0	Fair	Poor	Good	Good	n/a	400	
107	Src	SRC	6714	Good	98	Good	Good	Good	Good	5	140	
108	Srms/Ptk70	SRMS	6725	n/a	n/a	Fair	Good	Good	Good	15	146	
109	Brdg1	STAP1	26228	Good	82	Poor	Good	GOOD	Good	13	409	[43]
110	Bks/Stap2	STAP2	55620	n/a	n/a	Good	Good	Good	n/a	n/a	468	
111	Stat1	STAT1	6772	Poor	0	Poor	Poor	Poor	n/a	n/a	714	[44]
112	Stat2	STAT2	6773	Good	0	Poor	Poor	Poor	n/a	n/a	1036	[44]
113	Stat3	STAT3	6774	Poor	39	Poor	Poor	Poor	n/a	n/a	1395	[5, 45, 46]
114	Stat4	STAT4	6775	Poor	0	Poor	Poor	n/a	n/a	n/a	725	
115	Stat5A	STAT5A	6776	Poor	0	Poor	Poor	Poor	n/a	n/a	1051	[47]
116	Stat5B	STAT5B	6777	n/a	n/a	Poor	Poor	Poor	n/a	n/a	1440	[48]
117	Stat6	STAT6	6778	Poor	0	Fair	Poor	n/a	n/a	n/a	992	

118	Supt6h	SUPT6H	6830	n/a	n/a	Poor	Good	Good	Good	2	185	[49, 50]
119	Syk (C)	SYK	6850	Good	97	Good	Good	Good	Good	14	425	
120	Syk(N)	SYK	6850	Good	65	Good	Good	Good	Good	9	388	
121	Syk (NC)	SYK	6850	Good	58	Good	n/a	n/a	n/a	n/a	n/a	
122	Tec	TEC	7006	Good	99	Fair	Good	Low exp	Good	6	164	
123	Tenc1/Tns2	TENC1	23371	Good	100	Good	Good	Good	Good	6	153	
124	Tensin/Tns1	TNS1	7145	Good	86	Good	Good	Good	Good	37	246	
125	Tem6/Tns3/ Ten s1	TNS3	64759	Good	100	Fair	Low exp	Good	n/a	n/a	430	
126	Cten/Tns4	TNS4	84951	Good	100	Fair	Good	Good	Good	9	132	
127	Txk	TXK	7294	Good	91	Poor	Good	Good	Good	6	604	[51]
128	Tyk2	TYK2	7297	n/a	n/a	Poor	Poor	Low exp	n/a	n/a	285	[52]
129	Vav1	VAV1	7409	Good	100	Good	Good	Good	GOOD	18	218	
130	Vav2	VAV2	7410	Good	100	Good	Good	Good	Good	14	76	
131	Vav3	VAV3	10451	Good	87	Fair	Good	Poor	Good	22	58	
132	Yes	YES1	7525	Good	99	Good	Good	Good	Good	9	217	
133	Zap70 (C)	ZAP70	7535	Good	88	Good	Low exp	n/a	n/a	n/a	696	
134	Zap70 (N)	ZAP70	7535	Good	99	Good	Good	n/a	Good	21	513	
135	Zap70 (NC)	ZAP70	7535	Good	99	Good	n/a	Good	n/a	n/a	n/a	[53]

(continued)

Table 1
(continued)

(b)				
No.	***Protein***	***Gene symbol***	***Gene ID***	***Amino acid sequence of SH2 domains (bold: SH2 domain; *Arg betaB5 position)***
1	Abl	ABL1	25	PSNYITPVNSLEKHS**WYHGPVSRNAAEYLLSSGINGSFLVR*ESESSPGQRSISLRYEGRVYHYRIN TASDGKLYVSSESRFNTLAELVHHHSTVADGLITTLHYPA**PKRNKPTVYGVSPNY
2	Arg	ABL2	27	VPSNYITPVNSLEKHS**WYHGPVSRSAAEYLLSSLINGSFLVR*ESESSPGQLSISLRYEGRVYHYRINT TADGKVYVTAESRFSTLAELVHHHSTVADGLVTTLHYPA**PKCNKPTVYGVSPIHD
3	Nsp2/SH2D3B	BCAR3	8412	KELEEELLLSSEDLRS**HAWYHGRIPRQVSENLVQRDGDFLVR*DSLSSPGNFVLTCQWKNLAQHF KINRTVLRLSEAYSRVQYQFEMESFDSIPGLVRCYVGNRRPISQQSGAIIFQPI**NRTVPLR
4	Blk	BLK	640	FVARVESLEMER**WFFRSQGRKEAERQLLAPINKAGSFLIR*ESETNKGAFSLSVKDVTTQGE LIKHYKIRCLDEGGYYISPRITFPSLQALVQHYSKKGDGLCQRLTLPC**VRPAPQNPWAQ
5	Blnk	BLNK	29760	TISEQEAGVLCKP**WYAGACDRKSAEEALHRSNKDGSFLIR*KSSGHDSKQPYTLVVFFNKR VYNIPVRFIEATKQYALGRKKNGEEYFGSVAEIIRNHQHSPLVLIDSQNNTKDSTRLKY AV**KVS
6	Bmx	BMX	660	SSEEEENLDDYD**WFAGNISRSQSEQLLRQKGKEGAFMVR*NSSQVGMYTVSLFSKAVND KKGTVKHYHVHTNAENKLYLAENYCFDSIPKLIHYHQHNSAGMITRLRHPV**STKAN KVPDSVS
7	Btk	BTK	695	PSNYVTEAEDSIEMYE**WYSKHMTRSQAEQLLKQEGKEGGFIVR*DSSKAGKYTVSVFAKS TGDPQGVIRHYVVCSTPQSQYYLAEKHLFSTIPELINYHQHNSAGLISRLKYPV**SQQN KNAPSTAGLGYGSWE
8	CblA (SH2)	CBL	867	NDYISVFEFDIFTRLF**QPWSSLLRNWNSLAVTHPGYMAFLTYDEVKARLQKFIHKPGSYIFR *LSCTRLGQWAIGYVTADGNILQTIPHNKPLFQALIDGFREGFYLFPDGRNQNPDLTG** LCEPTPQD

9	CblA (TKB)	CBL	867	DAFQPHHHHHHLSPH**PPGTVDKKMVEKCWKLMDKVVRLCQNPKLALKNSPPYIL DLLPDTYQHLRTILSRYEGKMETLGENEYFRVFM ENLMKKTKQTISLFKEGKERMYEENSQPR RNLTKLSLIFSHMLAELKGIFPSGLF QGDTFRITKADAAEFWRKAFGEKTIVPWKSFRQALHEVHPISSGLEA MALKSTIDLTCNDYISVFEFDIFTRLFQPW SSLLRNWNSLAVTHPGYMAFLTYDEVKARLQKFIHKP GSYIFR*LSCTRLGQWAIGYVTADGNILQTIPHNKPLFQALIDGFREG FYLFPDGRNQNPDLTG**LCEPTPQD
10	CblB (SH2)	CBLB	868	NDYISVFEFDIFTRLF**QPWGSILRNWNFLAVTHPGYMAFLTYDEVKARLQKYSTKPGSYIFR *LSCTRLGQWAIGYVTGDGNILQTIPHNKPLFQALIDGSREGFYLYPDGRSYNPDLTG**L CEPTPHD
11	CblB (TKB)	CBLB	868	DAIQDAVG**PPKQAAADRRTVEKTWKLMDKVVRLCQNPKLQLKNSPPYILDILPDTYQHLR LILSKYDDNQKLAQLSENEYFKIYIDSLMKKSKRAIRLFKEGKERMYEEQSQDRRNLTKL SLIFSHMLAEIKAIFPNGQFQGDNFRITKADAAEFWRKFFGDKTIVPWKVFRQCLHEVHQ ISSSLEAMALKSTIDLTCNDYISVFEFDIFTRLFQPWGSILRNWNFLAVTHPGYMAFLTYDE VKARLQKYSTKPGSYIFR*LSCTRLGQWAIGYVTGDGNILQTIPHNKPLFQALIDGSREGFY LYPDGRSYNPDLTG**LCEPTPHD
12	CblC (SH2)	CBLC	23624	DLTCSGHVSIFEFDVFTRLF**QPWPTLLKNWQLLAVNHPGYMAFLTYDEVQERLQACRDKPG SYIFR*PSCTRLGQWAIGYVSSDGSILQTIPANKPLSQVLLEGQKDGFYLYPDGKTHNPDL TE**LGQAEPQQR
13	CblC (TKB)	CBLC	23624	MALAVA**PWGRQWEEARALGRAVRMLQRLEEQCVDPRLSVSPPSLR DLLPRTAQLLREVAHS RRAAGGGGPGGPGGSGDFLLIYL ANLEAKSRQVAALLPPRGRRSA N DELFRAGSRLRRQLAKLAIIFSH MHAELHALFPGGKYCGHMYQLTKAPAHTFW RESCGARCVLPWAEFESLLGTCHPVEPGCTALALRTTIDLTCS GHVSIFEFDVFTRLFQPWPT LLKNWQLLAVNHPGYMAFLTYDEVQ ERLQACRDKPGSYIFR*PSCTRLGQWAIGYVSSDGSILQTIPANKPLSQVLLEG QKDGFYLYPDGKTHNPDLTE**LGQAEPQQR

(continued)

Table 1
(continued)

(b)				
No.	***Protein***	***Gene symbol***	***Gene ID***	***Amino acid sequence of SH2 domains (bold: SH2 domain; *Arg betaB5 position)***
14	Chimerin	CHN1	1123	TCEVENRPKYYGR**EFHGMISREAADQLLIVAEGSYLIR*ESQRQPGTYTLALRFGSQTRNFRL YYDGKHFVGEKRFESIHDLVTDGLITLYIETKAAEYIAKMTINPIYEHV**GYTTLNRE
15	Chimerin2	CHN2	1124	VENRPKYYGR**EFHGIISREQADELLGGVEGAYILR*ESQRQPGCYTLALRFGNQTLNYRLFHDG KHFVGEKRFESIHDLVTDGLITLYIETKAAEYISKMTTNPIYEHI**GYATLLREK
16	Cis1/Cish	CISH	1154	IAKTFSYLRESG**WYWGSITASEARQHLQKMPEGTFLVR*DSTHPSYLFTLSVKTTRGPTNVRIE YADSSFRLDSNCLSRPRILAFPDVVSLVQHY**VASCTADTRSDSPDPAPTPAL
17	Mist/Clnk	CLNK	116449	KRSDRKDVQH**NEWYIGEYSRQAVEEAFMKENKDGSFLVR*DCSTKSKEEPYVLAVFYENKVYN VKIRFLERNQQFALGTGLRGDEKFDSVEDIIEHYKNFPIILIDGKDKTGVHRKQCHLTQPL** PLTRHLLPL
18	Crk	CRK	1398	MAGNFDSEERSS**WYWGRLSRQEAVALLQGQRHGVFLVR*DSSTSPGDYVLSVSENSRVSHYII NSSGPRPPVPPSPAQPPPGVSPSRLRIGDQEFDSLPALLEFYKIHYLDTTTLIEPV**SRSRQGSGVIL
19	CrkL	CRKL	1399	MSSARFDSSDRSA**WYMGPVSRQEAQTRLQGQRHGMFLVR*DSSTCPGDYVLSVSENSRVSHYII NSLPNRRFKIGDQEFDHLPALLEFYKIHYLDTTTLIEPA**PRYPSPPMGSVS
20	Csk	CSK	1445	GVKAGTKLSLMP**WFHGKITREQAERLLYPPETGLFLVR*ESTNYPGDYTLCVSCDGKVEHYR IMYHASKLSIDEEVYFENLMQLVEHYTSDADGLCTRLIKPK**VMEGTVAAQ
21	Dapp1	DAPP1	27071	WSRSDGEAELLQDLG**WYHGNLTRHAAEALLLSNGCDGSYLLR*DSNETTGLYSLSVRAKDSV KHFHVEYTGYSFKFGFNEFSSLKDFVKHFANQPLIGSETGTLMVLKHPY**PRKVEEPSIYESVRV
22	Fer	FER	2241	DMISISEKPLAEQD**WYHGAIPRIEAQELLKKQGDFLVR*ESHGKPGEYVLSVYSDGQRRHFIIQ YVDNMYRFEGTGFSNIPQLIDHHYTTKQVITKKSGVVLLNPI**PKDKKWILS
23	Fes/Fps	FES	2242	EVQKPLHEQL**WYHGAIPRAEVAELLVHSGDFLVR*ESQGKQEYVLSVLWDGLPRHFIIQSLDNL YRLEGEGFPSIPLLIDHLLSTQQPLTKKSGVVLHRAV**PKDKWVLNHED
24	Fgr/Src2	FGR	2268	PSNYVAPVDSIQAEE**WYFGKIGRKDAERQLLSPGNPQGAFLIR*ESETTKGAYSLSIRDWDQTRG DHVKHYKIRKLDMGGYYITTRVQFNSVQELVQHYMEVNDGLCNLLIAPC**TIMKPQTLGLAKDA

25	Frk	FRK	2444	YVAEDRSLQAEP**WFFGAIGRSDAEKQLLYSENKTGSFLIR*ESESQKGEFSLSVLDGAVVKHYRIKRLDEGGFFLTRRRIFSTLNEFVSHYTKTSDGLCVKLGKPC**LKIQVPAPFDLSY
26	FynA	FYN	2534	IPSNYVAPVDSIQAEE**WYFGKLGRKDAERQLLSFGNPRGTFLIR*ESETTKGAYSLSIRDWDDMKGDHVKHYKIRKLDNGGYYITTRAQFETLQQLVQHYSGTWNGNTKVAIKTLKPG**TMSPESFLEEAQ
27	Grap	GRAP	10750	KNYIRVKP**HPWYSGRISRQLAEEILMKRNHLGAFLIR*ESESSPGEFSVSVNYGDQVQHFKVLREASGKYFLWEEKFNSLNELVDFYRTTTIAKKRQIFLRDEEPL**LKSPGA
28	GrapL	GRAPL	400581	KNYIRVKP**HPWYSGRISRQLAEEILMKRNHLGAFLIR*ESESSPGEFSVSVNNRAQRGPCLGPKSHSRLG**
29	Grap2/Gads	GRAP2	9402	PKNYIRVKP**HPWYSGRISRQLAEEILMKRNHLGAFLIR*ESESSPGEFSVSVNYGDQVQHFKVLREASGKYFLWEEKFNSLNELVDFYRTTTIAKKRQIFLRDEEPL**LKSPGA
30	Grb10	GRB10	2887	STLSTVIHRT**QHWFHGRISREESHRIIKQQGLVDGLFLLR*DSQSNPKAFVLTLCHHQKIKNFQILPCEDDGQTFFSLDDGNTKFSDLIQLVDFYQLNKGVLPCKLKHHC**IRVAL
31	Grb14	GRB14	2888	SQSSATNMAIHRS**QPWFHHKISRDEAQRLIIQQGLVDGVFLVR*DSQSNPKTFVLSMSHGQKIKHFQIIPVEDDGEMFHTLDDGHTRFTDLIQLVEFYQLNKGVLPCKLKHYC**ARIAL
32	Grb2/Ash	GRB2	2885	IPKNYIEMKP**HPWFFGKIPRAKAEEMLSKQRHDGAFLIR*ESESAPGDFSLSVKFGNDVQHFKVLRDGAGKYFLWVVKFNSLNELVDYHRSTSVSRNQQIFLRDIE**QVPQQPTY
33	Grb7	GRB7	2886	SGTSLSAAIHRT**QLWFHGRISREESQRLIGQQGLVDGLFLVR*ESQRNPQGFVLSLCHLQKVKHYLILPSEEEGRLYFSMDDGQTRFTDLLQLVEFHQLNRGILPCLLRHCC**TRVAL
34	Hck	HCK	3055	ARVDSLET**EEWFFKGISRKDAERQLLAPGNMLGSFMIR*DSETTKGSYSLSVRDYDPRQGDTVKHYKIRTLDNGGFYISPRSTFSTLQELVDHYKKGNDGLCQKLSVPC**MSSKPQKPWEKD
35	Hsh2	HSH2D	84941	DWFVHTQMGQLAQDGV**PEWFHGAISREDAENLLESQPLGSFLIR*VSHSHVGYTLSYKAQSSCCHFMVKLLDDGTFMIPGEKVAHTSLDALVTFHQQKPIEPRRELLTQPC**RQKDPAN
36	Ship1/Inpp5d	INPP5D	3635	MV**PCWNHGNITRSKAEELLSRTGKDGSFLVR*ASESISRAYALCVLYRNCVYTYRILPNEDDKFTVQASEGVSMRFFTKLDQLIEFYKKENMGLVTHLQYPV**PLEEEDTGDDPEEDT
37	Ship2/Inppl1	INPPL1	3636	MASACGAPGPGGALGSQA**PSWYHRDLSRAAAEELLARAGRDGSFLVR*DSESVAGAFALCVLYQKHVHTYRILPDGEDFLAVQTSQGVPVRRFQTLGELIGLYAQPNQGLVCALLLPV**EGEREPDPPDDRD

(continued)

Table 1
(continued)

(b)				
No.	Protein	Gene symbol	Gene ID	Amino acid sequence of SH2 domains (bold: SH2 domain; *Arg betaB5 position)
38	Emt/Itk	ITK	3702	PSSYLVEKSPNNLETYE**WYNKSISRDKAEKLLLDTGKEGAFMVR*DSRTAGTYTVSVFTKAVVSENNPCIKHYHIKETNDNPKRYYVAEKYVFDSIPLLINYHQHNGGGLVTRLRYPV**CFGRQKAPVTAG
39	Jak1	JAK1	3716	DVAPPLIVHNI**QNGCHGPICTEYAINKLRQEGSEEGMYVLR*WSCTDFDNILMTVTCFEKSEQVQGAQKQFKQIEVQKGRYSLHGSDRSFPSLGDLMSHLKKQILRTDNISFMLKRCCQPKPREIS**NLLVATKKAQEWQ
40	Jak2	JAK2	3717	PPAVLENI**QSNCHGPISMDFAISKLKKAGNQTGLYVLR*CSPKDFNKYFLTFAVERENVIEYKHCLITKNENEEYNLSGTKKNFSSLKDLLNCYQMETVRSDNIIFQFTKCCPPKPKDKS**NLLVFRTNGVSD
41	Jak3	JAK3	3718	CKEVAPP**RLLEEVAEQCHGPITLDFAINKLKTGGSRPGSYVLR*RSPQDFDSFLLTVCVQNPLGPDYKGCLIRRSPTGTFLLVGLSRPHSSLRELLATCWDGGLHVDGVAVTLTSCCIP**RPKEKSNLI
42	Lck	LCK	3932	PFNFVAKANSLEP**EPWFFKNLSRKDAERQLLAPGNTHGSFLIR*ESESTAGSFSLSVRDFDQNQGEVVKHYKIRNLDNGGFYISPRITFPGLHELVRHYTNASDGLCTRLSRPC**QTQKPQKPWWED
43	Slp76/Lcp2	LCP2	3937	SPAEEENSLN**EEWYVSYITRPEAEAALRKINQDGTFLVR*DSSKKTTTNPYVLMVLYKDKVYNIQIRYQKESQVYLLGTGLRGKEDFLSVSDIIDYFRKMPLLLIDGKNRGSRYQCTLTHA**AGYP
44	Lyn	LYN	4067	PSNYVAKLNTLET**EEWFFKDITRKDAERQLLAPGNSAGAFLIR*ESETLKGSFSLSVRDFDPVHGDVIKHYKIRSLDNGGYYISPRITFPCISDMIKHYQKQADGLCRRLEKAC**ISPKPQKPWD
45	Matk/Lsk	MATK	4145	REALSADPKLSL**MPWFHGKISGQEAVQQLQPPEDGLFLVR*ESARHPGDYVLCVSFGRDVIHYRVLHRDGHLTIDEAVFFCNLMDMVEHYSKDKGAICTKLVRPK**RKHGTKSAEEELA
46	Nck1	NCK1	4690	RPSLTGKFAG**NPWYYGKVTRHQAEMALNERGHEGDFLIR*DSESSPNDFSVSLKAQGKNKHFKVQLKETVYCIGQRKFSTMEELVEHYKKAPIFTSEQGEKLYLVKHL**S
47	Nck2	NCK2	8440	TGPSSSGRFAG**REWYYGNVTRHQAECALNERGVEGDFLIR*DSESSPSDFSVSLKASGKNKHFKVQLVDNVYCIGQRRFHTMDELVEHYKKAPIFTSEHGEKLYLVRAL**Q

48	p85a (C)	PIK3R1	5295	DDEDLPHHDE**KTWNVGSSNRNKAENLLRGKRDGTFLVR*ESSKQGCYACSVVVDGEVKHCVINKTATGYGFAEPYNLYSSLKELVLHYQHTSLVQHNDSLNVTLAYPV**YAQQRR
49	p85a (N)	PIK3R1	5295	GMNNNMSLQD**AEWYWGDISREEVNEKLRDTADGTFLVR*DASTKMHGDYTLTLRKGGNNKLIKIFHRDGKYGFSDPLTFSSVVELINHYRNESLAQYNPKLDVKLLYPV**SKYQQD
50	p85a (NC)	PIK3R1	5295	GMNNNMSLQD**AEWYWGDISREEVNEKLRDTADGTFLVR*DASTKMHGDYTLTLRKGGNNKLIKIFHRDGKYGFSDPLTFSSVVELINHYRNESLAQYNPKLDVKLLYPV**SKYQQDQVVKEDNIEAVGKKLHEYNTQFQEKSREYDRLYEEYTRTSQEIQMKRTAIEAFNETIKIFEEQCQTQERYSKEYIEKFKREGNEKEIQRIMHNYDKLKSRISEIIDSRRRLEEDLKKQAAEYREIDKRMNSIKPDLIQLRKTRDQYLMWLTQKGVRQKKLNEWLGNENTEDQYSLVEDDEDLPHHDE**KTWNVGSSNRNKAENLLRGKRDGTFLVR*ESSKQGCYACSVVVDGEVKHCVINKTATGYGFAEPYNLYSSLKELVLHYQHTSLVQHNDSLNVTLAYPV**YAQQRR
51	p85b (C)	PIK3R2	5296	DDLPHHEE**RTWYVGKINRTQAEEMLSGKRDGTFLIR*ESSQRGCYACSVVVDGDTKHCVIYRTATGFGFAEPYNLYGSLKELVLHYQHASLVQHNDALTVTLAHPV**RAPGPGPPPAAR
52	p85b (N)	PIK3R2	5296	GGSPPSLQD**AEWYWGDISREEVNEKLRDTPDGTFLVR*DASSKIQGEYTLTLRKGGNNKLIKVFHRDGHYGFSEPLTFCSVVDLINHYRHESLAQYNAKLDTRLLYPV**SKYQQD
53	p85b (NC)	PIK3R2	5296	GGSPPSLQD**AEWYWGDISREEVNEKLRDTPDGTFLVR*DASSKIQGEYTLTLRKGGNNKLIKVFHRDGHYGFSEPLTFCSVVDLINHYRHESLAQYNAKLDTRLLYPV**SKYQQDQIVKEDSVEAVGAQLKVYHQQYQDKSREYDQLYEEYTRTSQELQMKRTAIEAFNETIKIFEEQGQTQEKCSKEYLERFRREGNEKEMQRILLNSERLKSRIAEIHESRTKLEQQLRAQASDNREIDKRMNSLKPDLMQLRKIRDQYLVWLTQKGARQKKINEWLGIKNETEDQYALMEDEDDLPHHEE**RTWYVGKINRTQAEEMLSGKRDGTFLIR*ESSQRGCYACSVVVDGDTKHCVIYRTATGFGFAEPYNLYGSLKELVLHYQHASLVQHNDALTVTLAHPV**RAPGPGPPPAAR
54	p55g (C)	PIK3R3	8503	EEDENLPHYDE**KTWFVEDINRVQAEDLLYGKPDGAFLIR*ESSKKGCYACSVVADGEVKHCVIYSTARGYGFAEPYNLYSSLKELVLHYQQTSLVQHNDSLNVRLAYPV**HAQMPSLCR
55	p55g (N)	PIK3R3	8503	MKDSSVSLQD**AEWYWGDISREEVNDKLRDMPDGTFLVR*DASTKMQGDYTLTLRKGGNNKLIKIYHRDGKYGFSDPLTFNSVVELINHYHHESLAQYNPKLDVKLMYPV**SRYQQD

(continued)

Table 1
(continued)

(b)				
No.	***Protein***	***Gene symbol***	***Gene ID***	***Amino acid sequence of SH2 domains (bold: SH2 domain; *Arg betaB5 position)***
56	p55g (NC)	PIK3R3	8503	DSSVSLQD**AEWYWGDISREEVNDKLRDMPDGTFLVR*DASTKMQGDYTLTLRKGGNNKLIKIYHRDGKYGFSDPLTFNSVVELINHYHHESLAQYNPKLDVKLMYPVSRYQQD**QLVKEDNIDAVGKKLQEYHSQYQEKSKEYDRLYEEYTRTSQEIQMKRTAIEAFNETIKIFEEQCHTQEQHSKEYIERFRREGNEKEIERIMMNYDKLKSRLGEIHDSKMRLEQDLKNQALDNREIDKKMNSIKPDLIQLRKIRDQHLVWLNHKGVRQKRLNVWLGIKNEDADENYFINEEDENLPHYDE**KTWFVEDINRVQAEDLLYGKPDGAFLIR*ESSKKGCYACSVVADGEVKHCVIYSTARGYGFAEPYNLYSSLKELVLHYQQTSLVQHNDSLNVRLAYPV**HAQMPSLCR
57	Plcg1 (C)	PLCG1	5335	PVPQTNAHES**KEWYHASLTRAQAEHMLMRVPRDGAFLVR*KRNEPNSYAISFRAEGKIKHCRVQQEGQTVMLGNSEFDSLVDLISYYEKHPLYRKMKLRYPI**NEEALEKIGT
58	Plcg1 (N)	PLCG1	5335	SSSTELHSN**EKWFHGKLGAGRDGRHIAERLLTEYCIETGAPDGSFLVR*ESETFVGDYTLSFWRNGKVQHCRIHSRQDAGTPKFFLTDNLVFDSLYDLITHYQQVPLRCNEFEMRLSEPV**PQTNAHE
59	Plcg1 (NC)	PLCG1	5335	SSSTELHSN**EKWFHGKLGAGRDGRHIAERLLTEYCIETGAPDGSFLVR*ESETFVGDYTLSFWRNGKVQHCRIHSRQDAGTPKFFLTDNLVFDSLYDLITHYQQVPLRCNEFEMRLSEPV**PQTNAHES**KEWYHASLTRAQAEHMLMRVPRDGAFLVR*KRNEPNSYAISFRAEGKIKHCRVQQEGQTVMLGNSEFDSLVDLISYYEKHPLYRKMKLRYPI**NEEALEKIGT
60	Plcg2 (C)	PLCG2	5336	DPVPNPNPHES**KPWYYDSLSRGEAEDMLMRIPRDGAFLIR*KREGSDSYAITFRARGKVKHCRINRDGRHFVLGTSAYFESLVELVSYYEKHSLYRKMRLRYPV**TPELLERYNMERDI
61	Plcg2 (N)	PLCG2	5336	DIPPTELHFG**EKWFHKKVEKRTSAEKLLQEYCMETGGKDGTFLVR*ESETFPNDYTLSFWRSGRVQHCRIRSTMEGGTLKYYLTDNLTFSSIYALIQHYRETHLRCAEFELRLTDPV**PNPNPHESKP
62	Plcg2 (NC)	PLCG2	5336	DIPPTELHFG**EKWFHKKVEKRTSAEKLLQEYCMETGGKDGTFLVR*ESETFPNDYTLSFWRSGRVQHCRIRSTMEGGTLKYYLTDNLTFSSIYALIQHYRETHLRCAEFELRLTDPV**PNPNPHES**KPWYYDSLSRGEAEDMLMRIPRDGAFLIR*KREGSDSYAITFRARGKVKHCRINRDGRHFVLGTSAYFESLVELVSYYEKHSLYRKMRLRYPV**TPELLERYNMERD

63	Brk/Ptk6	PTK6	5753	VPHNYLAERETVESEP**WFFGCISRSEAVRRLQAEGNATGAFLIR*VSEKPSADYVLSVRDTQA** **VRHYKIWRRAGGRLHLNEAVSFLSLPELVNYHRAQSLSHGLRLAAP**CRKHEPEPLPHWD
64	SHP-2 (C)	PTPN11	5781	ELKYPLNCADPTS**ERWFHGHLSGKEAEKLLTEKGKHGSFLVR*ESQSHPGDFVLSVRTGDD** **KGESNDGKSKVTHVMIRCQELKYDVGGGERFDSLTDLVEHYKKNPMVETLGTVLQLK** **QPL**NTTRINAAEIESRVRE
65	SHP-2 (N)	PTPN11	5781	MTS**RRWFHPNITGVEAENLLLTRGVDGSFLAR*PSKSNPGDFTLSVRRNGAVTHIKIQNTG** **DYYDLYGGEKFATLAELVQYYMEHHGQLKEKNGDVIELKYPL**NCADPTS
66	SHP-2 (NC)	PTPN11	5781	MTS**RRWFHPNITGVEAENLLLTRGVDGSFLAR*PSKSNPGDFTLSVRRNGAVTHIKIQNTG** **DYYDLYGGEKFATLAELVQYYMEHHGQLKEKNGDVIELKYPL**NCADPTS**ERWFHGH** **LSGKEAEKLLTEKGKHGSFLVRESQSHPGDFVLSVRTGDD** **KGESNDGKSKVTHVMIRCQELKYDVGGGERFDSL** **TDLVEHYKKNPMVETLGTVLQLKQPL**NTTRINAAEIESRVRE
67	SHP-1 (C)	PTPN6	5777	TSER**WYHGHMSGGQAETLLQAKGEPWTFLVR*ESLSQPGDFVLSVLSDQPKAGPGSPLR** **VTHIKVMCEGGRYTVGGLETFDSLTDLVEHFKKTGIEEASGAFVYLRQPY**YATRVNAAD
68	SHP-1 (N)	PTPN6	5777	**MVRWFHRDLSGLDAETLLKGRGVHGSFLAR*PSRKNQGDFSLSVRVGDQVTHIRIQNSG** **DFYDLYGGEKFATLTELVEYYTQQQGVLQDRDGTIIHLKYPL**NCSDPTSER
69	SHP-1 (NC)	PTPN6	5777	**MVRWFHRDLSGLDAETLLKGRGVHGSFLAR*PSRKNQGDFSLSVRVGDQVTHIRIQNS** **GDFYDLYGGEKFATLTELVEYYTQQQGVLQDRDGTIIHLKYPLN**CSDPTSER**WYHGHM** **SGGQAETLLQAKGEPWTFLVR*ESLSQPGDFVLSVLSDQPKAGPGSPLRVTHIKVMCEG** **GRYTVGGLETFDSLTDLVEHFKKTGIEEASGAFVYLRQPY**YATRVNAAD
70	RasGAP (C)	RASA1	5921	VGREEDPHEGKI**WFHGKISKQEAYNLLMTVGQVCSFLVR*PSDNTPGDYSLYFRTNENIQR** **FKICPTPNNQFMMGGRYYNSIGDIIDHYRKEQIVEGYYLKEPV**PMQDQEQVLND
71	RasGAP (N)	RASA1	5921	EVAIPLTAPPTNQ**WYHGKLDRTIAEERLRQAGKSGSYLIR*ESDRRPGSFVLSFLSQMNVV** **NHFRIIAMCGDYYIGGRRFSSLSDLIGYYSHVSCLLKGEKLLYPV**APPEAIEDRRR
72	RasGAP (NC)	RASA1	5921	EVAIPLTAPPTNQ**WYHGKLDRTIAEERLRQAGKSGSYLIR*ESDRRPGSFVLSFLSQMNVVN** **HFRIIAMCGDYYIGGRRFSSLSDLIGYYSHVSCLLKGEKLLYPV**APPEPVEDRRRVRAILPYT KVPDTDEISFLKGDMFIVHNELEDGWMWVTNLRTDEQGLIVEDLVEEVGREEDPHEGKI**WF** **HGKISKQEAYNLLMTVGQVCSFLVR*PSDNTPGDYSLYFRTNENIQRFKICPTPNNQFM** **MGGRYYNSIGDIIDHYRKEQIVEGYYLKEPV**PMQDQEQVLND
73	Rin1	RIN1	9610	RVVSLRERLLLTR**PVWLQLQANAAAALHMLRTEPPGTFLVR*KSNTRQCQALCMRLPEAS** **GPSFVSSHYILESPGGVSLEGSELMFPDLVQLICAYCHTRDILLLPLQLP**RAIHHAATHKELE

(continued)

Table 1
(continued)

(b)				
No.	***Protein***	***Gene symbol***	***Gene ID***	***Amino acid sequence of SH2 domains (bold: SH2 domain; *Arg betaB5 position)***
74	Rin2	RIN2	54453	LSILDRLLHTH**PIWLQLSLSEEEAAEVLQAQPPGIFLVH*KSTKMQKKVLSLRLPCEFGAPLKEFAIKESTYTFSLEGSGISFADLFRLIAFYCISRDVLPFTLKLPY**AISTAKS
75	Rin3	RIN3	79890	SILEKLIKTC**PVWLQLSLGQAEVARILHRVVAGMFLVR*RDSSSKQLVLCVHFPSLNESSAEVLEYTIKEEKSILYLEGSALVFEDIFRLIAFYCVSRDLLPFTLRLPQ**AILEASSFT
76	Sh2-B	SH2B1	25970	EPEGGEGDQPLSG**YPWFHGMLSRLKAAQLVLTGGTGSHGVFLVR*QSETRRGEYVLTFNFQGKAKHLRLSLNEEGQCRVQHLWFQSIFDMLEHFRVHPIPLESGGSSDVVLVSYV**PSSQRQQGEQ
77	Aps	SH2B2	10603	PEAEPELELSDYP**WFHGTLSRVKAAQLVLAGGPRNHGLFVIR*QSETRPGEYVLTFNFQGKAKASHLRLSLNGHGQCHVQHLWFQSVLDMLRHFHTHPIPLESGGSADITLRSYV**RAQDPPPE
78	Lnk	SH2B3	10019	LDPACQKTDHFLSC**YPWFHGPISRVKAAQLVQLQGPDAHGVFLVR*QSETRRGEYVLTFNFQGIAKHLRLSLTERGQCRVQHLHFPSVVDMLHHFQRSPIPLECGAACDVRLSSYV**VVVSQPPGSCNT
79	Sap/SH2D1A	SH2D1A	4068	MDAVA**VYHGKISRETGEKLLLATGLDGSYLLR*DSESVPGVYCLCVLYHGYIYTYRVSQTETGSWSAETAPGVHKRYFRKIKNLISAFQKPDQGIVIPLQYPV**EKKSARSTQGTTGIREDPDVCLKAP
80	Eat2/SH2D1B	SH2D1B	117157	MDLP**YYHGRLTKQDCETLLLKEGVDGNFLLR*DSESIPGVLCLCVSFKNIVYTYRIFREKHGYYRIQTAEGSPKQVFPSLKELISKFEKPNQGMVVHLLKPI**KRTSPSLRW
81	Vrap/SH2D2A	SH2D2A	9047	KTQAHWLLQHGAA**PAWFHGFITRREAERLLEPKPQGCYLVR*FSESAVTFVLTYRSRTCCRHFLLAQLRDGRHVVLGEDSAHARLQDLLLHYTAHPLSPYGETLTEP**LARQTPEPAGLSLRTEES
82	Nsp1/SH2D3A	SH2D3A	10045	DGEDLAG**QPWYHGLLSRQKAEALLQQNGDFLVR*ASGSRGGNPVISCRWRGSALHFEVFRVALRPRPGRPTALFQLEDEQFPSIPALVHSYMTGRRPLSQATGAVVSRPV**TWQGPLRRSFSEDT

83	Nsp3/SH2D3C	SH2D3C	10044	ELKLSSTDLRS**HAWYHGRIPREVSETLVQRNGDFLIR*DSLTSLGDYVLTCRWRNQALHFKINKVVVKAGESYTHIQYLFEQESFDHVPALVRYHVGSRKAVSEQSGAIIYCPV**NRTFPLR
84	Sh2A/SH2D4A	SH2D4A	63898	RAGYQKTSDTI**APWFHGILTLKKANELLLSTGMPGSFLIR*VSERIKGYALSYLSEDGCKHFLIDASADAYSFLGVDQLQHATLADLVEYHKEEPITSLGKELLLYPC**GQQDQLPD
85	SH2D4B	SH2D4B	387694	RRAGFERNTKFI**APWFHGIISREDAEALLENMTEGAFLVR*VSEKIWGYTLSYRLQKGFKHFLVDASGDFYSFLGVDPNRHATLTDLVDFHKEEIITVSGGELLQEPC**GQRDSPPDYHLLFE
86	SH2D5	SH2D5	400745	HSCLVESEGSLTE**NIWAFAGISRPCALALLRRDVLGAFLLW*PELGASGQWCLSVRTQCGVVPHQVFRNHLGRYCLEHLPAEFPSLEALVENHAVTERSLFCPLDMGRLNPTY**EEQDCGPPGRPPRT
87	SH2D7	SH2D7	646892	APFILQNGAL**PPWFHGFITRKQTEQLLRDKALGSFLIR*LSDRATGYILSYRGSDRCRHFVINQLRNRRYIISGDTQSHSTLAELVHHYQEAQL**EPFKEMLTAACPRPED
88	Slnk	SH2D6	284948	GSASAAEDSDLLT**QPWYSGNCDRYAVESALLHLQKDGAYTVR*PSSGPHGSQPFTLAVLLRGRVFNIPIRRLDGGRHYALGREGRNREELFSSVAAMVQHFMWHPLPLVDRHSGSRELTCLLFPT**KP
89	SH3BP2	SH3BP2	6452	GKPGLLQVPLP**NSVFVNTTESCEVERLFKATSPRGEPQDGLYCIR*NSSTKSGKVLVVWDETSNKVRNYRIFEKDSKFYLEGEVLFVSVGSMEHYHTHVLPSHQSLLLRHPY**GYTGPR
90	Shb	SHB	6461	GERVDPAVPLEK**QIWYHGAISRGDAENLLRLCKECSYLVR*NSQTSKHDYSLSLRSNQGFMHMKLAKTKEKYVLGQNSPPFDSVPEVIHYYTTRKLPIKGAEHLSLLYPV**AVRTL
91	ShcA/Shc1	SHC1	6464	SVSMAEQLRG**EPWFHGKLSRREAEALLQLNGDFLVR*ESTTTPGQYVLTGLQSGQPKHLLLVDPEGVVRTKDHRFESVSHLISYHMDNHLPIISAGSELCLQQPV**ERKL
92	Sck/Shc2/ShcB	SHC2	25759	TEEQL**RQEPWYHGRMSRRAAERMLRADGDFLVR*DSVTNPGQYVLTGMHAGQPKHLLLVDPEGVVRTKDVLFESISHLIDHHLQNGQPIVAAESELHLRGVV**SREP
93	ShcC/Shc3	SHC3	53358	PDAKMLEELQA**ETWYQGEMSRKEAEGLLEKDGDFLVR*KSTTNPGSFVLTGMHNGQAKHLLLVDPEGTIRTKDRVFDSISHLINHHLESSLPIVSAGSELCLQQPV**ERKQ
94	ShcD/Shc4/ RaLP	SHC4	399694	SSHSLPHIKQQLWSE**ECYHGKLSRKAAESLLVKDGDFLVR*ESATSPGQYVLSGLQGGQAKHLLLVDPEGKVRTKDHVFDNVGHLIRYHMDNSLPIISSGSEVSLKQPV**RKDNNPALLHSNK
95	SHD	SHD	56961	DPALPLEK**QPWFHGPLNRADAESLLSLCKEGSYLVR*LSETNPQDCSLSLRSSQGFLHLKFARTRENQVVLGQHSGPFPSVPELVLHYSSRPLPVQGAEHLALLYPV**VTQTP

(continued)

Table 1
(continued)

(b)				
No.	***Protein***	***Gene symbol***	***Gene ID***	***Amino acid sequence of SH2 domains (bold: SH2 domain; *Arg betaB5 position)***
96	SHE	SHE	126669	SEGEKVDPGLPLEK**QPWYHGAISRAEAESRLQPCKEAGYLVR*NSESGNSRYSIALKTSQGCVHIVAQTKDNKYTLNQTSAVFDSIPEVVHYYSNEKLPFKGAEHMTLLYPV**HSKLH
97	SHF	SHF	90525	TDPALPLEN**QVWYHGAISRTDAENLLRLCKEASYLVR*NSETSKNDFSLSLKSSQGFMHMKLSRTKEHKYVLGQNSPPFSSVPEIVHHYASRKLPIKGAEHMSLLYPV**AIRTL
98	Slap	SLA	6503	IPGICVARVY**HGWLFEGLGRDKAEELLQLPDTKVGSFMIR*ESETKKGFYSLSVRHRQVKHYRIFRLPNNWYYISPRLTFQCLEDLVNHYSEVADGLCCVLTTPC**LTQSTAAPAV
99	Slap2	SLA2	84174	SGREYNIPSVHVAKVS**HGWLYEGLSREKAEELLLLPGNPGGAFLIR*ESQTRRGSYSLSVRLSRPASWDRIRHYRIHCLDNGWLYISPRLTFPSLQALVDHYSELADDICCLLKEPC**VLQRAGPLP
100	SOCS1	SOCS1	8651	DYRRITRASALLDA**CGFYWGPLSVHGAHERLRAEPVGTFLVR*DSRQRNCFFALSVKMASGPTSIRVHFQAGRFHLDGSRESFDCLFELLEHYVAAPRRMLGAPLRQRRVRP**LQELCRQ
101	SOCS2	SOCS2	8835	QAARLAKALRELGQ**TGWYWGSMTVNEAKEKLKEAPEGTFLIR*DSSHSDYLLTISVKTSAGPTNLRIEYQDGKFRLDSIICVKSKLKQFDSVVHLIDYYVQMCKDKRTGPEAPRNGTVHLYLTKP**
102	SOCS3	SOCS3	9021	SSKSEYQLVVNAVRKLQE**SGFYWSAVTGGEANLLLSAEPAGTFLIR*DSSDQRHFFTLSVKTQSGTKNLRIQCEGGSFSLQSDPRSTQPVPRFDCVLKLVHHY**MPPPGAPSFPSPPTEPSSE
103	SOCS4	SOCS4	122809	DLLQINNN**PCYWGVMDKYAAEALLEGKPEGTFLLR*DSAQEDYLFSVSFRRYSRSLHARIEQWNHNFSFDAHDPCVFHSPDITGLLEHYKDPSACMFFEPLLSTPL**IRTFPFS
104	SOCS5	SOCS5	9655	DYIHCLVPDLLQITG**NPCYWGVMDRYEAEALLEGKPEGTFLLR*DSAQEDYLFSVSFRRYNRSLHARIEQWNHNFSFDAHDPCVFHSSTVTGLLEHYKDPSSCMFFEPLLTISL**NRTFPFS
105	SOCS6	SOCS6	9306	TEELKKLAK**QGWYWGPITRWEAEGKLANVPDGSFLVR*DSSDDRYLLSLSFRSHGKTLHTRIEHSNGRFSFYEQPDVEGHTSIVDLIEHSIRDSENGAFCYSRSRLPGSATYPV**RLTNPVSR

106	SOCS7	SOCS7	30837	DSSSFAASLRELEK**CGWYWGPMNWEDAEMKLKGKPDGSFLVR*DSSDPRYILSLSFRSQGITHHTRMEHYRGTFSLWCHPKFEDRCQSVVEFIKRAIMHSKNGKFLYFLRSRVPGLPPTPVQLLYPVS**RFSNVKS
107	Src	SRC	6714	YIPSNYVAPSDSIQA**EEWYFGKITRRESERLLLNAENPRGTFLVR*ESETTKGAYCLSVSDFDNAKGLNVKHYKIRKLDSGGFYITSRTQFNSLQQLVAYYSKHADGLCHRLTTVC**PTSKPQTQGL
108	Srms/Ptk70	SRMS	6725	SPETLSD**QPWYFSGVSRTQAQQLLLSPPNEPGAFLIR*PSESSLGGYSLSVRAQAKVCHYRVSMAADGSLYLQKGRLFPGLEELLTYYKANWKLIQNPLLQPC**MPQKAPRQ
109	Brdg1	STAP1	26228	DYVDVLNPMPACFYT**VSRKEATEMLQKNPSLGNMILR*PGSDSRNYSITIRQEIDIPRIKHYKVMSVGQNYTIELEKPVTLPNLFSVIDYFVKETRGNLRPFICSTDENTGQEPS**MEGRS
110	Bks/Stap2	STAP2	55620	TDLTLLPGHL**YMMSEVLAKEEARRALETPSCFLKVSRLEAQLLLERYPECGNLLLR*PSGDGADGVSVTTRQMHNGTHVVRHYKVKREGPKYVIDVEQPFSCTSLDAVVNYFVSHTKKALVPFLLD**EDYEKVLGYVEA
111	Stat1	STAT1	6772	ESILELIKKHLL**PLWNDGCIMGFISKERERALLKDQQPGTFLLR*FSESSREGAITFTWVERSQNGGEPDFHAVEPYTKKELSAVTFPDIIRNYKVMAAENIPENPLKYLYPNIDKDHAFGKYYSRPKEAPEP**MELDGPKG
112	Stat2	STAT2	6773	DKILELVHDHLK**DLWNDGRIMGFVSRSQERRLLKKTMSGTFLLR*FSESSEGGITCSWVEHQDDDKVLIYSVQPYTKEVLQSLPLTEIIRHYQLLTEENIPENPLRFLYPR**IPRDEAFGCYYQEKVNLQERRK
113	Stat3	STAT3	6774	DLVKKYIL**ALWNEGYIMGFISKERERAILSTKPPGTFLLR*FSESSKEGGVTFTWVEKDISGKTQIQSVEPYTKQQLNNMSFAEIIMGYKIMDATNILVSPL**VYLYPDIPKEEAFGKYCRPESQE
114	Stat4	STAT4	6775	KKHILPLWI**DGYVMGFVSKEKERLLLKDKMPGTFLLR*FSESHLGGITFTWVDHSESGEVRFHSVEPYNKGRLSALPFADILRDYKVIMAENIPENPLKYLYPDIPKDKAFGKHYSSQPCE**VSRPTERGDKG
115	Stat5A	STAT5A	6776	DGVMEVLKKHHK**PHWNDGAILGFVNKQQAHDLLINKPDGTFLLR*FSDSEIGGITIAWKFDSPERNLWNLKPFTTRDFSIRSLADRLGDLSYLIYVFPDRPKDEVFSKYYTPV**LAKAVDGYVKPQIKQ
116	Stat5B	STAT5B	6777	DGVMEVLKKHLK**PHWNDGAILGFVNKQQAHDLLINKPDGTFLLR*FSDSEIGGITIAWKFDSQERMFWNLMPFTTRDFSIRSLADRLGDLNYLIYVFPDRPKDEVYSKYYTPV**PCESATAKAVDG
117	Stat6	STAT6	6778	TKRCLR**SYWSDRLIIGFISKQYVTSLLLNEPDGTFLLR*FSDSEIGGITIAHVIRGQDGSPQIENIQPFSAKDLSIRSLGDRIRDLAQLKNLYPKKPKDEAFRSHYKPE**QMGKDGRGYVPATIKMTVERDQ

(continued)

Table 1
(continued)

(b)				
No.	*Protein*	*Gene symbol*	*Gene ID*	*Amino acid sequence of SH2 domains (bold: SH2 domain; *Arg betaB5 position)*
118	Supt6h	SUPT6H	6830	DMKRKQQRTT**YIKRVIAHPSFHNINFKQAEKMMETMDQGDVIIR*PSSKGENHLTVTWKVSDGIYQHVDVREEGKENAFSLGATLWINSEEFEDLDEIVARYVQPMASFARDLLNHKY**YQDCSGGDRKKLEE
119	Syk (C)	SYK	6850	TTAHEKMP**WFHGKISREESEQIVLIGSKTNGKFLIR*ARDNNGSYALCLLHEGKVLHYRIDKDKTGKLSIPEGKKFDTLWQLVEHYSYKADGLLRVLTVPC**QKIGTQGNVNFGGR
120	Syk (N)	SYK	6850	DSANH**LPFFFGNITREEAEDYLVQGGMSDGLYLLR*QSRNYLGGFALSVAHGRKAHHYTIERELNGTYAIAGGRTHASPADLCHYHSQESDGLVCLLKKPF**NRPQGVQPKTG
121	Syk (NC)	SYK	6850	MASSGMADSANH**LPFFFGNITREEAEDYLVQGGMSDGLYLLR*QSRNYLGGFALSVAHGRKAHHYTIERELNGTYAIAGGRTHASPADLCHYHSQESDGLVCLLKKPF**NRPQGVQPKTGPFEDLKENLIREYVKQTWNLQGQALEQAIISQKPQLEKLIATTAHEKMP**WFHGKISREESEQIVLIGSKTNGKFLIR*ARDNNGSYALCLLHEGKVLHYRIDKDKTGKLSIPEGKKFDTLWQLVEHYSYKADGLLRVLTVP**CQKIGTQGNVNFGGR
122	Tec	TEC	7006	TGKKSNNLDQ**YEWYCRNMNRSKAEQLLRSEDKEGGFMVR*DSSQPGLYTVSLYTKFGGEGSSGFRHYHIKETTTSPKKYYLAEKHAFGSIPEIIEYHKHNAAGLVTRLRYP**VSVKGKNAPTTAG
123	Tenc1/Tns2	TENC1	23371	TQESQSNVKFVQDTS**KFWYKPHLSRDQAIALLKDKDPGAFLIR*DSHSFQGAYGLALKVATPPPSAQPWKGDPVEQLVRHFLIETGPKGVKIKGCPSEPYFGSLSALVSQHSISPISLPCCLRIPS**KDPLEETPEAPVPTNMSTAAD
124	Tensin/Tns1	TNS1	7145	RAKVKFVQDTS**KYWYKPEISREQAIALLKDQEPGAFIIR*DSHSFRGAYGLAMKVSSPPPTIMQQNKKGDMTHELVRHFLIETGPRGVKLKGCPNEPNFGSLSALVYQHSIIPLALPCKLVIPN**RDPTDESKDSS

125	Tem6/Tns3/ Tens 1	TNS3	64759	GDKLVIVKFVQDTS**KFWYKADISREQAIAMLKDKEPGSFIVR*DSHSFRGAYGLAMKVATPPPSVL QLNKKAGDLANELVRHFLIECTPKGVRLKGCSNEPYFGSLTALVCQHSITPLALPCKLLIPE**RD PLEEIAESS
126	Cten/Tns4	TNS4	84951	TMKFVMDTSKY**WFKPNITREQAIELLRKEEPGAFVIR*DSSSYRGSFGLALKVQERHFLIESSAKG VHLKGADEEPYFGSLSAFVCQHSIMALALPCKLTIPQ**RELGGADG
127	Txk	TXK	7294	TENKITNLEI**YEWYHRNITRNQAEHLLRQESKEGAFIVR*DSRHLGSYTISVFMGARRSTEAAIKH YQIKKNDSGQWYVAERHAFQSIPELIWYHQHNAAGLMTRLRYPV**GLMGSCLPAT
128	Tyk2	TYK2	7297	RLVMSIRD**GIHGPLLEPFVQAKLRPEDGLYLIH*WSTSHPYRLILTVAQRSQAPDGMQSLRLRKF PIEQQDGAFVLEGWGRSFPSVRELGAALQGCLLRAGDDCFSLRRCCLP**QPGETSNLIIMRGARASPR
129	Vav1	VAV1	7409	HGPPQDLSV**HLWYAGPMERAGAESILANRSDGTFLVR*QRVKDAAEFAISIKYNVEVKHIKIMTAE GLYRITEKKAFRGLTELVEFYQQNSLKDCFKSLDTTLQFPF**KEPEKR
130	Vav2	VAV2	7410	PPSREIDYTA**YPWFAGNMERQQTDNLLKSHASGTYLIR*ERPAEAERFAISIKFNDEVKHIKVVEKD NWIHITEAKKFDSLLELVEYYQCHSLKESFKQLDTTLKYPY**KSRERSA
131	Vav3	VAV3	10451	CVPKPVDYSC**QPWYAGAMERLQAETELINRVNSTYLVR*HRTKESGEYAISIKYNNEAKHIKILTRD GFFHIAENRKFKSLMELVEYYKHHSLKEGFRTLDTTLQFPY**KEPEHSAG
132	Yes	YES1	7525	SNYVAPADSIQA**EEWYFGKMGRKDAERLLLNPGNQRGIFLVR*ESETTKGAYSLSIRDWDEIRGDN VKHYKIRKLDNGGYYITTRAQFDTLQKLVKHYTEHADGLCHKLTTVC**PTVKPQTQGLAKD
133	Zap70 (C)	ZAP70	7535	SQAPQVEKLIATTAHER**MPWYHSSLTREEAERKLYSGAQTDGKFLLR*PRKEQGTYALSLIYGKTV YHYLISQDKAGKYCIPEGTKFDTLWQLVEYLKLKADGLIYCLKEAC**PNSSASNASG
134	Zap70 (N)	ZAP70	7535	MPDPAAH**LPFFYGSISRAEAEEHLKLAGMADGLFLLR*QCLRSLGGYVLSLVHDVRFHHFPIERQ LNGTYAIAGGKAHCGPAELCEFYSRDPDGLPCNLRKPC**NRPSGLEPQPGVFDCLRD
135	Zap70 (NC)	ZAP70	7535	MPDPAAH**LPFFYGSISRAEAEEHLKLAGMADGLFLLR*QCLRSLGGYVLSLVHDVRFHHFPIERQL NGTYAIAGGKAHCGPAELCEFYSRDPDGLPCNLRKPC**NRPSGLEPQPGVFDCLRDAMVRDYVRQ TWKLEGEALEQAIISQAPQVEKLIATTAHER**MPWYHSSLTREEAERKLYSGAQTDGKFLLR*PRKEQ GTYALSLIYGKTVYHYLISQDKAGKYCIPEGTKFDTLWQLVEYLKLKADGLIYCLKEAC**PNSSASNASG

(continued)

Table 1
(continued)

(b)				
No.	***Protein***	***Gene symbol***	***Gene ID***	***Amino acid sequence of SH2 domains (bold: SH2 domain; *Arg betaB5 position)***
tag1	GST	GSTM1	n/a	MSPILGYWKIKGLVQPTRLLLEYLEEKYEEHLYERDEGDKWRNKKFELGLEFPNLPYYIDGDVKLTQSMAII RYIADKHNMLGGCPKERAEISMLEGAVLDIRYGVSRIAYSKDFETLKVDFLSKLPEMLKMFEDRLCHKTYL NGDHVTHPDFMLYDALDVVLYMDPMCLDAFPKLVCFKKRIEAIPQIDKYLKSSKYIAWPLQGWQATFGG GDHPPK
tag2	MBP	MALE	n/a	MKIKTGARILALSALTTMMFSASALAKIEEGKLVIWINGDKGYNGLAEVGKKFEKDTGIKVTVEHPDKLE EKFPQVAATGDGPDIIFWAHDRFGGYAQSGLLAEITPDKAFQDKLYPFTWDAVRYNGKLIAYPIAVEAL SLIYNKDLLPNPPKTWEEIPALDKELKAKGKSALMFNLQEPYFTWPLIAADGGYAFKYENGKYDIKDVG VDNAGAKAGLTFLVDLIKNKHMNADTDYSIAEAAFNKGETAMTINGPWAWSNIDTSKVNYGVTVLPT FKGQPSKPFVGVLSAGINAASPNKELAKEFLENYLLTDEGLEAVNKDKPLGAVALKSYEEELVKDPRIAA TMENAQKGEIMPNIPQMSAFWYAVRTAVINAASGRQTVDEALKDAQT

(a) Summary of reported protein solubility of SH2 domain proteins. Evaluations are based on genome-wide studies using purified SH2 domain proteins [2, 4, 7–9]. In addition, aggregation scores based on TANGO algorithm are shown [19]. The solubility is categorized into either good (soluble), poor (insoluble), or fair (relatively insoluble) according to the original publications. Additional references for individual SH2 domains are provided to the right. (b) Gene ID and amino acid sequences. This list contains a total of 122 different human SH2 domains from 121 validated and 1 provisional (i.e., GRAPL) genes at the NCBI Gene site (http://goo.gl/X4r2kP). Bold text indicates estimated boundaries of SH2 domains and the letter "R*" indicates the position equivalent to Arg βB5 in the FLVRES motif [20, 21]

vide a standard method for detection of your protein. In other cases, when the removal of the tag is opted for studies such as X-ray crystallography or NMR, proteinase digestion is performed through a specific cleavage site located between the tag and domain. For example, the N-terminal GST moiety can be removed by treatment with thrombin or PreScission protease for pGEX-based clones [13]. The ideal tag system for your biochemical studies should be determined on a case-by-case basis. Representative vectors containing affinity and epitope tags that are compatible for inducible expression of SH2 proteins in *E. coli* are presented in Fig. 1.

The most commonly used affinity tags for SH2 domains are GST and poly-His [2, 4, 9, 14]. The GST tag derived from the *Schistosoma japonicum* GST protein has been one of the most popular tag systems for expression and purification of SH2 proteins produced in *E. coli*. N-terminal or C-terminal GST-tagged SH2 protein is expressed as ~40 kDa protein with high solubility and is known to form dimers. GST-SH2 proteins are purified from bacterial lysates by affinity chromatography using immobilized glutathione. GST-SH2 proteins are eluted under non-denaturing conditions using reduced glutathione (*see* Subheading 3.4). The collection of N-terminal GST tagged SH2 constructs are available through Addgene [4]. The polyhistidine tags, commonly a 6×His tag, is a small tag placed at N- or C-terminal of proteins and have a high affinity for Ni^{2+} and several other metal ions. His-tagged proteins can be purified with immobilized metal ions, e.g., Ni-NTA resin. As being small in size, the poly-His tag is less likely to interfere with function SH2 domains and may not need to be removed prior to the downstream assay. An advantage of the poly-His tag over GST tag is that purification can be performed under denaturing conditions, allowing for solubilization of inclusion bodies and subsequent on-column refolding [15]. Many other expression tags such as MBP, Strep, and SUMO are available; see review for descriptions of other tag systems [16], and comprehensive technical guide from GE Healthcare about protein purification and selection of affinity tags (https://goo.gl/KFp9hZ).

In the following sections, we present two representative protocols for cloning and purification of N-terminal GST-tagged SH2 domains (via pGEX bacterial expression vector) and N-terminal His-tagged SH2 domains (via Gateway cloning and pDEST bacterial expression vector). Both protocols are based on a simple affinity purification without requiring analytical chromatography, and are suitable for production of 1–10 mg of functional SH2 protein, sufficient for typical biochemical and proteomic studies. Readers should follow Subheadings 2.1–2.4 and 3.1–3.4 for purification of GST-tagged SH2 domains and Subheadings 2.5–2.6 and 3.5–3.7 for purification of His-tagged SH2 domains. Additional approaches for evaluation of purified protein quality and guidance for troubleshooting are described in Subheadings 2.7 and 3.8.

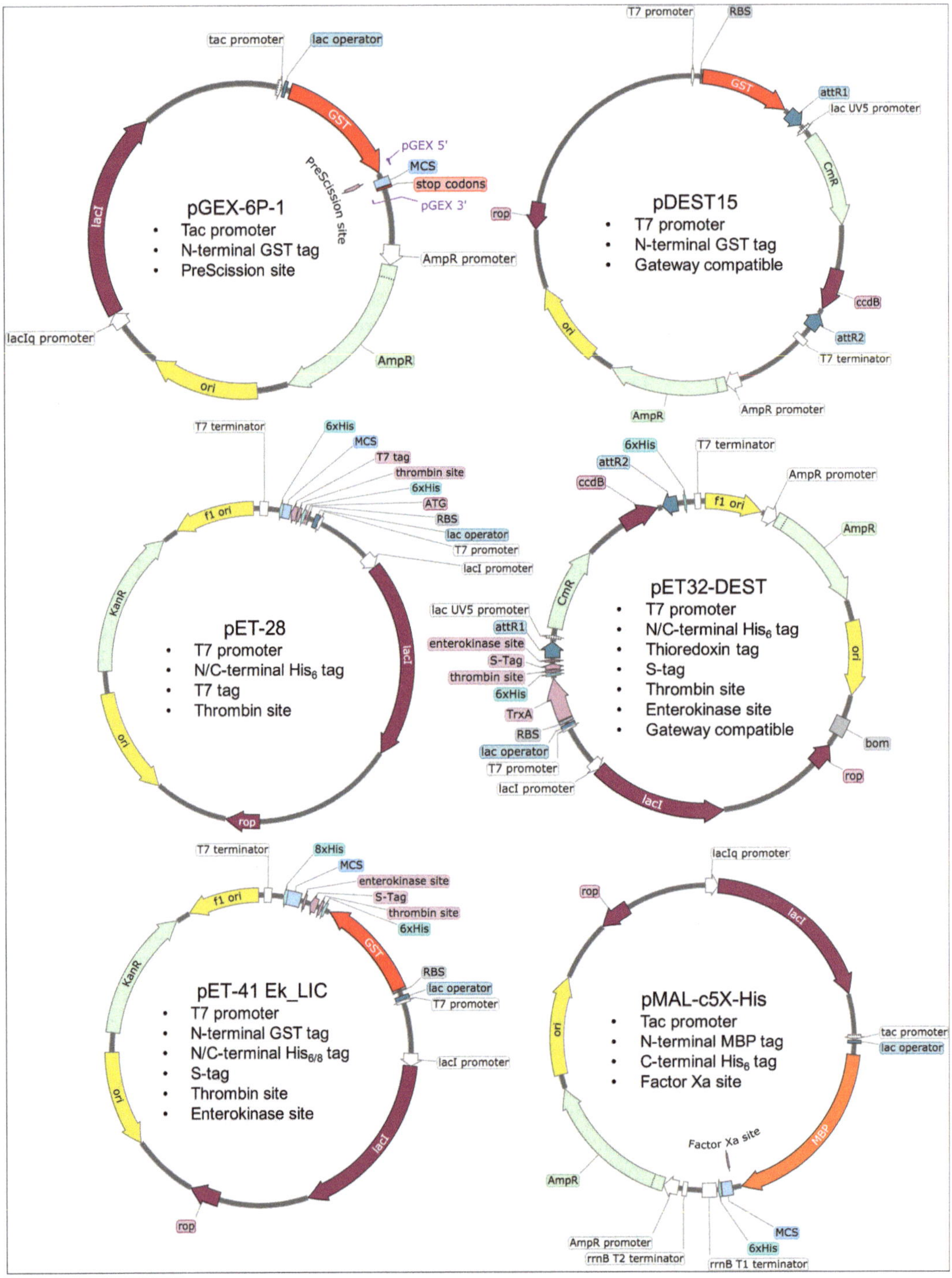

Fig. 1 Characteristics of bacterial expression vectors. An SH2 DNA fragment can be inserted into bacterial expression vectors with various features. Expression of tagged SH2 domain is induced with IPTG and the protein is purified on a tag-specific affinity column. The affinity tag can then be removed using a proteases cleavage site. The pGEX-6P-1 has an N-terminal GST tag and PreScission protease site. pDEST15 has a N-terminal GST tag and Gateway cloning site. pET-28 contains N- and C-terminal hexahistidine (His_6)-tags. Other vectors containing multiple affinity tags may help solubility and provide multiple purification options: pET32-DEST (His_6, thioredoxin, and S tags), pET41 Ek_LIC (GST, $His_{6/8}$, and S tags), and pMAL-c5X-His (MBP and His_6 tags). *GST* glutathione S-transferase, *MBP* maltose binding protein, *TrxA* thioredoxin, *MCS* multiple cloning site, *Ni-NTA* nickel-nitrilotriacetic acid

2 Materials

2.1 Cloning of pGEX-SH2

1. pGEX series bacterial expression vector (Fig. 1) (*see* **Note 1**).
2. Luria-Bertani (LB)-ampicillin agar plate: LB agar plate with 100 μg/mL ampicillin.
3. DNA polymerase (*see* **Note 2**).
4. Custom-designed oligonucleotide primers.
5. 1–2 % agarose gel.
6. TAE (Tris-acetate-EDTA) buffer: 40 mM Tris-acetate, 1 mM EDTA.
7. SYBR Safe DNA gel stain.
8. T4 DNA ligase with 10× ligation buffer.
9. Restriction enzymes.
10. Alkaline phosphatase.
11. Competent bacteria (DH5α, BL21, etc.) (*see* **Note 3**).
12. LB (−): LB media without antibiotics.
13. Water bath at 42 and 65 °C.

2.2 Evaluation of pGEX-SH2 Clones

1. LB-ampicillin: LB broth with 50 μg/mL ampicillin.
2. Isopropyl β-D-1-thiogalactopyranoside (IPTG) (10 mM).
3. Bacteria Triton X-100 lysis buffer (BXB): Phosphate buffered saline (PBS), 100 mM ethylene diamine tetraacetic acid (EDTA), 1 % Triton X-100, 10 % glycerol; add phenylmethyl sulfonyl fluoride (PMSF) to 1 mM, aprotinin solution to 1 % v/v, and dithiothreitol (DTT) to 1 mM just before use.
4. Sonicator with microtip probe (e.g., Branson Sonifier 450 or equivalent).
5. 5× SDS sample buffer: 0.3 M Tris–HCl pH 6.8, 10 % sodium dodecyl sulfate (SDS), 25 % β-mercaptoethanol, 0.1 mg/mL bromophenol blue, 45 % glycerol.
6. Glutathione Sepharose 4B.
7. 12 % SDS-polyacrylamide gel electrophoresis (PAGE) mini gel.
8. CBB gel staining solution: 40% methanol, 10% acetic acid, 0.25 % Coomassie Brilliant Blue R-250.
9. Destaining solution: 20 % methanol, 10 % acetic acid.
10. Miniprep Kit.
11. Restriction enzymes.
12. 1–2 % agarose gel.
13. TAE buffer: 40 mM Tris-acetate, 1 mM EDTA.
14. SYBR Safe DNA gel stain.

15. Sequencing primers: pGEX 5′ (GGGCTGGCAAGCCACG TTTGGTG) and pGEX 3′ (CCGGGAGCTGCATGTG TCAGAGG).
16. Bacteria stock solution: 50% glycerol (autoclaved).
17. Cryogenic tubes.

2.3 Large Scale Expression of GST-SH2

1. LB-ampicillin: LB broth with 50 μg/mL ampicillin.
2. IPTG (1 M).
3. PBS with 10% glycerol.
4. 50 mL Falcon tube.

2.4 Affinity Purification of GST-SH2

1. Bacteria Triton X-100 Lysis buffer (BXB).
2. Sonicator with microtip probe.
3. 5× SDS sample buffer.
4. Glutathione Sepharose 4B.
5. 12% SDS-PAGE mini gel.
6. CBB gel staining solution.
7. Destaining solution.
8. Elution buffer: 20 mM glutathione (reduced), 150 mM NaCl, 10% glycerol, 50 mM pH 8.8 Tris–HCl, 1 mM EDTA; add PMSF to 1 mM, aprotinin to 1% v/v, and DTT to 1 mM just before use.
9. Centrifuge tube filter (e.g., Spin-X Plastic Centrifuge Tube Filter).
10. Sephadex G-25 PD-10 column.
11. PBS with 10% glycerol.
12. Bradford assay reagent.
13. PreScission Protease (GE Healthcare).
14. Cleavage buffer: 50 mM Tris–HCl, pH 7.0, 150 mM NaCl, 1 mM EDTA, 1 mM DTT.
15. Empty chromatography column, e.g., Bio-Rad Poly-Prep columns.

2.5 Cloning of pDEST-His-SH2

1. TOPO cloning Kit (Invitrogen): Vector (e.g., pENTR-D-TOPO vector), water, salt solution.
2. One-Shot MACH1 or TOP10 chemically competent cells (Invitrogen).
3. *Escherichia coli* expression vector (example pET-32-DEST).
4. SH2 domain cDNA.
5. LR Clonase II Plus (Invitrogen).
6. TE buffer pH 8.0.

7. Proteinase K.
8. 1.7 mL microcentrifuge tube.
9. 14 mL round-bottom tube.
10. Miniprep Kit.
11. Luria broth (LB).
12. Ampicillin sodium salt.
13. Kanamycin.
14. LB Amp: 100 μg/mL Ampicillin in LB.
15. LB Kan: 50 μg/mL Kanamycin in LB.
16. Chloramphenicol.
17. DH5α competent cells.
18. LB-Kan plate: LB agar plates containing kanamycin (50 ng/mL).
19. One Shot BL21 (DE3) pLysS Chemically Competent *E. coli* (Invitrogen).
20. LB-Amp plate: LB agar plates containing 100 μg/mL ampicillin.

2.6 Expression of His-SH2 in E. coli

1. Isopropyl β-D-1-thiogalactopyranoside (IPTG).
2. Centrifuge bottles.
3. Oakridge tube.
4. Vortexer.
5. Ni-NTA His binding resin.
6. 50 mL Falcon tube.
7. 15 mL Falcon tubes.
8. Benzonase nuclease.
9. Sodium azide.
10. Glycerol.
11. NAP-10 columns.
12. Phosphate-buffered saline (PBS): 100 mM KCl, 20 mM sodium phosphate, pH 7.4.
13. Blocking solution: 100 mM KCl, 20 mM HEPES, 1% BSA (wt/vol), pH 7.4.
14. Probing buffer: 100 mM KCl, 20 mM HEPES, 0.1% Tween 20, pH 7.4.
15. 10× lysis buffer: 500 mM NaH_2PO_4, 3 M NaCl, 50 mM imidazole, pH 8.0.
16. Wash buffer: 50 mM NaH_2PO_4, 300 mM NaCl, 20 mM imidazole, pH 8.0.
17. Elution buffer: 50 mM NaH_2PO_4, 300 mM NaCl, 250 mM imidazole, pH 8.0.

18. Sonicator.
19. Bug buster.
20. Protease inhibitors: PMSF and benzamidine hydrochloride.
21. Bacterial shaker (temperature should be variable).
22. 15 mL Falcon tube.
23. Freezing buffer (300 mM NaCl, 50 mM Na_2PO_4, pH 8, 20% glycerol v/v).

2.7 Assessing Quality of SH2 Protein

1. 8–15% gradient SDS-page gels.
2. Coomassie GelCode Blue staining.
3. Superdex 200 10/300 GL column.
4. Freezing buffer: 300 mM NaCl, 50 mM Na_2PO_4, pH 8, 20% glycerol v/v.
5. SYPRO orange: 5000× in DMSO.
6. PCR plate for qPCR.
7. PCR optical plate seal.
8. Centrifuge with a 96-well plate adaptor.
9. qPCR machine (e.g., Bio-Rad CFX 96).
10. Chaperone Competent Cell BL21 Series (Clontech).

3 Method

Follow Subheadings 3.1–3.4 and 3.8 for purification of GST-tagged SH2 domains and Subheadings 3.5–3.8 for His-tagged SH2 domains.

3.1 Cloning of pGEX-SH2

1. Retrieve DNA sequence of SH2 domains from public databases such as NCBI Gene and Human Protein Reference Database. All human and mouse SH2 domain sequences have been identified [17]. As a quick reference, Table 1 provides NCBI Gene ID and amino acid sequences of human SH2 domains that can be utilized for cloning (*see* **Note 4**).
2. Find academic or industry sources of cDNA templates corresponding to the SH2 domains or perform molecular cloning (*see* **Note 5**).
3. Determine amino acid sequence of the SH2 domain insert (*see* **Note 6**).
4. Design PCR primers for the SH2 domain insert (*see* **Note 7**).
5. Amplify the SH2 fragment by PCR using the designed primers and the cDNA template.
6. Run the PCR product on a 1–2% agarose gel and extract the amplified DNA fragment.

7. Digest the SH2 DNA with appropriate restriction enzymes, run the digested DNA on an agarose gel and then purify the DNA fragment from the gel.
8. Digest a pGEX vector with appropriate restriction enzymes.
9. Incubate the digested vector with alkaline phosphatase for 1 h at 37 °C.
10. Purify the linearized and dephosphorylated pGEX vector from a 1 % agarose gel.
11. Mix 1 μL 10× ligation buffer, 50 ng linearized, dephosphorylated vector, equimolar amount of insert (*see* **Note 8**) and sterile water to 9.5 μL (*see* **Note 9**).
12. Incubate for 2 min at 65 °C to melt any ends that might have annealed, then place on ice.
13. Add 0.5 μL T4 ligase.
14. Incubate at 16 °C overnight or at room temperature for 1 h.
15. Heat-inactivate the ligase for 10 min at 65 °C then place on ice.
16. Thaw competent bacteria (50 μL) on ice and add 5 μL of ligation product.
17. Gently mix the ligation product with competent cells and incubate for 20 min on ice tapping occasionally.
18. Heat-shock at 42 °C for 45 s, then immediately transfer the tubes to ice and cool down for 2 min.
19. Add 500 μL LB (−) and incubate at 37 °C for 1 h.
20. Warm up LB-ampicillin plate in 37 °C incubator before plating.
21. Centrifuge at low speed (800 × *g*) for 3 min to pellet bacteria.
22. Remove the supernatant and leave ~50 μL pellet and vortex.
23. Plate the transformed cell mixture on an LB-ampicillin plate and incubate at 37 °C overnight.

3.2 Evaluation of pGEX-SH2 Clones

To determine whether the cloning of the SH2 domain was a success, protein expression screening is a fast and informative method. It will provide whether the pGEX-SH2 clones are producing protein of the expected size, and whether those proteins are soluble. In addition, a pull-down assay can be performed concurrently which will provide a quick test if the GST-SH2 is functional based on pTyr-dependent binding activity. The selected clones will be verified by sequencing and cryopreserved.

1. Inoculate colonies from LB plates into 4 mL LB-ampicillin and grow at 37 °C overnight. Mark the colonies on the plate for future use.

2. Inoculate 0.4 mL dense liquid culture of bacteria into 1.6 mL fresh pre-warmed LB-ampicillin (*see* **Note 10**).
3. Shake at 37 °C for 1.5–2 h.
4. Add IPTG to 0.1 mM (20 μL of 10 mM IPTG) and shake at 37 °C for 2 h to induce protein expression.
5. Transfer 1.5 mL bacteria to microcentrifuge tubes and spin at 10,000 × *g* at 4 °C for 3 min in microcentrifuge.
6. Remove supernatant and resuspend bacterial pellet in 0.2–0.4 mL BXB.
7. Vortex to resuspend, then sonicate briefly (e.g., 2–5 s at relatively low power) on ice to break cells. Try to avoid foaming.
8. Take 10 μL of total lysate and add 2.5 μL of 5× sample buffer for SDS-PAGE (whole cell fraction *W*, Fig. 2).
9. Spin the rest of the total lysate for 5 min in a microcentrifuge at 10,000 × *g* at 4 °C, transfer supernatant to a new tube.
10. Take 10 μL of the cleared supernatant lysate and add 2.5 μL of 5× sample buffer for SDS-PAGE (soluble fraction *S*, Fig. 2).

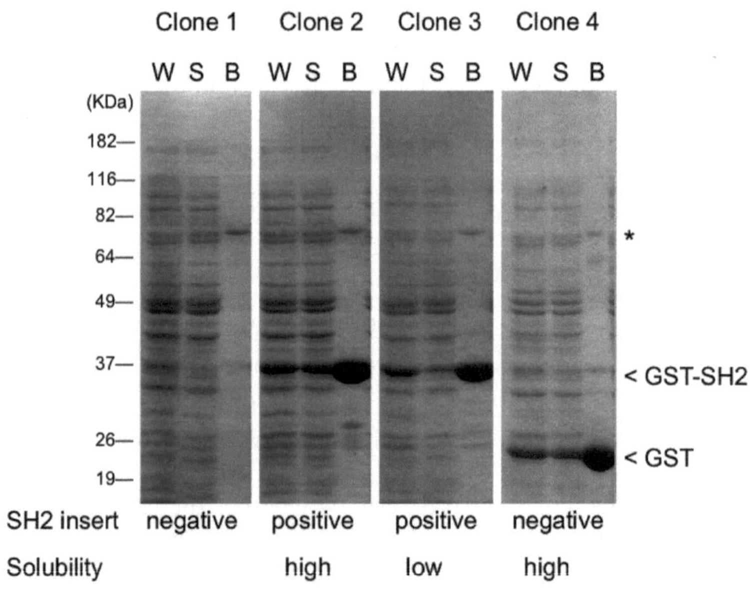

Fig. 2 Evaluation of GST-SH2 clone by small scale protein expression. Three purification fractions of four bacterial clones were prepared as described in Subheading 3.2 and their Coomassie-stained gels are shown: *W*, whole cell fraction; *S*, soluble fraction; *B*, glutathione bead-bound fraction. In the presence of ~37 kDa GST-SH2 size band, Clone 2 and 3 are considered ligation positive clones. Clone 1 (no expression) and Clone 4 (expression of ~26 kDa GST size band) are considered negative. By comparing the GST-SH2 band in *W* and *S* lanes, Clone 2 suggests good solubility ($W \approx S$) and Clone 3 shows relatively insoluble ($W > S$), respectively. *Asterisk* indicates a co-purified bacterial protein (*see* Note 22). Data shown are representative gels from multiple experiments

11. Take 100–150 μL of the cleared lysate and add 10 μL glutathione agarose bead slurry (cut end off pipet tip with razor blade to accurately pipet beads).
12. Rotate at 4 °C for 30 min.
13. Spin out briefly, and wash beads three times with 1 mL cold BXB.
14. Remove most of the supernatant (leave 20 μL in the tube), vortex, and add 2.5 μL 5× sample buffer for SDS-PAGE (bead-bound fraction *B*, Fig. 2) (*see* **Note 11**).
15. Boil all the samples collected (from **steps 8**, **10**, **14**) and run samples on a 12% SDS-PAGE gel.
16. When the gel is finished running, remove the SDS-PAGE gel and stain for 15 min with CBB gel staining solution, destain with destaining solution, and then dry. Check the size, purity, and solubility of the purified fusion protein (Fig. 2).
17. Optional (17–20): pTyr binding activity of the clones can be tested by pull down. Incubate remaining beads from **step 14** with 10 μg of pervanadate treated cell lysate (*see* **Note 12**).
18. Rotate for 1 h at 4 °C, and wash three times with BXB.
19. Add 5x sample buffer, boil and run on 12% SDS gel.
20. Perform Western blotting with anti-phosphotyrosine antibody (*see* **Note 13**).
21. Subject positive clones to DNA sequencing and store bacteria in 25% (v/v) sterile glycerol at –80 °C.

3.3 Large Scale Expression of GST-SH2

The following protocol typically yields 5–10 mg of GST-SH2 fusion protein for soluble SH2 domains and lesser amount for insoluble SH2 domains (Table 1). Solubility is also one of the major determining factors of SH2 protein binding activity. Several groups reported highly insoluble SH2 domains that give poor yields and tend to lack detectable binding activity [2, 4, 7–9].

1. Inoculate frozen stock culture of a verified GST-SH2 clone in 20 mL LB-ampicillin and incubate at 37 °C overnight.
2. Inoculate 20 mL of dense overnight culture to 400 mL LB-ampicillin. Shake at 37 °C for about 2 h until OD at 600 nm reaches to 0.4–0.6.
3. Add IPTG to 0.1 mM (40 μL of 1 M IPTG). Shake at 37 °C for 2–4 h (*see* **Note 14**).
4. Centrifuge the bacteria culture at 5,000 ×*g* at 4 °C for 10 min and decant the supernatant media.
5. Resuspend the pellet with 20 mL ice-cold PBS with 10% glycerol and transfer to 50 mL Falcon tube.
6. Centrifuge at 5,000 ×*g* at 4 °C for 20 min and decant the supernatant.
7. Store the bacterial pellet at -80 °C.

3.4 Affinity Purification of GST-SH2

1. Resuspend the pellet with 10–20 mL ice-cold BXB buffer and sonicate on ice until cells are broken (*see* **Note 15**).
2. Take 10 μL of total lysate and add 2.5 μL of 5× sample buffer for SDS-PAGE (whole cell fraction *W*, Fig. 3).
3. Centrifuge at 5000 ×*g* at 4 °C for 10 min to remove debris and transfer supernatant to a new 50 mL Falcon tube (*see* **Note 16**).
4. Take 10 μL of supernatant and add 2.5 μL of 5× sample buffer for SDS-PAGE (soluble fraction *S*, Fig. 3).
5. Add glutathione-agarose to the supernatant: 500 μL bead slurry per 400 mL of the original bacterial culture.
6. Rotate at 4 °C for about 1 h (up to 2 h).
7. Spin out beads at low speed (200–400 ×*g*), remove supernatant, and wash beads in cold BXB buffer. Repeat 3–5 times.
8. Transfer beads to an Eppendorf tube.
9. Take 10 μL of the beads solution and add 2.5-μL 5× sample buffer for SDS-PAGE (bead-bound fraction *B*, Fig. 3).
10. Spin out beads and remove the supernatant.
11. Optional: To cleave the SH2 domain from GST with PreScission Protease, follow **steps 12–18**. To elute GST-SH2 protein from beads without cleavage, go to **step 19**.

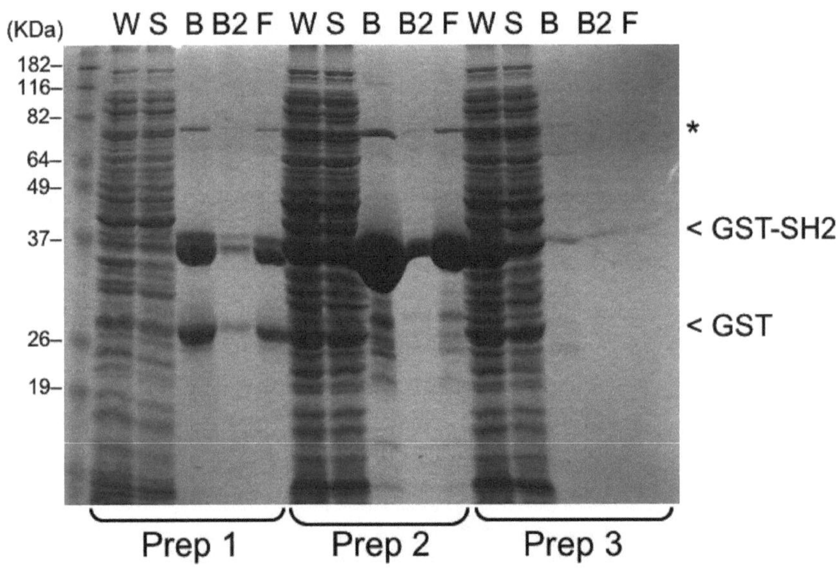

Fig. 3 Evaluation of purification of GST-SH2 proteins. Representative purification fractions of three different GEX-SH2 clones. Twelve percentage SDS PAGE gel were stained with CBB. *W*, whole cell fraction; *S*, soluble fraction; *B*, glutathione bead-bound fraction; *B2*, post elution bead fraction; *F*, final elution fraction. Prep 1 showed ~37 kDa GST-SH2 band and GST size degradation product. Prep 2 shows high expression of soluble fusion protein while Prep 3 shows low yield of the final product due to low solubility. *Asterisk* indicates a co-purified bacterial protein (*see* **Note 22**)

12. Wash beads several times with cleavage buffer.
13. Mix 40 μL (80 units) of PreScission Protease with 960 μL of cleavage buffer at 4 °C.
14. Add PreScission Protease mixture (about 1 bead volume) to the GST-SH2-bound beads and gently resuspend. Incubate at 4 °C for 4 h.
15. Collect cleaved SH2 protein either by pouring into the chromatography column or by centrifugation at 500 × *g* for 5 min.
16. Add ~500 μL cleavage buffer and collect the flow-through. Repeat twice. If the eluate is collected by centrifugation. Quickly check the relative protein concentration of the eluates using the Bradford assay. Pool the fractions that show a bright blue color.
17. Keep the eluates at 4 °C.
18. Go to **step 26** for buffer exchange (*see* **Note 17**).
19. Elution without cleavage (19–26): Add 500 μL of elution buffer to the beads and rotate at 4 °C for 15 min.
20. Spin out beads and transfer supernatant to a new Eppendorf tube on ice. Try not to suck up or remove the beads when removing the supernatant or wash. Alternatively, you can remove beads by centrifuge tube filter at **step 25**.
21. Repeat **steps 19** and **20** for the second elution.
22. Repeat **steps 19** and **20** for the third elution.
23. Take 10 μL of beads after elution and add 2.5 μL 5× sample buffer for SDS-PAGE (post elution bead fraction *B2*, Fig. 3).
24. Quickly check relative protein concentration of the eluates using the Bradford assay. Pool the fractions that show a bright blue color (*see* **Note 18**).
25. Remove accidentally transferred beads using a centrifuge tube filter.
26. Buffer exchange by gel filtration on a Sephadex G-25 PD-10 column according to the supplier's instructions. Briefly, equilibrate column with approximately 25 mL PBS-10% glycerol. Discard the flow through. Add your eluted protein sample to the column and allow the whole sample to flow into the column. Elute with PBS-10% glycerol (collect sequential 0.5 mL fractions of the elute in separate tubes (*see* **Note 19**).
27. Quickly check the relative protein concentration of each tube with the Bio-Rad Bradford dye reagent. Combine fractions that show a bright blue color into a single tube (*see* **Note 20**).
28. Take 10 μL of combined eluate and add 2.5 μL of 5× sample buffer for SDS-PAGE (final product fraction *F*, Fig. 3).
29. Determine protein concentration using the Bradford assay or absorbance at 280 nm and store at −80 °C (*see* **Note 21**).

30. Boil all the SDS-PAGE samples (from **steps 2**, **4**, **9**, **23**, **28**) and evaluate the purification fractions by 12 % SDS gel (Fig. 3) (*see* **Note 22**).

3.5 Cloning of pDEST-His-SH2

The protocol below describes cloning of SH2 domains using the Invitrogen TOPO entry vectors, which are part of Invitrogen's Gateway system. This section also describes the standard purification protocols for expression and purification of His-tagged SH2 domains using pDEST-His (Subheadings 3.5–3.7) (*see* **Note 23**). When deciding on a vector consider the downstream application and determine whether a protease cleavage site between the tag and the SH2 domain is needed (*see* **Note 24**).

1. Identify the SH2 domain family members of interest and determine consistent truncation parameters in setting boundaries (*see* Table 1b for boundaries). To do this, use the SMART database (*see* Chapter 4) and typically include 10–15 residues upstream and 10–15 residues downstream of the domain sequence. SH2 domains can be cloned from cDNA, subcloned ORFs from previously cloned genes or obtained ORFs through de novo gene synthesis.
2. Design the appropriate primers to clone out the SH2 domains of interest (*see* **Note 25**).
3. Amplify the SH2 domain by PCR.
4. Mix 1–4 μL of the PCR product, 1 μL of the salt solution, 1 μL of the pENTR-D-TOPO vector. Add water to a final volume of 6 μL.
5. Incubate for 5 min at room temperature.
6. Thaw competent cells on ice.
7. Mix the TOPO reaction in with the competent cells. Leave on ice for 15–30 min.
8. Heat-shock cells by placing tubes at 42 °C for 30 s and place back on ice.
9. Add 250 μL of room temperature S.O.C. Medium.
10. Cap the tube tightly and shake the tube horizontally (200 rpm) at 37 °C for 1 h.
11. Plate 50–200 μL from each transformation on a pre-warmed LB-Kan plate and incubate overnight at 37 °C.
12. Pick colony(s) and culture in 3 mL of LB-Kan media overnight at 37 °C.
13. Miniprep clone(s).
14. Recommended: verify samples by DNA sequencing
15. Mix Entry clone (50–150 ng), pDEST-His (150 ng), TE buffer, up to 8 μL in a microcentrifuge tube. Add 1 μL LR Clonase II Plus and mix.
16. Incubate reaction for 1 h at 25 °C.

17. Add 1 μL of Proteinase K solution to the sample, mix briefly and incubate for 10 min at 37 °C.
18. Transform competent DH5α *E. coli* cells with 2 μL of the reaction mixture and plate cells onto LB-Amp plate. Incubate plates at 37 °C overnight.
19. Inoculate 3 mL of LB-Amp media in a 14 mL round-bottom tube with a single colony and shake at 200 rpm for 16 h at 37 °C.
20. Purify plasmids using a DNA Miniprep Kit.
21. Store vectors at −20 °C.

3.6 Expression of His-SH2 in E. coli

1. Transform BL21(DE3)pLysS *E. coli* cells with 1 μL of the purified plasmid. Plate cells onto LB agar plates containing 100 μg/mL ampicillin and 50 μg/mL chloramphenicol.
2. Inoculate 2 mL of LB broth containing 100 μg/mL ampicillin and 50 μg/mL chloramphenicol in a 14 mL round-bottom tube with a single colony and shake at 200 rpm, 37 °C for 16 h.
3. Add 1 mL culture to 500 mL LB broth containing 100 μg/mL ampicillin and 50 μg/mL chloramphenicol in a 2 L baffled flask. Shake at 200 rpm at 37 °C until the absorbance of the cultures at 600 nm is 0.8 (*A*600 = 0.8; 1 cm path length). This step will take approximately 4 h.
4. Cool the cultures to room temperature in a sink filled with cool water. Once at room temperature, move the flasks to a shaker at room temperature and add IPTG to a final concentration of 1.0 mM. Continue to shake at approximately 25 °C for 12–16 h (overnight).
5. Transfer media to 500 mL centrifuge bottles and pellet bacteria at 11,150 × *g* for 15 min.
6. Discard the supernatant and add 5 mL of 0.5× lysis buffer containing protease inhibitors (1 mM PMSF and 1 mM benzamidine hydrochloride).
7. Bacteria pellets can be stored at –80 °C.

3.7 Purification of His-SH2

1. Break apart cell pellet using a 2 mL pipette and short periods of agitation with a benchtop vortex. Transfer suspension to a 30 mL Oakridge tube.
2. Add 500 μL of 10× BugBuster and 5 μL benzonase nuclease. Incubate for 1.5 h with rocking at 4 °C. If pellet lysates are particularly aggregated, incubation can be increased to 2.5 h.
3. Add 5 mL of 1.5× lysis buffer containing protease inhibitors (1 mM PMSF and 1 mM benzamidine hydrochloride) to each sample.
4. Centrifuge each sample at 35,800 × *g* for 30 min, 4 °C.

5. While the samples are in the centrifuge, prepare the Ni-NTA resin. For each sample, transfer 600 μL of resin to a 50 ml Falcon tube and mark the level of the resin on the tube. Centrifuge at 3000 × *g* for 1 min and remove the supernatant. Fill the Falcon tube with 1× lysis buffer and centrifuge at 3000 × *g* for 1 min. Discard supernatant. Add 1× lysis buffer to the original marked volume and resuspend the resin.
6. Decant the supernatant from the cell lysates into 15 mL Falcon tubes and store on ice. Add 600 μL of the resuspended resin to each sample. Allow proteins to bind to the resin by rotating samples at 4 °C for 30 min.
7. Set disposable fritted columns on a rack suspended above a waste collection container. Moisten each column by passing 1 ml of ddH_2O through, followed by 1 mL of 1× lysis buffer.
8. Transfer the cell lysate/resin mixture to each column and let the flow through drain.
9. Add 10 mL of wash buffer to each column and let the flow through drain.
10. Plug each column and add 0.5 mL of elution buffer to resin. Incubate for 5 min and then unplug the column and collect the flow-through in a 1.7 mL microcentrifuge tube. Repeat with an additional 0.5 mL of elution buffer.
11. Perform a buffer exchange to transfer the purified proteins into PBS. Equilibrate the NAP-10 columns with PBS by flushing three times with 5 ml of PBS (*see* **Note 26** and **27**).
12. Add 1 mL of purified protein and drain to waste. If volume is less than 1 mL, add PBS to make up the difference.
13. Add 1.5 mL of PBS to the column and collect the flow-through in a 1.7 mL microcentrifuge tube. This fraction contains the purified protein in PBS buffer.
14. Centrifuge samples on a tabletop centrifuge at 14,000 × *g* for 2 min to remove any precipitate.
15. Determine protein concentration. Protein concentrations were determined based on their absorbance at 280 nm.
16. Adjust concentration to 75 μM. If the protein concentration is less than 75 μM, use the sample for subsequent experiments, but note its actual concentration. This information is useful if unexpected negatives are observed on the arrays.
17. Prepare protein domains for long-term storage at −80 °C. If the domains will be used to prepare protein microarrays, mix three parts protein with one part 80 % glycerol and divide into 100 μL aliquots. If proteins will be used for fluorescent polarization assays, do not add glycerol and divide into 100–1000 μL aliquots.

18. Flash freeze the aliquots in liquid nitrogen.
19. Protein domains can typically be stored at −80 °C for up to 6 months in buffer. Proteins can be stored for longer periods at −80 °C when stabilized in 20% glycerol.

3.8 Assessing Quality of SH2 Protein

Protein quality and stability is extremely critical for downstream applications such as solving crystal structures, biophysical assays, and biochemical assays. To assess the purity and stability of the SH2 domain proteins several methods are listed below. A simple way to assess the purity is through SDS-PAGE and coomassie staining. This quick and easy method may reveal impurities or degradation products during the purification process (**step 1**). To determine if the SH2 domain is monomeric or dimeric is by using size-exclusion chromatography (**step 2**, *see* Chapter 14 for details on this method). Another example of determining the stability of the protein is by using differential scanning fluorimetry (DSF) that measures the thermal denaturation temperature of a protein [18] (**step 3–7**). SYPRO orange binds nonspecifically to hydrophobic surfaces, and water strongly quenches its fluorescence. When the protein unfolds, the exposed hydrophobic surfaces bind the dye, resulting in an increase in fluorescence by excluding water. This is extremely useful if you are unsure whether the SH2 domain is folded properly and can aid in identifying suitable buffers to enhance the stability of the protein. For some SH2 proteins that are more challenging to express (e.g., inclusion bodies, degradation by protease, improper folding), commercial chaperone expressing BL21 cells are available by Clontech (Chaperone Competent Cell BL21 Series) and other manufacturers (**step 8**).

1. Assess the purity of the protein by SDS–polyacrylamide gel electrophoresis. Restrict proteins with >80% purity for studies. Purified proteins were assessed for appropriate molecular weights and integrity on an 8–15% gradient gels and subsequent coomassie GelCode Blue staining.
2. Assess the aggregation state of each domain by analytical size exclusion column chromatography. We use a Superdex 200 10/300 GL column run in PBS at 0.75 ml/min and typically observe monomeric domains of 30–40 kDa eluting at ~18 min after injection. We restrict our subsequent studies to domains that are predominantly monomeric or dimeric.
3. Make fresh 200× SYPRO orange solution in the PBS or in the dialysis buffer. Add 5 μL of 200× SYPRO orange with 45 μL of 5 μM SH2 domain protein in triplicate into a 96-well plate (*see* **Note 28**).
4. Seal plate, use a roller to ensure a good seal on to the plate.
5. Load plate into the qPCR machine.

6. Setup machine to “FRET” mode, Fast Melt (25 °C for 30 s, Hold at temperature for 30 s, read sample, increase temp up 0.5 °C, go from 25 to 99 °C and return to 25 °C).
7. Run PCR plate. Export data when done and calculate the melting temperature (Tm) (*see* **Note 29**).
8. If the SH2 domain has poor solubility and expression one can transform DNA into chaperone expressing competent BL21 cells. Repeat steps in subheadings 3.2 and 3.3 for GST-tag expression or Subheadings 3.6 and 3.7 for His-tag expression.

4 Notes

1. This section describes a protocol using the pGEX GST expression system. Various other vector systems (e.g., pDEST) are available as shown in the Fig. 1. *See* **Note 3** for selection of a bacterial host.
2. Since an SH2 PCR amplicon is small (300–500 bp), the probability of mutation is relatively low regardless of the type of DNA polymerase used.
3. In general, DH5α is suitable for subcloning of pGEX-SH2 clones and BL21 is suitable for protein expression. In the presence of internal tac promoter, lac operator, and lacI^q, pGEX vectors are compatible with general *E. coli* strains (Fig. 1). For pET vectors, host cells harboring the T7 RNA polymerase gene are required, e.g., BL21 (DE3).
4. More information about SH2 domains including phylogenetic relationships and multiple alignment is available on the SH2 Domain Site (sites.google.com/site/sh2domain/). To determine the presence of SH2 domains in unknown sequences, use SMART site (smart.embl-heidelberg.de) or ScanProsite (us.expasy.org/cgi-bin/scanprosite) or *see* Chapter 3.
5. SH2 domain expression clones or plasmids containing human SH2 domain cDNA are available from Addgene (e.g., Mayer lab clones), DNASU, GE Dharmacon, the Pawson Lab, etc.
6. The amino acids surrounding the SH2 domain can significantly affect solubility and expression of recombinant SH2 proteins. Empirically, including additional amino acids at the boundaries of the domain is more beneficial than removing a few residues. We usually add 5–10 amino acids to the N- and C-terminal boundaries of SH2 domains (*see* Table 1). SH2 domain structural information (e.g., Sequence Chain View/Protein Data Bank) can provide clues to the boundaries and in designing the GST-SH2 fusion (for example a flexible loop region might be better than a rigid secondary structure segment).

7. PCR primers should have approximately 15–20 nucleotides of homology to the template; the 5′-end should have five C's (this clamps the end, allowing efficient digestion with restriction nucleases) followed by a restriction site to be used for cloning in-frame into the pGEX expression vector, followed by the region with template homology.
8. If the vector is 5.0 kb and the insert is 0.4 kb, equimolar amount would be $0.4/5.0 \times 50$ ng = 4 ng. This does not have to be exact. In some cases, more insert or less insert works better.
9. Always set up a ligation with the digested vector without the insert as a control. This will tell you how much background occurred such as self-ligation of the vector. If the background is high, redigest and phosphatase treat your vector for a longer period of time to ensure the vector does not re-ligate to itself.
10. Alternately, colonies picked up from a plate can be directly inoculated into 1.6 ml of pre-warmed LB-ampicillin and incubated for 3–4 h before IPTG induction (wait until the medium becomes cloudy).
11. If an optional pull down assay is intended (**steps 17–20**), save half of the beads at 4 °C before adding 5× sample buffer.
12. Pervanadate inhibits endogenous protein tyrosine phosphatases, thus strongly enhancing tyrosine phosphorylation in vivo. Ideally aliquots of a pervanadate-treated cell lysate are prepared in advance and stored at –80 °C. For the pull-down experiment, it is recommended to include positive (known functional GST-SH2 protein) and negative control proteins (GST only).
13. The expected result is readily detectable pTyr signal for the positive control and some sample clones, but not the GST tag only. If the result from the small scale assay is uncertain, e.g., weak signal, it does not necessarily mean the GST-SH2 is nonfunctional. Repeat the experiment using a larger quantity or proceed to the large scale purification.
14. If the protein is insoluble, consider protein expression at a lower temperature (30 or 16 °C). When the OD reaches 0.4–0.6, cool down the culture to 30 or 16 °C with ice before adding IPTG, then add IPTG to 0.1 mM, and shake at 30 °C for 4 h or 16 °C for 24 h, respectively. If the protein is still highly insoluble, consider purifying inclusion bodies and perform refolding (*see* Chapter 10).
15. Sonication time is 2–3 min in total with microtip at the medium power level. Avoid foaming and overheating the lysate (always keep tubes on ice).
16. Supernatant does not need to be absolutely clear at this step.
17. As PreScission Protease is also a GST fusion protein, the PreScission Protease should remain bound to the GSH beads.

18. If elution efficiency is low, incubate bacterial lysate with glutathione beads for a shorter period of time (~30 min) or use elution buffer with higher pH (up to 9.6), higher glutathione concentration (up to 50 mM), and detergent to 0.1% (e.g., Triton X, Tween 20, or NP40). If all the native elution solutions are ineffective, consider a denaturing solution (3–6 M guanidine hydrochloride, 20 mM sodium phosphate, 500 mM NaCl, pH 7.8) though this has the potential risk of loss of function.
19. Alternatively, the eluate can be dialyzed at 4 °C overnight in several large volumes of PBS-10% glycerol.
20. To quickly check the relative protein concentration, add 2 μL of each PD-10 elution fraction to 50 μL diluted Bradford dye and vortex. Pool the top three or four fractions with the brightest blue color. If the color change is not obvious (this occurs when protein concentration is less than 0.2 μg/μL), take three to five fractions and proceed to ultrafiltration for concentration (below).
21. To calculate the protein concentration via OD at 280 nm, use the SH2 domain and GST sequences in Table 1. ProtParam (web.expasy.org/protparam/) has a calculator for molecular weight and extinction coefficient values. If protein concentration is low (<0.5 μg/μL), sample can be concentrated by ultrafiltration (e.g., Amicon Ultra-4 Centrifugal Filter Unit).
22. Protein concentration of the final fraction can be estimated by including 1–10 μg of a BSA control sample loaded on the same SDS-PAGE gel. If the final product is found to be highly degraded, select a bacterial host with protease-deficiency and reevaluate purification conditions such as prep scale, temperature of induction, protease inhibitors, EDTA concentration, and sonication strength. The 65–70 kDa protein band co-eluted with GST-SH2 is most likely the bacterial chaperone DnaK [18]. It can be removed by additional washing in the presence of 5 mM ATP and 5 mM $MgCl_2$ prior to elution.
23. Multiple tags can be used such as His-GST, His-Thioredoxin or His-MBP (e.g, pDEST-His-MBP) to improve expression and solubility (*see* Fig. 1). This also allows for tandem affinity purification which can increase the purity of your protein.
24. Also determine whether the chosen tag system is ideal for your biochemical studies. For example, in peptide microarray studies, tags are good for antibody detection of your protein. For studies requiring removal of your tag, a thrombin cleavage site or a TEV (Tobacco Etch Virus) cleavage site should be included. If there is a structured region close to the cleavage site it may affect cleavage, so be sure to leave a flexible linker between the cleavage site and the SH2 domain. Many pDEST vectors containing tags and protease cleavage sites are available from Addgene.

25. When using Invitrogen's TOPO entry vectors (e.g., pENTR-D-TOPO), which are part of Invitrogen's Gateway system, a TOPO sequence (CACC) must be included at the 5′ end of the linear DNA insert being cloned. Check the Invitrogen website for more details and instructions (https://goo.gl/MKUeqK).
26. We use NAP-10 columns for this step, rather than dialysis membranes, as this procedure is quick and can be performed on many samples simultaneously.
27. Purified proteins can also be dialyzed in 300 mM NaCl, 50 mM Na_2PO_4, pH 8. Protein concentrations were determined based on their absorbance at 280 nm.
28. A phosphotyrosine peptide can be added to the protein and SYPRO-orange mixture to determine if the domain is functional. Binding of a ligand may induce a shift in the melting temperature.
29. Check if the qPCR software is compatible with melting curves estimation. The qPCR machine exports data (in Excel format) for both the raw fluorescence signal and the derivative of the fluorescence signal as a function of the temperature. Take the average of the three replicates (either fluorescence signal or derivative). Subtract the average of the three replicates of buffer + dye (no protein control). Plot the corrected average of the fluorescence signal or derivative as a function of temperature.

Acknowledgments

We thank Kin Leung, Richard B. Jones, Piers Nash, and Brett Engelmann for sharing their protocols on SH2 domains. We thank Joshua Jadwin for assistance with editing the manuscript, and Bruce Mayer for his continuous encouragement and support. This study was partly supported by grant CA1154966 from the National Institutes of Health and Quest for CURES (QFC) grant from the Leukemia and Lymphoma Society (to K.M.).

References

1. Jadwin JA, Ogiue-Ikeda M, Machida K (2012) The application of modular protein domains in proteomics. FEBS Lett 586(17):2586–2596. doi:10.1016/j.febslet.2012.04.019
2. Jones RB, Gordus A, Krall JA, MacBeath G (2006) A quantitative protein interaction network for the ErbB receptors using protein microarrays. Nature 439(7073):168–174. doi:10.1038/nature04177
3. Liu BA, Engelmann BW, Jablonowski K, Higginbotham K, Stergachis AB, Nash PD (2012) SRC homology 2 domain binding sites in insulin, IGF-1 and FGF receptor mediated signaling networks reveal an extensive potential interactome. Cell Commun Signal 10(1):27. doi:10.1186/1478-811x-10-27
4. Machida K, Thompson CM, Dierck K, Jablonowski K, Karkkainen S, Liu B, Zhang H, Nash PD,

Newman DK, Nollau P, Pawson T, Renkema GH, Saksela K, Schiller MR, Shin DG, Mayer BJ (2007) High-throughput phosphotyrosine profiling using SH2 domains. Mol Cell 26(6):899–915. doi:10.1016/j.molcel.2007.05.031

5. Takakuma K, Ogo N, Uehara Y, Takahashi S, Miyoshi N, Asai A (2013) Novel multiplexed assay for identifying SH2 domain antagonists of STAT family proteins. PLoS One 8(8), e71646. doi:10.1371/journal.pone.0071646
6. Hause RJ Jr, Leung KK, Barkinge JL, Ciaccio MF, Chuu CP, Jones RB (2012) Comprehensive binary interaction mapping of SH2 domains via fluorescence polarization reveals novel functional diversification of ErbB receptors. PLoS One 7(9), e44471. doi:10.1371/journal.pone.0044471
7. Liu H, Li L, Voss C, Wang F, Liu J, Li SS (2015) A comprehensive immunoreceptor phosphotyrosine-based signaling network revealed by reciprocal protein-peptide array screening. Mol Cell Proteomics 14(7):1846–1858. doi:10.1074/mcp.M115.047951
8. Tinti M, Kiemer L, Costa S, Miller ML, Sacco F, Olsen JV, Carducci M, Paoluzi S, Langone F, Workman CT, Blom N, Machida K, Thompson CM, Schutkowski M, Brunak S, Mann M, Mayer BJ, Castagnoli L, Cesareni G (2013) The SH2 domain interaction landscape. Cell Rep 3(4):1293–1305. doi:10.1016/j.celrep.2013.03.001
9. Huang H, Li L, Wu C, Schibli D, Colwill K, Ma S, Li C, Roy P, Ho K, Songyang Z, Pawson T, Gao Y, Li SS (2008) Defining the specificity space of the human SRC homology 2 domain. Mol Cell Proteomics 7(4):768–784. doi:10.1074/mcp.M700312-MCP200
10. Jarvis DL (2009) Baculovirus-insect cell expression systems. Methods Enzymol 463:191–222. doi:10.1016/s0076-6879(09)63014-7
11. Endo Y, Sawasaki T (2006) Cell-free expression systems for eukaryotic protein production. Curr Opin Biotechnol 17(4):373–380. doi:10.1016/j.copbio.2006.06.009
12. Costa S, Almeida A, Castro A, Domingues L (2014) Fusion tags for protein solubility, purification and immunogenicity in Escherichia coli: the novel Fh8 system. Front Microbiol 5:63. doi:10.3389/fmicb.2014.00063
13. Harper S, Speicher DW (2008) Expression and purification of GST fusion proteins. Curr Protoc Protein Sci 6:6. doi:10.1002/0471140864.ps0606s52
14. Liu BA, Jablonowski K, Shah EE, Engelmann BW, Jones RB, Nash PD (2010) SH2 domains recognize contextual peptide sequence information to determine selectivity. Mol Cell Proteomics 9(11):2391–2404. doi:10.1074/mcp.M110.001586
15. Lemercier G, Bakalara N, Santarelli X (2003) On-column refolding of an insoluble histidine tag recombinant exopolyphosphatase from Trypanosoma brucei overexpressed in Escherichia coli. J Chromatogr B Analyt Technol Biomed Life Sci 786(1-2):305–309
16. Esposito D, Chatterjee DK (2006) Enhancement of soluble protein expression through the use of fusion tags. Curr Opin Biotechnol 17(4):353–358. doi:10.1016/j.copbio.2006.06.003
17. Liu BA, Jablonowski K, Raina M, Arce M, Pawson T, Nash PD (2006) The human and mouse complement of SH2 domain proteins-establishing the boundaries of phosphotyrosine signaling. Mol Cell 22(6):851–868. doi:10.1016/j.molcel.2006.06.001
18. Schafer F, Seip N, Maertens B, Block H, Kubicek J (2015) Purification of GST-tagged proteins. Methods Enzymol 559:127–139. doi:10.1016/bs.mie.2014.11.005
19. Andersen TC, Lindsjo K, Hem CD, Koll L, Kristiansen PE, Skjeldal L, Andreotti AH, Spurkland A (2014) Solubility of recombinant Src homology 2 domains expressed in E. coli can be predicted by TANGO. BMC Biotechnol 14:3. doi:10.1186/1472-6750-14-3
20. Mayer BJ, Jackson PK, Van Etten RA, Baltimore D (1992) Point mutations in the abl SH2 domain coordinately impair phosphotyrosine binding in vitro and transforming activity in vivo. Mol Cell Biol 12(2):609–618
21. Kuriyan J, Cowburn D (1997) Modular peptide recognition domains in eukaryotic signaling. Annu Rev Biophys Biomol Struct 26:259–288. doi:10.1146/annurev.biophys.26.1.259
22. Peschard P, Ishiyama N, Lin T, Lipkowitz S, Park M (2004) A conserved DpYR motif in the juxtamembrane domain of the Met receptor family forms an atypical c-Cbl/Cbl-b tyrosine kinase binding domain binding site required for suppression of oncogenic activation. J Biol Chem 279(28):29565–29571. doi:10.1074/jbc.M403954200
23. de Jong R, ten Hoeve J, Heisterkamp N, Groffen J (1995) Crkl is complexed with tyrosine-phosphorylated Cbl in Ph-positive leukemia. J Biol Chem 270(37):21468–21471
24. Wang J, Dai H, Yousaf N, Moussaif M, Deng Y, Boufelliga A, Swamy OR, Leone ME, Riedel H (1999) Grb10, a positive, stimulatory signaling adapter in platelet-derived growth factor BB-, insulin-like growth factor I-, and insulin-mediated mitogenesis. Mol Cell Biol 19(9):6217–6228
25. Ambaye ND, Gunzburg MJ, Traore DA, Del Borgo MP, Perlmutter P, Wilce MC, Wilce JA (2014) Preparation of crystals for characterizing the Grb7 SH2 domain before and after

complex formation with a bicyclic peptide antagonist. Acta Crystallogr Sect F Struct Biol Commun 70(Pt 2):182–186. doi:10.1107/s2053230x13033414
26. Yap MY, Wilce MC, Clayton DJ, Perlmutter P, Aguilar MI, Wilce JA (2010) Preparation and crystallization of the Grb7 SH2 domain in complex with the G7-18NATE nonphosphorylated cyclic inhibitor peptide. Acta Crystallogr Sect F Struct Biol Cryst Commun 66(Pt 12):1640–1643. doi:10.1107/s1744309110041850
27. Porter CJ, Matthews JM, Mackay JP, Pursglove SE, Schmidberger JW, Leedman PJ, Pero SC, Krag DN, Wilce MC, Wilce JA (2007) Grb7 SH2 domain structure and interactions with a cyclic peptide inhibitor of cancer cell migration and proliferation. BMC Struct Biol 7:58. doi:10.1186/1472-6807-7-58
28. Prasad N, Topping RS, Decker SJ (2001) SH2-containing inositol 5′-phosphatase SHIP2 associates with the p130(Cas) adapter protein and regulates cellular adhesion and spreading. Mol Cell Biol 21(4):1416–1428. doi:10.1128/mcb.21.4.1416-1428.2001
29. Pesesse X, Moreau C, Drayer AL, Woscholski R, Parker P, Erneux C (1998) The SH2 domain containing inositol 5-phosphatase SHIP2 displays phosphatidylinositol 3,4,5-trisphosphate and inositol 1,3,4,5-tetrakisphosphate 5-phosphatase activity. FEBS Lett 437(3):301–303
30. Lupardus PJ, Skiniotis G, Rice AJ, Thomas C, Fischer S, Walz T, Garcia KC (2011) Structural snapshots of full-length Jak1, a transmembrane gp130/IL-6/IL-6Ralpha cytokine receptor complex, and the receptor-Jak1 holocomplex. Structure (Lond 1993) 19(1):45–55. doi:10.1016/j.str.2010.10.010
31. Erdmann D, Allard B, Chène P (2008) Kinetic study of human full-length wild-type JAK2 and V617F mutant proteins. Open Enzyme Int J 1:80–84
32. Mishra J, Karanki SS, Kumar N (2012) Identification of molecular switch regulating interactions of Janus kinase 3 with cytoskeletal proteins. J Biol Chem 287(49):41386–41391. doi:10.1074/jbc.C112.363507
33. Wang LL, Blasioli J, Plas DR, Thomas ML, Yokoyama WM (1999) Specificity of the SH2 domains of SHP-1 in the interaction with the immunoreceptor tyrosine-based inhibitory motif-bearing receptor gp49B. J Immunol 162(3):1318–1323
34. Cao X, Tanis KQ, Koleske AJ, Colicelli J (2008) Enhancement of ABL kinase catalytic efficiency by a direct binding regulator is independent of other regulatory mechanisms. J Biol Chem 283(46):31401–31407. doi:10.1074/jbc.M804002200
35. Tall GG, Barbieri MA, Stahl PD, Horazdovsky BF (2001) Ras-activated endocytosis is mediated by the Rab5 guanine nucleotide exchange activity of RIN1. Dev Cell 1(1):73–82
36. Kajiho H, Saito K, Tsujita K, Kontani K, Araki Y, Kurosu H, Katada T (2003) RIN3: a novel Rab5 GEF interacting with amphiphysin II involved in the early endocytic pathway. J Cell Sci 116(Pt 20):4159–4168. doi:10.1242/jcs.00718
37. Naudin C, Sirvent A, Leroy C, Larive R, Simon V, Pannequin J, Bourgaux JF, Pierre J, Robert B, Hollande F, Roche S (2014) SLAP displays tumour suppressor functions in colorectal cancer via destabilization of the SRC substrate EPHA2. Nat Commun 5:3159. doi:10.1038/ncomms4159
38. Bullock AN, Debreczeni JE, Edwards AM, Sundstrom M, Knapp S (2006) Crystal structure of the SOCS2-elongin C-elongin B complex defines a prototypical SOCS box ubiquitin ligase. Proc Natl Acad Sci U S A 103(20):7637–7642. doi:10.1073/pnas.0601638103
39. Greenhalgh CJ, Rico-Bautista E, Lorentzon M, Thaus AL, Morgan PO, Willson TA, Zervoudakis P, Metcalf D, Street I, Nicola NA, Nash AD, Fabri LJ, Norstedt G, Ohlsson C, Flores-Morales A, Alexander WS, Hilton DJ (2005) SOCS2 negatively regulates growth hormone action in vitro and in vivo. J Clin Invest 115(2):397–406. doi:10.1172/jci22710
40. Kershaw NJ, Murphy JM, Liau NP, Varghese LN, Laktyushin A, Whitlock EL, Lucet IS, Nicola NA, Babon JJ (2013) SOCS3 binds specific receptor-JAK complexes to control cytokine signaling by direct kinase inhibition. Nat Struct Mol Biol 20(4):469–476. doi:10.1038/nsmb.2519
41. Babon JJ, McManus EJ, Yao S, DeSouza DP, Mielke LA, Sprigg NS, Willson TA, Hilton DJ, Nicola NA, Baca M, Nicholson SE, Norton RS (2006) The structure of SOCS3 reveals the basis of the extended SH2 domain function and identifies an unstructured insertion that regulates stability. Mol Cell 22(2):205–216. doi:10.1016/j.molcel.2006.03.024
42. Linossi EM, Chandrashekaran IR, Kolesnik TB, Murphy JM, Webb AI, Willson TA, Kedzierski L, Bullock AN, Babon JJ, Norton RS, Nicola NA, Nicholson SE (2013) Suppressor of cytokine signaling (SOCS) 5 utilises distinct domains for regulation of JAK1 and interaction with the adaptor protein Shc-1. PLoS One 8(8), e70536. doi:10.1371/journal.pone.0070536
43. Li L, Wu C, Huang H, Zhang K, Gan J, Li SS (2008) Prediction of phosphotyrosine signaling networks using a scoring matrix-assisted ligand identification approach. Nucleic Acids Res 36(10):3263–3273. doi:10.1093/nar/gkn161

44. Gupta S, Yan H, Wong LH, Ralph S, Krolewski J, Schindler C (1996) The SH2 domains of Stat1 and Stat2 mediate multiple interactions in the transduction of IFN-alpha signals. EMBO J 15(5):1075–1084
45. Zhang X, Yue P, Fletcher S, Zhao W, Gunning PT, Turkson J (2010) A novel small-molecule disrupts Stat3 SH2 domain-phosphotyrosine interactions and Stat3-dependent tumor processes. Biochem Pharmacol 79(10):1398–1409. doi:10.1016/j.bcp.2010.01.001
46. Haan S, Hemmann U, Hassiepen U, Schaper F, Schneider-Mergener J, Wollmer A, Heinrich PC, Grotzinger J (1999) Characterization and binding specificity of the monomeric STAT3-SH2 domain. J Biol Chem 274(3):1342–1348
47. Mahajan S, Vassilev A, Sun N, Ozer Z, Mao C, Uckun FM (2001) Transcription factor STAT5A is a substrate of Bruton's tyrosine kinase in B cells. J Biol Chem 276(33):31216–31228. doi:10.1074/jbc.M104874200
48. Varco-Merth B, Feigerlova E, Shinde U, Rosenfeld RG, Hwa V, Rotwein P (2013) Severe growth deficiency is associated with STAT5b mutations that disrupt protein folding and activity. Mol Endocrinol (Baltimore, MD) 27(1):150–161. doi:10.1210/me.2012-1275
49. Dengl S, Mayer A, Sun M, Cramer P (2009) Structure and in vivo requirement of the yeast Spt6 SH2 domain. J Mol Biol 389(1):211–225. doi:10.1016/j.jmb.2009.04.016
50. Yoh SM, Cho H, Pickle L, Evans RM, Jones KA (2007) The Spt6 SH2 domain binds Ser2-P RNAPII to direct Iws1-dependent mRNA splicing and export. Genes Dev 21(2):160–174. doi:10.1101/gad.1503107
51. Maruyama T, Nara K, Yoshikawa H, Suzuki N (2007) Txk, a member of the non-receptor tyrosine kinase of the Tec family, forms a complex with poly(ADP-ribose) polymerase 1 and elongation factor 1alpha and regulates interferon-gamma gene transcription in Th1 cells. Clin Exp Immunol 147(1):164–175. doi:10.1111/j.1365-2249.2006.03249.x
52. Wallweber HJ, Tam C, Franke Y, Starovasnik MA, Lupardus PJ (2014) Structural basis of recognition of interferon-alpha receptor by tyrosine kinase 2. Nat Struct Mol Biol 21(5):443–448. doi:10.1038/nsmb.2807
53. Northrop JP, Pustelnik MJ, Lu AT, Grove JR (1996) Characterization of the roles of SH2 domain-containing proteins in T-lymphocyte activation by using dominant negative SH2 domains. Mol Cell Biol 16(5):2255–2263

Chapter 9

Expression and Purification of Soluble STAT5b/STAT3 Proteins for SH2 Domain Binding Assay

Akira Asai and Kazuyuki Takakuma

Abstract

When a large hydrophobic full-length protein is expressed in bacteria, it is often challenging to obtain recombinant proteins in the soluble fraction. One way to overcome this challenge is expression of deletion mutants that have improved solubility while maintaining biological activity. In this chapter, we describe a protocol for expression of truncated forms of STAT5b and STAT3 proteins that are soluble and retain SH2-mediated activity for phospho-Tyr peptide recognition.

Key words STAT5b, STAT3, SH2 domain, Purification, Soluble protein

1 Introduction

Signal transducer and activator of transcription (STAT) family members were originally discovered as latent cytoplasmic transcription factors that transmit signals from cytokine receptors and growth factor receptors to the nucleus [1]. These signaling pathways involve the activation of receptor tyrosine kinases, such as epidermal growth factor and platelet-derived growth factor receptors, and Janus kinases (JAKs). Following phosphorylation at a conserved tyrosine residue, two STAT monomers dimerize through a reciprocal interaction between phosphotyrosine and the Src homology 2 (SH2) domain. The STAT dimers subsequently translocate to the nucleus, where they regulate gene expression by binding to specific DNA sequences. Among the seven members of STAT family transcription factors, particularly STAT3 and STAT5b are thought to be important therapeutic targets for anticancer chemotherapy because of the activation in many types of cancer [2–4]. There are several approaches including inhibition of the activation process to abrogate the function of STAT. However, there are numerous activating pathways that have been reported to converge on STAT [2–4]. Direct inhibition of STAT therefore would be

Kazuya Machida and Bernard A. Liu (eds.), *SH2 Domains: Methods and Protocols*, Methods in Molecular Biology, vol. 1555, DOI 10.1007/978-1-4939-6762-9_9,

effective for the treatment of STAT related diseases. Targeting the SH2 domain by which diminishes the formation of functional dimer of STAT is one of the most investigated strategies. To identify new SH2 antagonists for STAT3 and STAT5b, we developed a multiplexed assay of STAT3- and STAT5b-SH2 binding by combining AlphaLISA and AlphaScreen beads, which produce different colors, in a single-well assay. For the purpose of development of such multiplexed assay system, we therefore performed large-scale preparation of recombinant human STAT3 and STAT5b.

We have previously prepared soluble recombinant human STAT3 and STAT1 proteins, which were suitable for the SH2 binding assay [5]. However, full length STAT5b was not expressed in the soluble fraction in *Escherichia coli*. One of the truncated forms of STAT5b, STAT5b(136–703), in which the N- and C-terminal domains were deleted, was successfully expressed in the soluble fraction, although some of the protein still remained in the insoluble fraction. The truncated form of STAT3, i.e., STAT3(136–705), was also constructed. The amino acid sequence of STAT3(136–705) is homologous to that of STAT5b(136–703). Those soluble forms of the truncated mutants could be used for the multiplexed SH2 binding assay [6] (*see* Chapter 19).

2 Materials

Prepare all solutions with Milli-Q water (prepared with Milli-Q® Water Purification System, Merck Millipore) and the analytical grade reagents. Prepare and store all reagents at room temperature unless described otherwise.

2.1 Expression Plasmid Construction

1. Primers for PCR amplification of STAT5b(136–703) cDNA: 5′-gatcgaagctttggccatgtcccagaaacacctc-3′ and 5′-gatcgctcgagtcactgtggcttcacgtatccatc-3′ (*see* **Note 1**).
2. KOD plus.
3. Agarose gel.
4. Gel electrophoresis system.
5. *Hin*dIII.
6. *Xho*I.
7. pET28a(+) plasmid with 6×His and Avi-tag (*see* **Note 2**).
8. Ligation High (T4 Ligase).
9. *E. coli* DH5α competent cells.
10. The plate of LB medium: Dissolve one pack of LB broth with DAIGO (Wako Pure Chemical Industries) or equivalent agar with 100 mL water and autoclaved at 121 °C for 20 min. Add 60 μL of 50 mg/mL kanamycin, spread the sterilized plate, and store at 4 °C.

2.2 Protein Expression Components

1. Expression plasmid of SH2-containing truncated STAT5b, STAT5b(136–703) were prepared as described in Subheading 3.1 and store at −20 °C (*see* **Note 3**).
2. *E. coli* strains AVB99 (Avidity) and BL-21 Star (DE3) pLysS (Life Technologies Corporation). Isolate pBirAcm plasmid from AVB99 cells. Co-transform BL-21 Star (DE3) pLysS cells with pBirAcm and the above-mentioned expression plasmids. Store the bacteria in the presence of glycerol at −80 °C or plate at 4 °C (*see* **Note 4**).
3. LB medium: Dissolve one pack of DAIGO (Wako Pure Chemical Industries) or equivalent LB broth with 100 mL water, autoclaved at 121 °C for 20 min and store at 4 °C.
4. 10 mg/mL chloramphenicol stock solution: weigh 100 mg chloramphenicol, dissolve with 10 mL 100% ethanol and sterilize through the prepared 0.22 μm syringe filter. It is stored at −20 °C.
5. 50 mg/mL kanamycin stock solution: weigh 500 mg kanamycin, dissolve with 10 mL water and sterilize through the prepared 0.22 μm syringe filter. It is stored at −20 °C.
6. 100 mM isopropyl-β-D-thiogalactopyranoside (IPTG) stock solution: weigh 238 mg of IPTG, dissolve with 10 mL of water and sterilize through the prepared 0.22 μm syringe filter. It is stored at −20 °C.
7. 5 M biotin stock solution: weigh 1.2 g of biotin, dissolve with 1 mL of water and heat gradually on block incubator until dissolved.
8. Lysis buffer: 50 mM Tris–HCl (pH 7.4), 500 mM NaCl, 5 mM β-mercaptoethanol, 0.1% (w/v) Nonidet P-40 (NP-40), and one tablet of complete protease inhibitor cocktail tablets per 50 mL (*see* **Note 5**).
9. Ultrasonicator.

2.3 Protein Purification Components

1. HisTrap HP column (GE Healthcare Bio-Sciences).
2. ÄKTA FPLC system (GE Healthcare Bio-Sciences).
3. Wash buffer: 50 mM Tris–HCl (pH 7.4), 300 mM NaCl, 0.2 mM EDTA, 0.1% (w/v) NP-40, 5 mM β-mercaptoethanol, 10 mM imidazole, and 10% (v/v) glycerol (*see* **Note 6**).
4. Elution buffer: 50 mM Tris–HCl (pH 7.4), 300 mM NaCl, 0.2 mM EDTA, 0.1% (w/v) NP-40, 5 mM β-mercaptoethanol, 10% (v/v) glycerol, and 10–300 mM imidazole (*see* **Note 7**).
5. 500 mM imidazole solution: weigh 34 g of imidazole and dissolved with 1 L of water.
6. Dialysis buffer: 10 mM HEPES-NaOH (pH 7.4), 50 mM NaCl, 10 mM β-mercaptoethanol, 0.1 mM EGTA, 0.02%

(w/v) NP-40, 0.2 mM phenylmethylsulfonyl fluoride (PMSF), and 10% (v/v) glycerol (*see* **Note 8**).

7. 0.5 M EGTA (pH 8.0) solution: weigh 19 g of EGTA and dissolve with about 80 mL of water. After pH was adjusted to 7.95–8.05 with 5 or 1 N NaOH solution, the solution was diluted to 100 mL with water.
8. 200 mM PMSF solution: weigh 0.35 g of PMSF, dissolve with 10 mL of dimethyl sulfoxide and store at −20 °C.
9. Standard SDS-PAGE system with Coomassie brilliant blue (CBB) staining solution.

3 Methods

Carry out all procedures at room temperature unless described otherwise.

3.1 Expression Plasmid Construction

1. Amplify STAT5b(136–703) cDNA by PCR with KOD plus, primers and the full length of STAT5b cDNA as a template (*see* **Note 9**).
2. Load the PCR product on an agarose gel and run a gel electrophoresis.
3. Cut the part of the amplified cDNA from the agarose gel and extract the cDNA with the gel extraction kit provided (*see* **Note 10**).
4. Digest the cDNA with *Hin*dIII and *Xho*I.
5. Also digest the pET28a(+) plasmid with 6×His and Avi-tag with *Hin*dIII and *Xho*I.
6. Ligate them with Ligation High at 16 °C for 1 h (*see* **Note 11**).
7. Transform *E. coli* DH5α competent cell with the ligated plasmid.
8. Plate transformed cells on the plate of LB medium with 30 μg/mL kanamycin.
9. Inoculate and grow some colonies in 2 mL of LB medium with 100 μg/mL kanamycin at 37 °C overnight.
10. Centrifuge the *E. coli* cultures at 10,000 × *g* for 5 min.
11. Extract the plasmid with QIAprep Spin Miniprep kit (*see* **Note 12**).
12. Check that the objective cDNA is inserted at the correct site and direction by the digestion with restriction enzymes and sequencing.
13. Store the plasmid at −20 °C and use the following study.

3.2 Protein Expression of SH2 Domain-Containing Truncated STAT5b

1. Grow the *E. coli* BL-21 Star (DE3) pLysS carrying pBirAcm and STAT5b(136–703) expression plasmids at 37 °C in 2 mL of LB medium containing 10 μg/mL chloramphenicol and 30 μg/mL kanamycin overnight.
2. Transfer the precultured *E. coli* to 100 mL of LB medium containing 10 μg/mL chloramphenicol and 30 μg/mL kanamycin in conical flask (*see* **Note 13**).
3. Culture at 37 °C at 180 rpm to the OD600 of 0.6 (*see* **Note 14**).
4. Cool the *E. coli* to 30 °C in the incubator that is set at 30 °C.
5. Apply 0.2 mL of 100 mM IPTG and 0.2 mL of 5 M biotin to 100 mL culture (*see* **Note 15**).
6. Culture for 3 h at 30 °C at 180 rpm.
7. Transfer the *E. coli* to two vials of 50 mL conical tube.
8. Centrifuge at 5000 × *g* for 10 min and eliminate supernatant.
9. Suspend the bacterial pellet with 5 mL lysis buffer and lyse with ten cycles of sonication, each consisting of a constant pulse for 1 min on ice (*see* **Note 16**).
10. Centrifuge the cell lysate at 12,000 × *g* for 20 min at 4 °C (*see* **Note 17**) (Fig. 1).

3.3 Protein Purification of SH2 Domain-Containing Truncated STAT5b

In this procedure, all solutions are cooled to 4 °C prior to use. The overview of this procedure was described in Fig. 2a.

1. Transfer the supernatant to a 15 mL conical tube.
2. Load the supernatant onto a HisTrap HP column using AKTA system at 4 °C (*see* **Note 18**).
3. Wash out the nonspecifically bound proteins with 2 mL of wash buffer.
4. Elute the His-tagged STAT5b(136–703) protein with elution buffer by imidazole gradient (*see* **Note 19**).
5. Correct all eluents to a volume of 1 mL in a fresh tube.

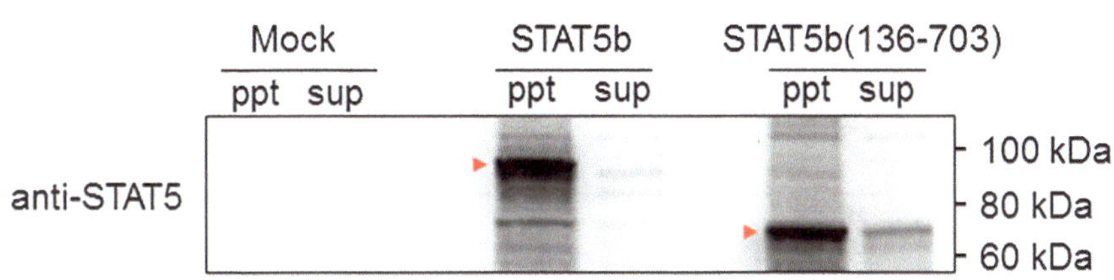

Fig. 1 STAT5b protein expression [6]. Western blot analysis of the soluble (sup) and insoluble (ppt) fractions of *E. coli* expressing full-length STAT5b or STAT5b(136–703). The proteins (5 μg) were analyzed by SDS-PAGE, and then probed with an antibody. The *arrows* indicate the expressed proteins

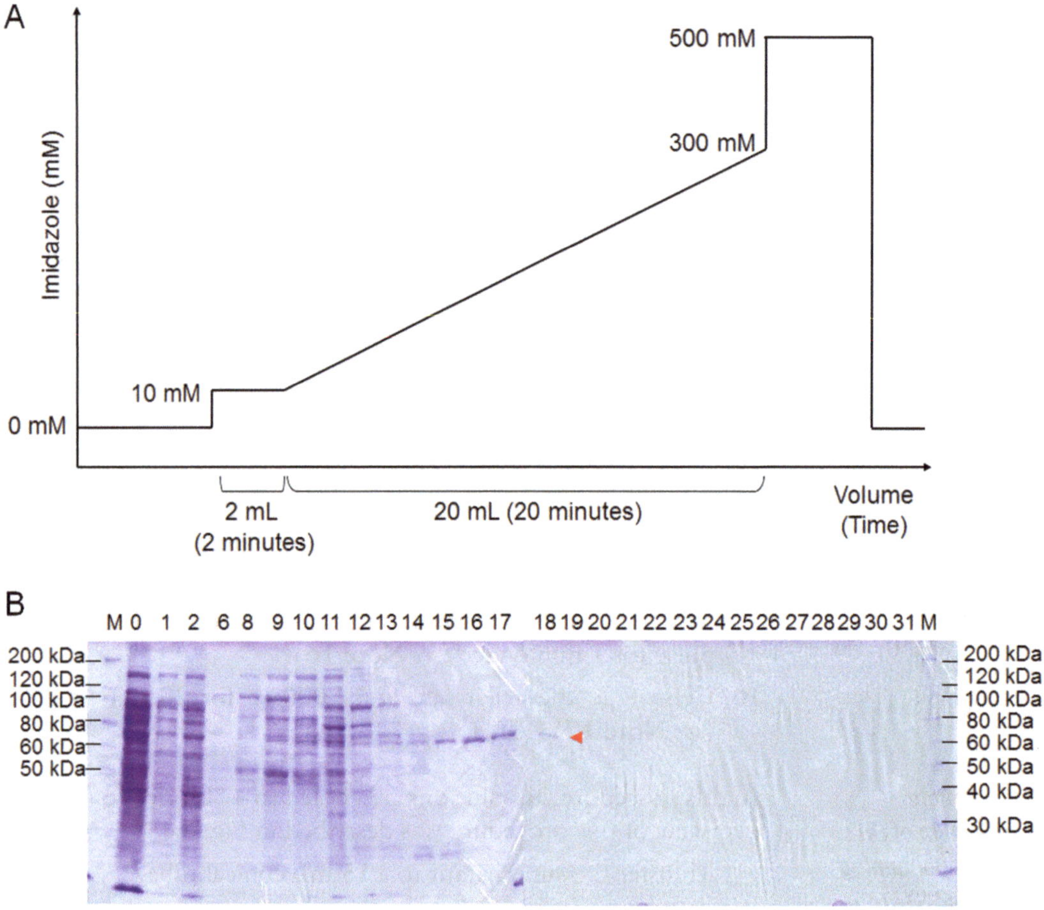

Fig. 2 The purification of the STAT5b(136–703) protein. (**a**) The overview of the purification procedure using a chromatography system. The soluble fraction of the STAT5b(136–703) protein was loaded in His Trap column, washed with 2 mL of 10 mM imidazole, and eluted the proteins by imidazole gradient. (**b**) SDS-PAGE and CBB staining of the eluted proteins. *Lanes*: *0* non-purified sample, *1*, *2* column unbound fractions, *6* eluents with wash buffer, *8–27* eluents by imidazole gradient, *28–31* eluents with 500 mM imidazole, *M* molecular marker

6. Number respective tubes of eluents and check the expression of the STAT5b(136–703) protein by CBB staining and western blotting (*see* **Note 20**) (Fig. 2b).
7. Dialyze the eluted proteins in dialysis buffer while mixing with a magnetic stir bar. Exchange into a fresh buffer several times at 4 °C overnight (*see* **Note 21**).

4 Notes

1. The sites of the restriction enzymes *Hin*dIII and *Xho*I are contained in the primers for N-terminal and C-terminal sides, respectively.

2. Avi-tag is the amino acid sequence necessary for biotinylation [7]. Biotinylation of the proteins is necessary for the dimerization assay by using streptavidin-conjugated AlphaScreen beads (PerkinElmer Life Sciences).
3. Thus far, the full length of STAT5b protein is difficult to express as a soluble fraction. In order to improve the efficiency of STAT5b expression, the truncated STAT5b protein, STAT5b(136–703) was constructed. It contains an SH2 domain and tyrosine phosphorylation site (Tyr699), which is necessary for protein-protein interaction. On the other hand, the parts not involved in protein-protein interaction were eliminated, which were the regions for polymerization, DNA binding and transactivation. We constructed the truncated STAT3 protein, STAT3(136–705) in a similar manner (Fig. 3). We succeeded to improve the expression and purification by constructing a truncated STAT3(136–705) and STAT5b proteins as compared to the full length proteins.
4. The pBirAcm plasmid has IPTG-inducible biotin ligase and chloramphenicol resistance genes. The pBirAcm was used to ligate the biotin to STAT5b protein expressed in an IPTG-inducible fashion. *E. coli* BL21(DE3) pLys strain could efficiently express the target proteins because the background protein expression is suppressed by T7 lysozyme associating with T7 promoter without IPTG.

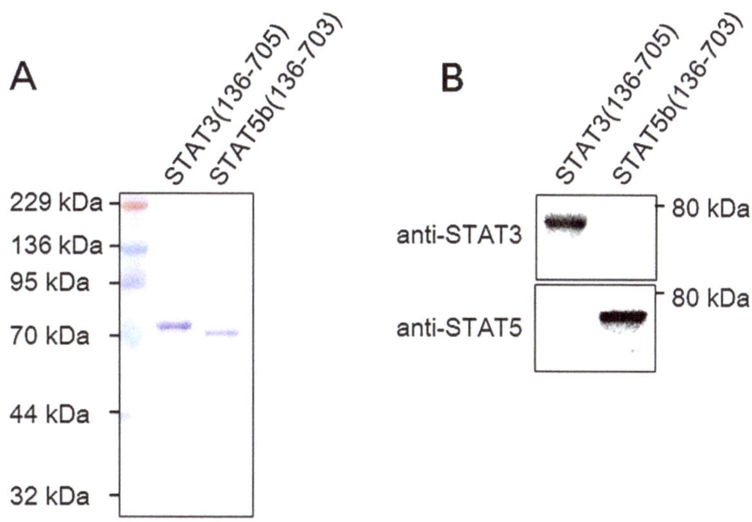

Fig. 3 STAT3(136–705) and STAT5b(136–703) protein expression [6]. (**a**) CBB staining of the STAT3(136–705) and STAT5b(136–703) proteins. The purified soluble proteins (0.5 μg) were analyzed by SDS-PAGE. (**b**) Western blot analysis of the STAT3(136–705) and STAT5b(136–703) proteins. The blotted proteins were detected with an antibody

5. Simple method of preparing lysis buffer: weigh 0.39 g of Tris (hydroxymethyl)aminomethane (Tris), 1.5 g of NaCl, 20 mg of β-mercaptoethanol, 50 μL of NP-40 and one tablet of complete protease inhibitor cocktail tablets, and dissolve with about 40 mL of water. Adjust pH to 7.35–7.45 with 6 or 1 N HCl solution and dilute to 50 mL with water. We use Nacalai Tesque's Protease Inhibitor Cocktail (100X, EDTA free). First roughly adjust the pH using concentrated HCl (6 N) solution and then use diluted HCl (1 N) for the fine adjustment.
6. Simple method of preparing wash buffer: weigh 7.9 g of Tris, 18 g of NaCl, 0.4 mL of 500 mM EDTA solution, 0.39 g of β-mercaptoethanol, 1 mL of NP-40, 40 mL of 500 mM imidazole solution, and 100 mL of glycerol, and dissolve with about 800 mL of water. After pH was adjusted to 7.35–7.45 with HCl solution, the buffer was diluted to 1 L with water.
7. Simple method of preparing elution buffer: weigh 7.9 g of Tris, 18 g of NaCl, 0.4 mL of 500 mM EDTA solution, 0.39 g of β-mercaptoethanol, 1 mL of NP-40, and 100 mL of glycerol, and dissolve with about 800 mL of water. After the pH was adjusted to 7.35–7.45 with HCl solution, the buffer was diluted to 1 L with water. Imidazole gradient was prepared in the AKTA system in eluting protein.
8. Simple method of preparing dialysis buffer: weigh 2.4 g of HEPES, 2.9 g of NaCl, 0.78 g of β-mercaptoethanol, 0.2 mL of 0.5 M EGTA (pH 8.0) solution, 0.2 mL of NP-40, 1.0 mL of 200 mM PMSF, and 100 mL of glycerol, and dissolve with about 800 mL of water. After the pH was adjusted to 7.35–7.45 with 5 N or 1 N NaOH solution, the buffer was diluted to 1 L with water.
9. 50 μL of PCR reaction mixture was prepared by mixing 100 ng of template DNA, 1 μL of 10 mM primers, 5 μL of 2 mM dNTPs, 5 μL of 10× buffer, and 1 U of PCR enzyme. The PCR was performed by GeneAmp PCR System 9700 (Applied Biosystems) as following three steps: (a) Initial denaturation at 95 °C for 5 min, (b) 30 cycles of 95 °C for 30 s, 60 °C for 30 s, and 68 °C for 2 min, (c) Final extension 68 °C for 7 min. We amplified STAT5b(136–703) cDNA using a full length STAT5b cDNA template which was isolated from a human cDNA library. Similarly, the truncated STAT5b can be amplified using a full-length STAT5b cDNA from other sources and PCR systems following the manufacturer's instructions.
10. We often use QIAprep Spin Miniprep Kit (QIAGEN) as a gel extraction kit. You could use the gel extraction kit you like.
11. It is favorable that over threefold amount of the insert DNA to the vector is applied in ligation procedure. You could use the

ligation reagent you like. If you so, follow the manufacturer's instruction.

12. You could use the plasmid extraction kit you like. If you so, follow the manufacturer's instruction.
13. In culturing with 100 mL LB medium, 300 mL conical flask could be used.
14. It is important to check OD600 of *E. coli* suspension. It is recommended to check the OD600 frequently. The protein could be expressed into the insoluble fraction if OD600 is over 0.6. The culture time depends on the proliferation of precultured *E. coli*.
15. Final concentrations of IPTG and biotin are 0.2 mM and 10 μM, respectively.
16. We used the Branson Sonifier (power setting 3, ten cycles of 30-s bursts and 30-s rest intervals).
17. The centrifuge and the rotor are precooled to 4 °C prior to use. The supernatant and the precipitate were considered as the soluble and the insoluble fractions, respectively (Fig. 1).
18. ÄKTA FPLC system is preconditioned by flowing with water at first and then with the wash buffer. 6×His-tagged STAT5b(136–703) protein is bound to the column. In ÄKTA system, the flow rate is 1 mL/min. If you do not have ÄKTA system or equivalent FPLC systems, you can use nickel magnetic beads. When using a nickel magnetic beads-based system, follow the manufacturer's instruction.
19. The protein was eluted with 20 mL of the elution buffer that the flow rate is 1 mL/min and is set the imidazole gradient with 10–300 mM. The eluents are collected by 1 mL in the respective tube.
20. In eluting with around 100–150 mM imidazole, the objective protein is almost eluted. All corrected eluents should be subjected to CBB staining and western blotting.
21. During the dialysis step, the dialysis buffer should be exchanged into fresh buffer at least three times. The tube for dialysis or the disposable dialysis cassette could be used. The pore size of the tube should be less than 10 kDa. You can select the dialysis buffer composition, dependent on the following assay.

Acknowledgment

This work was supported by National Institute of Biomedical Innovation (NIBO) grant number 06–02, JSPS KAKENHI Grant Number 23510281/26430166, and Pharma Valley Center (PVC) Shizuoka, Japan.

References

1. Levy DE, Darnell JE Jr (2002) Stats: transcriptional control and biological impact. Nat Rev Mol Cell Biol 3:651–662
2. Turkson J, Jove R (2000) STAT proteins: novel molecular targets for cancer drug discovery. Oncogene 19:6613–6626
3. Yu H, Jove R (2004) The STATs of cancer—new molecular targets come of age. Nat Rev Cancer 4:97–105
4. Turkson J (2004) STAT proteins as novel targets for cancer drug discovery. Expert Opin Ther Targets 8:409–422
5. Uehara Y, Mochizuki M, Matsuno K, Haino T, Asai A (2009) Novel high-throughput screening system for identifying STAT3-SH2 antagonists. Biochem Biophys Res Commun 380:627–631
6. Takakuma K, Ogo N, Uehara Y, Takahashi S, Miyoshi N, Asai A (2013) Novel multiplexed assay for identifying SH2 domain antagonists of STAT family proteins. PLoS One 8:71646. doi:10.1371/journal.pone.0071646
7. Athavankar S, Peterson BR (2003) Control of gene expression with small molecules: biotin-mediated acylation of targeted lysine residues in recombinant yeast. Chem Biol 10:1245–1253

Chapter 10

Purification of SOCS (Suppressor of Cytokine Signaling) SH2 Domains for Structural and Functional Studies

Nicholas P.D. Liau, Artem Laktyushin, and Jeffrey J. Babon

Abstract

Src Homology 2 (SH2) domains are protein domains which have a high binding affinity for specific amino acid sequences containing a phosphorylated tyrosine residue. The Suppressors of Cytokine Signaling (SOCS) proteins use an SH2 domain to bind to components of certain cytokine signaling pathways to downregulate the signaling cascade. The recombinantly produced SH2 domains of various SOCS proteins have been used to undertake structural and functional studies elucidating the method of how such targeting occurs. Here, we describe the protocol for the recombinant production and purification of SOCS SH2 domains, with an emphasis on SOCS3.

Key words Src homology 2 (SH2) domains, SOCS

1 Introduction

The SOCS family of proteins are negative regulators of cytokine signaling [1–3]. SOCS proteins are able to specifically downregulate signaling by the intracellular receptor associated kinases, Janus Kinases (JAKs), which are critical elements of cytokine signaling [4, 5].

1.1 The JAK Proteins

The JAK proteins consist of four family members, JAK1, JAK2, JAK3, and TYK2 [6]. The JAKs are tyrosine kinases, and are constitutively attached to the intracellular portion of a cytokine receptor [7]. Upon receptor activation by cytokine binding, JAK is activated, and is able to phosphorylate downstream substrate molecules, including the STAT family transcription factors. Once STATs are activated by phosphorylation, they are able to enter the nucleus and upregulate the expression of certain gene sets, such as those involved in hematopoietic growth and differentiation [4]. Overactive JAKs can cause uncontrolled growth and differentiation of affected cells, so the regulation of JAK catalytic activity is crucial [8–12].

Kazuya Machida and Bernard A. Liu (eds.), *SH2 Domains: Methods and Protocols*, Methods in Molecular Biology, vol. 1555,
DOI 10.1007/978-1-4939-6762-9_10, © Springer Science+Business Media LLC 2017

1.2 The SOCS Proteins

The SOCS proteins are identifiable by a common SOCS Box motif at the C terminus of the protein [3, 13]. The SOCS box acts as an adapter domain, which recruits the necessary cellular machinery for the ubiquitination of SOCS substrates, resulting in downregulation by proteasomal degradation of the ubiquitinated target [14, 15]. We have specifically investigated SOCS3 because of its ability to directly inhibit cytokine signaling through a separate method, namely, by inhibiting JAK catalytic activity by a domain known as the KIR, present only in SOCS1 and SOCS3 [16–18].

An important factor for obtaining specificity for the downregulation of certain cytokine signaling pathways, common to all SOCS family members, is their SH2 domain [19]. The SH2 domain is able to bind to specific amino acid sequences flanking a phosphorylated tyrosine residue [20]. In the SOCS context, the SH2 domain is crucial for obtaining specificity for the downregulation of certain cytokine signaling pathways. For example, the SOCS3 SH2 domain targets SOCS3 only to those JAKs which are bound to specific phosphorylated cytokine receptors [18]. The important function of the SH2 domains of SOCS proteins makes them a target of great interest for further study.

All of the SOCS proteins also contain unique N terminal domains of varying lengths (Fig. 1). In some case, function can be attributed to at least a portion of the N terminal domain—for instance, the SOCS1 and SOCS3 N terminal domains contain a region which is able to directly inhibit JAK catalytic kinase activity [21]. In other cases, the function of the N terminal domain is largely unknown.

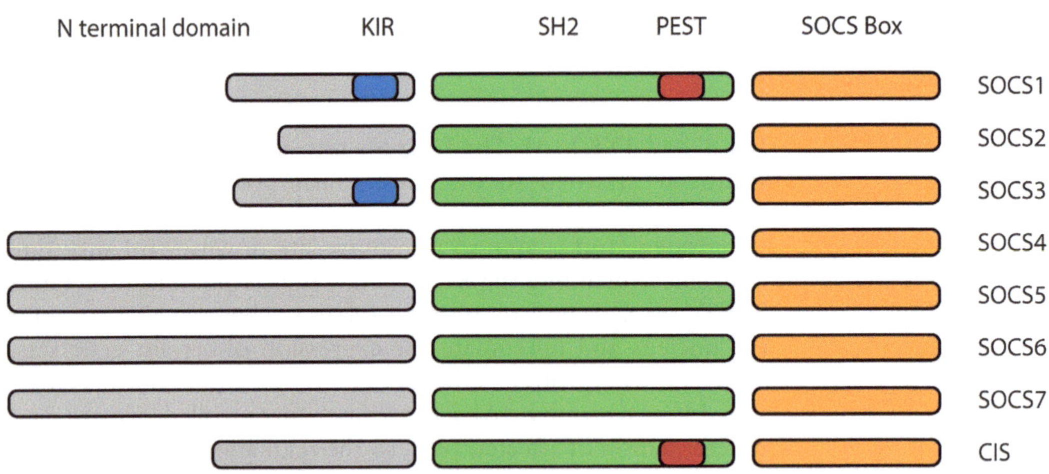

Fig. 1 Domain architecture diagram of the SOCS family of proteins. The SOCS family consists of eight members. The N terminal domains differ in length, and are largely unstructured. The SH2 domain and the SOCS Box are common features shared between all SOCS family members

1.3 Overview of Protocol

Here, we describe a method we have used to successfully purify the SH2 domain of SOCS3 using an *E. coli* expression system [22, 23]. The SH2 domain of SOCS3 is expressed as an unstructured protein inside insoluble inclusion bodies of the *E. coli* cytoplasm [24]. Although this results in unfolded protein, extremely high yields are obtained. The pure protein is refolded into its native state by the gradual removal of guanidine by dialysis. We have previously shown that SOCS3 produced in this way is structured [23], inhibits JAK1, JAK2, and TYK2 [25], binds phosphopeptide and forms a stable complex with JAK [18].

2 Materials

2.1 Cloning

1. Standard reagents for cloning (e.g., restriction enzymes and buffers, miniprep kit, competent cells).
2. pET-15b vector (or other suitable *E. coli* expression vector).

2.2 Protein Expression

1. Superbroth: 35 g tryptone, 20 g yeast extract, 5 g NaCl, 2.5 mL 2 M NaOH to 1 L with deionized water.
2. Ampicillin.
3. Isopropylthio-β-D-galactoside (IPTG).
4. 50 mL Falcon tubes.
5. 2 L baffled flask.
6. Centrifuge (capable of 5000 × *g* with capacity for 1 L bottles).

2.3 Purification of Inclusion Bodies

1. Resuspension buffer: Dulbecco's phosphate buffered saline (138 mM NaCl, 2.7 mM KCl, 1.5 mM KH_2PO_4, 8.1 mM Na_2HPO_4 pH 7.2), 1 mM phenylmethanesulfonyl fluoride, 5 mM dithiothreitol (DTT), 0.5 mg/mL lysozyme, one EDTA-free Complete protease inhibitor cocktail tablet per 100 mL, 5 U/mL DNAseI.
2. Sonicator.
3. High-speed centrifuge (capable of >20,000 × *g* with capacity for at 40 mL tubes).
4. Guanidine buffer A: 7 M guanidine HCl, 10 mM Tris, 100 mM NaH_2PO_4 pH 8.0.
5. Guanidine buffer B: 7 M guanidine HCl, 10 mM Tris, 100 mM NaH_2PO_4 pH 6.3.
6. Guanidine buffer C: 7 M guanidine HCl, 10 mM Tris, 100 mM NaH_2PO_4 pH 4.0.
7. Nickel agarose affinity resin (e.g., Ni-NTA Agarose). Matrix must be resistant to 7 M guanidine HCl.
8. Fritted column.

2.4 Refolding of Solubilized Protein

1. Guanidine buffer D: 7 M guanidine HCl, 10 mM Tris, 100 mM NaH_2PO_4 pH 8.0.
2. Dialysis tubing, 10 kDa cut-off.
3. Dialysis buffer: 20 mM Tris pH 8.5, 2 mM DTT.
4. SH2 domain binding phosphopeptide (e.g., for the SOCS3 SH2 domain, STASTVEpYSTVVHSG; pY = phosphotyrosine [26]) (*see* **Note 1**).
5. 50 mL polypropylene Falcon tubes.
6. Centrifugal concentrator, 10 kDa cutoff.
7. Equipment for sodium dodecyl sulfate–polyacrylamide gel electrophoresis (SDS-PAGE).

2.5 Cleavage of His_6 Tag

1. Gel filtration buffer: 20 mM Tris pH 7.5, 150 mM NaCl, 2 mM DTT, 5 mM sodium phenyl phosphate.
2. Fast Protein Liquid Chromatography apparatus (e.g., ÄKTA pure, GE Healthcare Life Sciences).
3. Gel filtration column (e.g., Superdex 200 HR 10/30, GE Healthcare Life Sciences).
4. Centrifugal concentrator, 10 kDa cutoff.
5. Liquid nitrogen.

3 Methods

3.1 Cloning

This section describes the design and preparation of DNA constructs for expression of SH2 domains.

1. For SOCS3, we use the following domain boundaries (*see* **Notes 2** and **3**):

   SOCS3 22-185 (*see* **Note 4**) (GenBank: Protein, NP_003946; cDNA, NG_016851).
2. Clone the SH2 domain-containing gene into pET-15b vector using standard techniques (*see* **Note 5**).
3. Using standard techniques, transform the plasmid into an expression line of *E. coli* (e.g., BL21(DE3)) and freeze as a glycerol stock for later use.

3.2 Expression of Protein

This section describes the method for expressing SH2 domains in *E. coli* into inclusion bodies. Inclusion bodies are insoluble aggregates of unfolded protein.

1. Take a scraping of glycerol stock and inoculate into 3 mL Superbroth containing 100 μg/mL ampicillin.
2. Incubate overnight at 37 °C with 180 rpm shaking.

3. Inoculate 1 L of Superbroth containing 100 μg/mL ampicillin with the 3 mL overnight culture.
4. Incubate at 37 °C with 180 rpm shaking until A_{600} reaches 0.8–1.0.
5. Induce expression by adding IPTG to a concentration of 1 mM.
6. Incubate 3 h at 37 °C with 180 rpm shaking to express protein.
7. Harvest cells by centrifugation at 6000 × *g*, 20 min, 4 °C. Discard supernatant and transfer cell pellet to a 50 mL Falcon tube.
8. The cell pellet can be frozen at < −30 °C indefinitely for later purification, or purified immediately.

3.3 Purification of Inclusion Bodies

This section describes the method for the purification under denaturing conditions of insoluble inclusion bodies containing SOCS SH2 domain. Tagged protein is able to bind to a Nickel affinity resin by coordination of immobilized nickel ions through histidine residues in the His_6 tag. A decrease in pH results in the protonation of histidine residues, decreasing affinity to the nickel resin.

Nonspecifically bound proteins are washed by decreasing the pH of the wash buffer, and an elution buffer with a further reduced pH elutes SOCS3 SH2 from the column.

1. Resuspend the cell pellet in 40 mL of resuspension buffer (*see* **Note 6**).
2. Sonicate with 15 × 10s bursts followed by 10 s rest at a power of 40–60 W. Perform sonication in an ice-water bath to ensure that pellet does not overheat.
3. Pellet the insoluble material, including inclusion bodies, by centrifugation (20,000 × *g*, 30 min, 4 °C) (*see* **Note 7**).
4. Add 50 mL guanidine buffer A to the pelleted material. Solubilize the pellet by inversion for at least 1 h at room temperature.
5. In order to speed solubilization of the pellet, sonicate with 15 × 10 s bursts followed by 10 s rest at a power of 40–60 W (*see* **Note 8**).
6. Remove material which remains insoluble by centrifugation (20,000 × *g*, 10 min). Keep the supernatant.
7. Transfer 2 mL of a 50% Ni-agarose slurry into a fritted column. This will give a packed volume of 1 mL of resin.
8. Wash the resin with 5-10 mL of guanidine buffer A (*see* **Note 9**).
9. Pass the solubilized inclusion bodies over the resin (*see* **Note 10**).
10. Wash the resin with 30 mL guanidine buffer A.
11. Wash the resin with 15 mL guanidine buffer B.

12. Elute the sample with 5 mL guanidine buffer C. The protein will be in the flow through (*see* **Note 11**).
13. Test the concentration of the protein by spectrophotometric analysis of absorbance at 280 nm. The molar extinction coefficient (ε) of murine SOCS3 (22–185) is 14,565 M^{-1}. The concentration of protein is typically >1 mg/mL at this stage.

3.4 Refolding of Purified Solubilized Protein

After purification, the denatured protein is refolded into its native state through dilution and the gradual removal of guanidine by dialysis.

1. Measure an appropriate volume of eluted protein solution containing approximately 5–10 mg of protein. Mix with 50 mL guanidine buffer D (*see* **Note 12**).
2. Add 50 mL dialysis buffer to the solution.
3. Transfer this solution into dialysis tubing with a molecular weight cut-off of not more than 10 kDa (*see* **Note 13**).
4. Immerse the dialysis tubing in 4 L of dialysis buffer which has been prechilled to 4 °C. Stir the dialysis buffer at 4 °C for at least 2 h with a magnetic stirrer. Discard the used dialysis buffer (*see* **Note 14**).
5. Replace the dialysis buffer an additional three times, with at least one of the dialysis steps taking place overnight.
6. Remove solution from the dialysis bag (*see* **Note 15**).
7. Centrifuge the solution at to remove any precipitated protein (5000 × *g*, 10 min, 4 °C). Keep the supernatant and discard any pelleted aggregate.
8. If available, add a 1.5× molar excess of SH2 binding phosphopeptide to the supernatant (*see* **Note 16**).
9. Concentrate the supernatant in a 10 kDa centrifugal concentrator to >5 mg/mL (*see* **Note 17**). Agitate the solution with a pipette regularly during concentration (approximately every 5 min) and inspect for precipitation (*see* **Notes 18** and **19**).
10. We typically analyze fractions by reducing SDS-PAGE to check for purity (*see* **Note 20**).

3.5 Removal of His6 Tag by Protease Cleavage

Depending upon the desired downstream application, it is often desirable to remove the His_6 tag from the SOCS3 SH2 domain. This section describes such removal by protease cleavage. After cleavage, the SH2 domain must be separated from free tag and protease by gel filtration.

1. Dilute 1–5 mg protein in 20 mL gel filtration buffer (*see* **Note 21**).
2. Add 30 μg/mL thrombin and incubate overnight at 4 °C.
3. Concentrate the mixture to 500 μL volume (*see* **Note 22**).

4. Separate the cleaved SH2 domain from protease and free tag by running on a gel filtration column in gel filtration buffer.
5. Take fractions corresponding to expected elution volume for the molecular weight of your SH2 domain on your chosen gel filtration column.
6. Analyse fractions by reducing SDS-PAGE.
7. Pool the most pure fractions (as determined by SDS-PAGE) and concentrate to >5 mg/mL in a 10 kDa centrifugal concentrator.
8. To store protein long term, glycerol may be added to 10% (v/v), and aliquots may be snap frozen on liquid nitrogen and stored at −80 °C.

4 Notes

1. This is a synthetic peptide containing a phosphorylated tyrosine residue flanked by an amino acid sequence known to bind the SOCS3 SH2 domain. Such peptides can be custom designed and commercially synthesized by many commercial suppliers (e.g., GenScript). If no SH2 domain binding phosphopeptides are known for a particular SH2 domain, the phosphotyrosine mimetic phenyl phosphate may be used instead (*see* **Note 16**).
2. The design of domain boundaries is critical to the ability of the protein to successfully refold. Examples of stabilizing mutations include removing unstructured regions of the protein, and removing other domains exogenous to the SH2 domain. For example, we found that removing the PEST motif found in the SOCS3 SH2 domain (residues 129–163) and replacing it with $(\text{Gly-Ser})_4$ yields a more stable protein [22].
3. The SOCS proteins also contain a domain known as the SOCS Box, this binds to an adapter complex (elonginBC) with high affinity. Co-expressing constructs of numerous SOCS proteins with elonginBC results in a stable ternary complex that is typically more stable and more easily purified and concentrated than the SH2 domain alone [14, 15]. In addition, such ternary complexes can be purified under native, rather than denaturing conditions (*See* ref [14] for a full protocol on how to achieve this.). Examples of SOCS proteins purified in this manner are SOCS2, SOCS4, and SOCS5 [14, 27–29].
4. The SH2 domain for SOCS3 begins at residue 45. However, residues 22–44 are optimal for the stability of the final construct. The N terminal boundary can be further truncated to residue 30, but no further. SOCS3 SH2 domains are unusual in requiring the presence of a short α-helix (the ESS) immediately N terminal to the SH2 domain proper, for stability.

5. This vector encodes an N terminal His_6 tag followed by a thrombin protease site. Alternatively, you may use another vector suitable for *E. coli* expression which encodes a His_6 tag and a protease cleavage site adjacent to the protein of interest. Unlike most affinity tags, a His_6 tag does not rely on secondary or tertiary structural motifs for binding to its substrate, so it is suitable for use under the denaturing conditions used in this procedure.
6. This purification procedure can be scaled up for multiple litres of cell culture by increasing buffer and resin volumes accordingly.
7. Some SOCS SH2 domain proteins are partially expressed as soluble proteins in *E. coli*, for example SOCS2, and will therefore not all be pelleted in the inclusion body fraction (although to our knowledge, the majority of all human SOCS SH2 constructs are directed to inclusion bodies). However, for this reason it is prudent to bind a small volume (~1 mL) of supernatant to Nickel affinity resin (~20 μL) and assay the affinity resin by reducing SDS-PAGE. If there is a significant amount of protein in the soluble fraction, this protein may be isolated using standard techniques for purifying soluble His_6 tagged proteins.
8. Since the protein is denatured at this stage, there is no need to keep the solution cool.
9. It is important that the resin is pre-equilibrated with guanidine buffer A before the addition of protein solution to the resin.
10. Be careful to pour the solution down the side of the column to avoid disturbing the packed affinity resin.
11. The eluted protein solution can be stored indefinitely at room temperature.
12. It is important that the protein concentration is <0.1 mg/mL (and preferably <0.05 mg/mL) after the addition of guanidine buffer D. The protein will aggregate if the concentration of protein is too high at this stage.
13. Upon dialysis, the sample volume will increase as a large amount of buffer moves into the dialysis tubing. Allow plenty of spare capacity in the dialysis tubing for this to occur.
14. This dialysis buffer is of low ionic strength and relatively high pH, as we found SOCS3 to be most stable in these conditions.
15. As the protein is now refolded, it is important to keep the sample cold after this point in the protocol to prevent precipitation.
16. Ideally, the SH2 domain will be stabilized by a known binding partner phosphopeptide. However, if no binding phosphopeptides are known, the phosphotyrosine mimetic phenyl

phosphate may be used instead. Phenyl phosphate binds the phosphotyrosine binding site of SH2 domains with relatively low affinity. In this case, omit SH2 binding phosphopeptide, but include phenyl phosphate to a concentration of 5 mM in all preparations of the dialysis buffer and the gel filtration buffer. If required, phenyl phosphate may be removed before downstream applications by a desalting column (e.g., PD-10, GE Healthcare Life Sciences). For example, we have successfully expressed SOCS1 in this manner.

17. We typically find that the protein is pure enough after refolding that it does not need further downstream purification steps. Other purification steps such as size exclusion chromatography or ion-exchange could be performed if the protein is not of sufficient purity as judged by SDS-PAGE analysis.
18. If significant precipitation starts to occur, cease concentrating the protein if possible. Often, the majority of protein still remains in solution even when precipitate is visible; however, this cannot be guaranteed.
19. The SOCS3 SH2 domain precipitates over several days at 4 °C. We suggest snap-freezing in liquid nitrogen if the protein is not to be used immediately.
20. Guanidine present in the pre-dialysis samples will immediately precipitate if loaded onto an SDS-PAGE gel, so we typically dilute our samples one in ten with water before loading. Protein concentrations are usually high enough to be easily visualized with Coomassie staining, even when diluted at this level.
21. The amount of protein which can be cleaved at this stage is limited by the capacity of your chosen gel filtration column. This protocol is for a 10/30 sized column; however, you may scale up the protocol accordingly by using a larger column.
22. Scale this volume according to the manufacturer's recommended loading volume for your chosen gel filtration column.

Acknowledgments

The authors' original work was supported by grants from the National Health and Medical Research Council, Australia (program grants 461219, 1016647; The Independent Research Institutes Infrastructure Scheme 361646) the Australian Research Council (future fellowship FT110100169 JJB), the Cancer Council of Victoria (1065180) the National Institutes of Health, Bethesda, MD (Grant RO1CA22556) and Operational Infrastructure Support grants from the State Government of Victoria.

References

1. Endo TA et al (1997) A new protein containing an SH2 domain that inhibits JAK kinases. Nature 387(6636):921–4
2. Naka T et al (1997) Structure and function of a new STAT-induced STAT inhibitor. Nature 387(6636):924–9
3. Starr R et al (1997) A family of cytokine-inducible inhibitors of signalling. Nature 387(6636):917–21
4. Shuai K et al (1993) Polypeptide signalling to the nucleus through tyrosine phosphorylation of Jak and Stat proteins. Nature 366(6455):580–3
5. Wilks AF et al (1991) Two novel protein-tyrosine kinases, each with a second phosphotransferase-related catalytic domain, define a new class of protein kinase. Mol Cell Biol 11(4):2057–65
6. Ihle JN (1995) Cytokine receptor signalling. Nature 377(6550):591–4
7. Haan C et al (2006) Jaks and cytokine receptors—an intimate relationship. Biochem Pharmacol 72(11):1538–46
8. Baxter EJ et al (2005) Acquired mutation of the tyrosine kinase JAK2 in human myeloproliferative disorders. Lancet 365(9464):1054–61
9. James C et al (2005) A unique clonal JAK2 mutation leading to constitutive signalling causes polycythaemia vera. Nature 434(7037):1144–8
10. Kralovics R et al (2005) A gain-of-function mutation of JAK2 in myeloproliferative disorders. N Engl J Med 352(17):1779–90
11. Levine RL et al (2005) Activating mutation in the tyrosine kinase JAK2 in polycythemia vera, essential thrombocythemia, and myeloid metaplasia with myelofibrosis. Cancer Cell 7(4):387–97
12. Zhao R et al (2005) Identification of an acquired JAK2 mutation in polycythemia vera. J Biol Chem 280(24):22788–92
13. Hilton DJ et al (1998) Twenty proteins containing a C-terminal SOCS box form five structural classes. Proc Natl Acad Sci U S A 95(1):114–9
14. Babon JJ et al (2008) The SOCS box domain of SOCS3: structure and interaction with the elonginBC-cullin5 ubiquitin ligase. J Mol Biol 381(4):928–40
15. Babon JJ et al (2009) The SOCS box encodes a hierarchy of affinities for Cullin5: implications for ubiquitin ligase formation and cytokine signalling suppression. J Mol Biol 387(1):162–74
16. Yasukawa H et al (1999) The JAK-binding protein JAB inhibits Janus tyrosine kinase activity through binding in the activation loop. EMBO J 18(5):1309–20
17. Sasaki A et al (1999) Cytokine-inducible SH2 protein-3 (CIS3/SOCS3) inhibits Janus tyrosine kinase by binding through the N-terminal kinase inhibitory region as well as SH2 domain. Genes Cells 4(6):339–51
18. Kershaw NJ et al (2013) SOCS3 binds specific receptor-JAK complexes to control cytokine signaling by direct kinase inhibition. Nat Struct Mol Biol 20(4):469–76
19. Hilton DJ et al (1998) Twenty proteins containing a C-terminal SOCS box form five structural classes. Proc Natl Acad Sci U S A 95(1):114–9
20. Koch CA et al (1991) SH2 and SH3 domains: elements that control interactions of cytoplasmic signaling proteins. Science 252(5006):668–74
21. Alexander WS (2002) Suppressors of cytokine signalling (SOCS) in the immune system. Nat Rev Immunol 2(6):410–6
22. Babon JJ et al (2005) Secondary structure assignment of mouse SOCS3 by NMR defines the domain boundaries and identifies an unstructured insertion in the SH2 domain. FEBS J 272(23):6120–30
23. Babon JJ et al (2006) The structure of SOCS3 reveals the basis of the extended SH2 domain function and identifies an unstructured insertion that regulates stability. Mol Cell 22(2):205–16
24. Lilie H, Schwarz E, Rudolph R (1998) Advances in refolding of proteins produced in E. coli. Curr Opin Biotechnol 9(5):497–501
25. Varghese LN et al (2014) Mechanistic insights into activation and SOCS3-mediated inhibition of myeloproliferative neoplasm-associated JAK2 mutants from biochemical and structural analyses. Biochem J 458(2):395–405
26. Nicholson SE et al (2000) Suppressor of cytokine signaling-3 preferentially binds to the SHP-2-binding site on the shared cytokine receptor subunit gp130. Proc Natl Acad Sci U S A 97(12):6493–8
27. Kershaw NJ et al (2014) Reconstruction of an active SOCS3-based E3 ubiquitin ligase complex in vitro: identification of the active components and JAK2 and gp130 as substrates. Growth Factors 32(1):1–10
28. Bullock AN et al (2006) Crystal structure of the SOCS2-elongin C-elongin B complex defines a prototypical SOCS box ubiquitin ligase. Proc Natl Acad Sci U S A 103(20):7637–42
29. Bullock AN et al (2007) Structure of the SOCS4-ElonginB/C complex reveals a distinct SOCS box interface and the molecular basis for SOCS-dependent EGFR degradation. Structure (London 1993) 15(11):1493–504

Chapter 11

Expression and Purification of SH2 Domains Using Baculovirus Expression System

Mari Ogiue-Ikeda and Kazuya Machida

Abstract

Recombinant proteins expressed in bacteria are sometimes insoluble, aggregated, and incorrectly folded. For those Src homology 2 (SH2) domains that are insoluble in bacteria, baculovirus-insect cell expression systems can be an alternative to produce soluble and functionally active proteins. We describe a protocol for cloning and purification of GST-tagged SH2 domains using the Bac-to-Bac baculovirus expression system.

Key words Baculovirus expression system, Bac-to-Bac, Insect cells, GST fusion protein, SH2 domain

1 Introduction

SH2 and other modular protein interaction domains can be expressed and purified as isolated protein modules without loss of function. Modular domain interactions have been widely used in both analytical and high-throughput studies. Due to its convenience and cost effectiveness, bacteria expression systems are the first choice for production of recombinant SH2 domain proteins. However, overexpression of exogenous proteins in bacteria is inherently susceptible to improper folding, aggregation, and degradation, resulting in insoluble and inactive recombinant proteins. According to genome-wide studies using bacterial expressed SH2 domains, 20–40% of SH2 domain proteins are either low expression, insoluble, or degraded (*see* **Chapter 8**). To address this issue, modifications to the standard method such as low temperature culture as well as denaturing and refolding are sometimes effective. However, when those efforts are unsuccessful, baculovirus-insect cell expression system may be considered.

Insect baculoviruses have been widely used as a tool for protein expression since the mid-1980s [1–3]. Baculovirus expression systems allow expression of larger proteins, co-purification of multiple proteins, and improvement of solubility and activity.

Kazuya Machida and Bernard A. Liu (eds.), *SH2 Domains: Methods and Protocols*, Methods in Molecular Biology, vol. 1555, DOI 10.1007/978-1-4939-6762-9_11, © Springer Science+Business Media LLC 2017

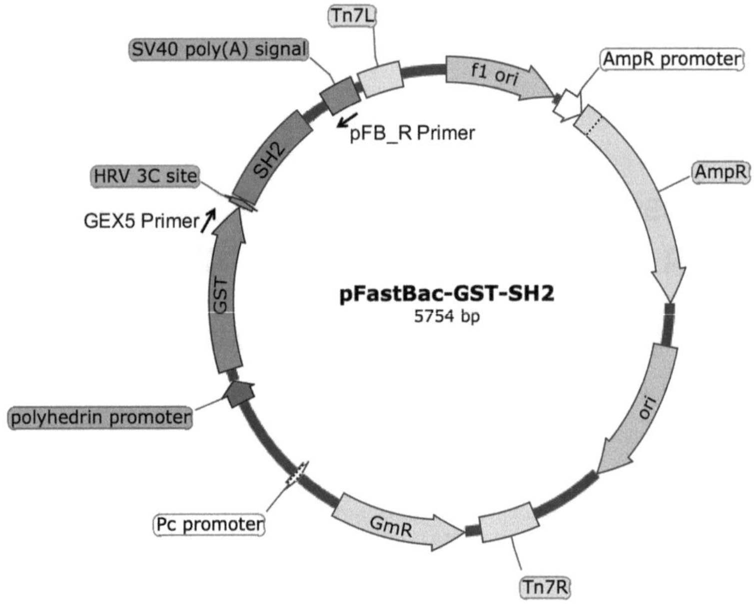

Fig. 1 Plasmid map of pFastBac-GST-SH2. The SH2 domain coding sequence is inserted into the pFastBac-GST insect cell expression vector. This vector contains a polyhedrin promoter and 5′ GST tag sequence and is compatible with the Bac-to-Bac Baculovirus Expression System. Binding sites for sequencing primers GEX5 (forward) and pFB_R (reverse) are indicated by *arrows*

Baculovirus expression systems are commonly used to produce mammalian full-length proteins and catalytic domains that require a eukaryotic host for proper posttranslational modifications [4–6]. Although posttranslational modifications of SH2 domains are not essential for their canonical binding, SH2 proteins expressed in insect cells are more likely to be processed and folded correctly than in bacterial cells.

In the following sections, we describe cloning and purification of a GST-tagged SH2 domain using the Bac-to-Bac baculovirus expression system (Figs. 1 and 2). Briefly, the SH2 domain cDNA is cloned into the pFastBac-GST donor plasmid (Subheadings 3.1 and 3.2), transferred into a bacmid (Subheadings 3.3 and 3.4), and transfected into insect cells to produce the recombinant baculovirus (Subheading 3.5). The baculovirus containing the GST-SH2 cDNA is then amplified and used for infection of insect cells (Subheadings 3.6–3.8). The insect cells expressing the GST-SH2 are then harvested and the protein is affinity-purified with a glutathione column (Subheading 3.9).

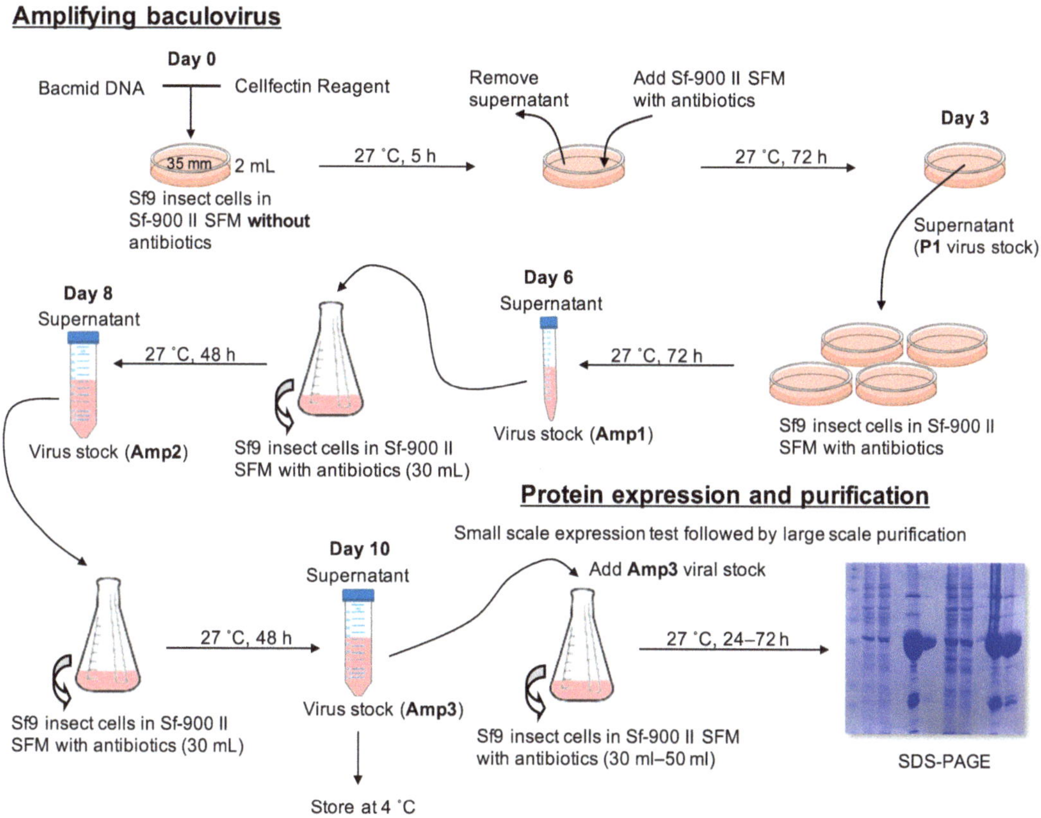

Fig. 2 Schematic diagram of baculovirus amplification and protein expression. Once the high molecular weight recombinant bacmid DNA containing the SH2 domain sequence is isolated, the bacmid DNA is transfected into Sf9 insect cells to generate the recombinant baculovirus. The baculoviral stock is amplified for several rounds, and the high-titer stock is used to infect insect cells for large-scale protein expression and purification of GST-SH2. Typically, the whole process outlined here takes 10–14 days

2 Materials

2.1 Subcloning of GST-SH2 Construct

1. SH2 domain cDNA (*see* **Note 1**).
2. Baculovirus expression vectors for insect cells: pFastBac1 (no tag), pFastBac-GST (N-terminal GST tag, custom made), pDEST8 (no tag), pDEST10 (N-terminal 6xHis tag), and pDEST20 (N-terminal GST tag) (*see* **Note 2**).
3. Luria-Bertani (LB)-ampicillin agar plate: LB agar plate with 100 μg/mL ampicillin.
4. LB (−): LB broth without ampicillin.
5. LB-ampicillin: LB broth with 50 μg/mL ampicillin.
6. DNA polymerase.
7. Custom-designed oligonucleotide primers (*see* **Note 3**).
8. 1–2 % agarose gel.

9. TAE (Tris–acetate–EDTA) buffer: 40 mM Tris–acetate, 1 mM ethylenediaminetetraacetic acid (EDTA).
10. SYBR Safe DNA gel stain.
11. T4 DNA ligase with 10× ligation buffer.
12. Restriction enzymes.
13. Alkaline phosphatase.
14. Competent bacteria (a *rec*A, *end*A *E. coli* strain such as TOP10, DH10B, or DH5α).

2.2 Evaluation of pFastBac-GST-SH2 Clones

1. LB-ampicillin: LB broth with 50 μg/mL ampicillin.
2. QIAprep Spin Miniprep Kit.
3. Restriction enzymes.
4. 1–2 % agarose gel.
5. TAE buffer: 40 mM Tris–acetate, 1 mM EDTA.
6. SYBR Safe DNA gel stain.
7. Sequencing primers: Forward 5′GGGCTGGCAAGCCACG TTTGGTG3′ (GEX5) and Reverse 5′ACAAATGTGGTATG GCTGATT3′ (pFB_R) (*see* **Note 4**).
8. Bacteria stock solution: 50 % glycerol (autoclaved).
9. Cryogenic tubes.

2.3 Generating GST-SH2 Recombinant Bacmid

1. Purified pFastBac-GST-SH2 construct (200 pg/μL).
2. DH10Bac competent cells (Life Technologies, Carlsbad, CA).
3. LB agar plates containing 50 μg/mL kanamycin, 7 μg/mL gentamicin, 10 μg/mL tetracycline, 100 μg/mL X-gal, and 40 μg/mL isopropyl-β-D-thiogalactoside (IPTG) (3 freshly prepared plates for each transformation).
4. S.O.C. medium.
5. 15 mL round-bottom polypropylene tubes.

2.4 Purification and Evaluation of GST-SH2 Recombinant Bacmid

1. Solution 1: 50 mM Tris–HCl pH 8.0, 10 mM EDTA, 100 μg/mL RNase A.
2. Solution 2: 200 mM NaOH, 1 % sodium dodecyl sulfate (SDS).
3. Solution 3: 3.3 M potassium acetate pH 5.5.
4. Isopropanol.
5. 70 % ethanol.
6. Tris–EDTA (TE) buffer: 10 mM Tris–HCl, 1 mM EDTA, pH 8.0.
7. Primers for PCR: GEX5 5′GGGCTGGCAAGCCACGTTT GGTG3′, pUC/M13 Forward 5′CCCAGTCACGACGTT GTAAAACG3′, pUC/M13 Reverse 5′AGCGGATAACAA TTTCACACAGG3′.
8. DNA polymerase.

9. 1% agarose gel.
10. TAE buffer: 40 mM Tris–acetate, 1 mM EDTA.
11. SYBR Safe DNA gel stain.
12. Bacteria stock solution: 50% glycerol (autoclaved).
13. Cryogenic tubes.

2.5 Bacmid Transfection and Virus Amplification

1. Sf9 insect cells.
2. CellFECTIN (Shake well before use) (Life Technologies, Carlsbad, CA).
3. Sf-900 II SFM, serum-free, protein-free insect cell culture medium (Life Technologies, Carlsbad, CA).
4. Sf-900 II SFM with antibiotics.
5. Bacmid DNA (500 ng/μL).
6. 35 mm dish.
7. 50 mL Falcon tube.
8. 15 mL Falcon tube.

2.6 Small Scale Expression Test

1. Sf9 insect cells: 3×10^6 cells/mL cell culture.
2. Sf-900 II SFM with antibiotics.
3. Viral stock solution (Amp3).
4. 5× sample buffer: 0.3 M Tris–HCl pH 6.8, 10% SDS, 25% β-mercaptoethanol, 0.1 mg/mL bromophenol blue, 45% glycerol.
5. 12% SDS-polyacrylamide gel electrophoresis (PAGE) mini gel.
6. Coomassie blue solution: 40% methanol, 10% acetic acid, 0.25% Coomassie Blue R-250.
7. Fixing solution: 20% methanol, 10% acetic acid.

2.7 Expression of GST-SH2 in Insect Cells

1. Sf9 insect cells: 5×10^6 cells/mL cell culture.
2. Sf-900 II SFM with antibiotics.
3. Viral stock solution (*Amp3*).
4. PBS with 10% glycerol.
5. 5× sample buffer.
6. 12% SDS-PAGE mini gel.
7. Coomassie blue solution.
8. Fixing solution.

2.8 Purification of GST-SH2 Protein

1. Lysis buffer (10–15 mL for 50 mL cell culture): Phosphate buffered saline (PBS) with 100 mM ethylene diamine tetraacetic acid (EDTA), 1% Triton X-100, 10% glycerol; add phenylmethyl sulfonyl fluoride (PMSF) to 1 mM, aprotinin to 1% v/v, and dithiothreitol (DTT) to 1 mM right before use.

2. 5× sample buffer.
3. Glutathione Sepharose 4B.
4. 12% SDS-PAGE mini gel.
5. Coomassie blue solution.
6. Fixing solution.
7. Elution buffer: 20 mM glutathione (reduced), 150 mM NaCl, 10% glycerol, 50 mM pH 8.8 Tris–HCl, 1 mM EDTA; add PMSF to 1 mM, aprotinin to 1% v/v, and DTT to 1 mM just before use; adjust pH to 8–9 with pH 9.6 Tris–HCl.
8. Centrifuge tube filter (Corning Costar Spin-X Plastic Centrifuge Tube Filters, Sigma-Aldrich, St. Louis, MO).
9. Sephadex G-25 PD-10 column.
10. PBS with 10% glycerol.
11. Bradford dye reagent.
12. Ultrafiltration membrane.

3 Methods

3.1 Subcloning of pFastBac GST-SH2 Construct

1. Retrieve SH2 domain DNA sequences from public databases such as NCBI Gene and Human Protein Reference Database (*see* **Note 5**).
2. Identify the location of the SH2 domain using the protein sequence at ScanProsite.
3. Identify the nucleotide sequence corresponding to the SH2 domain using a sequence viewer.
4. Obtain cDNA templates corresponding to the SH2 domains from academic or industry sources or perform molecular cloning (*see* **Note 1**).
5. Design primers for PCR.
6. Amplify the SH2 fragment by PCR using the designed primers and the cDNA template.
7. Check and purify the PCR product using 1–2% agarose gel.
8. Digest the PCR product with appropriate restriction enzymes and purify the digested PCR product from agarose gel.
9. Digest pFastBac-GST with appropriate restriction enzymes.
10. Incubate the digested vector with alkaline phosphatase for 1 h at 37 °C.
11. Purify the linearized and dephosphorylated pFastBac-GST vector from 1% agarose gel.
12. Mix 1 μL 10× ligation buffer, 50 ng linearized and dephosphorylated vector, equimolar amount of insert (*see* **Note 6**) and sterile water to 9.5 μL (*see* **Note 7**).

13. Incubate for 2 min at 65 °C to melt any ends that might have annealed, then place on ice.
14. Add 0.5 μL T4 ligase.
15. Incubate at 16 °C overnight or at room temperature for 1 h.
16. Heat-inactivate the ligase for 10 min at 65 °C then place on ice.
17. Thaw competent bacteria (50 μL) on ice.
18. Add 5 μL of ligation solution to the competent bacteria, gently mix, and incubate for 20 min on ice with occasional tapping.
19. Heat-shock at 42 °C for 45 s, then immediately transfer the tubes to ice and cool down for 2 min.
20. Add 500 μL LB (−) and incubate at 37 °C for 1 h.
21. Warm up LB-ampicillin plates in a 37 °C incubator before plating.
22. Centrifuge at low speed (800 × *g*) for 3 min to pellet bacteria.
23. Aspirate the supernatant, leaving ~50 μL with the pellet, and vortex.
24. Plate the transformed cell mixture on an LB-ampicillin plate and incubate at 37 °C overnight.

3.2 Evaluation of pFastBac-GST-SH2 Clones

1. Inoculate colonies from the LB plate into 4–6 mL LB-ampicillin and grow at 37 °C overnight. Mark the colonies on the plate for future use.
2. Purify the plasmids using miniprep kit (keep ~1 mL culture at 4 °C for **step 5**).
3. Digest the purified-plasmid DNA with appropriate restriction enzymes and check the size of DNA on agarose gel.
4. Subject positive clones to DNA sequencing.
5. After verifying the sequence of the pFastBac-GST-SH2 clone(s), mix the corresponding bacterial solutions (0.5–1 mL, from mid-logarithmic phase culture) with the equal amount of bacterial stock solution in cryogenic tubes and store them at −80 °C.

3.3 Generating Recombinant GST-SH2 Bacmid

Follow the procedure below to transform DH10Bac competent cells with your pFastBac-GST-SH2 construct.

1. Thaw DH10Bac competent cells on ice.
2. Gently mix and transfer 100 μL of the DH10Bac cells into a prechilled, 15 mL round-bottom polypropylene tube.
3. Add 1 ng (5 μL) of pFastBac-GST-SH2 to the cells and mix gently by tapping the tube (*see* **Note 8**).
4. Incubate the cells on ice for 30 min.
5. Heat-shock the cells for 45 s at 42 °C without shaking.

6. Immediately transfer the tubes to ice and cool down for 2 min. Add 900 μL of room temperature S.O.C. medium.
7. Shake tubes at 37 °C at 225 rpm for 4 h.
8. Prepare serial dilutions of the cells (10^{-1}, 10^{-2}, 10^{-3}) with S.O.C. medium. Plate 100 μL of **each** dilution on an LB agar plate containing 50 μg/mL kanamycin, 7 μg/mL gentamicin, 10 μg/mL tetracycline, 100 μg/mL X-gal, and 40 μg/mL IPTG.
9. Incubate the plates for 48 h at 37 °C. Pick several *white* colonies for analysis (*see* **Note 9**).

3.4 Purification and Evaluation of the GST-SH2 Recombinant Bacmid

1. Pick several white colonies and restreak them on fresh LB agar plates containing 50 μg/mL kanamycin, 7 μg/mL gentamicin, 10 μg/mL tetracycline, 100 μg/mL X-gal, and 40 μg/mL IPTG. Incubate the plates overnight at 37 °C.
2. Pick 1 colony and inoculate it into 2 mL of LB containing 50 μg/mL kanamycin, 7 μg/mL gentamicin, and 10 μg/mL tetracycline. Mark the colony on the plate for future use.
3. Incubate the culture at 37 °C with shaking (at 250 rpm) overnight.
4. Transfer 1.5 mL of bacterial culture to a 1.5 mL centrifuge tube and centrifuge at 9000 × *g* for 15 min (keep the remaining culture for **step 19**).
5. Remove the supernatant and resuspend the bacteria in 0.3 mL Solution 1.
6. Add 0.3 mL of Solution 2 and mix gently by inverting the tube several times. Incubate at room temperature for 5 min (*see* **Note 10**).
7. Add 0.3 mL of Solution 3 and mix immediately by inverting the tube until the mixture is homogeneous. Place the tube on ice for 5–10 min.
8. Centrifuge the mixture at 15,000 × *g* for 10 min.
9. Gently transfer the supernatant to a new microcentrifuge tube containing 0.8 mL of isopropanol (*see* **Note 11**). Invert the tube several times to mix and place it on ice for 5–10 min (*see* **Note 12**).
10. Centrifuge the sample for 15 min at 15,000 × *g* at room temperature.
11. Carefully remove the supernatant without disturbing the pellet. Add 0.5 mL of cold 70% ethanol. Invert the tube several times to wash the pellet.
12. Centrifuge at 15,000 × *g* for 5 min at room temperature.
13. Repeat **steps 11** and **12** to remove residual isopropanol.
14. Decant supernatant and air-dry pellet.

15. Resuspend the pellet in 30 μL TE buffer. Allow pellet to dissolve for at least 10 min on ice (*see* **Note 13**).
16. Store the bacmid at 4 °C. *Do not freeze* the bacmid to avoid shearing of the DNA.
17. To verify the presence of the GST-SH2 in the recombinant bacmid, run PCR using the following combinations of primers.
    (a) pUC/M13 Forward and Reverse primers.
    (b) GEX5 primer and pUC/M13 Reverse primer.
18. Run the PCR product on agarose gel (*see* **Notes 14** and **15**).
19. Mix the bacterial solution from the verified clone (0.5–1 mL, from mid-logarithmic phase culture) with the equal amount of bacterial stock solution in cryogenic tubes and store them at −80 °C.

3.5 Bacmid Transfection

As shown in Fig. 2, the verified bacmid containing the GST-SH2 gene is transfected into insect cells to produce the recombinant baculovirus. After several rounds of viral amplification, the high titer virus is stored. Refer to the Bac-to-Bac Baculovirus Expression System manual (Life Technologies) for more detailed protocols to perform the steps outlined below.

1. Dilute Sf9 cells from $1–2 \times 10^6$ to 4.5×10^5 cells/mL with Sf-900 II SFM without antibiotics.
2. Gently add 2 mL of diluted cell culture to a 35 mm dish.
3. Place the dish in a 27 °C incubator and incubate for 1 h without shaking.
4. 30 min after putting the dish in the incubator, start preparing the following solutions.
    (a) 1 μL DNA+ 100 μL Sf-900 II SFM without antibiotics.
    (b) 6 μL CellFECTIN + 100 μL Sf-900 II SFM without antibiotics.
5. Gently mix the DNA and CellFECTIN solutions and incubate for 15–45 min at room temperature.
6. Add 0.8 mL Sf-900 II SFM without antibiotics to the DNA/CellFECTIN mixture and mix gently.
7. Remove the supernatant from the 35 mm dish in **step 3** (cells should have settled down to the bottom of the plate) and gently add 2 mL Sf-900 II SFM without antibiotics.
8. Shake the dish gently and remove the supernatant again.
9. Gently add the DNA/CellFECTIN mixture to the dish.
10. Place the dish in a 27 °C incubator and incubate for 5 h without shaking.

11. Remove the supernatant and add 2 mL Sf-900 II SFM *with* antibiotics.
12. Incubate for 72 h in a 27 °C incubator without shaking and collect the media (*P1* virus stock).

3.6 Virus Amplification

1. Dilute Sf9 cells ($1–2 \times 10^6$ cells/mL) to 7.5×10^5 cells/mL with Sf-900 II SFM with antibiotics.
2. Gently add 1.5 mL of the diluted cell culture to four dishes (35 mm).
3. Add 0.5 mL of *P1* virus to each dish.
4. Incubate the plate in a 27 °C incubator without shaking for 72 h.
5. Collect the media from all four dishes and centrifuge at $500 \times g$ for 3 min to remove cells and large debris (*Amp1*).
6. Add *Amp1* virus and fresh insect cells to a 125 mL culture flask as outlined below.

3×10^6 cells/mL cell culture	10 mL
Virus containing medium (*Amp1*)	8 mL
Fresh Sf-900 II SFM with antibiotics	12 mL

7. Incubate the flask in a 27 °C incubator *with* shaking for 48 h.
8. Collect the virus containing medium and centrifuge at $500 \times g$ for 3 min to remove cells and large debris (*Amp2*).
9. Add *Amp2* virus and fresh insect cells to a 125 mL culture flask as outlined below.

3×10^6 cells/mL cell culture	10 mL
Virus containing medium (*Amp2*)	10 mL
Fresh Sf-900 II SFM with antibiotics	10 mL

10. Incubate the flask in a 27 °C incubator *with* shaking for 48 h.
11. Collect the virus containing medium and centrifuge at $500 \times g$ for 3 min to remove cells and large debris (*Amp3*).
12. Store all viral solutions at 4 °C (*see* **Note 16**).

3.7 Small Scale Expression Test

1. Add virus containing medium (*Amp3*) to 6 mL 1×10^6 cells/mL cell culture.

   Example:

3×10^6 cells/mL cell culture	2 mL
Viral stock solution (Amp3)	0.2 mL
Fresh Sf-900 II SFM with antibiotics	3.8 mL

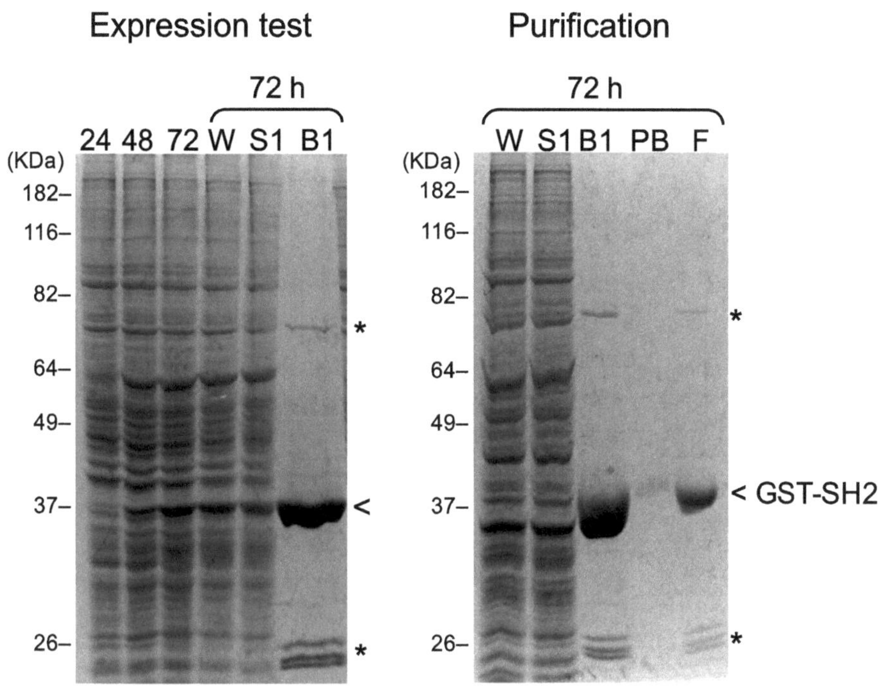

Fig. 3 Expression and purification of GST-SH2 using baculovirus-insect cell expression system. *Left*, insect cells were harvested at 24, 48, and 72 h and expression of GST-SH2 was determined using a small amount of cells and Coomassie blue gel staining. GST-SH2 domains appears as ~40 kDa proteins and co-purified proteins are indicated by *asterisks* [9]. *Right*, fractions from the large scale purification. *W* whole cell fraction, *S1* soluble fraction, *B1* glutathione bead-bound fraction, *PB* post elution bead fraction, *F* final product

2. Incubate the flask in a 27 °C incubator with shaking 24–72 h.
3. At 24 h, 48 h, and 72 h after infection, collect 100 μL of cell culture and centrifuge at 500 × *g* for 3 min. Remove the supernatant and add 100 μL of 1× sample buffer.
4. Boil the samples for 3 min, and check the expression by SDS-PAGE.
5. Identify optimal incubation time with an ideal expression level (Fig. 3).

3.8 Expression of GST-SH2 in Insect Cells

1. Add *Amp3* virus and fresh insect cells to a 125 mL culture flask as outlined below.

5×10^6 cells/mL cell culture	10 mL
Viral stock solution (*Amp3*)	1 mL
Fresh Sf-900 II SFM with antibiotics	39 mL

2. Incubate the flask in a 27 °C incubator with shaking for 48–72 h (use optimal incubation time from Subheading 3.7).

3. Harvest the cells and centrifuge at 500 × *g* for 3 min.
4. Add PBS with 10% glycerol and centrifuge again at 500 × *g* for 5 min.
5. Aspirate the supernatant and store the pellet at –80 °C.

3.9 Purification of GST-SH2 Protein

1. Resuspend insect cells in ice-cold lysis buffer (10–15 mL for 50 mL cell culture pellet).
2. Vortex and leave the sample on ice for 10 min.
3. Remove 10 μL of total lysates and add 2.5 μL of 5× sample buffer for SDS-PAGE (whole cell fraction *W*, Fig. 3).
4. Centrifuge the remaining sample at 5000 × *g* at 4 °C for 10 min to pellet the debris and transfer the supernatant to a new tube.
5. Take 10 μL of supernatant, add 2.5 μL 5× sample buffer for SDS-PAGE (soluble fraction *S1*, Fig. 3).
6. (Optional: **step 6–8**) Add 10–15 mL lysis buffer to the pellet from **step 4**, vortex, and leave for 10 min on ice.
7. Centrifuge the remaining sample at 5000 × *g* at 4 °C for 10 min and transfer the supernatant to a new tube. This fraction may contain a lower concentration of fusion protein (save for future use).
8. Take 10 μL of supernatant, add 2.5 μL 5× sample buffer for SDS-PAGE (soluble fraction *S2*).
9. Add Glutathione Sepharose 4B to the supernatant from **step 4**. Use 200–500 μL of bead slurry per 50 mL of original culture.
10. Rotate at 4 °C for about 1 h (up to 2 h).
11. Wash beads with cold lysis buffer: spin down the beads at low speed (200–400 × *g*), remove the supernatant, and resuspend the beads in fresh buffer. Repeat 3–5 times.
12. Transfer the beads to a new 1.5 mL tube.
13. Take 10 μL of beads solution and add 2.5 μL of 5× sample buffer for SDS-PAGE (glutathione bead-bound fraction *B1*, Fig. 3).
14. Spin down the remaining beads solution and remove the supernatant.
15. Add 500 μL of elution buffer to the beads and rotate at 4 °C for 15 min.
16. Spin out beads and transfer supernatant to a new Eppendorf tube on ice (*see* **Note 17**).
17. Repeat **steps 15** and **16** for the second elution.
18. Repeat **steps 15** and **16** for the third elution.

19. Take 10 μL of beads after elution and add 2.5 μL 5× sample buffer for SDS-PAGE (post elution bead fraction *PB*, Fig. 3).
20. Quickly check the relative protein concentration of the eluates using Bradford assay dye. Pool the fractions that show a bright blue color (*see* **Notes 18** and **20**).
21. Perform buffer exchange by gel filtration using a Sephadex G-25 PD-10 column according to the supplier's instructions. Briefly, equilibrate the column with approximately 25 mL PBS-10% glycerol. Discard the flow through. Add your eluted protein sample to the column and allow the whole sample to flow into the column. Elute with PBS-10% glycerol and collect sequential 0.5 mL fractions of the elute in separate tubes (*see* **Note 19**).
22. Quickly check the relative protein concentration of each tube with the Bradford dye reagent. Combine fractions that show a bright blue color into a single tube (*see* **Note 20**).
23. Take 10 μL of combined eluate and add 2.5 μL of 5× sample buffer for SDS-PAGE (final product fraction *F*, Fig. 3).
24. Determine the protein concentration using Bradford assay or absorbance at 280 nm and store at −80 °C (*see* **Note 21**).
25. Boil *W*, *S1*, *B1*, *PB*, and *F* sample buffer fractions (from **steps 3**, **5**, **13**, **19**, and **23**) and evaluate the purification by 12% SDS-PAGE and Coomassie staining (Fig. 3).

4 Notes

1. Sources for human SH2 domain cDNA include Addgene (e.g., Mayer lab clones), DNASU, GE Dharmacon, and the Pawson Lab.
2. Choose an appropriate expression vector for your purpose. This protocol is based on a custom-made pFastBac-GST vector, which is compatible with the Bac-to-Bac Baculovirus Expression System and purification of GST fusion protein. Alternatively, the GST-tagged SH2 gene can be transferred from other expression vectors such as pGEX-SH2 [7]. For Gateway expression vectors (pDEST8, 10, 20), Gateway entry clones containing SH2 domain ORFs should be prepared in advance.
3. Regarding the target sequence for PCR, the amino acids surrounding the SH2 domain should be carefully determined as they can significantly affect solubility and expression of recombinant SH2 proteins. Empirically, including additional amino acids at the boundaries of the domain is more beneficial than removing a few residues. We usually add 5–10 amino

acids to the N- and C-terminal boundaries of SH2 domains. SH2 domain structural information (e.g., Protein Data Bank) can provide clues about the location of boundaries and how to fuse the SH2 to GST (for example a flexible loop region might be better than a rigid secondary structure segment).

4. GEX5 primer recognizes the C-terminal region of GST. pFB_R primer recognizes the SV40 poly(A) signal region of the pFastBac vector (Fig. 1).
5. All human and mouse SH2 domain sequences have been identified [8]. For SH2 domain NCBI Gene IDs and amino acid sequences, *see* Chapter 8. More information about SH2 domains including phylogenetic relationships and multiple alignment SH2 domains is available at the SH2 Domain Site (sites.google.com/site/sh2domain/). To determine the presence of SH2 domains in unknown sequences, use the SMART web-based tool (smart.embl-heidelberg.de) or ScanProsite (us.expasy.org/cgi-bin/scanprosite) or *see* Chapter 3 or 4.
6. If the vector is 5.0 kb and the insert is 0.4 kb, equimolar amount would be $0.4/5.0 \times 50$ ng = 4 ng. This does not have to be exact. In some cases, more or less insert works better.
7. Always set up a ligation with the digested vector without the insert as a control. This will tell you how much background occurred as a result of self-ligation. If the background is high, re-digest and phosphatase treat your vector for a longer period of time to ensure the vector does not re-ligate to itself.
8. Do not pipette up and down or vortex to mix, this can damage cells and reduce the transformation efficiency.
9. In this system, transposition on the bacmid disrupts expression of the lacZα gene. As a result, white colonies indicate the presence of the recombinant bacmid (i.e., positive clone) and blue colonies indicate the unaltered parent bacmid (i.e., background).
10. The appearance of the suspension should change from turbid to almost translucent.
11. Do not transfer any white precipitate.
12. You may proceed directly to **step 10** or store sample at −20 °C overnight.
13. To avoid shearing the DNA, pipette only 1–2 times for resuspension.
14. The pUC/M13 Forward and Reverse PCR is performed to verify transposition and insert size. PCR of a positive clone will amplify a ~3400 bp fragment containing the GST-SH2 insert. A negative clone will amplify a ~300 bp fragment.

   (a) Bacmid transposed with GST-SH2: ~2300 bp + GST-SH2 (~1100 bp).
   (b) Bacmid only: ~300 bp.

15. The GEX5 and pUC/M13 Reverse PCR is performed to verify the presence of GST and the insert directionality. PCR of a positive clone will amplify a ~1150 bp fragment.
16. For long-term storage, add serum (final concentration 2 %) or BSA (final concentration 0.1–0.5 %) to the viral solution and keep at 4 °C. You can determine virus titer by plaque assay. We instead perform a small scale protein expression test as Subheading 3.7.
17. Try not to suck up or remove the beads when removing the supernatant or wash. Accidentally transferred beads can be removed using a centrifuge tube filter.
18. If elution efficiency is inefficient, incubate insect cell lysate with glutathione beads for a shorter period of time (~30 min) or use elution buffer with higher pH (up to 9.6), higher glutathione concentration (up to 50 mM), and add detergent to 0.1 % (e.g., Triton X, Tween 20, or NP40). If the native elution solution does not work, a denaturing solution (3–6 M guanidine hydrochloride, 20 mM sodium phosphate, 500 mM NaCl, pH 7.8) may be used at the expense of potential activity loss.
19. Alternatively, the eluate can be dialyzed at 4 °C overnight in several large volumes of PBS-10 % glycerol.
20. To quickly check the relative protein concentration, add 2 μL of each PD-10 elution fraction to 50 μL of diluted Bradford dye and vortex. Pool the top three or four fractions with the brightest blue color. If the color change is not obvious (this occurs when protein concentration is less than 0.2 μg/μL), take 3–5 fractions and concentrate by ultrafiltration (e.g., Amicon Ultra-4 Centrifugal Filter Unit).
21. If protein concentration is low (<0.5 μg/μL), the sample can be concentrated by ultrafiltration.

Acknowledgment

We thank Joshua Jadwin and Silas Khong Ng for assistance with editing the manuscript, and Bruce Mayer for his continuous encouragement and support. This study was partly supported by grant CA1154966 from the National Institutes of Health and Quest for CURES (QFC) grant from the Leukemia and Lymphoma Society (to K.M.).

References

1. Hitchman RB, Possee RD, King LA (2009) Baculovirus expression systems for recombinant protein production in insect cells. Recent Pat Biotechnol 3(1):46–54
2. Jarvis DL (2009) Baculovirus-insect cell expression systems. Methods Enzymol 463:191–222. doi:10.1016/s0076-6879(09)63014-7
3. van Oers MM, Pijlman GP, Vlak JM (2015) Thirty years of baculovirus-insect cell protein expression: from dark horse to mainstream technology. J Gen Virol 96(Pt 1):6–23. doi:10.1099/vir.0.067108-0
4. Shao H, Cheng HY, Cook RG, Tweardy DJ (2003) Identification and characterization of signal transducer and activator of transcription 3 recruitment sites within the epidermal growth factor receptor. Cancer Res 63(14):3923–3930
5. Joseph RE, Min L, Andreotti AH (2007) The linker between SH2 and kinase domains positively regulates catalysis of the Tec family kinases. Biochemistry 46(18):5455–5462. doi:10.1021/bi602512e
6. Zhao Y, Zhang X, Chen Y, Lu S, Peng Y, Wang X, Guo C, Zhou A, Zhang J, Luo Y, Shen Q, Ding J, Meng L, Zhang J (2014) Crystal structures of PI3Kalpha complexed with PI103 and its derivatives: new directions for inhibitors design. ACS Med Chem Lett 5(2):138–142. doi:10.1021/ml400378e
7. Machida K, Thompson CM, Dierck K, Jablonowski K, Karkkainen S, Liu B, Zhang H, Nash PD, Newman DK, Nollau P, Pawson T, Renkema GH, Saksela K, Schiller MR, Shin DG, Mayer BJ (2007) High-throughput phosphotyrosine profiling using SH2 domains. Mol Cell 26(6):899–915. doi:10.1016/j.molcel.2007.05.031
8. Liu BA, Jablonowski K, Raina M, Arce M, Pawson T, Nash PD (2006) The human and mouse complement of SH2 domain proteins-establishing the boundaries of phosphotyrosine signaling. Mol Cell 22(6):851–868. doi:10.1016/j.molcel.2006.06.001
9. Bichet P, Mollat P, Capdevila C, Sarubbi E (2000) Endogenous glutathione-binding proteins of insect cell lines: characterization and removal from glutathione S-transferase (GST) fusion proteins. Protein Express Purif 19(1):197–201. doi:10.1006/prep.2000.1239

Chapter 12

Functionally Altered SH2 Domains for Biochemical Studies: Loss-of-Function Mutant and Domain Concatenation

Mari Ogiue-Ikeda and Kazuya Machida

Abstract

Recombinant modular protein domains have been a convenient proteomics tool for deciphering protein-protein interactions and elucidating the role of protein modifications in cell signaling. To obtain reliable experimental data, these protein domain probes require sufficient specificity and sensitivity. Since naturally evolved protein domains do not always have optimal biochemical characteristics for in vitro assays, functional alterations such as improved affinity are sometimes needed. In this chapter, we describe preparation of loss-of-function and concatenated (tandem) SH2 domains that should be widely applicable to both high- and low-throughput phosphoproteomics studies.

Key words SH2 domain, Loss of function, Domain concatenation, Tandem SH2

1 Introduction

Eukaryotic proteins are modular in nature, built on independently folded globular domains such as kinase and Src Homology 2 (SH2). Purified modular protein domains, commonly obtained using bacterial expression systems, have been an inexpensive and convenient tool for cell signaling studies. In particular, their ability to function as affinity reagents has made them very useful for array-based and MS-based proteomics studies [1]. However, physiological interactions between signaling proteins mediated by domain–motif binding are often weak and transient which may hamper their unequivocal detection in modular domain binding assays [2]. For example, unlike antibodies that are purposely selected for high affinity, naturally occurring protein modules do not always provide optimal signal-to-noise characteristics as desired for in vitro assays. To address this issue, functional alteration of modular domains to gain optimal sensitivity and specificity can be considered.

SH2 domains, the most prevalent type of phosphotyrosine binding module in eukaryotic cells, have been used in multiple phosphoproteomics platforms including forward- and reverse-phase arrays,

Kazuya Machida and Bernard A. Liu (eds.), *SH2 Domains: Methods and Protocols*, Methods in Molecular Biology, vol. 1555,
DOI 10.1007/978-1-4939-6762-9_12, © Springer Science+Business Media LLC 2017

in-solution assays, and bead-based assays [3–9]. Purified SH2 domains have also been proposed for use in molecular diagnostics to profile cancer patient specimens based on their signaling states. The level of tyrosine phosphorylation in patient specimens, however, is inherently low likely due to instability of the phosphorylation state and high phosphatase activity in surgically removed tissues, posing a challenge to the control of signal and noise in the assay [10, 11]. To obtain dependable experimental data using recombinant SH2 domain proteins, several functionally altered SH2 domains can be engineered. First, a loss-of-function SH2 domain mutant is useful as a negative control to assess assay specificity. Next, concatenation of SH2 domains by aligning multiple domains in the same protein (engineered tandem domains) can be an effective way to improve signal [12]. In the following sections, we describe methods to create these modified SH2 domain proteins. In addition, another approach to increase the affinity of SH2 domains, i.e., superbinders, is described in Chapter 13.

2 Materials

2.1 Site Directed Mutagenesis

1. Plasmid containing SH2 domain cDNA (*see* **Note 1**). This protocol is based on pGEX vector containing an ampicillin-resistance gene.
2. Molecular biology grade water.
3. Phusion High-Fidelity DNA Polymerase and 5× Phusion Buffer (New England Biolabs).
4. 10× dNTP (adjust each to 2 mM).
5. Custom-designed oligonucleotide primers (*see* **Note 2**).
6. DpnI restriction enzyme.
7. Competent bacteria (dam + E. coli strain such as DH5α).
8. Luria-Bertani (LB)-ampicillin agar plate: LB agar plate with 100 μg/mL ampicillin.
9. LB (−): LB broth without antibiotics.
10. LB-ampicillin: LB broth with 50 μg/mL ampicillin.
11. Miniprep kit.
12. PCR purification kit.
13. Sequencing primers.

2.2 Domain Concatenation

1. Plasmid containing SH2 domain cDNA (*see* **Note 3**). This protocol is based on pGEX vector containing an ampicillin-resistance gene.
2. DNA polymerase.
3. Custom-designed oligonucleotide primers (*see* **Note 4**).

4. 1–2 % agarose gel.
5. TAE buffer: 40 mM Tris, 20 mM acetic acid, 1 mM EDTA.
6. SYBR Safe DNA gel stain (Life Technologies).
7. T4 DNA ligase with the reaction buffer.
8. Restriction enzymes.
9. Alkaline phosphatase.
10. Competent bacteria.
11. LB-ampicillin agar plate.
12. LB (−).
13. LB-ampicillin.
14. Miniprep kit.
15. Gel extraction kit.
16. Sequencing primers.

3 Methods

3.1 Site Directed Mutagenesis

The loss-of-function SH2 mutant can be generated by introducing a point mutation to the conserved βB5 arginine residue as part of the **FLVRES** motif which is crucial for the pTyr recognition [13, 14] (Table 1). Other residues of SH2 domains can be targeted for mutagenesis based on structural and genomic information although the functional consequences of other mutations need to be determined on a case by case basis [15–18].

1. Identify the conserved βB5 arginine residue in your SH2 domain sequence. Table 1 provides human SH2 domain amino acid and putative **FLVRES** motif sequences (*see* **Note 5**). Retrieve cDNA sequence from NCBI Gene site using either gene symbol or Gene ID.
2. Synthesize primers for PCR-based mutagenesis with the following characteristics (Fig. 1).
   (a) A total length of 25–45 bp with a melting temperature (Tm) ≥ 78 °C.
   (b) A point mutation for arginine to lysine substitution into both primers (*see* **Note 6**).
   (c) At least four bases between the mutation and the 5′ end.
   (d) At least eight bases between the mutation and the 3′ end.
   (e) At least eight non-overlapping bases at the 3′ end of each primer.
   (f) At least one G or C at the 3′ and 5′ ends of each primer.
3. Miniprep the SH2 domain plasmid.

Table 1
Amino acid sequence and the FLVRES motif of human SH2 domain proteins

No.	Protein	Gene symbol	Gene ID	Amino acid sequences of SH2 domains (bold: SH2 domain; *Arg betaB5 position)
1	Abl	ABL1	25	PSNYITPVNSLEKHS**WYHGPVSRNAAEYLLSSGINGSFLVR*ESESSPGQRSISLRYEGRVYHYRINTASDGKLYVSSESRFNTLAELVHHHSTVADGLITTLHYPA**PKRNKPTVYGVSPNY
2	Arg	ABL2	27	VPSNYITPVNSLEKHS**WYHGPVSRSAAEYLLSSLINGSFLVR*ESESSPGQLSISLRYEGRVYHYRINTTADGKVYVTAESRFSTLAELVHHHSTVADGLVTTLHYPA**PKCNKPTVYGVSPIHD
3	Nsp2/SH2D3B	BCAR3	8412	KELEEELLLSSEDLRS**HAWYHGRIPRQVSENLVQRDGDFLVR*DSLSSPGNFVLTCQWKNLAQHFKINRTVLRLSEAYSRVQYQFEMESFDSIPGLVRCYVGNRRPISQQSGAIIFQPI**NRTVPLR
4	Blk	BLK	640	FVARVESLEMER**WFFRSQGRKEAERQLLAPINKAGSFLIR*ESETNKGAFSLSVKDVTTQGELIKHYKIRCLDEGGYYISPRITFPSLQALVQHYSKKGDGLCQRLTLPC**VRPAPQNPWAQ
5	Blnk	BLNK	29760	TISEQEAGVLCKP**WYAGACDRKSAEEALHRSNKDGSFLIR*KSSGHDSKQPYTLVVFFNKRVYNIPVRFIEATKQYALGRKKNGEEYFGSVAEIIRNHQHSPLVLIDSQNNTKDSTRLKYAV**KVS
6	Bmx	BMX	660	SSEEEENLDDYD**WFAGNISRSQSEQLLRQKGKEGAFMVR*NSSQVGMYTVSLFSKAVNDKKGTVKHYHVHTNAENKLYLAENYCFDSIPKLIHYHQHNSAGMITRLRHPV**STKANKVPDSVS
7	Btk	BTK	695	PSNYVTEAEDSIEMYE**WYSKHMTRSQAEQLLKQEGKEGGFIVR*DSSKAGKYTVSVFAKSTGDPQGVIRHYVVCSTPQSQYYLAEKHLFSTIPELINYHQHNSAGLISRLKYPV**SQQNKNAPSTAGLGYGSWE
8	CblA (SH2)	CBL	867	NDYISVFEFDIFTRLF**QPWSSLLRNWNSLAVTHPGYMAFLTYDEVKARLQKFIHKPGSYIFR*LSCTRLGQWAIGYVTADGNILQTIPHNKPLFQALIDGFREGFYLFPDGRNQNPDLTG**LCEPTPQD
9	CblA (TKB)	CBL	867	DAFQPHHHHHHHLSPH**PPGTVDKKMVEKCWKLMDKVVRLCQNPKLALKNSPPYILDLLPDTYQHLRTILSRYEGKMETLGENEYFRVFMENLMKKTKQTISLFKEGKERMYEENSQPRRNLTKLSLIFSHMLAELKGIFPSGLFQGDTFRITKADAAEFWRKAFGEKTIVPWKSFRQALHEVHPISSGLEAMALKSTIDLTCNDYISVFEFDIFTRLFQPWSSLLRNWNSLAVTHPGYMAFLTYDEVKARLQKFIHKPGSYIFR*LSCTRLGQWAIGYVTADGNILQTIPHNKPLFQALIDGFREGFYLFPDGRNQNPDLTG**LCEPTPQD

10	CblB (SH2)	CBLB	868	NDYISVFEFDIFTRLFQPWGSILRNWNFLAVTHPGYMAFLTYDEVKARLQKYSTKPGSYIFR*LSCTRLGQWAIGYVTGDGNILQTIPHNKPLFQALIDGSREGFYLYPDGRSYNPDLTGLCEPTPHD
11	CblB (TKB)	CBLB	868	DAIQDAVGPPKQAAADRRTVEKTWKLMDKVVRLCQNPKLQLKNSPPYILDILPDTYQHLRLILSKYDDNQKLAQLSENEYFKIYIDSLMKKSKRAIRLFKEGKERMYEEQSQDRRNLTKLSLIFSHMLAEIKAIFPNGQFQGDNFRITKADAAEFWRKFFGDKTIVPWKVFRQCLHEVHQISSSLEAMALKSTIDLTCNDYISVFEFDIFTRLFQPWGSILRNWNFLAVTHPGYMAFLTYDEVKARLQKYSTKPGSYIFR*LSCTRLGQWAIGYVTGDGNILQTIPHNKPLFQALIDGSREGFYLYPDGRSYNPDLTGLCEPTPHD
12	CblC (SH2)	CBLC	23624	DLTCSGHVSIFEFDVFTRLFQPWPTLLKNWQLLAVNHPGYMAFLTYDEVQERLQACRDKPGSYIFR*PSCTRLGQWAIGYVSSDGSILQTIPANKPLSQVLLEGQKDGFYLYPDGKTHNPDLTELGQAEPQQR
13	CblC (TKB)	CBLC	23624	MALAVAPWGRQWEEARALGRAVRMLQRLEEQCVDPRLSVSPPSLRDLLPRTAQLLREVAHSRRAAGGGGPGGPGGSGDFLLIYLANLEAKSRQVAALLPPRGRRSANDELFRAGSRLRRQLAKLAIIFSHMHAELHALFPGGKYCGHMYQLTKAPAHTFWRESCGARCVLPWAEFESLLGTCHPVEPGCTALALRTTIDLTCSGHVSIFEFDVFTRLFQPWPTLLKNWQLLAVNHPGYMAFLTYDEVQERLQACRDKPGSYIFR*PSCTRLGQWAIGYVSSDGSILQTIPANKPLSQVLLEGQKDGFYLYPDGKTHNPDLTELGQAEPQQR
14	Chimerin	CHN1	1123	TCEVENRPKYYGREFHGMISREAADQLLIVAEGSYLIR*ESQRQPGTYTLALRFGSQTRNFRLYYDGKHFVGEKRFESIHDLVTDGLITLYIETKAAEYIAKMTINPIYEHVGYTTLNRE
15	Chimerin2	CHN2	1124	VENRPKYYGREFHGIISREQADELLGGVEGAYILR*ESQRQPGCYTLALRFGNQTLNYRLFHDGKHFVGEKRFESIHDLVTDGLITLYIETKAAEYISKMTTNPIYEHIGYATLLREK
16	Cis1/Cish	CISH	1154	IAKTFSYLRESGWYWGSITASEARQHLQKMPEGTFLVR*DSTHPSYLFTLSVKTTRGPTNVRIEYADSSFRLDSNCLSRPRILAFPDVVSLVQHYVASCTADTRSDSPDPAPTPAL
17	Mist/Clnk	CLNK	116449	KRSDRKDVQHNEWYIGEYSRQAVEEAFMKENKDGSFLVR*DCSTKSKEEPYVLAVFYENKVYNVKIRFLERNQQFALGTGLRGDEKFDSVEDIIEHYKNFPIILIDGKDKTGVHRKQCHLTQPLPLTRHLLPL
18	Crk	CRK	1398	MAGNFDSEERSSWYWGRLSRQEAVALLQGQRHGVFLVR*DSSTSPGDYVLSVSENSRVSHYIINSSGPRPPVPPSPAQPPPGVSPSRLRIGDQEFDSLPALLEFYKIHYLDTTTLIEPVSRSRQGSGVIL

(continued)

Table 1
(continued)

No.	Protein	Gene symbol	Gene ID	Amino acid sequences of SH2 domains (bold: SH2 domain; *Arg betaB5 position)
19	CrkL	CRKL	1399	MSSARFDSSDRSA**WYMGPVSRQEAQTRLQGQRHGMFLVR*DSSTCPGDYVLSVSENSRVSHYIINSLPNRRFKIGDQEFDHLPALLEFYKIHYLDTTTLIEPA**PRYPSPPMGSVS
20	Csk	CSK	1445	GVKAGTKLSLMP**WFHGKITREQAERLLYPPETGLFLVR*ESTNYPGDYTLCVSCDGKVEHYRIMYHASKLSIDEEVYFENLMQLVEHYTSDADGLCTRLIKPK**VMEGTVAAQ
21	Dapp1	DAPP1	27071	WSRSDGEAELLQDLG**WYHGNLTRHAAEALLLSNGCDGSYLLR*DSNETTGLYSLSVRAKDSVKHFHVEYTGYSFKFGFNEFSSLKDFVKHFANQPLIGSETGTLMVLKHPY**PRKVEEPSIYESVRV
22	Fer	FER	2241	DMISISEKPLAEQD**WYHGAIPRIEAQELLKKQGDFLVR*ESHGKPGEYVLSVYSDGQRRHFIIQYVDNMYRFEGTGFSNIPQLIDHHYTTKQVITKKSGVVLLNPI**PKDKKWILS
23	Fes/Fps	FES	2242	EVQKPLHEQL**WYHGAIPRAEVAELLVHSGDFLVR*ESQGKQEYVLSVLWDGLPRHFIIQSLDNLYRLEGEGFPSIPLLIDHLLSTQQPLTKKSGVVLHRAV**PKDKWVLNHED
24	Fgr/Src2	FGR	2268	PSNYVAPVDSIQAEE**WYFGKIGRKDAERQLLSPGNPQGAFLIR*ESETTKGAYSLSIRDWDQTRGDHVKHYKIRKLDMGGYYITTRVQFNSVQELVQHYMEVNDGLCNLLIAPC**TIMKPQTLGLAKDA
25	Frk	FRK	2444	YVAEDRSLQAEP**WFFGAIGRSDAEKQLLYSENKTGSFLIR*ESESQKGEFSLSVLDGAVVKHYRIKRLDEGGFFLTRRRIFSTLNEFVSHYTKTSDGLCVKLGKPC**LKIQVPAPFDLSY
26	FynA	FYN	2534	IPSNYVAPVDSIQAEE**WYFGKLGRKDAERQLLSFGNPRGTFLIR*ESETTKGAYSLSIRDWDDMKGDHVKHYKIRKLDNGGYYITTRAQFETLQQLVQHYSGTWNGNTKVAIKTLKPG**TMSPESFLEEAQ
27	Grap	GRAP	10750	KNYIRVKP**HPWYSGRISRQLAEEILMKRNHLGAFLIR*ESESSPGEFSVSVNYGDQVQHFKVLREASGKYFLWEEKFNSLNELVDFYRTTTIAKKRQIFLRDEEPL**LKSPGA
28	GrapL	GRAPL	400581	KNYIRVKP**HPWYSGRISRQLAEEILMKRNHLGAFLIR*ESESSPGEFSVSVNNRAQRGPCLGPKSHSRLG**
29	Grap2/Gads	GRAP2	9402	PKNYIRVKP**HPWYSGRISRQLAEEILMKRNHLGAFLIR*ESESSPGEFSVSVNYGDQVQHFKVLREASGKYFLWEEKFNSLNELVDFYRTTTIAKKRQIFLRDEEPL**LKSPGA

30	Grb10	GRB10	2887	STLSTVIHRTQHWFHGRISREESHRIIKQQGLVDGLFLLR*DSQSNPKAFVLTLCHHQKI KNFQILPCEDDGQTFFSLDDGNTKFSDLIQLVDFYQLNKGVLPCKLKHHCIRVAL
31	Grb14	GRB14	2888	SQSSATNMAIHRSQPWFHHKISRDEAQRLIIQQGLVDGVFLVR*DSQSNPKTFVLSMSHG QKIKHFQIIPVEDDGEMFHTLDDGHTRFTDLIQLVEFYQLNKGVLPCKLKHYCARIAL
32	Grb2/Ash	GRB2	2885	IPKNYIEMKPHPWFFGKIPRAKAEEMLSKQRHDGAFLIR*ESESAPGDFSLSVKFGNDVQH FKVLRDGAGKYFLWVVKFNSLNELVDYHRSTSVSRNQQIFLRDIEQVPQQPTY
33	Grb7	GRB7	2886	SGTSLSAAIHRTQLWFHGRISREESQRLIGQQGLVDGLFLVR*ESQRNPQGFVLSLCHLQ KVKHYLILPSEEEGRLYFSMDDGQTRFTDLLQLVEFHQLNRGILPCLLRHCCTRVAL
34	Hck	HCK	3055	ARVDSLETEEWFFKGISRKDAERQLLAPGNMLGSFMIR*DSETTKGSYSLSVRDYDPRQG DTVKHYKIRTLDNGGFYISPRSTFSTLQELVDHYKKGNDGLCQKLSVPCMSSKPQKPWEKD
35	Hsh2	HSH2D	84941	DWFVHTQMGQLAQDGVPEWFHGAISREDAENLLESQPLGSFLIR*VSHSHVGYTLSYKA QSSCCHFMVKLLDDGTFMIPGEKVAHTSLDALVTFHQQKPIEPRRELLTQPCRQKDPAN
36	Ship1/Inpp5d	INPP5D	3635	MVPCWNHGNITRSKAEELLSRTGKDGSFLVR*ASESISRAYALCVLYRNCVYTYRILPNED DKFTVQASEGVSMRFFTKLDQLIEFYKKENMGLVTHLQYPVPLEEEDTGDDPEEDT
37	Ship2/Inppl1	INPPL1	3636	MASACGAPGPGGALGSQAPSWYHRDLSRAAAEELLARAGRDGSFLVR*DSESVAGAFALC VLYQKHVHTYRILPDGEDFLAVQTSQGVPVRRFQTLGELIGLYAQPNQGLVCALLLPV EGEREPDPPDDRD
38	Emt/Itk	ITK	3702	PSSYLVEKSPNNLETYEWYNKSISRDKAEKLLLDTGKEGAFMVR*DSRTAGTYTVSVFTKA VVSENNPCIKHYHIKETNDNPKRYYVAEKYVFDSIPLLINYHQHNGGGLVTRLRYPVCF GRQKAPVTAG
39	Jak1	JAK1	3716	DVAPPLIVHNIQNGCHGPICTEYAINKLRQEGSEEGMYVLR*WSCTDFDNILMTVTCFEK SEQVQGAQKQFKQIEVQKGRYSLHGSDRSFPSLGDLMSHLKKQILRTDNISFMLKRCC QPKPREISNLLVATKKAQEWQ
40	Jak2	JAK2	3717	PPAVLENIQSNCHGPISMDFAISKLKKAGNQTGLYVLR*CSPKDFNKYFLTFAVERENVIE YKHCLITKNENEEYNLSGTKKNFSSLKDLLNCYQMETVRSDNIIFQFTKCCPPKPKDKS NLLVFRTNGVSD

(continued)

Table 1
(continued)

No.	Protein	Gene symbol	Gene ID	Amino acid sequences of SH2 domains (bold: SH2 domain; *Arg betaB5 position)
41	Jak3	JAK3	3718	CKEVAPP**RLLEEVAEQCHGPITLDFAINKLKTGGSRPGS**YVLR*RS**PQDFDSFLLTVCVQNPLGPDYKGCLIRRSPTGTFLLVGLSRPHSSLRELLATCWDGGLHVDGVAVTLTSCCIP**RPKEKSNLI
42	Lck	LCK	3932	PFNFVAKANSLEP**EPWFFKNLSRKDAERQLLAPGNTHGS**FLIR*ES**ESTAGSFSLSVRDFDQNQGEVVKHYKIRNLDNGGFYISPRITFPGLHELVRHYTNASDGLCTRLSRP**CQTQKPQKPWWED
43	Slp76/Lcp2	LCP2	3937	SPAEEENSLN**EEWYVSYITRPEAEAALRKINQDGT**FLVR*DS**SKKTTTNPYVLMVLYKDKVYNIQIRYQKESQVYLLGTGLRGKEDFLSVSDIIDYFRKMPLLLIDGKNRGSRYQCTLTHA**AGYP
44	Lyn	LYN	4067	PSNYVAKLNTLET**EEWFFKDITRKDAERQLLAPGNSAGA**FLIR*ES**ETLKGSFSLSVRDFDPVHGDVIKHYKIRSLDNGGYYISPRITFPCISDMIKHYQKQADGLCRRLEKAC**ISPKPQKPWD
45	Matk/Lsk	MATK	4145	REALSADPKLSL**MPWFHGKISGQEAVQQLQPPEDGL**FLVR*ES**ARHPGDYVLCVSFGRDVIHYRVLHRDGHLTIDEAVFFCNLMDMVEHYSKDKGAICTKLVRPK**RKHGTKSAEEELA
46	Nck1	NCK1	4690	RPSLTGKFAG**NPWYYGKVTRHQAEMALNERGHEGD**FLIR*DS**ESSPNDFSVSLKAQGKNKHFKVQLKETVYCIGQRKFSTMEELVEHYKKAPIFTSEQGEKLYLVKHLS**
47	Nck2	NCK2	8440	TGPSSSGRFAG**REWYYGNVTRHQAECALNERGVEGD**FLIR*DS**ESSPSDFSVSLKASGKNKHFKVQLVDNVYCIGQRRFHTMDELVEHYKKAPIFTSEHGEKLYLVRALQ**
48	p85a (C)	PIK3R1	5295	DDEDLPHHDE**KTWNVGSSNRNKAENLLRGKRDGT**FLVR*ESS**KQGCYACSVVVDGEVKHCVINKTATGYGFAEPYNLYSSLKELVLHYQHTSLVQHNDSLNVTLAYPV**YAQQRR
49	p85a (N)	PIK3R1	5295	GMNNNMSLQD**AEWYWGDISREEVNEKLRDTADGT**FLVR*DAS**TKMHGDYTLTLRKGGNNKLIKIFHRDGKYGFSDPLTFSSVVELINHYRNESLAQYNPKLDVKLLYPV**SKYQQD

50	p85a (NC)	PIK3R1	5295	GMNNNMSLQDAEWYWGDISREEVNEKLRDTADGTFLVR*DASTKMHGDYTLTLRKGGNNKLIKIFHRDGKYGFSDPLTFSSVVELINHYRNESLAQYNPKLDVKLLYPVSKYQQDQVVKEDNIEAVGKKLHEYNTQFQEKSREYDRLYEEYTRTSQEIQMKRTAIEAFNETIKIFEEQCQTQERYSKEYIEKFKREGNEKEIQRIMHNYDKLKSRISEIIDSRRRLEEDLKKQAAEYREIDKRMNSIKPDLIQLRKTRDQYLMWLTQKGVRQKKLNEWLGNENTEDQYSLVEDDEDLPHHDEKTWNVGSSNRNKAENLLRGKRDGTFLVR*ESSKQGCYACSVVVDGEVKHCVINKTATGYGFAEPYNLYSSLKELVLHYQHTSLVQHNDSLNVTLAYPVYAQQRR
51	p85b (C)	PIK3R2	5296	DDLPHHEERTWYVGKINRTQAEEMLSGKRDGTFLIR*ESSQRGCYACSVVVDGDTKHCVIYRTATGFGFAEPYNLYGSLKELVLHYQHASLVQHNDALTVTLAHPVRAPGPGPPPAAR
52	p85b (N)	PIK3R2	5296	GGSPPSLQDAEWYWGDISREEVNEKLRDTPDGTFLVR*DASSKIQGEYTLTLRKGGNNKLIKVFHRDGHYGFSEPLTFCSVVDLINHYRHESLAQYNAKLDTRLLYPVSKYQQD
53	p85b (NC)	PIK3R2	5296	GGSPPSLQDAEWYWGDISREEVNEKLRDTPDGTFLVR*DASSKIQGEYTLTLRKGGNNKLIKVFHRDGHYGFSEPLTFCSVVDLINHYRHESLAQYNAKLDTRLLYPVSKYQQDQIVKEDSVEAVGAQLKVYHQQYQDKSREYDQLYEEYTRTSQELQMKRTAIEAFNETIKIFEEQGQTQEKCSKEYLERFRREGNEKEMQRILLNSERLKSRIAEIHESRTKLEQQLRAQASDNREIDKRMNSLKPDLMQLRKIRDQYLVWLTQKGARQKKINEWLGIKNETEDQYALMEDEDDLPHHEERTWYVGKINRTQAEEMLSGKRDGTFLIR*ESSQRGCYACSVVVDGDTKHCVIYRTATGFGFAEPYNLYGSLKELVLHYQHASLVQHNDALTVTLAHPVRAPGPGPPPAAR
54	p55g (C)	PIK3R3	8503	EEDENLPHYDEKTWFVEDINRVQAEDLLYGKPDGAFLIR*ESSKKGCYACSVVADGEVKHCVIYSTARGYGFAEPYNLYSSLKELVLHYQQTSLVQHNDSLNVRLAYPVHAQMPSLCR
55	p55g (N)	PIK3R3	8503	MKDSSVSLQDAEWYWGDISREEVNDKLRDMPDGTFLVR*DASTKMQGDYTLTLRKGGNNKLIKIYHRDGKYGFSDPLTFNSVVELINHYHHESLAQYNPKLDVKLMYPVSRYQQD
56	p55g (NC)	PIK3R3	8503	DSSVSLQDAEWYWGDISREEVNDKLRDMPDGTFLVR*DASTKMQGDYTLTLRKGGNNKLIKIYHRDGKYGFSDPLTFNSVVELINHYHHESLAQYNPKLDVKLMYPVSRYQQDQLVKEDNIDAVGKKLQEYHSQYQEKSKEYDRLYEEYTRTSQEIQMKRTAIEAFNETIKIFEEQCHTQEQHSKEYIERFRREGNEKEIERIMMNYDKLKSRLGEIHDSKMRLEQDLKNQALDNREIDKKMNSIKPDLIQLRKIRDQHLVWLNHKGVRQKRLNVWLGIKNEDADENYFINEEDENLPHYDEKTWFVEDINRVQAEDLLYGKPDGAFLIR*ESSKKGCYACSVVADGEVKHCVIYSTARGYGFAEPYNLYSSLKELVLHYQQTSLVQHNDSLNVRLAYPVHAQMPSLCR

(continued)

Table 1
(continued)

No.	Protein	Gene symbol	Gene ID	Amino acid sequences of SH2 domains (bold: SH2 domain; *Arg betaB5 position)
57	Plcg1 (C)	PLCG1	5335	PVPQTNAHES**KEWYHASLTRAQAEHMLMRVPRDGAFLVR*KRNEPNSYAISFRAEGKIKHCRVQQEGQTVMLGNSEFDSLVDLISYYEKHPLYRKMKLRYPI**NEEALEKIGT
58	Plcg1 (N)	PLCG1	5335	SSSTELHSN**EKWFHGKLGAGRDGRHIAERLLTEYCIETGAPDGSFLVR*ESETFVGDYTLSFWRNGKVQHCRIHSRQDAGTPKFFLTDNLVFDSLYDLITHYQQVPLRCNEFEMRLSE**PVPQTNAHE
59	Plcg1 (NC)	PLCG1	5335	SSSTELHSN**EKWFHGKLGAGRDGRHIAERLLTEYCIETGAPDGSFLVR*ESETFVGDYTLSFWRNGKVQHCRIHSRQDAGTPKFFLTDNLVFDSLYDLITHYQQVPLRCNEFEMRLSEPV**PQTNAHES**KEWYHASLTRAQAEHMLMRVPRDGAFLVR*KRNEPNSYAISFRAEGKIKHCRVQQEGQTVMLGNSEFDSLVDLISYYEKHPLYRKMKLRYPI**NEEALEKIGT
60	Plcg2 (C)	PLCG2	5336	DPVPNPNPHES**KPWYYDSLSRGEAEDMLMRIPRDGAFLIR*KREGSDSYAITFRARGKVKHCRINRDGRHFVLGTSAYFESLVELVSYYEKHSLYRKMRLRYPV**TPELLERYNMERDI
61	Plcg2 (N)	PLCG2	5336	DIPPTELHFG**EKWFHKKVEKRTSAEKLLQEYCMETGGKDGTFLVR*ESETFPNDYTLSFWRSGRVQHCRIRSTMEGGTLKYYLTDNLTFSSIYALIQHYRETHLRCAEFELRLTDPV**PNPNPHESKP
62	Plcg2 (NC)	PLCG2	5336	DIPPTELHFG**EKWFHKKVEKRTSAEKLLQEYCMETGGKDGTFLVR*ESETFPNDYTLSFWRSGRVQHCRIRSTMEGGTLKYYLTDNLTFSSIYALIQHYRETHLRCAEFELRLTDPV**PNPNPHES**KPWYYDSLSRGEAEDMLMRIPRDGAFLIR*KREGSDSYAITFRARGKVKHCRINRDGRHFVLGTSAYFESLVELVSYYEKHSLYRKMRLRYPV**TPELLERYNMERD
63	Brk/Ptk6	PTK6	5753	VPHNYLAERETVESEP**WFFGCISRSEAVRRLQAEGNATGAFLIR*VSEKPSADYVLSVRDTQAVRHYKIWRRAGGRLHLNEAVSFLSLPELVNYHRAQSLSHGLRLAAPC**RKHEPEPLPHWD
64	SHP-2 (C)	PTPN11	5781	ELKYPLNCADPTS**ERWFHGHLSGKEAEKLLTEKGKHGSFLVR*ESQSHPGDFVLSVRTGDDKGESNDGKSKVTHVMIRCQELKYDVGGGERFDSLTDLVEHYKKNPMVETLGTVLQLKQPL**NTTRINAAEIESRVRE
65	SHP-2 (N)	PTPN11	5781	MTS**RRWFHPNITGVEAENLLLTRGVDGSFLAR*PSKSNPGDFTLSVRRNGAVTHIKIQNTGDYYDLYGGEKFATLAEVQYYMEHHGQLKEKNGDVIELKYPL**NCADPTS

66	SHP-2 (NC)	PTPN11	5781	MTSRRWFHPNITGVEAENLLLTRGVDGSFLAR*PSKSNPGDFTLSVRRNGAVTHIKIQNTGDYYDLYGGEKFATLAELVQYYMEHHGQLKEKNGDVIELKYPLNCADPTSERWFHGHLSGKEAEKLLTEKGKHGSFLVRESQSHPGDFVLSVRTGDDKGESNDGKSKVTHVMIRCQELKYDVGGGERFDSLTDLVEHYKKNPMVETLGTVLQLKQPLNTTRINAAEIESRVRE
67	SHP-1 (C)	PTPN6	5777	TSERWYHGHMSGGQAETLLQAKGEPWTFLVR*ESLSQPGDFVLSVLSDQPKAGPGSPLRVTHIKVMCEGGRYTVGGLETFDSLTDLVEHFKKTGIEEASGAFVYLRQPYYATRVNAAD
68	SHP-1 (N)	PTPN6	5777	MVRWFHRDLSGLDAETLLKGRGVHGSFLAR*PSRKNQGDFSLSVRVGDQVTHIRIQNSGDFYDLYGGEKFATLTELVEYYTQQQGVLQDRDGTIIHLKYPLNCSDPTSER
69	SHP-1 (NC)	PTPN6	5777	MVRWFHRDLSGLDAETLLKGRGVHGSFLAR*PSRKNQGDFSLSVRVGDQVTHIRIQNSGDFYDLYGGEKFATLTELVEYYTQQQGVLQDRDGTIIHLKYPLNCSDPTSERWYHGHMSGGQAETLLQAKGEPWTFLVR*ESLSQPGDFVLSVLSDQPKAGPGSPLRVTHIKVMCEGGRYTVGGLETFDSLTDLVEHFKKTGIEEASGAFVYLRQPYYATRVNAAD
70	RasGAP (C)	RASA1	5921	VGREEDPHEGKIWFHGKISKQEAYNLLMTVGQVCSFLVR*PSDNTPGDYSLYFRTNENIQRFKICPTPNNQFMMGGRYYNSIGDIIDHYRKEQIVEGYYLKEPVPMQDQEQVLND
71	RasGAP (N)	RASA1	5921	EVAIPLTAPPTNQWYHGKLDRTIAEERLRQAGKSGSYLIR*ESDRRPGSFVLSFLSQMNVVNHFRIIAMCGDYYIGGRRFSSLSDLIGYYSHVSCLLKGEKLLYPVAPPEAIEDRRR
72	RasGAP (NC)	RASA1	5921	EVAIPLTAPPTNQWYHGKLDRTIAEERLRQAGKSGSYLIR*ESDRRPGSFVLSFLSQMNVVNHFRIIAMCGDYYIGGRRFSSLSDLIGYYSHVSCLLKGEKLLYPVAPPEPVEDRRRVRAILPYTKVPDTDEISFLKGDMFIVHNELEDGWMWVTNLRTDEQGLIVEDLVEEVGREEDPHEGKIWFHGKISKQEAYNLLMTVGQVCSFLVR*PSDNTPGDYSLYFRTNENIQRFKICPTPNNQFMMGGRYYNSIGDIIDHYRKEQIVEGYYLKEPVPMQDQEQVLND
73	Rin1	RIN1	9610	RVVSLRERLLLTRPVWLQLQANAAAALHMLRTEPPGTFLVR*KSNTRQCQALCMRLPEASGPSFVSSHYILESPGGVSLEGSELMFPDLVQLICAYCHTRDILLLPLQLPRAIHHAATHKELE
74	Rin2	RIN2	54453	LSILDRLLHTHPIWLQLSLSEEEAAEVLQAQPPGIFLVH*KSTKMQKKVLSLRLPCEFGAPLKEFAIKESTYTFSLEGSGISFADLFRLIAFYCISRDVLPFTLKLPYAISTAKS
75	Rin3	RIN3	79890	SILEKLIKTCPVWLQLSLGQAEVARILHRVVAGMFLVR*RDSSSKQLVLCVHFPSLNESSAEVLEYTIKEEKSILYLEGSALVFEDIFRLIAFYCVSRDLLPFTLRLPQAILEASSFT
76	Sh2-B	SH2B1	25970	EPEGGEGDQPLSGYPWFHGMLSRLKAAQLVLTGGTGSHGVFLVR*QSETRRGEYVLTFNFQGKAKHLRLSLNEEGQCRVQHLWFQSIFDMLEHFRVHPIPLESGGSSDVVLVSYVPSSQRQQGEQ

(continued)

Table 1
(continued)

No.	Protein	Gene symbol	Gene ID	Amino acid sequences of SH2 domains (bold: SH2 domain; *Arg betaB5 position)
77	Aps	SH2B2	10603	PEAEPELELSDYP**WFHGTLSRVKAAQLVLAGGPRNHGLFVIR*QSETRPGEYVLTFNFQG KAKASHLRLSLNGHGQCHVQHLWFQSVLDMLRHFHTHPIPLESGGSADITLRSYV** RAQDPPPE
78	Lnk	SH2B3	10019	LDPACQKTDHFLSC**YPWFHGPISRVKAAQLVQLQGPDAHGVFLVR*QSETRRGEYVLTF NFQGIAKHLRLSLTERGQCRVQHLHFPSVVDMLHHFQRSPIPLECGAACDVRLSSYV** VVVSQPPGSCNT
79	Sap/SH2D1A	SH2D1A	4068	MDAVA**VYHGKISRETGEKLLLATGLDGSYLLR*DSESVPGVYCLCVLYHGYIYTYRVSQT ETGSWSAETAPGVHKRYFRKIKNLISAFQKPDQGIVIPLQYPV**EKKSSARSTQGTTGIRED PDVCLKAP
80	Eat2/SH2D1B	SH2D1B	117157	MDLP**YYHGRLTKQDCETLLLKEGVDGNFLLR*DSESIPGVLCLCVSFKNIVYTYRIFREK HGYYRIQTAEGSPKQVFPSLKELISKFEKPNQGMVVHLLKPI**KRTSPSLRW
81	Vrap/SH2D2A	SH2D2A	9047	KTQAHWLLQHGAA**PAWFHGFITRREAERLLEPKPQGCYLVR*FSESAVTFVLTYRSRTCCR HFLLAQLRDGRHVVLGEDSAHARLQDLLLHYTAHPLSPYGETLTEPL**ARQTPEPAGL SLRTEES
82	Nsp1/SH2D3A	SH2D3A	10045	DGEDLAG**QPWYHGLLSRQKAEALLQQNGDFLVR*ASGSRGGNPVISCRWRGSALHFE VFRVALRPRPGRPTALFQLEDEQFPSIPALVHSYMTGRRPLSQATGAVVSRPV** TWQGPLRRSFSEDT
83	Nsp3/SH2D3C	SH2D3C	10044	ELKLSSTDLRS**HAWYHGRIPREVSETLVQRNGDFLIR*DSLTSLGDYVLTCRWRNQALHFKI NKVVVKAGESYTHIQYLFEQESFDHVPALVRYHVGSRKAVSEQSGAIIYCPV**NRTFPLR
84	Sh2A/SH2D4A	SH2D4A	63898	RAGYQKTSDTI**APWFHGILTLKKANELLLSTGMPGSFLIR*VSERIKGYALSYLSEDGC KHFLIDASADAYSFLGVDQLQHATLADLVEYHKEEPITSLGKELLLYPC**GQQDQLPD
85	SH2D4B	SH2D4B	387694	RRAGFERNTKFI**APWFHGIISREDAEALLENMTEGAFLVR*VSEKIWGYTLSYRLQKGFKHF LVDASGDFYSFLGVDPNRHATLTDLVDFHKEEIITVSGGELLQEPC**GQRDSPPDYHLLFE
86	SH2D5	SH2D5	400745	HSCLVESEGSLTE**NIWAFAGISRPCALALLRRDVLGAFLLW*PELGASGQWCLSVRTQ CGVVPHQVFRNHLGRYCLEHLPAEFPSLEALVENHAVTERSLFCPLDMGRLNPTY** EEQDCGPPGRPPRT

87	SH2D7	SH2D7	646892	APFILQNGALPPWFHGFITRKQTEQLLRDKALGSFLIR*LSDRATGYILSYRGSDRCRHFVINQLRNRRYIISGDTQSHSTLAELVHHYQEAQLEPFKEMLTAACPRPED
88	Slnk	SH2D6	284948	GSASAAEDSDLLTQPWYSGNCDRYAVESALLHLQKDGAYTVR*PSSGPHGSQPFTLAVLLRGRVFNIPIRRLDGGRHYALGREGRNREELFSSVAAMVQHFMWHPLPLVDRHSGSRELTCLLFPTKP
89	SH3BP2	SH3BP2	6452	GKPGLLQVPLPNSVFVNTTESCEVERLFKATSPRGEPQDGLYCIR*NSSTKSGKVLVVWDETSNKVRNYRIFEKDSKFYLEGEVLFVSVGSMEHYHTHVLPSHQSLLLRHPYGYTGPR
90	Shb	SHB	6461	GERVDPAVPLEKQIWYHGAISRGDAENLLRLCKECSYLVR*NSQTSKHDYSLSLRSNQGFMHMKLAKTKEKYVLGQNSPPFDSVPEVIHYYTTRKLPIKGAEHLSLLYPVAVRTL
91	ShcA/Shc1	SHC1	6464	SVSMAEQLRGEPWFHGKLSRREAEALLQLNGDFLVR*ESTTTPGQYVLTGLQSGQPKHLLLVDPEGVVRTKDHRFESVSHLISYHMDNHLPIISAGSELCLQQPVERKL
92	Sck/Shc2/ShcB	SHC2	25759	TEEQLRQEPWYHGRMSRRAAERMLRADGDFLVR*DSVTNPGQYVLTGMHAGQPKHLLLVDPEGVVRTKDVLFESISHLIDHHLQNGQPIVAAESELHLRGVVSREP
93	ShcC/Shc3	SHC3	53358	PDAKMLEELQAETWYQGEMSRKEAEGLLEKDGDFLVR*KSTTNPGSFVLTGMHNGQAKHLLLVDPEGTIRTKDRVFDSISHLINHHLESSLPIVSAGSELCLQQPVERKQ
94	ShcD/Shc4/RaLP	SHC4	399694	SSHSLPHIKQQLWSEECYHGKLSRKAAESLLVKDGDFLVR*ESATSPGQYVLSGLQGGQAKHLLLVDPEGKVRTKDHVFDNVGHLIRYHMDNSLPIISSGSEVSLKQPVRKDNNPALLHSNK
95	SHD	SHD	56961	DPALPLEKQPWFHGPLNRADAESLLSLCKEGSYLVR*LSETNPQDCSLSLRSSQGFLHLKFARTRENQVVLGQHSGPFPSVPELVLHYSSRPLPVQGAEHLALLYPVVTQTP
96	SHE	SHE	126669	SEGEKVDPGLPLEKQPWYHGAISRAEAESRLQPCKEAGYLVR*NSESGNSRYSIALKTSQGCVHIIVAQTKDNKYTLNQTSAVFDSIPEVVHYYSNEKLPFKGAEHMTLLYPVHSKLH
97	SHF	SHF	90525	TDPALPLENQVWYHGAISRTDAENLLRLCKEASYLVR*NSETSKNDFSLSLKSSQGFMHMKLSRTKEHKYVLGQNSPPFSSVPEIVHHYASRKLPIKGAEHMSLLYPVAIRTL
98	Slap	SLA	6503	IPGICVARVYHGWLFEGLGRDKAEELLQLPDTKVGSFMIR*ESETKKGFYSLSVRHRQVKHYRIFRLPNNWYYISPRLTFQCLEDLVNHYSEVADGLCCVLTTPCLTQSTAAPAV
99	Slap2	SLA2	84174	SGREYNIPSVHVAKVSHGWLYEGLSREKAEELLLLPGNPGGAFLIR*ESQTRRGSYSLSVRLSRPASWDRIRHYRIHCLDNGWLYISPRLTFPSLQALVDHYSELADDICCLLKEPCVLQRAGPLP
100	SOCS1	SOCS1	8651	DYRRITRASALLDACGFYWGPLSVHGAHERLRAEPVGTFLVR*DSRQRNCFFALSVKMASGPTSIRVHFQAGRFHLDGSRESFDCLFELLEHYVAAPRRMLGAPLRQRRVRPLQELCRQ

(continued)

Table 1
(continued)

No.	Protein	Gene symbol	Gene ID	Amino acid sequences of SH2 domains (bold: SH2 domain; *Arg betaB5 position)
101	SOCS2	SOCS2	8835	QAARLAKALRELGQ**TGWYWGSMTVNEAKEKLKEAPEGTFLIR*DSSHSDYLLTISVKTSAGPTNLRIEYQDGKFRLDSIICVKSKLKQFDSVVHLIDYYVQMCKDKRTGPEAPRNGTVHLYLTKP**
102	SOCS3	SOCS3	9021	SSKSEYQLVVNAVRKLQE**SGFYWSAVTGGEANLLLSAEPAGTFLIR*DSSDQRHFFTLSVKTQSGTKNLRIQCEGGSFSLQSDPRSTQPVPRFDCVLKLVHHY**MPPPGAPSFPSPPTEPSSE
103	SOCS4	SOCS4	122809	DLLQINNN**NPCYWGVMDKYAAEALLEGKPEGTFLLR*DSAQEDYLFSVSFRRYSRSLHARIEQWNHNFSFDAHDPCVFHSPDITGLLEHYKDPSACMFFEPLLSTPL**IRTFPFS
104	SOCS5	SOCS5	9655	DYIHCLVPDLLQITG**NPCYWGVMDRYEAEALLEGKPEGTFLLR*DSAQEDYLFSVSFRRYNRSLHARIEQWNHNFSFDAHDPCVFHSSTVTGLLEHYKDPSSCMFFEPLLTISL**NRTFPFS
105	SOCS6	SOCS6	9306	TEELKKLAK**QGWYWGPITRWEAEGKLANVPDGSFLVR*DSSDDRYLLSLSFRSHGKTLHTRIEHSNGRFSFYEQPDVEGHTSIVDLIEHSIRDSENGAFCYSRSRLPGSATYPV**RLTNPVSR
106	SOCS7	SOCS7	30837	DSSSFAASLRELEK**CGWYWGPMNWEDAEMKLKGKPDGSFLVR*DSSDPRYILSLSFRSQGITHHTRMEHYRGTFSLWCHPKFEDRCQSVVEFIKRAIMHSKNGKFLYFLRSRVPGLPPTPVQLLYPVS**RFSNVKS
107	Src	SRC	6714	YIPSNYVAPSDSIQA**EEWYFGKITRRESERLLLNAENPRGTFLVR*ESETTKGAYCLSVSDFDNAKGLNVKHYKIRKLDSGGFYITSRTQFNSLQQLVAYYSKHADGLCHRLTTV**CPTSKPQTQGL
108	Srms/Ptk70	SRMS	6725	SPETLSD**QPWYFSGVSRTQAQQLLLSPPNEPGAFLIR*PSESSLGGYSLSVRAQAKVCHYRVSMAADGSLYLQKGRLFPGLEELLTYYKANWKLIQNPLLQPC**MPQKAPRQ
109	Brdg1	STAP1	26228	DYVDVLNPMPACFYT**VSRKEATEMLQKNPSLGNMILR*PGSDSRNYSITIRQEIDIPRIKHYKVMSVGQNYTIELEKPVTLPNLFSVIDYFVKETRGNLRPFICSTDENTGQEPS**MEGRS
110	Bks/Stap2	STAP2	55620	TDLTLLPGHL**YMMSEVLAKEEARRALETPSCFLKVSRLEAQLLLERYPECGNLLLR*PSGDGADGVSVTTRQMHNGTHVVRHYKVKREGPKYVIDVEQPFSCTSLDAVVNYFVSHTKKALVPFLLD**EDYEKVLGYVEA
111	Stat1	STAT1	6772	ESILELIKKHLL**PLWNDGCIMGFISKERERALLKDQQPGTFLLR*FSESSREGAITFTWVERSQNGGEPDFHAVEPYTKKELSAVTFPDIIRNYKVMAAENIPENPLKYLYPNIDKDHAFGKYYSRPKEAPEP**MELDGPKG

112	Stat2	STAT2	6773	DKILELVHDHLK**DLWNDGRIMGFVSRSQERRLLKKTMSGTFLLR*FSESSEGGITCSWVEHQDDDKVLIYSVQPYTKEVLQSLPLTEIIRHYQLLTEENIPENPLRFLYPR**IPRDEAFGCYYQEKVNLQERRK
113	Stat3	STAT3	6774	DLVKKYIL**ALWNEGYIMGFISKERERAILSTKPPGTFLLR*FSESSKEGGVTFTWVEKDISGKTQIQSVEPYTKQQLNNMSFAEIIMGYKIMDATNILVSPL**VYLYPDIPKEEAFGKYCRPESQE
114	Stat4	STAT4	6775	KKHILPLWI**DGYVMGFVSKEKERLLLKDKMPGTFLLR*FSESHLGGITFTWVDHSESGEVRFHSVEPYNKGRLSALPFADILRDYKVIMAENIPENPLKYLYPDIPKDKAFGKHYSSQPCE**VSRPTERGDKG
115	Stat5A	STAT5A	6776	DGVMEVLKKHHK**PHWNDGAILGFVNKQQAHDLLINKPDGTFLLR*FSDSEIGGITIAWKFDSPERNLWNLKPFTTRDFSIRSLADRLGDLSYLIYVFPDRPKDEVFSKYYTPV**LAKAVDGYVKPQIKQ
116	Stat5B	STAT5B	6777	DGVMEVLKKHLK**PHWNDGAILGFVNKQQAHDLLINKPDGTFLLR*FSDSEIGGITIAWKFDSQERMFWNLMPFTTRDFSIRSLADRLGDLNYLIYVFPDRPKDEVYSKYYTPV**PCESATAKAVDG
117	Stat6	STAT6	6778	TKRCLR**SYWSDRLIIGFISKQYVTSLLLNEPDGTFLLR*FSDSEIGGITIAHVIRGQDGSPQIENIQPFSAKDLSIRSLGDRIRDLAQLKNLYPKKPKDEAFRSHYKPE**QMGKDGRGYVPATIKMTVERDQ
118	Supt6h	SUPT6H	6830	DMKRKQQRTT**YIKRVIAHPSFHNINFKQAEKMMETMDQGDVIIR*PSSKGENHLTVTWKVSDGIYQHVDVREEGKENAFSLGATLWINSEEFEDLDEIVARYVQPMASFARDLLNHKY**YQDCSGGDRKKLEE
119	Syk (C)	SYK	6850	TTAHEKMP**WFHGKISREESEQIVLIGSKTNGKFLIR*ARDNNGSYALCLLHEGKVLHYRIDKDKTGKLSIPEGKKFDTLWQLVEHYSYKADGLLRVLTVPC**QKIGTQGNVNFGGR
120	Syk (N)	SYK	6850	DSANH**LPFFFGNITREEAEDYLVQGGMSDGLYLLR*QSRNYLGGFALSVAHGRKAHHYTIERELNGTYAIAGGRTHASPADLCHYHSQESDGLVCLLKKPF**NRPQGVQPKTG
121	Syk (NC)	SYK	6850	MASSGMADSANH**LPFFFGNITREEAEDYLVQGGMSDGLYLLR*QSRNYLGGFALSVAHGRKAHHYTIERELNGTYAIAGGRTHASPADLCHYHSQESDGLVCLLKKPF**NRPQGVQPKTGPFEDLKENLIREYVKQTWNLQGQALEQAIISQKPQLEKLIATTAHEKMP**WFHGKISREESEQIVLIGSKTNGKFLIR*ARDNNGSYALCLLHEGKVLHYRIDKDKTGKLSIPEGKKFDTLWQLVEHYSYKADGLLRVLTVPC**QKIGTQGNVNFGGR
122	Tec	TEC	7006	TGKKSNNLDQ**YEWYCRNMNRSKAEQLLRSEDKEGGFMVR*DSSQPGLYTVSLYTKFGGEGSSGFRHYHIKETTTSPKKYYLAEKHAFGSIPEIIEYHKHNAAGLVTRLRYPV**SVKGKNAPTTAG

(continued)

Table 1
(continued)

No.	Protein	Gene symbol	Gene ID	Amino acid sequences of SH2 domains (bold: SH2 domain; *Arg betaB5 position)
123	Tenc1/Tns2	TENC1	23371	TQESQSNVKFVQDTS**KFWYKPHLSRDQAIALLKDKDPGAFLIR*DSHSFQGAYGLALKVATPPPSAQPWKGDPVEQLVRHFLIETGPKGVKIKGCPSEPYFGSLSALVSQHSISPISLPCCLRIPS**KDPLEETPEAPVPTNMSTAAD
124	Tensin/Tns1	TNS1	7145	RAKVKFVQDTS**KYWYKPEISREQAIALLKDQEPGAFIIR*DSHSFRGAYGLAMKVSSPPPTIMQQNKKGDMTHELVRHFLIETGPRGVKLKGCPNEPNFGSLSALVYQHSIIPLALPCKLVIPN**RDPTDESKDSS
125	Tem6/Tns3/ Tensl	TNS3	64759	GDKLVIVKFVQDTS**KFWYKADISREQAIAMLKDKEPGSFIVR*DSHSFRGAYGLAMKVATPPPSVLQLNKKAGDLANELVRHFLIECTPKGVRLKGCSNEPYFGSLTALVCQHSITPLALPCKLLIPE**RDPLEEIAESS
126	Cten/Tns4	TNS4	84951	TMKFVMDTSKY**WFKPNITREQAIELLRKEEPGAFVIR*DSSSYRGSFGLALKVQERHFLIESSAKGVHLKGADEEPYFGSLSAFVCQHSIMALALPCKLTIPQ**RELGGADG
127	Txk	TXK	7294	TENKITNLEI**YEWYHRNITRNQAEHLLRQESKEGAFIVR*DSRHLGSYTISVFMGARRSTEAAIKHYQIKKNDSGQWYVAERHAFQSIPELIWYHQHNAAGLMTRLRYPV**GLMGSCLPAT
128	Tyk2	TYK2	7297	RLVMSIRD**GIHGPLLEPFVQAKLRPEDGLYLIH*WSTSHPYRLILTVAQRSQAPDGMQSLRLRKFPIEQQDGAFVLEGWGRSFPSVRELGAALQGCLLRAGDDCFSLRRCCLP**QPGETSNLIIMRGARASPR
129	Vav1	VAV1	7409	HGPPQDLSV**HLWYAGPMERAGAESILANRSDGTFLVR*QRVKDAAEFAISIKYNVEVKHIKIMTAEGLYRITEKKAFRGLTELVEFYQQNSLKDCFKSLDTTLQFPF**KEPEKR
130	Vav2	VAV2	7410	PPSREIDYTA**YPWFAGNMERQQTDNLLKSHASGTYLIR*ERPAEAERFAISIKFNDEVKHIKVVEKDNWIHITEAKKFDSLLELVEYYQCHSLKESFKQLDTTLKYPY**KSRERSA
131	Vav3	VAV3	10451	CVPKPVDYSC**QPWYAGAMERLQAETELINRVNSTYLVR*HRTKESGEYAISIKYNNEAKHIKILTRDGFFHIAENRKFKSLMELVEYYKHHSLKEGFRTLDTTLQFPY**KEPEHSAG

132	Yes	YES1	7525	SNYVAPADSIQA**EEWYFGKMGRKDAERLLLNPGNQRGIFLVR*ESETTKGAYSLSIRDWDEIRGDNVKHYKIRKLDNGGYYITTRAQFDTLQKLVKHYTEHADGLCHKLTTVC**PTVKPQTQGLAKD
133	Zap70 (C)	ZAP70	7535	SQAPQVEKLIATTAHER**MPWYHSSLTREEAERKLYSGAQTDGKFLLR*PRKEQGTYALSLIYGKTVYHYLISQDKAGKYCIPEGTKFDTLWQLVEYLKLKADGLIYCLKEAC**PNSSASNASG
134	Zap70 (N)	ZAP70	7535	MPDPAAH**LPFFYGSISRAEAEEHLKLAGMADGLFLLR*QCLRSLGGYVLSLVHDVRFHHFPIERQLNGTYAIAGGKAHCGPAELCEFYSRDPDGLPCNLRKPC**NRPSGLEPQPGVFDCLRD
135	Zap70 (NC)	ZAP70	7535	MPDPAAH**LPFFYGSISRAEAEEHLKLAGMADGLFLLR*QCLRSLGGYVLSLVHDVRFHHFPIERQLNGTYAIAGGKAHCGPAELCEFYSRDPDGLPCNLRKPC**NRPSGLEPQPGVFDCLRDAMVRDYVRQTWKLEGEALEQAIISQAPQVEKLIATTAHER**MPWYHSSLTREEAERKLYSGAQTDGKFLLR*PRKEQGTYALSLIYGKTVYHYLISQDKAGKYCIPEGTKFDTLWQLVEYLKLKADGLIYCLKEAC**PNSSASNASG

This list contains a total of 122 different human SH2 domains from 121 validated and one provisional (i.e., GRAPL) genes at the NCBI Gene site (http://goo.gl/X4r2kP). Estimated boundaries of SH2 domains are indicated by bold text, the FLVRES motif by red text, and the position equivalent to Arg βB5 by the letter "R*"

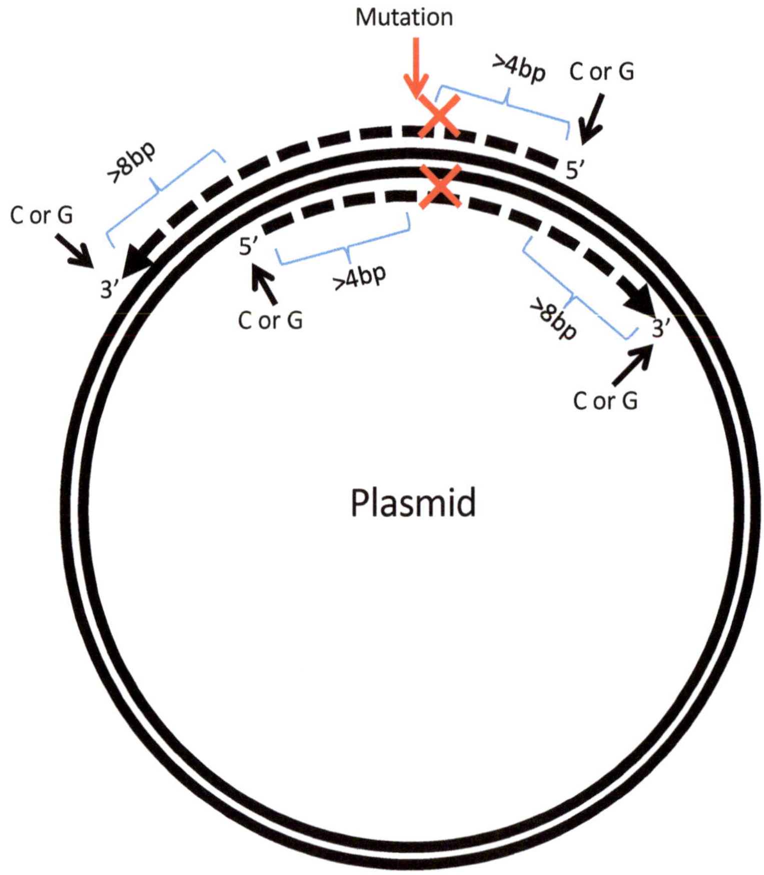

Fig. 1 Schematic diagram of the overlapped mutagenesis primers. The loss-of-function SH2 domain mutant is generated by site-directed mutagenesis using the primers outlined here. The mutation corresponding to the βB5 arginine to lysine substitution should be included in both primers. Red X indicates the location of the mutation. *See* Subheading 3.1 **step 2**

4. Make up a PCR mix for mutagenesis as below. PCR guideline using Phusion DNA Polymerase is available at New England Biolabs Web site (https://goo.gl/T6r3eQ).

10 μL	5× Phusion Buffer
5 μL (200 μM)	10× dNTPs
125 ng (0.2–0.5 μM)	Forward primer
125 ng (0.2–0.5 μM)	Reverse primer
5–50 ng	Template DNA
0.5 μL	Phusion DNA polymerase
Add water to 50 μL	

5. Run PCR with the following settings:

Step 1	98 °C	30 s
Step 2	98 °C	10 s
Step 3	60 °C	30 s
Step 4	72 °C	30 s per kb of plasmid length
Step 5	Repeat **steps 2–4** for 20 cycles	
Step 6	72 °C	5 min
Step 7	4 °C	Indefinite

6. Add 1 μL DpnI restriction enzyme directly to the PCR tube (*see* **Note 7**).
7. Incubate for 2–3 h at 37 °C (*see* **Note 8**).
8. Thaw competent bacteria (50 μL) on ice.
9. Add 1–5 μL DNA to competent bacteria, gently mix by tapping the tube, and incubate on ice for 20 min without shaking.
10. Heat shock at 42 °C for 45 s, then immediately cool down on ice for 2 min.
11. Add 500 μL LB (−) and incubate at 37 °C for 1 h with agitation.
12. Warm up LB-ampicillin agar plate at 37 °C.
13. Spin down the bacteria at 3000 × *g* for 3 min.
14. Remove the supernatant (leave 50 μL in the tube) and vortex.
15. Plate the bacteria on a prewarmed LB-ampicillin plate and incubate at 37 °C overnight.
16. Pick 7–8 colonies from the plate and inoculate into 4–6 mL fresh prewarmed LB-ampicillin.
17. Grow overnight at 37 °C.
18. Purify the plasmids using the miniprep kit.
19. Screen positive clones by restriction enzyme mapping or sequencing (*see* **Note 6**).
20. Express and purify the mutant SH2 domain and test binding activity using either pull-down or far-Western assays (*see* Chapters 8 and 17). Figure 3 shows expression and activity of wild type (Wt) and mutant (Mt) SH2 domain proteins in which binding activity of the mutant SH2 is undetectable (*see* **Note 9**).

3.2 Domain Concatenation

The domain concatenation method described here utilizes two restriction enzymes that generate complementary cohesive ends, such as BamHI and BglII (*see* **Note 4**). Upon successful ligation,

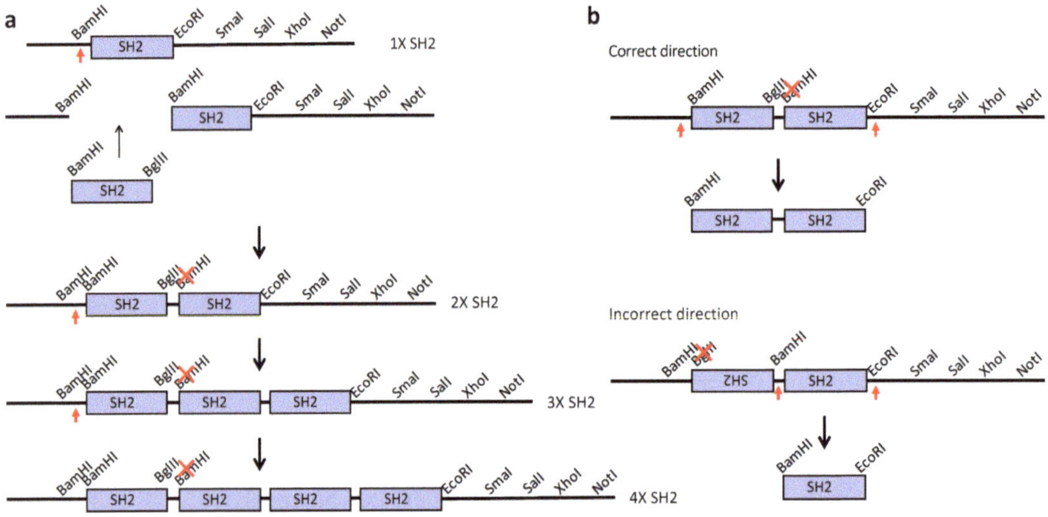

Fig. 2 Strategy for SH2 domain concatenation. An SH2 fragment flanked by BamHI and BglII sites are repeatedly inserted into the pGEX6P-1-SH2 plasmid vector in a stepwise fashion. (**a**) A BamHI/BglII SH2 fragment is inserted into a 1× SH2 plasmid vector that has been cut with BamHI. 2×, 3×, and 4× SH2 domain tandem constructs can be obtained by repeating this method. *Red arrow* indicates the site of digestion, and *red X* indicates the killed restriction site which is unable to be cut by neither BamHI nor BglII. (**b**) Orientation of the insert in ligation clones can be determined by restriction enzyme digestion. The presence of a ~600 bp fragment (i.e., 2× SH2) following digestion with BamHI and EcoRI indicates a clone with the correct orientation of the insert. The presence of ~300 bp size fragment (i.e., 1× SH2) indicates clones with incorrect orientation

the 3′ enzyme site will be killed allowing for repetitive insertion of the SH2 domain fragments which will result in multiple tandem domains (Fig. 2a). The procedure below describes a method to insert a BamHI–SH2–BglII fragment into an existing pGEX6P-1-SH2 construct in a stepwise fashion as depicted in Fig. 2a.

1. Amplify the SH2 fragment by PCR using primers containing restriction enzyme sites (e.g., BamHI-forward primer and BglII-reverse primer) with an SH2 cDNA template.
2. Run the PCR product on a 1–2% agarose gel and gel-extract the amplified DNA fragment.
3. Double digest the SH2 DNA fragment with both restriction enzymes (e.g., BamHI and BglII), and then purify the fragment from the agarose gel.
4. Digest a destination vector (maybe the same plasmid used as the PCR template such as pGEX) with the appropriate restriction enzyme (e.g., BamHI).
5. To dephosphorylate the N- and C- termini of the digested vector, add alkaline phosphatase and incubate at 37 °C for 1 h.
6. Purify the linearized and dephosphorylated pGEX vector from the agarose gel.

7. Mix 1 μL 10× ligation buffer, 50 ng linearized, dephosphorylated vector, equimolar amount of insert (*see* **Note 10**) and molecular grade water to 9.5 μL (*see* **Note 11**).
8. Heat the mixture to 65 °C for 2 min to melt sticky ends, then place on ice.
9. Add 0.5 μL T4 ligase.
10. Incubate at room temperature for 1 h, or at 16 °C overnight.
11. Heat inactivate the ligase at 65 °C for 10 min before transformation.
12. Thaw competent bacteria (50 μL) on ice.
13. Add 5 μL ligation mixture to competent bacteria, gently mix by tapping the tube, and incubate on ice for 20 min without shaking.
14. Heat shock at 42 °C for 45 s, then immediately transfer the tubes to ice and cool down for 2 min.
15. Add 500 μL LB (−) and incubate at 37 °C for 1 h with agitation.
16. Warm up LB-ampicillin agar plate at 37 °C.
17. Spin down bacteria at 3000 × *g* for 3 min.
18. Remove the supernatant (leaving about 50 μL in the tube) and vortex.
19. Plate the bacteria on the warmed LB-ampicillin plate and incubate at 37 °C overnight.
20. Next day, assess the ligation efficiency by comparing the ligated sample (vector + insert) to the negative control plate (vector only).
21. Pick 7–8 colonies from the plate and inoculate each into 4–6 mL fresh prewarmed LB-ampicillin.
22. Shake overnight at 37 °C.
23. Purify plasmid DNA using the miniprep kit.
24. Digest the plasmids with appropriate restriction enzymes to confirm that the plasmid has two SH2 domains (i.e., 2× SH2) with a correct orientation (*see* **Note 12** and Fig. 2b).
25. Sequence positive clones.
26. Repeat **steps 4–25** using the resultant tandem (2×) SH2 domain construct plasmid to generate a triple tandem (3×) construct.
27. Repeat **steps 4–25** using the resultant tandem (3×) SH2 domain construct plasmid to generate a quadruple tandem (4×) construct (*see* **Note 13**).
28. Express and purify the concatenated SH2 domains and test binding activity using either pull-down or far-Western assays

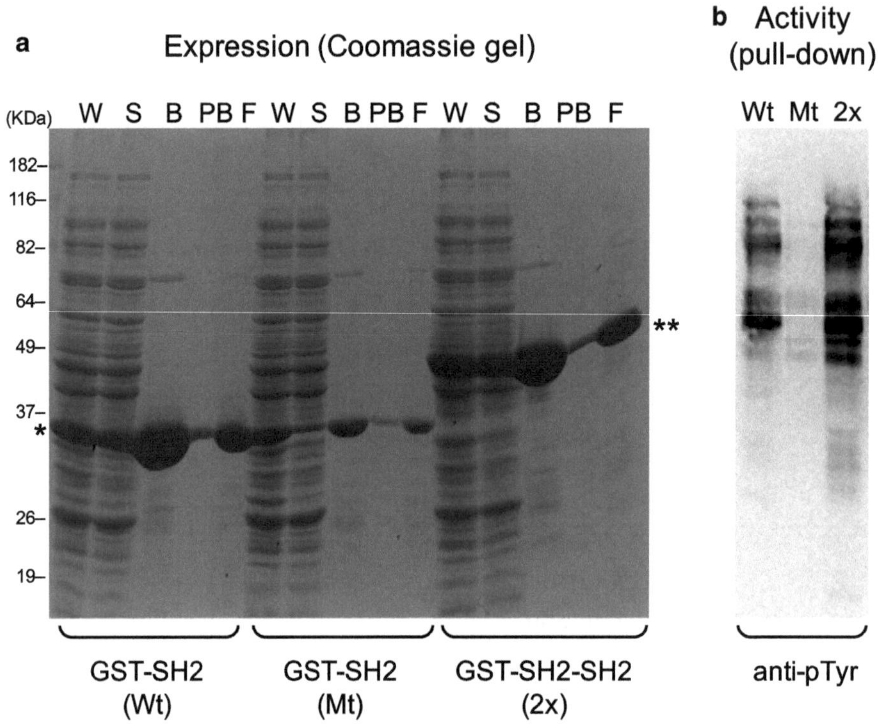

Fig. 3 Expression and activity of SH2 mutants. (**a**) Coomassie stained gel of GST-tagged wild type SH2 domain (Wt), βB5 arginine mutant (Mt), and tandem SH2 domain (2×) purified from bacteria. Wt and Mt GST-SH2 domains appear as ~35 kDa proteins (*single asterisk*) while the tandem SH2 domain appears as a ~45 kDa protein (*double asterisk*). (**b**) Binding activity of the GST-SH2 protein tested by pull-down assay. GST-SH2 proteins were bound to glutathione beads and incubated with a pervanadate-treated cell lysate. Bead-bound proteins were detected by pTyr-immunoblotting. *Lanes*: *W* whole cell fraction, *S* soluble fraction, *B* glutathione bead-bound fraction, *PB* post-elution bead fraction, *F* final product

(*see* Chapters 8 and 17). Figure 3 shows expression and activity of wild type (Wt) and tandem (2×) SH2 domain proteins in which the binding activity of the tandem SH2 is enhanced (*see* **Note 14**).

4 Notes

1. Sources for human SH2 domain cDNA include Addgene (e.g., Mayer lab clones), DNASU, GE Dharmacon, and the Pawson Lab Web site.
2. General procedures for designing primers and mutagenesis are described in [19] and the openwetware Web site (http://goo.gl/ITGxrf).
3. Ideally the SH2 domain fragment can be cloned into either BamHI or XhoI sites (*see* **Note 4**).

4. Design primers that contain: (a) 15–20 nucleotides of homology to the SH2 domain template; (b) Five C's at the 5′ end to allow efficient digestion with restriction enzymes; and (c) In-frame restriction enzyme sites for forward and reverse primers with compatible overhangs such as BamHI/BglII (Fig. 2) and XhoI/SalI. A list of enzymes with compatible cohesive ends is available on the New England Biolabs Web site (https://goo.gl/Kuqj8H).
5. The **FLVRES** motif in Table 1 was inferred from the multiple alignment of all SH2 domains (sh2domain.org) [20]. Some SH2 domains have divergent sequences in this region and a few do not even have a βB5 arginine, e.g., Rin 2. For these domains, structural information (e.g., Protein Data Bank) can be used to identify amino acids responsible for phosphotyrosine binding (like βB5 arginine) as the alternative target for the loss-of-function mutation.
6. It is convenient if mutation creates a unique restriction enzyme recognition site, which can be utilized for screening for positive clones [19]. If no positive clone is obtained with the primers, try modified PCR conditions or synthesize another set of primers or alternative mutagenesis using type IIs restriction enzymes can be considered [21].
7. DpnI is a restriction enzyme which digests methylated DNA. Purification of PCR product prior to DpnI addition is not necessary.
8. Purification of the PCR product before the transformation is recommended.
9. We often observe that mutated SH2 domains have relatively lower expression in bacteria (Fig. 3a).
10. If the vector is 5.0 kb and the insert is 0.4 kb, an equimolar amount would be $0.4/5.0 \times 50$ ng = 4 ng. This does not have to be exact. In some cases, more or less insert works better.
11. Always set up a ligation with the digested vector without the insert as a control. This will tell you how much background occurred as a result of self-ligation. In addition, this method can result in half of ligated clones having the reverse orientation.
12. Rarely some clones may contain three or more concatenated SH2 domains at this stage. The restriction digest product should contain the plasmid and a single insert with a length equal to some multiple of the SH2 fragment length (600, 900, 1200 bp, etc.).
13. Repeat until ideal number of domains are inserted.
14. In our experience, tandem constructs containing six or more SH2 domains tend to have lower expression and be insoluble.

Acknowledgements

We thank Joshua Jadwin and Silas Khong Ng for assistance with editing the chapter, and Bruce Mayer for his continuous encouragement and support. This study was partly supported by grant CA1154966 from the National Institutes of Health and Quest for CURES (QFC) grant from the Leukemia and Lymphoma Society (to K.M.).

References

1. Jadwin JA, Ogiue-Ikeda M, Machida K (2012) The application of modular protein domains in proteomics. FEBS Lett 586(17):2586–2596. doi:10.1016/j.febslet.2012.04.019
2. Mayer BJ (2015) The discovery of modular binding domains: building blocks of cell signalling. Nat Rev Mol Cell Biol 16(11):691–698. doi:10.1038/nrm4068
3. Jones RB, Gordus A, Krall JA, MacBeath G (2006) A quantitative protein interaction network for the ErbB receptors using protein microarrays. Nature 439(7073):168–174. doi:10.1038/nature04177
4. Machida K, Thompson CM, Dierck K, Jablonowski K, Karkkainen S, Liu B, Zhang H, Nash PD, Newman DK, Nollau P, Pawson T, Renkema GH, Saksela K, Schiller MR, Shin DG, Mayer BJ (2007) High-throughput phosphotyrosine profiling using SH2 domains. Mol Cell 26(6):899–915. doi:10.1016/j.molcel.2007.05.031
5. Chen X, Tan PH, Zhang Y, Pei D (2009) On-bead screening of combinatorial libraries: reduction of nonspecific binding by decreasing surface ligand density. J Comb Chem 11(4):604–611. doi:10.1021/cc9000168
6. Hause RJ Jr, Leung KK, Barkinge JL, Ciaccio MF, Chuu CP, Jones RB (2012) Comprehensive binary interaction mapping of SH2 domains via fluorescence polarization reveals novel functional diversification of ErbB receptors. PLoS One 7(9), e44471. doi:10.1371/journal.pone.0044471
7. Liu BA, Engelmann BW, Jablonowski K, Higginbotham K, Stergachis AB, Nash PD (2012) SRC homology 2 domain binding sites in insulin, IGF-1 and FGF receptor mediated signaling networks reveal an extensive potential interactome. Cell Commun Signal 10(1):27. doi:10.1186/1478-811x-10-27
8. Takakuma K, Ogo N, Uehara Y, Takahashi S, Miyoshi N, Asai A (2013) Novel multiplexed assay for identifying SH2 domain antagonists of STAT family proteins. PLoS One 8(8), e71646. doi:10.1371/journal.pone.0071646
9. Liu H, Li L, Voss C, Wang F, Liu J, Li SS (2015) A comprehensive immunoreceptor phosphotyrosine-based signaling network revealed by reciprocal protein-peptide array screening. Mol Cell Proteomics 14(7):1846–1858. doi:10.1074/mcp.M115.047951
10. Dierck K, Machida K, Voigt A, Thimm J, Horstmann M, Fiedler W, Mayer BJ, Nollau P (2006) Quantitative multiplexed profiling of cellular signaling networks using phosphotyrosine-specific DNA-tagged SH2 domains. Nat Methods 3(9):737–744. doi:10.1038/nmeth917
11. Machida K, Khenkhar M, Nollau P (2012) Deciphering phosphotyrosine-dependent signaling networks in cancer by SH2 profiling. Genes Cancer 3(5-6):353–361. doi:10.1177/1947601912459048
12. Saito Y, Furukawa T, Arano Y, Fujibayashi Y, Saga T (2010) Fusion protein based on Grb2-SH2 domain for cancer therapy. Biochem Biophys Res Commun 399(2):262–267. doi:10.1016/j.bbrc.2010.07.066
13. Mayer BJ, Jackson PK, Van Etten RA, Baltimore D (1992) Point mutations in the abl SH2 domain coordinately impair phosphotyrosine binding in vitro and transforming activity in vivo. Mol Cell Biol 12(2):609–618
14. Kuriyan J, Cowburn D (1997) Modular peptide recognition domains in eukaryotic signaling. Annu Rev Biophys Biomol Struct 26:259–288. doi:10.1146/annurev.biophys.26.1.259
15. Campbell SJ, Jackson RM (2003) Diversity in the SH2 domain family phosphotyrosyl peptide binding site. Protein Eng 16(3):217–227
16. Lappalainen I, Thusberg J, Shen B, Vihinen M (2008) Genome wide analysis of pathogenic SH2 domain mutations. Proteins 72(2):779–792. doi:10.1002/prot.21970
17. Filippakopoulos P, Muller S, Knapp S (2009) SH2 domains: modulators of nonreceptor tyrosine kinase activity. Curr Opin Struct Biol 19(6):643–649. doi:10.1016/j.sbi.2009.10.001

18. Gan W, Roux B (2009) Binding specificity of SH2 domains: insight from free energy simulations. Proteins 74(4):996–1007. doi:10.1002/prot.22209
19. Zheng L, Baumann U, Reymond JL (2004) An efficient one-step site-directed and site-saturation mutagenesis protocol. Nucleic Acids Res 32(14), e115. doi:10.1093/nar/gnh110
20. Liu BA, Jablonowski K, Raina M, Arce M, Pawson T, Nash PD (2006) The human and mouse complement of SH2 domain proteins-establishing the boundaries of phosphotyrosine signaling. Mol Cell 22(6):851–868. doi:10.1016/j.molcel.2006.06.001
21. Ko JK, Ma J (2005) A rapid and efficient PCR-based mutagenesis method applicable to cell physiology study. Am J Physiol Cell Physiol 288(6):C1273–C1278. doi:10.1152/ajpcell.00517.2004

Chapter 13

Creation of Phosphotyrosine Superbinders by Directed Evolution of an SH2 Domain

Haiming Huang, Tomonori Kaneko, Sachdev S. Sidhu, and Shawn S.C. Li

Abstract

Commercial antibodies raised against phosphotyrosine have been widely used as reagents to detect or isolate tyrosine-phosphorylated proteins from cellular samples. However, these antibodies are costly and are not amenable to in-house production in an academic lab setting. In this chapter, we describe a method to generate super-high affinity SH2 domains, dubbed the phosphotyrosine superbinders, by evolving a natural SH2 domain using the phage display technology. The superbinders are stable and can be easily produced in *Escherichia coli* in large quantities. The strategy presented here may also be applied to other protein domains to generate domain variants with markedly enhanced affinities for a specific post-translational modification.

Key words Phage display, Directed evolution, Src homology 2 (SH2) domain, Phosphotyrosine, Modular domain, Post-translational modification

1 Introduction

Rapid protein phosphorylation and dephosphorylation events dynamically regulate cell signaling that can proceed in the order of seconds [1–3]. Moderate binding affinities allow transient association and dissociation between signaling molecules, including the SH2 domain and its binding partner, so that the signal from an active kinase is promptly transmitted to the downstream proteins and the system can then be reset and ready for the next incoming signal. The dissociation constant (K_D) of SH2 domain-phosphotyrosine (pTyr) ligand complexes usually ranges between 0.1 and 10 μM, so an SH2 domain will readily dissociate from tyrosine-phosphorylated target sites instead of permanently masking the phosphotyrosine [3–6].

The SH2 domain specifically binds to phosphorylated tyrosine, but not nonphosphorylated tyrosine, with a few exceptions [7–9]. High affinity binding to other phosphorylated residues (e.g., phosphoserine or phosphothreonine) has not been observed.

Kazuya Machida and Bernard A. Liu (eds.), *SH2 Domains: Methods and Protocols*, Methods in Molecular Biology, vol. 1555, DOI 10.1007/978-1-4939-6762-9_13, © Springer Science+Business Media LLC 2017

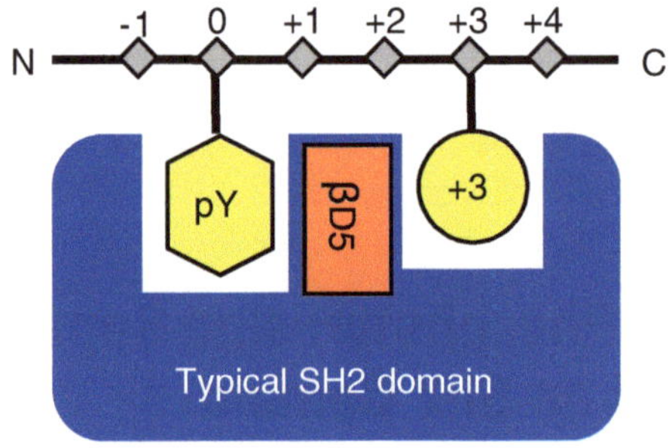

Fig. 1 The two pockets on an SH2 domain to capture a pTyr-containing ligand peptide. Many SH2 domains have two major binding pockets on the surface, the pTyr-binding pocket and the specificity-determining pocket. In the case of the Fyn SH2 domain, the two pockets are separated by a bulky tyrosine residue, called βD5 [40], and the specificity-determining pocket captures the third residue C-terminal to pTyr (pTyr+3). For the phage display experiment, we mutated pTyr-binding pocket residues, except for βD5 to preserve ligand recognition specificity (*see* **Note 3**)

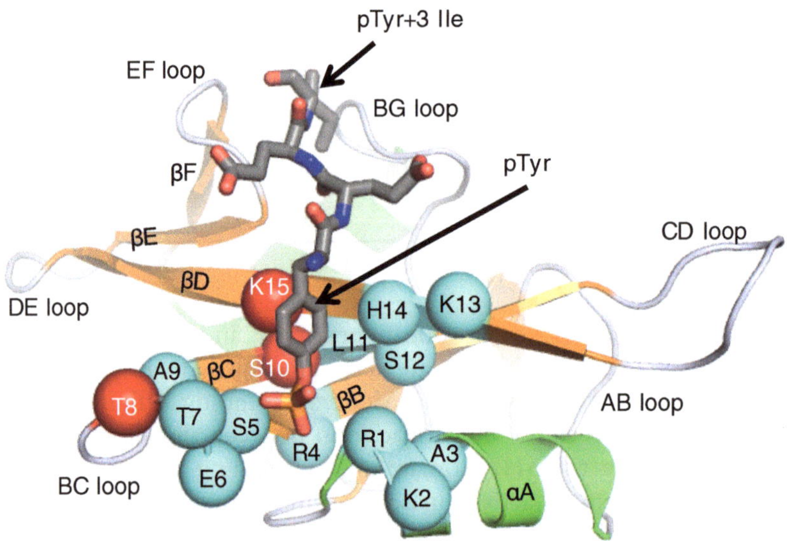

Fig. 2 Structure of the Fyn SH2 domain (PDB: 1AOT). The 15 positions mutated in our library are shown as balls on the Fyn SH2 domain in ribbon representation. The peptide is shown as *grey sticks*. The three positions (8, 10, 15), identified as key residues for superbinder creation, are shown as *red balls*. The remaining 12 residues are in *cyan*. α helices are in *green*, and β strands are in *orange*

The structures of SH2 domains in complex with pTyr targets reveal that the SH2 domain features a pTyr-binding pocket designed to specifically recognize the pTyr moiety [10] (Figs. 1 and 2). Therefore, SH2 domains are bona-fide pTyr-recognition modules and may be usable as tools to detect phosphotyrosine.

Commercial antibodies raised against phosphorylated tyrosine have been widely used in cellular and molecular biology experiments to study tyrosine phosphorylation events [12]. Although such antibodies are powerful tools, they are costly and are not ideal for researches that require large quantities. Antibody alternatives derived from single protein domains have potential advantages that include the ease of large-scale production in an established protein production system (e.g., overexpression in *Escherichia coli*) and introduction of the reagents into live cells, a feature not achievable with a multi-chain antibody assembled with disulfide bonds. For example, a few SH2 domains have been reported as Förster resonance energy transfer (FRET) biosensors to monitor tyrosine phosphorylation events in live cells [13]. On the other hand, a large-scale protein production and purification method from *E. coli* lysate has been established for the Src SH2 domain, without an affinity purification tag [14].

In this chapter, we describe a strategy we used to create phosphotyrosine superbinder SH2 domains that serve as alternatives to antibodies against phosphotyrosine. We found that simultaneous substitutions of three residues of the pTyr-binding pocket increased pTyr-binding affinity by a few hundredfold [15]. The superbinders can be used as a tool to isolate and detect phosphotyrosine or to monitor cell signaling events in live cells. The superbinders, which are specific for phosphotyrosine, have been successfully used to enrich tyrosine-phosphorylated peptides from cell lysate samples for subsequent identification by mass spectrometry (Bian et al., [16]). When introduced into mammalian cells stimulated with a growth factor, we observed inhibition of MAP kinase signaling and substantially reduced cell viability [15]. In mouse embryonic stem cells, an engineered version of the adaptor protein Grb2 that contains a superbinder SH2 domain caused precocious stem cell differentiation, presumably because of artificially enhanced affinity to transmembrane receptors [17].

It has been reported that a number of globular protein domains can be displayed on the surface of bacteriophages, including the SH3, PDZ, and FN3 domains, in addition to the antibody Fab fragment [18–21]. The SH2 domain is a small globular domain (Figs. 2 and 3). The majority of the 120 human SH2 domains are soluble when overexpressed in *E. coli* as a recombinant protein [4, 22–26]. Among the numerous available SH2 domains, we chose the SH2 domain of the kinase Fyn for phage display. The Fyn SH2 domain was an ideal choice because structures in complex with pTyr ligand peptides are available, and replacement of all three cysteine residues in the Fyn SH2 domain with serine residues does not affect ligand binding [27, 28]. Removal of cysteine from the domain is favorable for phage display screening to reduce false-positive binding.

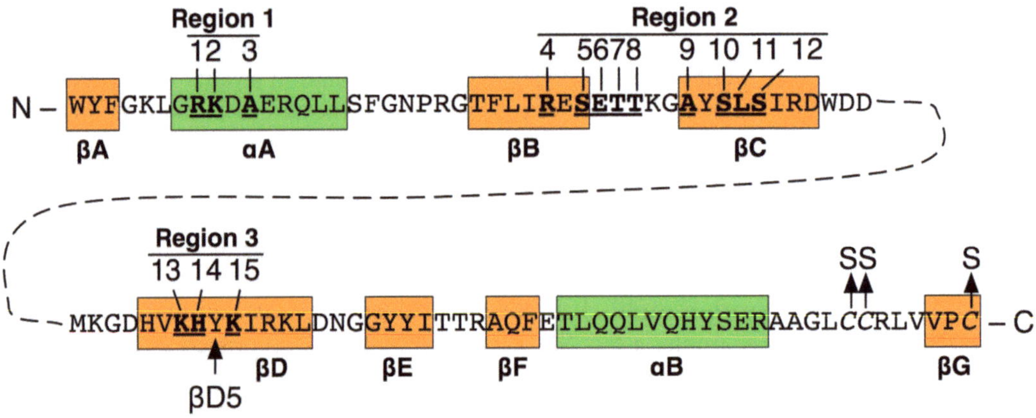

Fig. 3 The sequence of the Fyn SH2 domain. α Helices are colored *green*, while β strands are in *orange*. Three cysteine residues (*italicized*) were replaced with serine residues for phage display. The 15 residues mutated in the phage display experiment were numbered from 1 to 15, clustered into the three regions for the Kunkel mutagenesis

```
{========================= selections ============================}

{* selection for first atom set *}
{===>} set_select.1=(segid F and not(name c or name n or name ca or name o));

{* selection for second atom set *}
{===>} set_select.2=(segid P and residue 204);

{========================= parameters ============================}

{* distance cutoff value in Angstroms *}
{* only interactions which are closers than the specified distance
   cutoff are included *}
{===>} cutoff=100;

{* selection for contacting atoms *}
{* only interactions between selected atoms are included *}
{===>} contact_select=(known);
```

Fig. 4 A part of the contact.inp file used for distance calculation with the program CNS. This example uses the Fyn SH2 domain-ligand complex structure shown in Fig. 2. Residues 204 in Chain P (or segid P) is phosphotyrosine in the PDB file 1AOT

Below we describe the main steps by which to generate superbinders from the wild-type Fyn SH2 domain.

1. Identify pTyr-binding pocket residues of the Fyn SH2 domain based on the ligand complex structure (Figs. 3 and 4).
2. Pick 15 pocket residues to which simultaneous amino acid substitutions are applied (Fig. 5).

Secondary structure	Fyn (1AOT) Residue	Fyn (1AOT) Distance	Src (1FMK) Residue	Src (1FMK) Distance	Grb2 (1JYR) Residue	Grb2 (1JYR) Distance	Mutation Position	Region
αA1	GLY	n/a	THR	7.1	PRO	8.3		
αA2	ARG	3.0	ARG	2.8	ARG	2.8	1	Region 1
αA3	LYS	7.1	ARG	8.9	ALA	9.1	2	
αA4	ASP	8.7	GLU	10.9	LYS	10.9		
αA5	ALA	7.5	SER	6.7	ALA	7.8	3	
αA6	GLU	9.2	GLU	7.3	GLU	7.6		
αA7	ARG	12.4	ARG	13.4	GLU	13.4		
αA8	GLN	11.9	LEU	13.8	MET	13.7		
αA9	LEU	11.9	LEU	9.8	LEU	10.9		
αA10	LEU	14.5	LEU	12.6	SER	15.3		
AB1	SER	16.6	ASN	18.3	LYS	18.4		
AB2	PHE	20.4	ALA	21.3	–	–		
AB3	GLY	n/a	GLU	21.4	GLN	17.3		
AB4	ASN	17.9	ASN	17.4	ARG	22.9		
AB5	PRO	19.8	PRO	21.0	HIS	21.6		
AB6	ARG	17.3	ARG	17.0	ASP	18.3		
AB7	GLY	n/a	GLY	n/a	GLY	n/a		
βB1	THR	15.0	THR	16.2	ALA	16.8		
βB2	PHE	10.3	PHE	13.4	PHE	12.9		
βB3	LEU	9.4	LEU	9.3	LEU	9.2		
βB4	ILE	10.1	VAL	10.9	ILE	9.7		
βB5	ARG	3.2	ARG	2.7	ARG	2.7	4	Region 2
βB6	GLU	8.4	GLU	7.3	GLU	8.0		
βB7	SER	4.0	SER	3.2	SER	2.7	5	
BC1	GLU	5.2	GLU	3.1	GLU	4.2	6	
BC2	THR	3.3	THR	2.7	SER	2.6	7	
BC3	THR	7.6	THR	5.7	ALA	5.9	8	
BC4	LYS	11.5	LYS	10.2	PRO	8.3		
BC5	GLY	n/a	GLY	n/a	GLY	n/a		
βC1	ALA	7.7	ALA	7.5	ASP	6.5	9	
βC2	TYR	9.2	TYR	9.1	PHE	8.0		
βC3	SER	4.0	CYS	4.0	SER	2.6	10	
βC4	LEU	7.5	LEU	8.5	LEU	7.7	11	
βC5	SER	7.1	SER	6.0	SER	6.2	12	
βC6	ILE	8.2	VAL	10.4	VAL	10.4		
βC7	ARG	13.6	SER	14.5	LYS	14.9		
βC8	ASP	13.7	ASP	11.7	PHE	13.7		
CD1	TRP	18.6	PHE	17.3	GLY	n/a		
CD2	ASP	17.7	ASP	14.5	–	–		
CD3	ASP	23.0	ASN	20.4	–	–		
CD4	MET	20.0	ALA	17.4	–	–		
CD5	LYS	15.4	LYS	10.7	–	–		
CD6	GLY	n/a	GLY	n/a	–	–		
CD7	ASP	17.5	LEU	17.2	ASN	18.4		
βD1	HIS	13.8	ASN	10.7	ASP	11.5		
βD2	VAL	8.9	VAL	8.6	VAL	9.4		
βD3	LYS	7.0	LYS	7.4	GLN	7.5	13	Region 3
βD4	HIS	3.2	HIS	3.8	HIS	4.0	14	
βD5	TYR	5.5	TYR	6.1	PHE	5.6	(*)	
βD6	LYS	3.2	LYS	3.9	LYS	3.9	15	
βD7	ILE	8.1	ILE	8.7	VAL	8.5		

Fig. 5 Distances between SH2 domain residues and pTyr in the ligand complex structures. Calculated distances for three different SH2-peptide complex structures are shown. The values are distances in angstrom between pTyr and the side chain atom of each SH2 domain residue. Distances ≤3.2 Å (a cutoff distance commonly used for hydrogen bonds) is *shaded black*, whereas distances ≤8.0 Å are in *grey*. The distance values are largely comparable between Fyn, Src, and Grb2 SH2 domains, which indicate that pTyr-binding pockets are designed similarly among SH2 domain family members

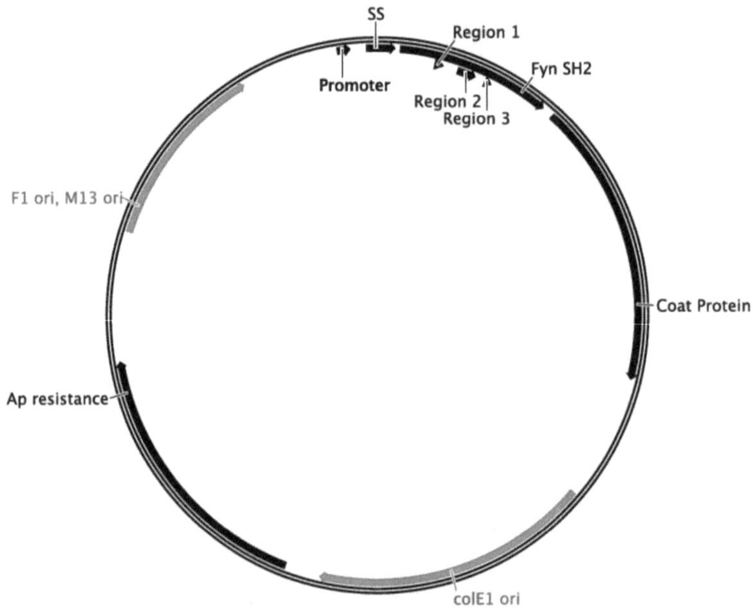

Fig. 6 A phagemid vector designed for the Fyn SH2 domain phage display. The phagemid vector contains origins of single-stranded (f1 ori) and double-stranded (colE1 ori) DNA replication. A β-lactamase gene confers resistance to ampicillin (Ap) and carbenicillin (Ap resistance). For SH2 domain display, the phagemid also contains a cassette consisting of a promoter that drives transcription of the SH2 domain fused to minor phage coat protein P3. The three regions to be mutated in the SH2 domain are also shown. A secretion signal (SS) directs the SH2-P3 fusion to the periplasm, where it is incorporated into phage particles that are secreted from the *Escherichia coli* host

3. Prepare a phagemid that encodes the Fyn SH2 domain, in which the three cysteine residues are replaced with serine residues (Figs. 3 and 6).
4. Apply the Kunkel method to the Fyn SH2 phagemid to introduce random amino acid substitutions at the 15 positions and obtain a library of mutant SH2 domains (the variants). We divided these 15 positions into three regions to facilitate the Kunkel reactions (Figs. 7 and 8) (*see* **Note 1**).
5. Produce phages that display a library of variant SH2 domains on the phage surface.
6. Screen the phage pool against immobilized pTyr bait peptides (Fig. 9).
7. Wash off unbound phages.
8. Elute positive (bound) phages from bait peptides and infect *E. coli* to amplify the phages.

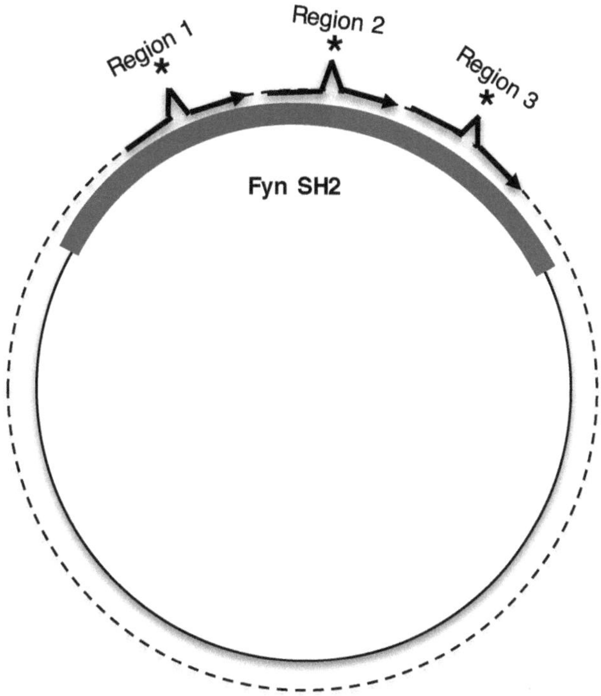

Fig. 7 Oligonucleotide-directed mutagenesis for library construction. Three synthetic oligonucleotides are annealed to the dU-ssDNA template (*solid circle*). The oligonucleotides are designed to introduce mutations (*asterisks*) at the 15 positions of the Fyn SH2 pTyr-binding pocket. Region1: Position 1–3, Region2: Position 4–12, Region3: Position 13–15. Heteroduplex CCC-dsDNA is enzymatically synthesized by T7 DNA polymerase and T4 DNA ligase (*dashed circle*) and is introduced into an *Escherichia coli* host where the mismatched region is repaired to either the wild-type sequence or the mutant sequence. The template contains stop codons in the regions to be mutated, and thus SH2 proteins are only displayed by clones in which all the stop codons have been repaired by the mutagenic oligonucleotides

9. Repeat the screening process (**steps 6–8**) for five rounds to enrich bound phages.
10. Isolate the positive phages and verify them by phage enzyme-linked immunosorbent assay (phage ELISA).
11. Sequence the positive clones (Figs. 10 and 11).
12. Analyze the sequences of positive variants to identify residue substitutions that may potentially increase pTyr-binding affinity (Fig. 12).
13. Evaluate binding affinities of mutants by in vitro binding assays (Fig. 13).

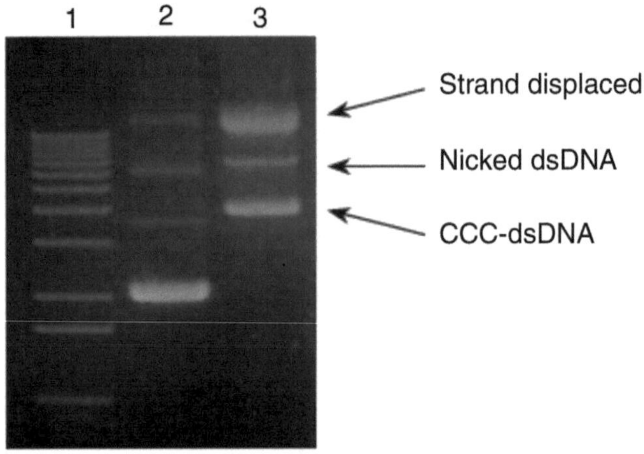

Fig. 8 In vitro synthesis of heteroduplex CCC-dsDNA. The reaction products were electrophoresed on a 1.0 % TAE/agarose gel for DNA visualization. The *lanes* were loaded as follows: (*1*) 1 kb DNA marker, (*2*) dU-ssDNA template, and (*3*) reaction product from the heteroduplex CCC-dsDNA synthesis reaction. The lower band is correctly extended and ligated CCC-dsDNA. The middle band is nicked dsDNA. The upper band is strand-displaced DNA

2 Materials

2.1 Identification of pTyr-Binding Pocket Residues

1. The protein data bank (PDB) file of your target SH2 domain or its homolog in complex with pTyr.
2. A program that can calculate atomic distances in a structure. We used the program Crystallography & NMR System (CNS, http://cns-online.org) [29].
3. Topology and parameter files that define phosphotyrosine (residue name PTR). The files can be obtained from the HIC-Up server (http://xray.bmc.uu.se/hicup/) [30].

2.2 Phage Display Experiments

1. 0.2-cm gap electroporation cuvette.
2. 1.0 M H_3PO_4.
3. 1.0 M Tris base, pH 8.0.
4. 1.0 mM HEPES, pH 7.4.
5. 3,3′,5,5′-tetramethylbenzidine/H_2O_2 peroxidase (TMB) substrate.
6. 10% (v/v) ultrapure glycerol.
7. 10 mM ATP.
8. 10× TM buffer (0.1 M $MgCl_2$, 0.5 M Tris, pH 7.5).
9. 100 mM dNTP mix.

		-3	-2	-1	0	+1	+2	+3	+4	+5	+6	
Bait-1	Biotin–linker–	E	P	Q	pY	E	N	E	E	E		–amide
Bait-2	Biotin–linker–	S	K	A	pY	G	N	G	A	S	S	–amide
Bait-3	Biotin–linker–	D	V	R	pY	V	N	A	A	K	F	–amide
Bait-4	Biotin–linker–	A	E	L	pY	S	N	A	A	P	V	–amide
Bait-5	Biotin–linker–	D	H	Q	pY	Y	N	D	A	P	G	–amide
Bait-6	Biotin–linker–	Q	P	E	pY	V	N	Q	A	D	V	–amide
Bait-7	Biotin–linker–	V	P	E	pY	I	N	Q	S	V	P	–amide
Bait-8	Biotin–linker–	N	P	V	pY	H	N	Q	P	L	N	–amide
Bait-9	Biotin–linker–	R	K	T	pY	V	N	T	T	L	Y	–amide
Bait-10	Biotin–linker–	N	P	E	pY	L	N	T	V	Q	P	–amide
Bait-11	Biotin–linker–	E	P	L	pY	L	N	T	F	A	N	–amide
Bait-12	Biotin–linker–	E	P	Q	pY	E	E	I	E	E		–amide
Bait-13	Biotin–linker–	T	T	E	pY	S	E	I	K	I	H	–amide
Bait-14	Biotin–linker–	P	Q	R	pY	L	V	I	Q	G	D	–amide
Bait-15	Biotin–linker–	D	R	V	pY	E	E	L	N	I	Y	–amide
Bait-16	Biotin–linker–	E	N	L	pY	E	G	L	N	L	D	–amide
Bait-17	Biotin–linker–	E	S	I	pY	E	V	L	G	M	Q	–amide
Bait-18	Biotin–linker–	N	A	E	pY	L	R	V	A	P	Q	–amide
Bait-19	Biotin–linker–	A	D	E	pY	L	I	P	Q	Q	G	–amide
Bait-20	Biotin–linker–	D	P	H	pY	Q	D	P	H	S	T	–amide
Bait-21	Biotin–linker–	S	K	P	pY	D	G	I	P	A	S	–amide
Bait-22	Biotin–linker–	S	N	F	pY	R	A	L	M	D	E	–amide
Bait-23	Biotin–linker–	I	D	V	pY	M	I	M	V	K	A	–amide
Bait-24	Biotin–linker–	E	P	Q	pY	E	E	E	L	E		–amide
Bait-25	Biotin–linker–	Q	D	T	pY	E	T	H	L	E	T	–amide
Bait-26	Biotin–linker–	Y	Q	H	pY	D	L	D	L	K	D	–amide
Bait-27	Biotin–linker–	Y	E	E	pY	T	S	K	L	D	A	–amide
Bait-28	Biotin–linker–	R	N	D	pY	D	D	T	I	P	I	–amide
Bait-29	Biotin–linker–	L	Q	R	pY	S	S	D	P	T	G	–amide
Bait-30	Biotin–linker–	N	P	D	pY	Q	Q	D	F	F	P	–amide
Bait-31	Biotin–linker–	E	K	E	pY	H	A	E	G	G	K	–amide
Bait-32	Biotin–linker–	P	P	I	pY	T	S	R	A	R	I	–amide
Bait-33	Biotin–linker–	Q	F	V	pY	R	D	G	G	F	A	–amide
Bait-34	Biotin–linker–	K	A	E	pY	L	K	N	N	I	L	–amide
Bait-35	Biotin–linker–	P	P	P	pY	R	H	R	N	T	V	–amide

Fig. 9 A list of bait peptidesimmobilized for phage display screening. Each peptide contains pTyr and N-terminal biotin. The linker is 6-aminohexanoic acid (ahx). The peptides are classified into three groups: (*1*) Asn at pTyr+2 position (colored *orange*), (*2*) a hydrophobic residue at pTyr+3 position (colored *cyan*), and (*3*) a hydrophobic residue at the pTyr+4 position (colored *magenta*)

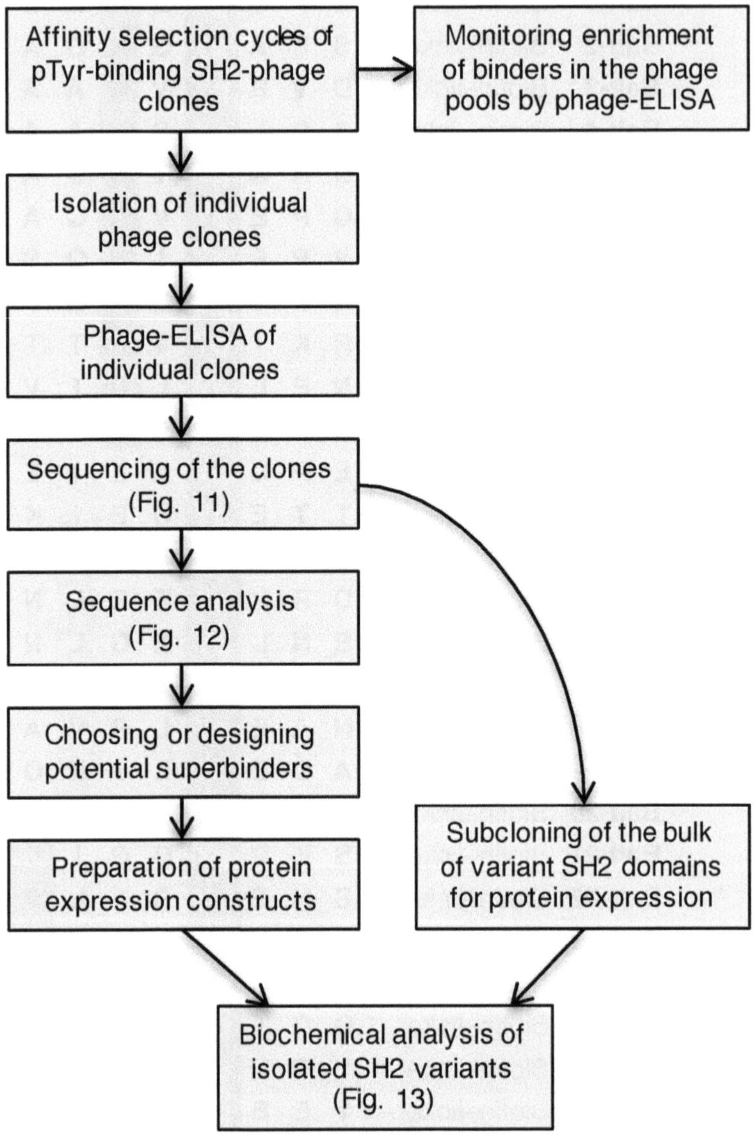

Fig. 10 Sequencing of the phage clones and preparation of protein expression constructs. Phage clones isolated and confirmed for peptide binding by phage ELISA are sent for DNA sequencing to identify promising mutations. Potential superbinders may be designed based on the sequence analysis. Alternatively, instead of predicting favorable mutations from the sequencing results, the bulk of variant SH2 domain genes can be directly amplified by PCR from the phage pool for subcloning into a protein overexpression vector. In this case, a number of mutant protein expression constructs can be obtained at once for further experimental analysis

	Region 1					Region 2													Region 3			
Position	1	2	–	3	–	4	–	5	6	7	8	–	9	–	10	11	12	–	13	14	–	15
Wild type	R	K	D	A	ERQLLSFGNPRGTFLI	R	E	S	E	T	T	KG	A	Y	S	L	S	IRDWDDMKGDHV	K	H	Y	K
Variant-1	R	K	D	A	ERQLLSFGNPRGTFLI	R	E	S	E	T	F	KG	E	Y	S	L	S	IRDWDDMKGDHV	K	H	Y	R
Variant-2	R	K	D	A	ERQLLSFGNPRGTFLI	R	E	S	E	R	S	KG	D	Y	V	L	S	IRDWDDMKGDHV	K	H	Y	K
Variant-3	R	K	D	A	ERQLLSFGNPRGTFLI	R	E	S	E	S	I	KG	A	Y	S	L	S	IRDWDDMKGDHV	K	H	Y	L
Variant-4	R	K	D	A	ERQLLSFGNPRGTFLI	R	E	S	E	T	T	KG	S	Y	V	L	S	IRDWDDMKGDHV	K	H	Y	L
Variant-5	R	K	D	A	ERQLLSFGNPRGTFLI	R	E	S	E	T	I	KG	S	Y	S	L	S	IRDWDDMKGDHV	K	H	Y	L
Variant-6	R	K	D	A	ERQLLSFGNPRGTFLI	R	E	S	E	T	T	KG	A	Y	A	L	S	IRDWDDMKGDHV	K	H	Y	I
Variant-7	R	I	D	A	ERQLLSFGNPRGTFLI	R	E	S	E	S	T	KG	A	Y	V	L	S	IRDWDDMKGDHV	K	H	Y	K
Variant-8	R	K	D	A	ERQLLSFGNPRGTFLI	R	E	S	E	T	I	KG	D	Y	S	L	S	IRDWDDMKGDHV	K	H	Y	L
Variant-9	R	K	D	A	ERQLLSFGNPRGTFLI	R	E	S	E	T	T	KG	A	Y	V	L	S	IRDWDDMKGDHV	K	H	Y	K
Variant-10	R	K	D	A	ERQLLSFGNPRGTFLI	R	E	S	P	R	T	KG	V	Y	V	L	S	IRDWDDMKGDHV	K	H	Y	K
Variant-11	R	K	D	A	ERQLLSFGNPRGTFLI	R	E	S	E	S	T	KG	A	Y	A	L	S	IRDWDDMKGDHV	K	H	Y	L
Variant-12	R	E	D	A	ERQLLSFGNPRGTFLI	R	E	S	E	T	V	KG	A	Y	A	L	S	IRDWDDMKGDHV	K	H	Y	I
Variant-13	R	K	D	A	ERQLLSFGNPRGTFLI	R	E	S	E	T	T	KG	A	Y	S	L	S	IRDWDDMKGDHV	K	H	Y	L
Variant-14	R	K	D	A	ERQLLSFGNPRGTFLI	R	E	S	E	T	E	KG	A	Y	A	L	S	IRDWDDMKGDHV	K	H	Y	K
Variant-15	R	Q	D	A	ERQLLSFGNPRGTFLI	R	E	S	E	T	F	KG	S	Y	A	L	S	IRDWDDMKGDHV	K	H	Y	L
Variant-16	R	E	D	A	ERQLLSFGNPRGTFLI	R	E	S	D	N	F	KG	G	Y	V	L	S	IRDWDDMKGDHV	K	H	Y	K
Variant-17	R	Q	D	A	ERQLLSFGNPRGTFLI	R	E	S	E	T	S	KG	N	Y	A	L	S	IRDWDDMKGDHV	K	H	Y	L
Variant-18	R	T	D	A	ERQLLSFGNPRGTFLI	R	E	S	E	S	V	KG	A	Y	V	L	S	IRDWDDMKGDHV	K	H	Y	L
Variant-19	R	K	D	A	ERQLLSFGNPRGTFLI	R	E	S	E	S	I	KG	G	Y	V	L	S	IRDWDDMKGDHV	K	H	Y	L
Variant-20	R	E	D	A	ERQLLSFGNPRGTFLI	R	E	S	E	S	R	KG	A	Y	A	L	S	IRDWDDMKGDHV	K	H	Y	L
Variant-21	R	K	D	A	ERQLLSFGNPRGTFLI	R	E	S	E	T	I	KG	A	Y	V	L	S	IRDWDDMKGDHV	K	H	Y	I
Variant-22	P	R	D	A	ERQLLSFGNPRGTFLI	R	E	R	V	Y	P	KG	P	Y	L	L	S	IRDWDDMKGDHV	K	P	Y	S
Variant-23	R	K	D	A	ERQLLSFGNPRGTFLI	R	E	S	E	T	I	KG	A	Y	A	L	S	IRDWDDMKGDHV	V	H	Y	L
Variant-24	R	S	D	A	ERQLLSFGNPRGTFLI	R	E	S	E	T	I	KG	A	Y	A	L	S	IRDWDDMKGDHV	K	H	Y	L
Variant-25	R	K	D	A	ERQLLSFGNPRGTFLI	R	E	S	E	S	I	KG	A	Y	V	L	S	IRDWDDMKGDHV	K	H	Y	I
Variant-26	R	E	D	A	ERQLLSFGNPRGTFLI	R	E	S	E	S	S	KG	A	Y	A	L	S	IRDWDDMKGDHV	K	H	Y	L
Variant-27	R	K	D	A	ERQLLSFGNPRGTFLI	R	E	S	E	T	I	KG	A	Y	S	L	S	IRDWDDMKGDHV	K	H	Y	L
Variant-28	R	K	D	A	ERQLLSFGNPRGTFLI	R	E	S	A	S	I	KG	A	Y	A	L	S	IRDWDDMKGDHV	K	H	Y	L
Variant-29	R	K	D	A	ERQLLSFGNPRGTFLI	R	E	S	E	T	V	KG	A	Y	S	L	S	IRDWDDMKGDHV	K	H	Y	L
Variant-30	R	K	D	A	ERQLLSFGNPRGTFLI	R	E	S	K	T	V	KG	A	Y	S	L	S	IRDWDDMKGDHV	K	H	Y	L
Variant-31	R	K	D	A	ERQLLSFGNPRGTFLI	R	E	R	E	S	V	KG	E	Y	V	L	S	IRDWDDMKGDHV	K	H	Y	R
Variant-32	R	E	D	A	ERQLLSFGNPRGTFLI	R	E	S	E	S	I	KG	A	Y	S	L	S	IRDWDDMKGDHV	K	H	Y	L
Variant-33	R	I	D	A	ERQLLSFGNPRGTFLI	R	E	S	E	S	I	KG	A	Y	S	L	S	IRDWDDMKGDHV	K	H	Y	L
Variant-34	R	K	D	A	ERQLLSFGNPRGTFLI	R	E	S	E	T	A	KG	A	Y	S	L	S	IRDWDDMKGDHV	K	H	Y	L
Variant-35	R	K	D	A	ERQLLSFGNPRGTFLI	R	E	S	G	T	T	KG	A	Y	S	L	S	IRDWDDMKGDHV	K	H	Y	L
Variant-36	R	K	D	A	ERQLLSFGNPRGTFLI	R	E	S	E	T	I	KG	A	Y	S	L	S	IRDWDDMKGDHV	I	H	Y	L
Variant-37	R	K	D	A	ERQLLSFGNPRGTFLI	R	E	S	E	T	V	KG	A	Y	A	L	S	IRDWDDMKGDHV	K	H	Y	L
Variant-38	R	D	D	A	ERQLLSFGNPRGTFLI	R	E	S	E	T	V	KG	A	Y	S	L	S	IRDWDDMKGDHV	T	H	Y	L
Variant-39	R	K	D	A	ERQLLSFGNPRGTFLI	R	E	S	E	T	I	KG	A	Y	V	L	S	IRDWDDMKGDHV	K	H	Y	L
Variant-40	R	K	D	A	ERQLLSFGNPRGTFLI	R	E	S	E	S	S	KG	D	Y	A	L	S	IRDWDDMKGDHV	K	H	Y	I
Variant-41	R	R	D	A	ERQLLSFGNPRGTFLI	R	E	S	E	S	R	KG	L	Y	S	L	S	IRDWDDMKGDHV	K	H	Y	L
Variant-42	R	E	D	A	ERQLLSFGNPRGTFLI	R	E	S	E	T	F	KG	D	Y	A	L	S	IRDWDDMKGDHV	K	H	Y	V
Variant-43	R	K	D	A	ERQLLSFGNPRGTFLI	R	E	S	E	S	E	KG	A	Y	A	L	S	IRDWDDMKGDHV	K	H	Y	K
Variant-44	R	K	D	A	ERQLLSFGNPRGTFLI	R	E	S	E	S	R	KG	A	Y	S	L	S	IRDWDDMKGDHV	K	H	Y	L
Variant-45	R	E	D	A	ERQLLSFGNPRGTFLI	R	E	S	E	S	R	KG	A	Y	S	L	S	IRDWDDMKGDHV	K	H	Y	L
Variant-46	R	K	D	A	ERQLLSFGNPRGTFLI	R	E	S	E	T	Y	KG	D	Y	S	L	S	IRDWDDMKGDHV	K	H	Y	R
Variant-47	R	K	D	A	ERQLLSFGNPRGTFLI	R	E	S	E	T	S	KG	D	Y	A	L	S	IRDWDDMKGDHV	K	H	Y	L
Variant-48	R	K	D	A	ERQLLSFGNPRGTFLI	R	E	S	E	S	I	KG	A	Y	A	L	S	IRDWDDMKGDHV	N	H	Y	L
Variant-49	R	R	D	A	ERQLLSFGNPRGTFLI	R	E	S	Q	S	I	KG	A	Y	V	L	S	IRDWDDMKGDHV	K	H	Y	I
Variant-50	R	K	D	A	ERQLLSFGNPRGTFLI	R	E	S	E	T	I	KG	A	Y	A	L	S	IRDWDDMKGDHV	K	H	Y	L
Variant-51	R	K	D	A	ERQLLSFGNPRGTFLI	R	E	S	E	N	V	KG	G	Y	V	L	S	IRDWDDMKGDHV	K	H	Y	K
Variant-52	R	K	D	A	ERQLLSFGNPRGTFLI	R	E	S	E	S	T	KG	A	Y	A	L	S	IRDWDDMKGDHV	K	H	Y	K
Variant-53	R	K	D	A	ERQLLSFGNPRGTFLI	R	E	S	E	T	A	KG	A	Y	V	L	S	IRDWDDMKGDHV	K	H	Y	L
Variant-54	R	K	D	A	ERQLLSFGNPRGTFLI	R	E	S	E	T	Y	KG	A	Y	S	L	S	IRDWDDMKGDHV	K	H	Y	L
Variant-55	R	L	D	A	ERQLLSFGNPRGTFLI	R	E	S	D	S	N	KG	G	Y	V	L	S	IRDWDDMKGDHV	K	H	Y	K
Variant-56	R	K	D	A	ERQLLSFGNPRGTFLI	R	E	S	E	R	T	KG	A	Y	V	L	S	IRDWDDMKGDHV	K	H	Y	K
Variant-57	R	K	D	A	ERQLLSFGNPRGTFLI	R	E	S	V	N	N	KG	A	Y	V	L	S	IRDWDDMKGDHV	K	H	Y	K
Variant-58	R	K	D	A	ERQLLSFGNPRGTFLI	R	E	S	K	N	S	KG	G	Y	V	L	S	IRDWDDMKGDHV	K	H	Y	R
Variant-59	R	K	D	A	ERQLLSFGNPRGTFLI	R	E	S	E	S	T	KG	A	Y	V	L	S	IRDWDDMKGDHV	K	H	Y	K
Variant-60	R	K	D	A	ERQLLSFGNPRGTFLI	R	E	S	E	S	I	KG	A	Y	A	L	S	IRDWDDMKGDHV	K	H	Y	L
Variant-61	R	K	D	A	ERQLLSFGNPRGTFLI	R	E	S	V	R	S	KG	G	Y	I	L	S	IRDWDDMKGDHV	K	H	Y	K
Variant-62	R	K	D	A	ERQLLSFGNPRGTFLI	R	E	S	K	T	T	KG	A	Y	A	L	S	IRDWDDMKGDHV	K	H	Y	K
Variant-63	R	K	D	A	ERQLLSFGNPRGTFLI	R	E	S	E	T	H	KG	A	Y	A	L	S	IRDWDDMKGDHV	K	H	Y	L

Fig. 11 Sequenced variants. Sequences of the 63 unique Fyn SH2 variants between Position 1 and 15 are shown here. Amino acids substituted from the wild type are colored *green*. Variant-37 (colored *red*) was identified to be a superbinder

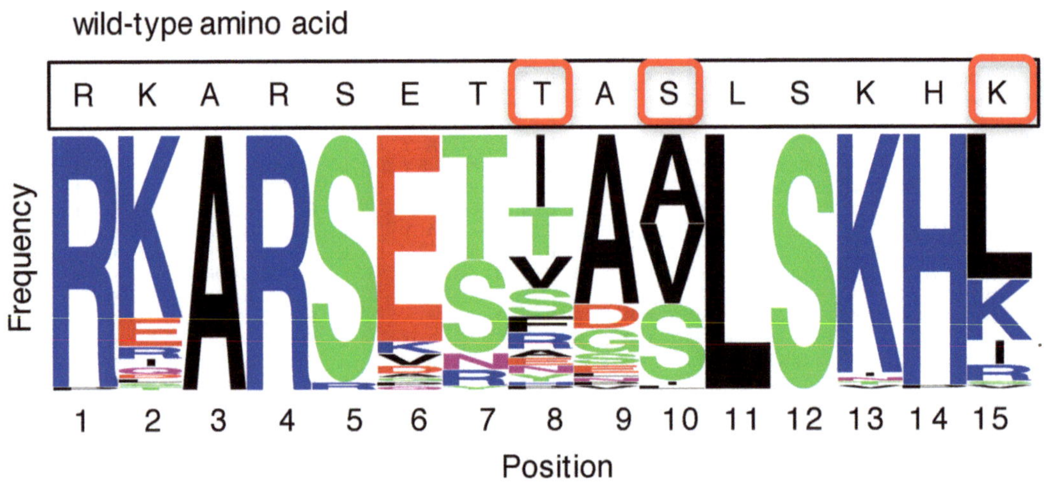

Fig. 12 Amino acid frequency observed from the 63 variants. At the three positions, 8, 10, and 15 *squared* in *red*, a non-wild-type residue is observed more frequently than the wild-type residue

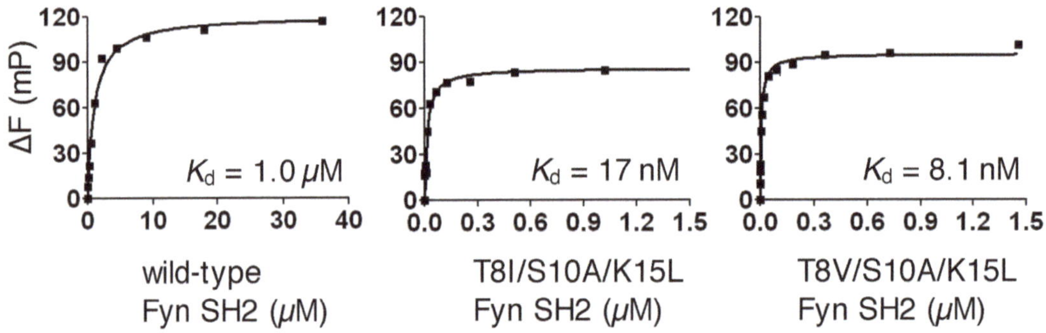

Fig. 13 Examples of fluorescence polarization assay. The SH2 domain is titrated. The vertical axis (Δ*F*) is fluorescence polarization after subtracting the blank value (at 0 μM SH2 domain). The peptide in this example is Shc1 pTyr239-derived peptide. Derived K_d values are also shown in each plot. Note the large differences in the presented titration range of the SH2 domains between the wild type and superbinders

10. 96-well Maxisorp immunoplates.
11. 100 mM HCl.
12. 100 mM dithiothreitol (DTT).
13. 2YT medium (10 g bacto-yeast extract, 16 g bacto-tryptone, 5 g NaCl. Add water to 1.0 L; adjust pH to 7.0 with NaOH; autoclave).
14. 2YT/carb/cmp medium (2YT, 50 μg/mL carbenicillin, 5 μg/mL chloramphenicol).
15. 2YT/carb/kan medium (2YT, 50 μg/mL carbenicillin, 25 μg/mL kanamycin).

16. 2YT/carb/kan/uridine medium (2YT, 50 μg/mL carbenicillin, 25 μg/mL kanamycin, 0.25 μg/mL uridine).
17. 2YT/carb/KO7 medium (2YT, 50 μg/mL carbenicillin, 10^{10} phage/mL of M13KO7).
18. 2YT/tet medium (2YT, 5 μg/mL tetracycline).
19. 2YT/carb/tet medium (2YT, 50 μg/mL carbenicillin, 10 μg/mL tetracycline).
20. 2YT/carb/tet/KO7 (2YT, 50 μg/mL carbenicillin, 10 μg/mL tetracycline, 10^{10} phage/mL of M13KO7).
21. 2YT/kan medium (2YT, 25 μg/mL kanamycin).
22. 2YT/kan/tet medium (2YT, 25 μg/mL kanamycin, 5 μg/mL tetracycline).
23. 2YT top agar (16 g tryptone, 10 g yeast extract, 5 g NaCl, 7.5 g granulated agar. Add water to 1.0 L and adjust pH to 7.0 with NaOH, heat to dissolve, autoclave).
24. Taq DNA polymerase.
25. Carbenicillin (5 mg/mL in water, filter sterilize).
26. Chloramphenicol (50 mg/mL in ethanol).
27. ECM-600 electroporator.
28. *E. coli* CJ236.
29. *E. coli* XL1-blue.
30. Horseradish peroxidase/anti-M13 antibody conjugate.
31. Kanamycin (5 mg/mL in water, filter sterilize).
32. LB/carb plates (LB agar, 50 μg/mL carbenicillin).
33. LB/tet plates (LB agar, 5 μg/mL tetracycline).
34. M13KO7 helper phage.
35. Magnetic stir bars (2 in.) soaked in ethanol.
36. Neutravidin.
37. Phosphate-buffered saline (PBS) (137 mM NaCl, 3 mM KCl, 8 mM Na_2HPO_4, 1.5 mM KH_2 PO_4; adjust pH to 7.2 with HCl, autoclave).
38. PBS, 0.2 % bovine serum albumin (BSA).
39. PBT buffer (PBS, 0.05 % Tween 20, 0.2 % BSA).
40. PT buffer (PBS, 0.05 % Tween 20).
41. PEG/NaCl (20% PEG-8000 (w/v), 2.5 M NaCl; mix and autoclave).
42. Qiagen QIAprep Spin M13 Kit (spin column, buffers MP, MLB, PE, EB).
43. Qiagen QIAquick DNA purification kit (spin column, buffers QG, PE).

44. SOC medium (5 g bacto-yeast extract, 20 g bacto-tryptone, 0.5 g NaCl, 0.2 g KCl; add water to 1 L and adjust pH to 7.0 with NaOH, autoclave; add 5.0 mL of autoclaved 2.0 M $MgCl_2$ and 20 mL of filter sterilized 1.0 M glucose).
45. T4 polynucleotide kinase.
46. T4 DNA ligase.
47. T7 DNA polymerase.
48. Tetracycline (5 mg/mL in water, filter sterilize).
49. Ultrapure irrigation USP water.
50. Uridine (25 mg/mL in water, filter sterilize).
51. Ultrapure glycerol.

2.3 In-Solution Peptide Binding Assay Using Fluorescence Polarization

1. A multilabel plate reader equipped with fluorescence polarization function (e.g., PerkinElmer 2103 Envision).
2. 96- or 384-well black flat bottom microplates.
3. Isopropyl β-D-1-thiogalactopyranoside (IPTG). Make a stock solution in water and filter sterilize.
4. *E. coli* lysis buffer for hexa-histidine tag affinity purification: 20 mM sodium phosphate, 20 mM imidazole, 0.1 M NaCl, pH 7.8, 0.5 % Triton-X 100, 1 mg/mL lysozyme.
5. Elution buffer for hexa-histidine tag affinity purification: 20 mM sodium phosphate, 0.25 M imidazole, 0.1 M NaCl, pH 7.8.
6. Dialysis buffer (also used for fluorescence polarization assay): 20 mM Tris–HCl, pH 7.0, 150 mM NaCl (*see* **Note 2**).
7. Peptides N-terminally labeled with fluorescein. We prepare peptides with an in-house peptide synthesizer. Alternatively, peptides may be purchased from companies that provide custom peptide synthesis service.

3 Methods

The following description is based on the Fyn SH2 domain that we used in our previous research [15]. A similar protocol or strategy may work well with other SH2 domains.

3.1 Identification of pTyr-Binding Pocket Residues to Mutate

The first step is to decide the number of residues to mutate simultaneously and choose which residues to mutate. We picked 15 residues from the Fyn SH2 domain with the intention to include all residues that form the pTyr-binding pocket in the mutagenesis.

The pocket residues can be identified from a list of distances between the pTyr residue and each SH2 domain residue. To determine the pocket residues of the Fyn SH2 domain, we used the NMR structure of the Fyn SH2 domain in complex with a pTyr-containing ligand peptide (PDB accession 1AOT) [27]. We calculated the minimal distance between each side chain of the SH2 domain and any atom from pTyr (Figs. 4 and 5), and identified pTyr-binding pocket residues whose side chains are located within 8 Å from the pTyr moiety. Hydrogen atoms were ignored for distance calculation. The distance calculation procedure using the program CNS is described below.

1. From the coordinates file 1AOT.pdb, run generate_easy.inp to produce PDB and MSF files that can be used as input files for contact.inp. You will need topology and parameter files that define phosphotyrosine (residue name PTR).
2. Edit necessary parameters in contact.inp. Figure 4 is a part of the input file used for 1AOT.pdb.
3. 16 Fyn SH2 domain residues are located within 8 Å from pTyr (Figs. 3 and 4). Exclude the βD5 residue to avoid affecting a second pocket (the specificity-determining pocket) (Fig. 1) (*see* **Note 3**). Choose the remaining 15 residues as targets for mutation.
4. Optionally, repeat the distance calculation using different SH2 domain-pTyr complex structures for comparison (Fig. 5).

3.2 Mutation of Cysteine Residues in the SH2 Domain

For phage display experiments, it is desirable to remove cysteine to minimize false positives during screening. In the case of the human Fyn SH2 domain (residues from Val138 to Gly249), the sequence contains three cysteine residues Cys239, 240, and 246 (Fig. 3). Literature suggested that replacement of these cysteines with serine residues did not damage the structure and pTyr-binding function of the SH2 domain [27, 28]. Using site-directed mutagenesis, we prepared the Fyn SH2 domain cDNA without cysteine for subsequent subcloning into the phagemid.

3.3 Purification of dU-ssDNA Template

We introduced mutations at the phosphotyrosine-binding pocket of the (cysteine-less) Fyn SH2 domain by site-directed mutagenesis. The Fyn SH2 domain was firstly subcloned into the phagemid pFN-OM6 to express the SH2 domain with phage coat protein P3 (Fig. 6). The pTyr-binding pocket of the SH2 domain was then incorporated with a stop codon (i.e., TAA). The pTyr-binding pocket is repaired by site-directed mutagenesis (Kunkel method) using degenerate oligonucleotides that anneal to the TAA-containing area of the template (Fig. 7). Mutagenesis efficiency depends on template purity, and thus the use of highly pure dU-ssDNA (Uracil-containing ssDNA) is critical for successful library construction. We use the Qiagen QIAprep Spin M13 Kit for dU-ssDNA purification, and the following is a modified version of the

Qiagen protocol. It yields at least 20 μg of dU-ssDNA for a medium copy number phagemid (e.g., pBR322 backbone), and this is sufficient for the construction of one library.

1. From a fresh LB/carb plate, pick a single colony of *E. coli* CJ236 (or another dut–/ung– strain) harboring the appropriate phagemid into 1 mL of 2YT medium supplemented with M13KO7 helper phage (10^{10} pfu/mL) and appropriate antibiotics to maintain the host F' episome and the phagemid. For example, 2YT/carb/cmp medium contains carbenicillin to select for phagemids that carry the β-lactamase gene and chloramphenicol to select for the F′ episome of *E. coli* CJ236.
2. Shake at 200 rpm and 37 °C for 2 h and add kanamycin (25 μg/mL) to select for clones that have been co-infected with M13KO7, which carries a kanamycin resistance gene.
3. Shake at 200 rpm and 37 °C for 6 h and transfer the culture to 30 mL of 2YT/carb/kan/uridine medium.
4. Shake for 20 h at 200 rpm and 37 °C.
5. Centrifuge for 10 min at 27,000 × *g* and 4 °C. Transfer the supernatant to a new tube containing 1/5 volume of PEG/NaCl and incubate for 5 min at room temperature.
6. Centrifuge for 10 min at 12,000 × *g* and 4 °C. Decant the supernatant; centrifuge briefly at 2000 × *g* and aspirate the remaining supernatant (*see* **Note 4**).
7. Resuspend the phage pellet in 0.5 mL of PBS and transfer to a 1.5-mL microcentrifuge tube.
8. Centrifuge for 5 min at 16,000 × *g* in a microcentrifuge to get rid of any insoluble matter, and transfer the supernatant to a 1.5-mL microcentrifuge tube.
9. Add 7.0 μL of buffer MP and mix. Incubate at room temperature for at least 2 min.
10. Apply the sample to a QIAprep spin column in a 2-mL microcentrifuge tube. Centrifuge for 30 s at 6000 × *g* in a microcentrifuge. Discard the flow-through. The phage particles remain bound to the column matrix.
11. Add 0.7 mL of buffer MLB to the column. Centrifuge for 30 s at 6000 × *g* and discard the flow-through.
12. Add 0.7 mL of buffer MLB. Incubate at room temperature for at least 1 min.
13. Centrifuge at 6000 × *g* for 30 s. Discard the flow-through. The DNA is separated from the protein coat and remains adsorbed to the matrix.
14. Add 0.7 mL of buffer PE. Centrifuge at 6000 × *g* for 30 s and discard the flow-through.

15. Repeat **step 14**. Residual proteins and salt are removed.
16. Centrifuge at 6000 ×*g* for 30 s in a fresh 1.5-mL microcentrifuge tube to remove residual PE buffer.
17. Transfer the column to a fresh 1.5-mL microcentrifuge tube.
18. Add 100 μL of buffer EB to the center of the column membrane. Incubate at room temperature for 10 min.
19. Centrifuge for 30 s at 6000 ×*g*. Save the eluant, which contains the purified dU-ssDNA.
20. Analyze the DNA by electrophoresing 1.0 μL on a TAE/agarose gel. The DNA should appear as a predominant single band, but faint bands with lower electrophoretic mobility are often visible. These are likely caused by secondary structure in the dU-ssDNA.
21. Determine the DNA concentration by measuring absorbance at 260 nm (*A*260 = 1.0 for 33 ng/μL of ssDNA). Typical DNA concentrations range from 200 to 500 ng/μL.

3.4 In Vitro Synthesis of CCC-dsDNA

A three-step procedure is used to incorporate the mutagenic oligonucleotides into heteroduplex covalently closed circular, double-strand DNA (CCC-dsDNA), using dU-ssDNA as a template. The protocol described here is an optimized, large-scale version of a published method [31]. The oligonucleotide is 5′-phosphorylated and annealed to a dU-ssDNA template. The oligonucleotide is enzymatically extended and ligated to form heteroduplex CCC-dsDNA, which is then purified and desalted. The protocol below produces ~20 μg of highly pure, low conductance CCC-dsDNA. This is sufficient for the construction of a library containing more than 10^{10} unique members.

3.5 Oligonucleotide Phosphorylation with T4 Polynucleotide Kinase

1. In a 1.5-mL microcentrifuge tube, combine 0.6 μg of the mutagenic oligonucleotide, 2.0 μL 10× TM buffer, 2.0 μL 10 mM ATP, and 1.0 μL 100 mM DTT. Add water to a total volume of 20 μL.
2. Add 20 units of T4 polynucleotide kinase. Incubate for 1 h at 37 °C and use immediately for annealing.

3.6 Annealing of the Oligonucleotides to the Template

1. To 20 μg of dU-ssDNA template, add 25 μL 10× TM buffer, 20 μL of each phosphorylated oligonucleotide, and water to a final volume of 250 μL.
2. Analogous to PCR amplification, perform a single cycle of denaturation and annealing to anneal the phosphorylated mutagenic oligonucleotide to the dU-ssDNA template, according to the following protocol: incubate at 90 °C for 3 min, 50 °C for 3 min, and 20 °C for 5 min.

3.7 Enzymatic Synthesis of CCC-dsDNA

1. To the annealed oligonucleotide/template mixture, add 10 μL 10 mM ATP, 10 μL 100 mM dNTP mix, 15 μL 100 mM DTT, 30 Weiss units T4 DNA ligase, and 30 units T7 DNA polymerase.
2. Incubate overnight at 20 °C.
3. Affinity-purify and desalt the DNA using the Qiagen QIAquick DNA purification kit. Add 1.0 mL of buffer QG and mix.
4. Apply the sample to two QIAquick spin columns placed in 2-mL microcentrifuge tubes. Centrifuge at 16,000 × *g* for 1 min in a microcentrifuge. Discard the flow-through.
5. Add 750 μL buffer PE to each column, and centrifuge at 16,000 × *g* for 1 min.
6. Transfer the column to a fresh 1.5-mL microcentrifuge tube, and centrifuge at 16,000 × *g* for 1 min.
7. Transfer the column to a fresh 1.5-mL microcentrifuge tube, and add 35 μL of ultrapure irrigation USP water to the center of the membrane. Incubate at room temperature for 2 min.
8. Centrifuge at 16,000 × *g* for 1 min to elute the DNA. Combine the eluants from the two columns. The DNA can be used immediately for *E. coli* electroporation, or it can be frozen for later use.
9. Electrophorese 1.0 μL of the eluted reaction product alongside the ssDNA template. Use a TAE/agarose gel with ethidium bromide for DNA visualization. A successful reaction results in the complete conversion of ssDNA to dsDNA, which has a lower electrophoretic mobility (Fig. 8). Usually, at least two product bands are visible and there should be no remaining ssDNA. The product band with higher electrophoretic mobility represents the desired product: correctly extended and ligated CCC-dsDNA, which transforms *E. coli* efficiently and provides a high mutation frequency (~80 %). The product band with lower electrophoretic mobility is a strand-displaced product resulting from an intrinsic, unwanted activity of T7 DNA polymerase. Although the strand-displaced product provides a low mutation frequency (~20 %), it also transforms *E. coli* at least 30-fold less efficiently than CCC-dsDNA. If a significant proportion of the single-stranded template is converted to CCC-dsDNA, a highly diverse library with high mutation frequency will result. Sometimes a third band is visible, with an electrophoretic mobility between the two product bands described previously. This intermediate band is correctly extended but unligated dsDNA (nicked dsDNA), which results from either insufficient T4 DNA ligase activity or incomplete oligonucleotide phosphorylation.

3.8 Conversion of CCC-dsDNA into a Phage-Displayed Library

To complete the library construction, the heteroduplex CCC-dsDNA is introduced into an *E. coli* host that contains an F′ episome to enable M13 bacteriophage infection and propagation. Phage-displayed library diversities are limited by methods for introducing DNA into *E. coli*, with the most efficient method being high-voltage electroporation.

We have constructed an *E. coli* strain (SS320) that is ideal for both high efficiency electroporation and phage production [32]. The following optimized protocols allow for the large-scale preparation of electrocompetent *E. coli* SS320, and subsequently, for electroporation to produce libraries with high diversity.

3.9 Preparation of Electrocompetent E. coli SS320

The following protocol yields approximately 12 mL of highly concentrated, electrocompetent *E. coli* SS320 (~3×10^{11} cfu/mL) infected by M13KO7 helper phage. The cells can be stored indefinitely at −70 °C in 10% glycerol. The use of *E. coli* infected by helper phage ensures that, once transformed with a phagemid, each cell will be able to produce phage particles.

1. Inoculate 25 mL 2YT/tet medium with a single colony of *E. coli* SS320 from a fresh LB/tet plate. Incubate at 37 °C with shaking at 200 rpm to mid-log phase ($OD_{550}=0.8$).
2. Make tenfold serial dilutions of M13K07 by diluting 20 μL into 180 μL of PBS (use a new pipette tip for each dilution).
3. Mix 500 μL of *E. coli* SS320 at exponential phase with 200 μL of each M13K07 dilution and 4 mL of 2YT top agar.
4. Pour the mixtures onto prewarmed LB/tet plates and grow overnight at 37 °C.
5. Pick a well-separated, single plaque and place in 1 mL of 2YT/kan/tet medium. Incubate for 8 h at 37 °C.
6. Transfer the culture to 250 mL of 2YT/kan medium in a 2-L baffled flask (*see* **Note 5**). Grow overnight at 37 °C with shaking at 200 rpm.
7. Inoculate six 2-L baffled flasks containing 900 mL of superbroth/tet/kan medium with 5 mL of the overnight culture. Incubate at 37 °C with shaking at 200 rpm to the mid-log phase ($OD_{550}=0.8$).
8. Chill three of the flasks on ice for 5 min with occasional swirling. The following steps should be done in a cold room, on ice, with prechilled solutions and equipments.
9. Centrifuge at $5000\times g$ and 4 °C for 10 min.
10. Decant the supernatant and add culture from the remaining flasks (these should be chilled while the first set is centrifuging) to the same tubes.
11. Repeat the centrifugation and decant the supernatant.

12. Fill the tubes with 1.0 mM Hepes, pH 7.4, and add sterile magnetic stir bars to facilitate pellet resuspension. Swirl to dislodge the pellet from the tube wall and stir at a moderate rate to resuspend the pellet completely.
13. Centrifuge at 5000 × *g* and 4 °C for 10 min. Decant the supernatant, being careful to retain the stir bar. To avoid disturbing the pellet, maintain the position of the centrifuge tube when removing from the rotor.
14. Repeat **steps 12** and **13**.
15. Resuspend each pellet in 150 mL of 10% ultrapure glycerol. Use stir bars and do not combine the pellets.
16. Centrifuge at 5000 × *g* and 4 °C for 15 min. Decant the supernatant and remove the stir bar. Remove remaining traces of supernatant with a pipette.
17. Add 3.0 mL of 10% ultrapure glycerol to one tube and resuspend the pellet by pipetting. Transfer the suspension to the next tube and repeat until all of the pellets are resuspended.
18. Transfer 350 μL aliquot into 1.5 mL microcentrifuge tubes.
19. Flash freeze with liquid nitrogen and store at −70 °C.

3.10 *E. coli* Electroporation and Phage Propagation

1. Chill the purified CCC-dsDNA (20 μg in a minimum volume) and a 0.2 cm gap electroporation cuvette on ice.
2. Thaw a 350 μL aliquot of electrocompetent *E. coli* SS320 on ice. Add the cells to the DNA and mix by pipetting several times (avoid introducing bubbles).
3. Transfer the mixture to the cuvette and electroporate. For electroporation, follow the manufacturer's instructions, preferably using a BTX ECM-600 electroporation system with the following settings: 2.5 kV field strength, 129 Ω resistance, and 50 μF capacitance. Alternatively, a Bio-Rad Gene Pulser can be used with the following settings: 2.5 kV field strength, 200 Ω resistance, and 25 μF capacitance.
4. Immediately rescue the electroporated cells by adding 1 mL SOC medium and transferring to 10 mL SOC medium in a 250-mL baffled flask. Rinse the cuvette twice with 1 mL SOC media. Add SOC medium to a final volume of 25 mL.
5. Incubate for 30 min at 37 °C with shaking at 200 rpm.
6. To determine the library diversity, plate serial dilutions on LB/carb plates to select for the phagemid.
7. Transfer the culture to a 2-L baffled flask containing 500 mL 2YT medium, supplemented with antibiotics for phagemid and M13KO7 helper phage selection (e.g., 2YT/carb/kan medium).
8. Incubate overnight at 37 °C with shaking at 200 rpm.

9. Centrifuge the culture for 10 min at 16,000×*g* and 4 °C.
10. Transfer the supernatant to a fresh tube and add 1/5 volume of PEG/NaCl solution to precipitate the phage. Incubate for 5 min at room temperature.
11. Centrifuge for 10 min at 16,000×*g* and 4 °C. Decant the supernatant. Spin again briefly and remove the remaining supernatant with a pipette.
12. Resuspend the phage pellet in 20 mL of PBT buffer.
13. Pellet insoluble matter by centrifuging for 5 min at 27,000×*g* and 4 °C. Transfer the supernatant to a clean tube.
14. Estimate the phage concentration spectrophotometrically (OD_{268} = 1.0 for a solution of 5×10^{12} phage/mL).
15. The library can be used immediately for selection experiments (*see* **Note 6**).

3.11 Affinity Selection of Superbinder SH2 Domains Against pTyr-Peptides

1. Coat Maxisorp immunoplate wells with 100 μL of Neutravidin (5 μg/mL PBS) overnight at 4 °C. The number of wells required depends on the diversity of the library. Ideally, the phage concentration should not exceed 10^{13} phage/mL, and the total number of phage should exceed the library diversity by 1000-fold. Thus for a diversity of 10^{10}, 10^{13} phage should be used, and using a concentration of 10^{13} phage/mL, 10 wells will be required. At the same time, coat an equal number of wells as a negative control.
2. Remove the coating solution and block for 1 h with 200 μL of PBS, 0.2% BSA.
3. Remove the block solution and wash two times with PT buffer.
4. Add 100 μL biotinylated, phosphotyrosine-containing peptides (e.g., biotin-linker-EPQ**pY**EEIEE) in PBS buffer to each of the coated wells (Fig. 9). In parallel, add 100 μL biotinylated, nonphosphorylated peptides (e.g., biotin-linker-EPQ**Y**EEIEE) in PBS buffer to each of the negative control wells. Incubate at room temperature for 30 min with gentle shaking (*see* **Note 7**).
5. Wash two times with PBS buffer.
6. Add 100 μL of library phage solution in PBT buffer to each of the negative control wells. Incubate at room temperature for 1 h with gentle shaking.
7. Transfer the phage solution to the wells immobilized with phosphotyrosine-containing peptides. Incubate at room temperature for 1 h with gentle shaking.
8. Remove the phage solution and wash ten times with PT buffer.

9. To elute bound phage, add 100 μL of 100 mM HCl. Incubate for 5 min at room temperature. Transfer the HCl solution to a 1.5-mL microfuge tube. Adjust to neutral pH by adding 1/8 volume, 1.0 M Tris–HCl, pH 11.
10. Add half the eluted phage solution to 10 volumes of actively growing *E. coli* XL1-Blue ($OD_{550} < 1.0$) in 2YT/tet medium. Incubate for 20 min at 37 °C with shaking at 200 rpm.
11. Plate serial dilutions on LB/carb plates to determine the number of phage eluted. Determine the enrichment ratio by dividing the number of phage eluted from a well coated with pTyr-peptides by the number of phage eluted from a well coated with Tyr peptide.
12. Add M13KO7 helper phage to a final concentration of 10^{10} phage/mL. Incubate for 45 min at 37 °C with shaking at 200 rpm.
13. Transfer the culture from the pTyr-coated wells to 25 volumes of 2YT/carb/kan medium and incubate overnight at 37 °C with shaking at 200 rpm.
14. Isolate phages by precipitation with 1/5 volume PEG/NaCl solution, resuspend in 1.0 mL of PBT buffer, and estimate phage concentration spectrophotometrically.
15. Repeat the selection cycle until the enrichment ratio has reached a maximum (*see* **Note 8**). The enrichment can be monitored by phage enzyme-linked immunosorbent assay (ELISA) using the phage pool from each cycle.
16. After completion of the enrichment cycles, isolate and pick individual clones for sequence analysis and phage ELISA.

3.12 Analysis of Binding Clones by Phage ELISA

The selection process described above produces a phage pool that is enriched for pTyr-binding clones. However, the population may also contain nonbinding clones and clones that bind nonspecifically. Binding of phage-displayed SH2 domains to immobilized pTyr-peptides can be detected by using antibodies raised against the phage particle, and ELISA methods have been adapted for the analysis of specific binding of phage-displayed SH2 domains to peptides. The culture medium into which phage particles are secreted can be used for direct-binding phage ELISAs to rapidly identify peptide-specific binding clones.

The affinities of specific pTyr-peptide binding clones can be estimated by competitive phage ELISAs. In this type of ELISA, binding of phage to immobilized peptide is competed by defined concentrations of peptide in solution. The assay can be done in a relatively high-throughput manner as a single-point competitive phage ELISA in which inhibition of binding is measured at a single peptide concentration. This allows for rank ordering of many unique clones on the basis of percent inhibition of binding, but does not provide an accurate measure of affinity.

Alternatively, a more accurate multipoint competitive phage ELISA can be used to accurately estimate the affinity as an IC_{50} value, which is defined as the concentration of peptide that blocks 50% of SH2-phage binding to immobilized peptide. The multipoint phage ELISA comprises two steps. First, the SH2-phage solution is titrated by serial dilution to determine a suitable concentration that provides a detectable but subsaturating signal on peptide-coated wells. Second, the binding of the subsaturating phage concentration to immobilized peptide is competed by serial dilutions of peptide in solution, and this produces a dose-response inhibition curve that is used to determine the IC_{50} value.

3.13 Direct-Binding Phage ELISA

1. Inoculate 450 μL aliquots of 2YT/carb/KO7 medium in 96-well microtubes with single colonies harboring phagemids and grow overnight at 37 °C with shaking at 200 rpm.
2. Centrifuge at 2000 × *g* for 10 min and transfer phage supernatants to fresh tubes.
3. Dilute phage supernatant threefold with PBT buffer.
4. Transfer 100 μL of diluted phage supernatant to 96-well Maxisorp immunoplates coated with pTyr-peptide and blocked with BSA, as described previously. In addition, binding should also be assayed using plates coated with Tyr-peptide as negative controls.
5. Incubate for 15 min with gentle shaking.
6. Wash eight times with PT buffer.
7. Add 100 μL of horseradish peroxidase/anti-M13 antibody conjugate (diluted 1:3000 in PBT buffer). Incubate for 30 min with gentle shaking.
8. Wash six times with PT buffer and two times with PBS.
9. Add 100 μL of freshly prepared TMB substrate. Allow color to develop for 5–10 min.
10. Stop the reaction with 100 μL of 1.0 M H_3PO_4 and read spectrophotometrically at 450 nm in a microtiter plate reader.
11. To assess specific binding, compare the signal strength on plates coated with pTyr-peptide to that on negative plates coated with Tyr-peptide.

3.14 Single-Point Competitive Phage ELISA

1. Grow and harvest phages as described previously.
2. Dilute phage supernatant fivefold, using PBT buffer alone or PBT buffer containing pTyr-peptide (nonbiotinylated) at a concentration close to the expected affinity.
3. Incubate for 1 h and transfer 100 μL of each solution to Maxisorp immunoplates coated with pTyr-peptide (biotinylated) and blocked with BSA, as described previously.

4. Incubate for 15 min with gentle shaking.
5. Wash, develop, and read the plates as described previously.
6. For each clone, calculate the fraction of SH2-phage uncomplexed with solution-phase pTyr-peptide by dividing the A_{450} in the presence of the solution-phase peptide by the A_{450} in the absence of the solution-phase pTyr-peptide. This ratio is inversely proportional to the affinity of the interaction and allows for the rank ordering of clones.

3.15 Multipoint Competitive Phage ELISA

1. Inoculate 1 mL 2YT/carb/tet/KO7 medium with a single colony of *E. coli* XL1-blue harboring phagemid. Incubate for 2 h at 37 °C with shaking at 200 rpm.
2. Add kanamycin (25 μg/mL) and continue incubation to mid-log phase ($OD_{550}=0.6$).
3. Transfer the culture to 25 mL of 2YT/carb/kan medium in a 250-mL baffled flask. Grow overnight at 37 °C with shaking at 200 rpm.
4. Isolate phage by precipitation with PEG/NaCl solution, resuspend in 1.0 mL of PBT buffer, and estimate phage concentration spectrophotometrically, as described previously.
5. Prepare threefold serial dilutions of phage stock in PBT buffer.
6. Transfer 100 μL of phage solution to Maxisorp immunoplates coated with pTyr-peptide and blocked as described previously.
7. Incubate for 15 min with gentle shaking.
8. Wash, develop, and read the plates as described previously.
9. For the subsequent steps, use a subsaturating concentration of the phage stock which gives ~50% of the ELISA signal at saturation.
10. Aliquot 135 μL of subsaturating phage solution into each of 12 wells of a 96-well plate.
11. Add 15 μL of serially diluted pTyr-peptide to each well. Use twofold dilutions, and ideally, start with a pTyr-peptide concentration 100-fold greater than the expected IC_{50} value.
12. Incubate for 1 h and transfer 100 μL to Maxisorp immunoplates coated with peptide and blocked as described previously.
13. Incubate for 15 min with gentle shaking.
14. Wash, develop, and read the plates as described previously.
15. Plot the OD_{450} reading as a function of the pTyr-peptide concentration and determine the IC_{50} value by standard curve fitting.

3.16 Sequencing and Subcloning of the Variant SH2 Domains

The phage ELISA experiment identifies variant SH2 domain clones that may have gained enhanced affinities to pTyr-containing ligand peptides. Next, we sequence the variant SH2 domain genes [33], and analyze the sequences to pick up or design promising variants for further biochemical studies, as we describe in the following sections (Fig. 10). Alternatively, instead of focusing on particular mutations from the sequence analysis, one can prepare the bulk of variant protein expression constructs at once. In this case, a pool of phage clones can be used as the PCR template for subcloning. The PCR products amplified from the phage pool constitute a mixture of SH2 variant genes for subsequent ligation into a protein expression vector, followed by transformation of *E. coli* (Fig. 10). Individual protein expression variant clones will be isolated as colonies.

3.17 Identifying Potential Superbinders from the Sequenced Variants

The phage display screening using the Fyn SH2 domain yielded 63 unique variants (Fig. 11). The arginine residue at Position 4 is indispensable for pTyr binding by an SH2 domain [34, 35]. Indeed, the arginine is absolutely conserved in the obtained variants that add a confidence to the screening results.

The next step is analysis of mutation frequency. If a non-wild-type residue occupies a position more frequently than the wild-type residue, the mutation will be particularly interesting since the observation may indicate that the wild-type residue is not optimal for ligand binding. A sequence logo is a convenient way to visualize mutation frequency.

1. Extract amino acid residues at the 15 positions from the list of variant sequences.
2. Submit the extracted pocket residue sequences to the WebLogo server (http://weblogo.berkeley.edu) [36].
3. Identify residues potentially interesting for further investigation (Fig. 12). In our experiments with the Fyn SH2 domain, positions 8, 10, and 15 looked interesting because a non-wild-type residue occupied the positions more frequently than the wild-type residue.
4. Look into variant sequences to find whether there are variants that contain a combination of interesting mutations at the three positions. From the sequence list in Fig. 11, Variant-37 contains a combination of Val-8, Ala-10, and Leu-15 mutations and reflects the high frequency amino acid substitutions (Fig. 12).

3.18 In-Solution pTyr-Peptide Binding Assay Using Purified SH2 Domains

There are a few popular methods, including surface plasmon resonance and isothermal titration calorimetry, that may be used to determine binding affinities of a protein to synthetic peptides. For simplicity, we used fluorescence polarization assay to determine in-solution binding affinities of the selected SH2 variants. This required purified SH2 domains and peptides labeled with a fluorophore (e.g., fluorescein).

1. Subclone superbinder candidates into an *E. coli* protein expression vector. The protein may be expressed as a fusion with an affinity purification tag (e.g., hexahistidine) (*see* **Note 9**).
2. Transform an *E. coli* strain such as BL21(DE3) with the plasmid for protein overexpression.
3. Induce protein expression at 37 °C for 4–5 h, or 18 °C overnight, with 0.3 mM IPTG (*see* **Note 10**).
4. Perform affinity purification of the SH2 domains from *E. coli* lysate. Follow manufacturer's instruction associated with commercial affinity beads.
5. Dialyze the eluted SH2 domain against dialysis buffer at 4 °C, overnight (*see* **Note 2**).
6. On next day, filter the protein solution to remove precipitates.
7. Make a series of twofold serial dilution of the SH2 domain using the dialysis buffer.
8. Perform fluorescence polarization assay by mixing protein and fluorescence-labeled pTyr-peptide solution. Fluorescence polarization caused by tumbling of the ligand peptide is titrated with different concentrations of the SH2 domain (*see* **Note 11**).
9. Subtract the blank (reading at 0 μM) from the fluorescence polarization data points.
10. Fit data into the equation: $\Delta F = \Delta F_{max} \times [SH2]/(K_d + [SH2])$, where ΔF is fluorescence polarization (after blank subtraction), [SH2] is SH2 domain concentration, and K_d is the dissociation constant [26] (Fig. 13) (*see* **Note 12**).

3.19 Transfer of the Observation to Other SH2 Domains

The sequence and structure of the phosphotyrosine-binding pocket is highly conserved between SH2 domains [37] (Fig. 5). Therefore, there is a good chance that the mutation information obtained from screening the Fyn SH2 domain can be transferred to other SH2 domains to enhance pTyr-binding affinity. Indeed, we created superbinders based on the Src and Grb2 SH2 domains, by applying mutations equivalent to the Fyn SH2-based superbinder [15].

1. Prepare a sequence alignment between the Fyn SH2 domain and the target SH2 domain (*see* **Note 13**). Refer to literature for an alignment between SH2 domains [8, 38], or use a sequence alignment program or Web server.
2. Identify the three key positions (Position 8, 10, and 15) in the target SH2 domain (Fig. 5).

3. Mutate Position 8 to Val, Position 10 to Ala, and Position 15 to Leu, by site-directed mutagenesis.
4. Perform peptide binding assay and compare affinities with the wild-type SH2 domain.

4 Notes

1. If more than one region of the DNA is to be mutated, two or more mutagenic oligonucleotides can be added simultaneously as long as sequences within the oligonucleotides do not overlap with each other.
2. Phosphate buffer is not recommended for in-solution binding assay involving SH2 domains. In our experience, in-solution binding assay in phosphate buffer gives lower affinities than in Tris buffer, presumably due to a competition between high concentration (millimolar) phosphate ion in phosphate buffer and a pTyr ligand peptide at a much lower (nanomolar) concentration.
3. An SH2 domain normally comprises two pockets, one for pTyr-binding and the other for specificity (two-pronged binding) [11, 39] (Fig. 1). To evolve only the pTyr-binding pocket for affinity, we avoided including residues that contribute to the specificity-determining pocket. Therefore, the residue βD5 that contributes to both of the two pockets was not mutated in the library.
4. To avoid contamination, always use pipette tips with filter to manipulate phage.
5. We recommend the use of baffled culture flasks to amplify phage as they increase aeration and cell growth.
6. The phage-displayed SH2 variant library can be frozen and stored at −80 °C, following the addition of glycerol to a final concentration of 20%. In general, it is best to use the libraries immediately as levels of displayed proteins can be reduced over time due to denaturation or proteolysis.
7. The amount of the coated phosphotyrosine-containing peptides should be titrated against wild-type Fyn SH2 domain beforehand. Typically, the immobilized peptide should be between 10 nM to 10 mM.
8. Typically, phage enrichment peaks at round 4 in selection and sorting beyond round 5 is seldom necessary.
9. Fusion with an affinity tag will not normally affect interaction between an SH2 domain and ligand peptides. However, artificial multivalent interaction may happen depending on affinity measurement methods. When using the surface plasmon

resonance method and SH2 domains fused with the GST tag, Ladbury et al. observed avidity effects and overestimates of affinities presumably caused by dimerization of the GST tag [40]. In-solution binding assay techniques such as isothermal titration calorimetry and fluorescence polarization should be free from artificial effects derived from immobilization of ligand materials on solid surface.

10. Try 37 °C first for protein overexpression. Variant proteins that show excellent production in *E. coli* with high stability will be ideal for application purposes.
11. We typically adjust final concentration of fluorescein-labeled peptides between 1 and 5 nM. Use fluorescence intensity of fluorescein (or 5(6)-carboxyfluorescein) at a known concentration as the standard to estimate concentration of the peptides.
12. The parameter [SH2] is the concentration of unbound, free fraction of the SH2 domain in solution, which is actually unknown. However, the initial (total) concentration of the SH2 domain can be used for [SH2] to obtain approximate K_d values [26]. For more precise calculation that involves peptide concentration, refer to Kaushansky et al. [41].
13. The mutation transfer may not work well if there is an alignment gap in the BC loop region.

Acknowledgements

Work described here was supported, in part, by grants from the Canadian Cancer Society Research Institute and the Ontario Research Fund. SSCL holds a Canada Research Chair in Functional Genomics and Cellular Proteomics.

References

1. Humphrey SJ, Azimifar SB, Mann M (2015) High-throughput phosphoproteomics reveals in vivo insulin signaling dynamics. Nat Biotechnol 33:990–995
2. Zheng Y, Zhang C, Croucher DR, Soliman MA, St-Denis N, Pasculescu A, Taylor L, Tate SA, Hardy WR, Colwill K et al (2013) Temporal regulation of EGF signalling networks by the scaffold protein Shc1. Nature 499:166–171
3. Mayer BJ (2012) Perspective: dynamics of receptor tyrosine kinase signaling complexes. FEBS Lett 586:2575–2579
4. Hause RJ, Leung KK, Barkinge JL, Ciaccio MF, Chuu C-P, Jones RB (2012) Comprehensive binary interaction mapping of SH2 domains via fluorescence polarization reveals novel functional diversification of ErbB receptors. PLoS One 7, e44471
5. Ladbury JE, Arold ST (2011) Energetics of Src homology domain interactions in receptor tyrosine kinase-mediated signaling. Meth Enzymol 488:147–183
6. Haslam NJ, Shields DC (2012) Peptide-binding domains: are limp handshakes safest? Sci Signal 5:40
7. Pawson T (2004) Specificity in signal transduction: from phosphotyrosine-SH2 domain interactions to complex cellular systems. Cell 116:191–203

8. Liu BA, Jablonowski K, Raina M, Arcé M, Pawson T, Nash PD (2006) The human and mouse complement of SH2 domain proteins-establishing the boundaries of phosphotyrosine signaling. Mol Cell 22:851–868
9. Kaneko T, Joshi R, Feller SM, Li SS-C (2012) Phosphotyrosine recognition domains: the typical, the atypical and the versatile. Cell Commun Signal 10:32
10. Waksman G, Kominos D, Robertson SC, Pant N, Baltimore D, Birge RB, Cowburn D, Hanafusa H, Mayer BJ, Overduin M et al (1992) Crystal structure of the phosphotyrosine recognition domain SH2 of v-src complexed with tyrosine-phosphorylated peptides. Nature 358:646–653
11. Eck MJ, Shoelson SE, Harrison SC (1993) Recognition of a high-affinity phosphotyrosyl peptide by the Src homology-2 domain of p56lck. Nature 362:87–91
12. Tinti M, Nardozza AP, Ferrari E, Sacco F, Corallino S, Castagnoli L, Cesareni G (2012) The 4G10, pY20 and p-TYR-100 antibody specificity: profiling by peptide microarrays. N Biotechnol 29:571–577
13. Aoki K, Kiyokawa E, Nakamura T, Matsuda M (2008) Visualization of growth signal transduction cascades in living cells with genetically encoded probes based on Förster resonance energy transfer. Philos Trans R Soc Lond B Biol Sci 363:2143–2151
14. Lesuisse D, Lange G, Deprez P, Bénard D, Schoot B, Delettre G, Marquette J-P, Broto P, Jean-Baptiste V, Bichet P et al (2002) SAR and X-ray. A new approach combining fragment-based screening and rational drug design: application to the discovery of nanomolar inhibitors of Src SH2. J Med Chem 45:2379–2387
15. Kaneko T, Huang H, Cao X, Li X, Li C, Voss C, Sidhu SS, Li SSC (2012) Superbinder SH2 domains act as antagonists of cell signaling. Sci Signal 5:68
16. Bian Y, Li L, Dong M, Liu X, Kaneko T, Cheng K, Liu H, Voss C, Cao X, Wang Y, Litchfield D, Ye M, Li SS, Zou H (2016) Ultra-deep tyrosine phosphoproteomics enabled by a phosphotyrosine superbinder. Nat Chem Biol 12(11):959–966. doi:10.1038/nchembio.2178
17. Findlay GM, Smith MJ, Lanner F, Hsiung MS, Gish GD, Petsalaki E, Cockburn K, Kaneko T, Huang H, Bagshaw RD et al (2013) Interaction domains of sos1/grb2 are finely tuned for cooperative control of embryonic stem cell fate. Cell 152:1008–1020
18. Kärkkäinen S, Hiipakka M, Wang J-H, Kleino I, Vähä-Jaakkola M, Renkema GH, Liss M, Wagner R, Saksela K (2006) Identification of preferred protein interactions by phage-display of the human Src homology-3 proteome. EMBO Rep 7:186–191
19. Tonikian R, Zhang Y, Sazinsky SL, Currell B, Yeh J-H, Reva B, Held HA, Appleton BA, Evangelista M, Wu Y et al (2008) A specificity map for the PDZ domain family. PLoS Biol 6, e239
20. Bloom L, Calabro V (2009) FN3: a new protein scaffold reaches the clinic. Drug Discov Today 14:949–955
21. Bradbury ARM, Sidhu S, Dübel S, McCafferty J (2011) Beyond natural antibodies: the power of in vitro display technologies. Nat Biotechnol 29:245–254
22. Huang H, Li L, Wu C, Schibli D, Colwill K, Ma S, Li C, Roy P, Ho K, Songyang Z et al (2008) Defining the specificity space of the human SRC homology 2 domain. Mol Cell Proteomics 7:768–784
23. Machida K, Thompson CM, Dierck K, Jablonowski K, Kärkkäinen S, Liu B, Zhang H, Nash PD, Newman DK, Nollau P et al (2007) High-throughput phosphotyrosine profiling using SH2 domains. Mol Cell 26:899–915
24. Zhao B, Tan PH, Li SS-C, Pei D (2013) Systematic characterization of the specificity of the SH2 domains of cytoplasmic tyrosine kinases. J Proteomics 81:56–69
25. Jones RB, Gordus A, Krall JA, MacBeath G (2006) A quantitative protein interaction network for the ErbB receptors using protein microarrays. Nature 439:168–174
26. Leung KK, Hause RJ, Barkinge JL, Ciaccio MF, Chuu C-P, Jones RB (2014) Enhanced prediction of SH2 domain binding potentials using a fluorescence polarization-derived c-Met, c-Kit, ErbB, and androgen receptor interactome. Mol Cell Proteomics 13:1705–1723
27. Mulhern TD, Shaw GL, Morton CJ, Day AJ, Campbell ID (1997) The SH2 domain from the tyrosine kinase Fyn in complex with a phosphotyrosyl peptide reveals insights into domain stability and binding specificity. Structure 5:1313–1323
28. Arold ST, Ulmer TS, Mulhern TD, Werner JM, Ladbury JE, Campbell ID, Noble ME (2001) The role of the Src homology 3-Src homology 2 interface in the regulation of Src kinases. J Biol Chem 276:17199–17205
29. Brunger AT (2007) Version 1.2 of the crystallography and NMR system. Nat Protoc 2:2728–2733
30. Kleywegt GJ, Jones TA (1998) Databases in protein crystallography. Acta Crystallogr D Biol Crystallogr 54:1119–1131

31. Kunkel TA, Roberts JD, Zakour RA (1987) Rapid and efficient site-specific mutagenesis without phenotypic selection. Meth Enzymol 154:367–382
32. Sidhu SS, Lowman HB, Cunningham BC, Wells JA (2000) Phage display for selection of novel binding peptides. Meth Enzymol 328:333–363
33. Huang H, Sidhu SS (2011) Studying binding specificities of peptide recognition modules by high-throughput phage display selections. Methods Mol Biol 781:87–97
34. Bradshaw JM, Mitaxov V, Waksman G (1999) Investigation of phosphotyrosine recognition by the SH2 domain of the Src kinase. J Mol Biol 293:971–985
35. Mayer BJ, Jackson PK, Van Etten RA, Baltimore D (1992) Point mutations in the abl SH2 domain coordinately impair phosphotyrosine binding in vitro and transforming activity in vivo. Mol Cell Biol 12:609–618
36. Crooks GE, Hon G, Chandonia J-M, Brenner SE (2004) WebLogo: a sequence logo generator. Genome Res 14:1188–1190
37. Kaneko T, Sidhu SS, Li SSC (2011) Evolving specificity from variability for protein interaction domains. Trends Biochem Sci 36: 183–190
38. Kaneko T, Huang H, Zhao B, Li L, Liu H, Voss CK, Wu C, Schiller MR, Li SS-C (2010) Loops govern SH2 domain specificity by controlling access to binding pockets. Sci Signal 3:34
39. Waksman G, Shoelson SE, Pant N, Cowburn D, Kuriyan J (1993) Binding of a high affinity phosphotyrosyl peptide to the Src SH2 domain: crystal structures of the complexed and peptide-free forms. Cell 72:779–790
40. Ladbury JE, Lemmon MA, Zhou M, Green J, Botfield MC, Schlessinger J (1995) Measurement of the binding of tyrosyl phosphopeptides to SH2 domains: a reappraisal. Proc Natl Acad Sci U S A 92:3199–3203
41. Kaushansky A, Allen JE, Gordus A, Stiffler MA, Karp ES, Chang BH, MacBeath G (2010) Quantifying protein-protein interactions in high throughput using protein domain microarrays. Nat Protoc 5:773–790

Part IV

Structural Analysis and Binding Assays

Chapter 14

Structural Characterization of Monomeric/Dimeric State of p59^{fyn} SH2 Domain

Radu Huculeci, Fabien Kieken, Abel Garcia-Pino, Lieven Buts, Nico van Nuland, and Tom Lenaerts

Abstract

Src homology 2 (SH2) domains are key modulators in various signaling pathways allowing the recognition of phosphotyrosine sites of different proteins. Despite the fact that SH2 domains acquire their biological functions in a monomeric state, a multitude of reports have shown their tendency to dimerize. Here, we provide a technical description on how to isolate and characterize by gel filtration, circular dichroism (CD), and nuclear magnetic resonance (NMR) each conformational state of p59^{fyn} SH2 domain.

Key words SH2 domain, Monomer/dimer, Gel filtration chromatography, Circular dichroism (CD), Nuclear magnetic resonance (NMR)

1 Introduction

Src homology 2 (SH2) domains are well-conserved protein domains, which allow for effective signal transduction of receptor tyrosine kinase. These domains can recognize phosphorylated tyrosine in a wide range of proteins. Structural characterization of SH2 domains is well documented and shown to perform their binding function under a monomeric conformation [1]. Yet a challenge of studying these domains is their tendency to dimerize or aggregate [2]. This tendency is mainly observed in NMR studies requiring a high concentration of soluble protein. As recently indicated by systematic analysis, intertwined structures of SH2 homomers are part of a vast category of intertwined of homo-oligomeric structures within the PDB [3]. The potential to aggregate or dimerize can be also be derived from sequence comparison and predictive methods like Waltz [4], which determines whether an amino-acid sequence can adopt a cross-β arrangement typically observed in amyloid fibrils. Our analysis for Src-related SH2 domains in [2] reveals that all eight Src-related SH2 domains may produce the same intertwined dimer. The biological importance of

Kazuya Machida and Bernard A. Liu (eds.), *SH2 Domains: Methods and Protocols*, Methods in Molecular Biology, vol. 1555,
DOI 10.1007/978-1-4939-6762-9_14, © Springer Science+Business Media LLC 2017

SH2 homo-oligomerization remains unclear; however, it has been shown that dimers can adopt different structural organizations that can alter the ligand binding affinity at different degrees [5]. Moreover, as with some kinases such as those belonging to the P21 activated kinases (PAK) I family [6], dimerization may have a relevant regulatory role.

Thus, it is critical to separate the different SH2 conformers (monomer/dimer). In this chapter, we use $p59^{fyn}$ SH2 domain to illustrate how to express and purify the monomer/dimer and how to characterize each form by gel filtration chromatography, circular dichroism (CD), and nuclear magnetic resonance (NMR).

2 Materials

1. Expression vectors: pColdII (TaKaRa), pET15b (Novagen).
2. *E. coli* strains: DH5α, BL21(DE3).
3. Ex Taq polymerase (Takara).
4. Agarose gel electrophoresis system.
5. Gel and PCR clean-up.
6. Restriction enzyme and T4 ligase.
7. Nucleospin® Plasmid (Macherey Nagel).
8. LB medium (liquid, agar).
9. 2× YT medium (liquid).
10. M9 Minimal Medium.
11. Antibiotics (ampicillin, carbenicillin).
12. ^{13}C-glucose and $^{15}NH_4Cl$ (Cambridge Isotope Laboratories).
13. Isopropyl β-D-1-thiogalactopyranoside (IPTG).
14. French pressure cell press.
15. Reducing agents: dithiothreitol (DTT), tris(2-carboxyethyl) phosphine (TCEP).
16. Thrombin.
17. FPLC chromatography columns: HisTrap HP, HiLoad 16/60 Superdex 75 pg, Superdex 75 10/300 GL (GE Healthcare).
18. AKTA protein purification system (GE Healthcare).
19. Sodium dodecyl sulfate–polyacrylamide gel electrophoresis (SDS-PAGE) system.
20. Centrifugal concentrator 5 kDa.
21. Protein standards for gel filtration chromatography.
22. J-715 spectro-polarimeter (Jasco).
23. 0.1 cm cuvettes (Hellma Analytics).

24. Deuterium oxide (D_2O).
25. NMR Shigemi 5 mm tube (Shigemi Inc.).
26. Peptide (sequence: EPQ-pY-EEIPIYL) with purity >90%.
27. Lysis buffer A: 20 mM HEPES pH 8.0, 1 M NaCl, 2 mM $MgCl_2$, 20 mM imidazole, 1 μg/mL leupeptin, 0.1 mg/mL AEBSF, 4 U/mL DNAse.
28. Lysis buffer B: 20 mM HEPES pH 8.0, 1 M NaCl, 2 mM $MgCl_2$, 1 M imidazole.
29. Gel filtration buffer C: 50 mM phosphate pH 6.5 and 5 mM DTT.

3 Methods

In this section, we describe methods used in our lab to clone, express, purify the Fyn-SH2 domain, providing technical details and tips on how to separate and characterize the SH2 monomeric and dimeric form by gel filtration, nuclear magnetic resonance (NMR), and circular dichroism (CD).

3.1 Cloning

3.1.1 Expression Plasmids

pCold II is an expression vector from TaKaRa using a promoter derived from the cold shock gene cspA. It also contains a 5′ untranslated region (5′ UTR) and a translation enhancing element (TEE). pET15b can be purchased from Novagen and used as a T7 promoter (*see* **Note 1**). Both plasmids are ampicillin resistant (Amp^R), have a lac promoter to control the expression, and can be repressed by the lac repressor ($lacI^q$). Addition of IPTG inactivates the lac repressor and allows the activation of the promoter. In case of the pColdII expression vector, the induction is driven by both IPTG addition and cold shock.

3.1.2 Cloning Strategy

1. Amplify the gene of interest (SH2 domain from human $p59^{fyn}$) using Ex Taq polymerase per manufacturer's instructions.
2. Separate the amplified DNA by agarose gel electrophoresis, followed by gel extraction and purification the PCR fragment using Nucleospin® Gel and PCR cleanup.
3. Digest expression vectors and genes of interest with appropriate enzymes per manufacturer's instructions (*see* **Note 2**).
4. Purify vectors and inserts by Nucleospin® PCR cleanup and ligated together using T4 ligase.
5. Transform ligation mixture into *E. coli* DH5α using standard methods [7] and plate on LB agar plate with 100 μg ampicillin or carbenicillin overnight at 37 °C.
6. Pick individual colony and inoculate them into 3 mL 2× YT liquid medium with 100 μg ampicillin and grow them overnight at 37 °C.

7. Centrifuge the cultures the next day at 4000 × *g* for 10 min, discard the supernatant and extract plasmids from pellet with Nucleospin® Plasmid.
8. Analyze plasmid sequence by Sanger DNA sequencing.

3.2 Protein Expression

3.2.1 Protein Expression Using pColdII Vector

1. Transform pColdII expression vector into BL21(DE3) and plate on LB plate with 100 μg ampicillin overnight at 37 °C.
2. Pick individual colonies and inoculate them into LB liquid medium with 100 μg ampicillin and grow overnight at 37 °C.
3. Inoculate 1 L of M9 medium (supplement with 4 g/L glucose and 1 g/L NH4Cl) with 10 mL of preculture and grow at 37 °C.
4. When OD reaches 0.5, place the culture in an ice–water bath to reach quickly 15 °C and incubate at the same temperature for 45 min to induce cold shock.
5. Harvest cells by centrifugation at 4000 × *g* for 20 min at 4 °C , wash with M9 medium and resuspend in 250 mL of M9 medium (supplement with 4 g/L ^{13}C-glucose and 1 g/L $^{15}NH_4Cl$) with 1 mM IPTG (*see* **Note 3**).
6. Allow for cells to grow overnight at 15 °C and harvest by centrifugation at 7000 × *g* for 10 min. Store cell pellets at −80 °C.

3.2.2 Protein Expression Using pET15b Vector

1. Transform pET15b expression vector into BL21(DE3) and plate on LB plate with 100 μg ampicillin overnight at 37 °C.
2. Pick individual colonies and inoculate them into LB liquid medium with 100 μg ampicillin and grow overnight in a 37 °C shaker at 220 rpm.
3. Inoculate 1 L of M9 media (supplement with 4 g/L ^{13}C-glucose and 1 g/L $^{15}NH_4Cl$) with 10 mL of preculture and grow at 37 °C.
4. When OD reaches 0.7, induce cultures with 1 mM IPTG.
5. Allow for cells to grow overnight at 30 °C and harvest by centrifugation at 7000 × *g* for 10 min. Store cell pellets at −80 °C until the next step.

3.3 Protein Purification

1. Resuspend cells into 40 mL chilled lysis buffer A (*see* **Note 4**).
2. Lyse cells in three passages through a French Press.
3. Centrifuge lysate by centrifugation at 35,000 × *g* for 1 h at 4 °C.
4. Equilibrate a 5 mL HisTrap HP column with lysis buffer A using AKTA purifier (*see* **Note 5**).
5. Load sample into the column, wash with 10× column volume (CV) lysis buffer A and elute protein with a linear gradient of lysis buffer B from 0 to 60 % over 10× CV and 60 to 100 % over 4× CV.

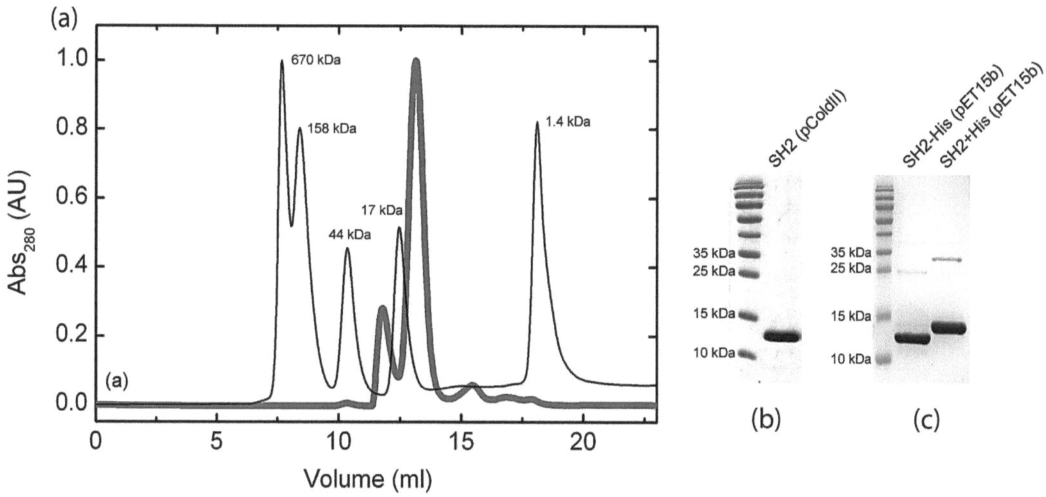

Fig. 1 Purification of the human Fyn SH2 domain. (**a**) Analytical gel filtration profile on a Superdex 75 HR column (SH2 loaded onto the column at 1 mM). The molecular-weight standards and protein chromatograms are shown in *black* and *grey lines*, respectively. (**b–c**) 20 % SDS-PAGE under reducing conditions of the molecular weight marker and the purified Fyn SH2 domain from pColdII (**b**) and pET15b (**c**) are illustrated. In the case of the pET15b vector, two different SH2 constructs are obtained with and without cleaving the His-tag. Figure 1a is reprinted with permission from Protein Science *24*, 1964–1978. © 2015 The Protein Society published by Wiley Blackwell

6. Pool the fractions containing the protein together and add 5 mM of dithiothreitol (DTT) to prevent protein cysteine-related dimerization (*see* **Note 4**). For the purification of the His-tagged construct go directly to **step 11**.
7. Add 2 U of bovine thrombin per mg of protein. Incubate in a shaker for 16 h at 26 °C.
8. Dialyze the sample against 4 L of lysis buffer A.
9. Equilibrate a 5 mL HisTrap HP column with lysis buffer using AKTA purifier.
10. Load sample into the column and collect the flow-through.
11. Load protein onto a Superdex 75 16/60 gel filtration column and elute using gel filtration buffer C (Fig. 1).
12. Pool fractions containing the protein together and check purity by SDS-PAGE (Fig. 1).
13. Quantify the protein by spectrophotometry using the protein extinction coefficient.

Table 1 reviews the advantages and disadvantages of each construct and their applications.

3.4 Gel Filtration Monomer/Dimer Fyn-SH2 Domain

SH2 domains are prone to aggregation. Therefore it is critical to separate the different conformers and assess their stability over time. The stability studies are important during a long experiment

Table 1
Three versions of human Fyn SH2 domain constructs used in this chapter with their advantages and limitations

Expression vector	Suitable for	Advantages	Limitations
pCold II (with His-tag)	Assignment	Production of ^{13}C	Partial isotope incorporation
	General NMR studies	^{15}N samples at low cost	Not suitable for NMR dynamics
pET15b (with His-tag)	Assignment General NMR studies Side-chain and backbone dynamics	Stability for more than 20 days in the monomeric form	Long His-tag
pET15b (without His-tag)	Dimerization studies	No His-tag	Highly unstable in monomeric form

(i.e., NMR structure determination and/or dynamics) and it is essential to keep only one conformer at a time in solution. By allowing molecules to move through a bed of porous beads, gel filtration chromatography permits for quick separation of molecules by molecular weight. As such gel filtration chromatography represents a very useful tool to confirm sample stability. We checked the stability of various Fyn SH2 constructs over time after incubation at 25 °C (Fig. 2). The two constructs of pET15b with and without His-tag behave differently in solution. While the Fyn SH2 (−His) dimerizes significantly at a 1:3 ratio after 48 h, and the Fyn SH2 (+His) forms a stable monomeric form over time. The gel filtration procedure for both forms is as follows:

1. The monomer fraction (Subheading 3.3) of Fyn SH2 is concentrated to the required value with a centrifugal concentrator 5 kDa (i.e., 1 mM for NMR structural studies) (*see* **Note 5**).
2. The sample is loaded to a superdex 75 10/300 GL column and eluted using gel filtration buffer C (*see* **Note 6**) (sample volume 100 μL, flow rate 0.5 mL/min) (Fig. 2).

3.5 Structural Comparison Monomer/Dimer Fyn-SH2 Domain

3.5.1 Circular Dichroism (CD)

Since the CD signal in the far UV region (190–260 nm) is due to the amides in the protein backbone, CD spectroscopy represents a powerful tool for assessing the general conformation of a protein. We recorded the far-UV CD for both Fyn SH2 conformations. Spectra indicate that both conformations are folded with a predominance of β-sheets (minimum between 215 and 220 nm) (Fig. 3). The dimeric signal in the 215–220 nm range is more

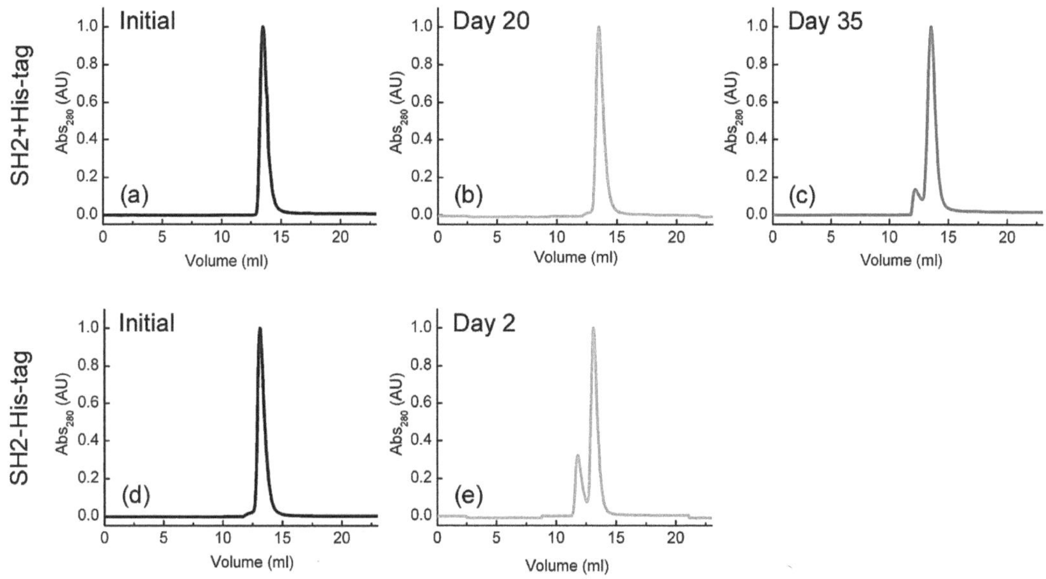

Fig. 2 Stability test of the human Fyn SH2 domain using size exclusion chromatography. Analytical gel-filtration profiles obtained from SH2 domain with (**a–c**) and without His-tag (**d**, **e**) expressed in pET15b vector. Figure 2d,e are reprinted with permission from Protein Science *24*, 1964–1978. © 2015 The Protein Society published by Wiley Blackwell

pronounced compared to the monomer, indicating that the dimerization induces a structural reorganization with an increase in β-sheet content. Sample preparation and CD data acquisition parameters are described below.

1. The monomer and dimer fractions (Fig. 1) are collected and the protein concentration is adjusted to 0.2 mg/mL with 50 mM phosphate pH 6.5 and 5 mM DTT (*see* **Note 7**).
2. The sample (300 μL) is added to 0.1 cm cell and the far UV CD spectra are recorded on a J-715 spectro-polarimeter with a range between 200 and 250 nm. Data are acquired at a scanning speed of 50 nm/min, at a 2 s integration time and 1 nm bandwidth.

3.5.2 Nuclear Magnetic Resonance (NMR)

Unlike CD, which provides information on the general conformation of a molecule, NMR spectroscopy enables insights into its structure and dynamics at the atomic level. Studies of the transverse relaxation rates (R_2) by NMR can be very useful for determining the state of aggregation of a system and dynamic changes during domain dimerization or binding to their targets. R_2 is in direct correlation with the size of the complex in solution. Similar to the size exclusion chromatography (SEC), the evolution of the dimerization can be estimated by the study of the transverse relaxation rates. Data in Fig. 4a illustrates R_2 rates of each conformer (monomer/dimer) alone and in complex with a specific phospho-tyrosine

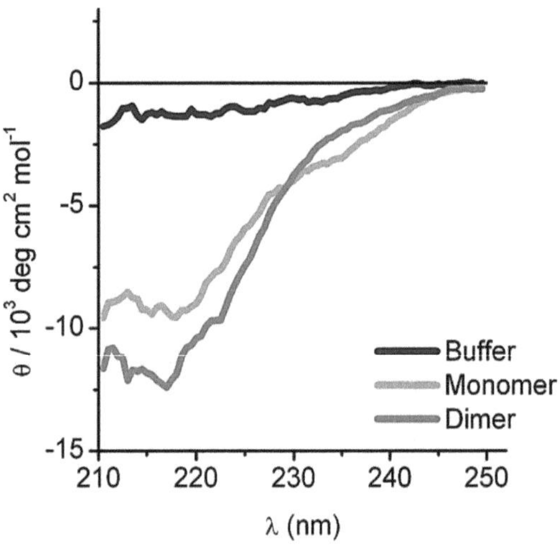

Fig. 3 Far-UV CD spectra of the monomeric and dimeric form of human Fyn SH2 domain. Reprinted with permission from Protein Science *24*, 1964–1978. © 2015 The Protein Society published by Wiley Blackwell

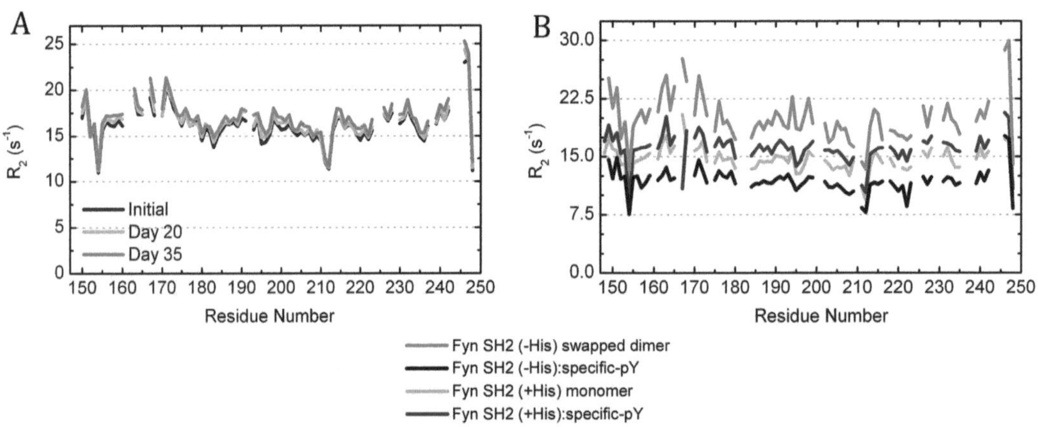

Fig 4 Transverse relaxation rates (R_2) of the human Fyn SH2 domain. (**a**) Stability of SH2 domain with His-tag over time. (**b**) Transverse relaxation rates (R_2) of the human Fyn SH2 domain monomer/dimer alone and in complex with specific phospho-tyrosine peptide (specific-pY). Figure 4b is reprinted with permission from Protein Science *24*, 1964–1978. © 2015 The Protein Society published by Wiley Blackwell

peptide (specific-pY). The specific-PY sequence has the following sequence: EPQ-pY-EEIPIYL. The swapped dimer shows the highest transverse relaxation rates, indicating the highest mass amongst the different complexes. Upon peptide addition, the swapped dimer R_2 values decrease, suggesting a loss in mass, and the formation of phospho-tyrosine peptide-SH2 monomer complex (Fig. 4b). R_2 rates of the monomer in its free form are as expected lower than for the swapped dimer. They show a slight increase upon addition of phospho-tyrosine peptide confirming the formation of a phospho-tyrosine peptide-SH2 monomer complex (Fig. 4b). Sample preparation for NMR analysis and data acquisition steps are indicated below.

1. The ^{15}N Fyn SH2 monomer and dimer fraction are collected and concentrated to reach 1 mM in 50 mM phosphate pH 6.5, 5 mM DTT with 10% D_2O. Samples (320 μL) were loaded into a Shigemi tube (*see* **Note 8**).
2. ^{15}N transverse relaxation rates are recorded on a Varian NMR Direct-Drive System 800 MHz spectrometer (*see* **Note 9**). R_2 are obtained from a series of 2D experiments [8]. Data are acquired with 1024×48 complex data points, and relaxation delays of 10, 30, 50, 70, 90, 110, 130, 150, and 170 ms (*see* **Note 10**).

4 Notes

1. Unlike the pCold system, which can be used with most *E. coli* strains, the pET system requires the presence of T7 RNA polymerase for expression of recombinant protein. Prokaryotic cells do not produce this polymerase and a recombinant plasmid has to be transferred to the host bacteria. We are using bacterial strains containing the bacteriophage DE3 that carries a lacI gene, a lacUV5 promoter, and the T7 RNA polymerase gene (i.e., BL21(DE3), BL21(DE3)pLys).
2. The digested vector can religate back on itself creating a high background during the ligation step. To overcome this issue we suggest adding phosphatase such as recombinant shrimp alkaline phosphatase (rSAP) on the digested vectors. This step can significantly reduce the background during the ligation step, as dephosphorylated DNA cannot be ligated by the T4 DNA ligase.
3. The expression of Fyn SH2 domain in pCold has been optimized for a maximum of protein expression. However, mass spectrometry tests of the purified protein show that a heterogenous isotopic incorporation, limiting sample suitability for NMR purposes. Please refer to Table 1 for the experiment suitability for the pCold expression system.
4. Fyn SH2 tends to precipitate during the early steps of purification requiring work under dilute (1 mg/mL) and reducing (TCEP or DTT) conditions throughout the purification process. The protein can be concentrated after the last gel filtration.
5. All samples and solutions for gel filtration must be filtered through 0.45 μm filter to remove any insoluble material and degassed by sonication or vacuum.
6. To optimize the separation between monomeric and dimeric form, it is imperative to load a minimum volume onto the gel

filtration column. For SH2 monomer/dimer separation, we load 1 mL of sample at a concentration of 15–20 mg/mL onto a Superdex 75 16/90. Analytical Superdex 75 HR 10/30 may also be used, when loading 200 µL of sample at a similar concentration.

7. DTT has a high absorbance in the far-UV region making the reading below 210 nm unreliable. Phosphate buffer 10 mM at pH 6.5 and 1 mM TCEP can be use to acquire CD between 190 and 210 nm.
8. Shigemi NMR microtubes should be magnetic susceptibility-matched to D_2O. These tubes allow for use of low sample volume without affecting the quality of the NMR experiment.
9. Transverse relaxation rates (R_2) experiment requires samples to be isotopically ^{15}N-labeled. If such sample is unavailable, pulse field gradient NMR (PFG-NMR) can be acquired to estimate the state of aggregation of a protein. The diffusion coefficient D can be calculated by PFG-NMR and is directly correlated to the size of the protein.
10. For error estimation, 10 ms relaxation delay is repeated. All NMR data are processed using NMRpipe [9]. R^2 values are obtained by fitting the connectivity intensity to a single exponential decay using CCPNMR software [10]. Errors are estimated from variation between corresponding experiments with different relaxation delays.

Acknowledgments

R.H., F.K., Nv.N., L.B., A.G., and T.L. were supported by Fonds Wetenschappelijk Onderzoek (FWO) via research grants G.0.116.09.N.10 and G025915N. Additional funding for T.L. was provided by Fondation de la Recherche Scientifique (FRS-FNRS) via the FRFC grant 2.4606.11. The VIB and the Hercules Foundation support the research in Nv.N. group.

References

1. Kaneko T, Joshi R, Feller SM, Li SS (2012) Phosphotyrosine recognition domains: the typical, the atypical and the versatile. Cell Commun Signal 10(1):32. doi:10.1186/1478-811X-10-32
2. Huculeci R, Garcia-Pino A, Buts L, Lenaerts T, van Nuland N (2015) Structural insights into the intertwined dimer of fyn SH2. Protein Sci. doi:10.1002/pro.2806
3. Mackinnon SS, Malevanets A, Wodak SJ (2013) Intertwined associations in structures of homooligomeric proteins. Structure 21(4):638–649. doi:10.1016/j.str.2013.01.019
4. Maurer-Stroh S, Debulpaep M, Kuemmerer N, Lopez de la Paz M, Martins IC, Reumers J, Morris KL, Copland A, Serpell L, Serrano L, Schymkowitz JW, Rousseau F (2010) Exploring the sequence determinants of amyloid structure

using position-specific scoring matrices. Nat Methods 7(3):237–242. doi:10.1038/nmeth.1432

5. Schiering N, Casale E, Caccia P, Giordano P, Battistini C (2000) Dimer formation through domain swapping in the crystal structure of the Grb2-SH2-Ac-pYVNV complex. Biochemistry 39(44):13376–13382
6. Rane CK, Minden A (2014) P21 activated kinases: structure, regulation, and functions. Small GTPases 5:28003. doi:10.4161/sgtp.28003
7. Sambrook J, Fritsch EF, Maniatis F (1989) Molecular cloning, a laboratory manual. Cold Spring Harbor Press, Cold Spring Harbor, NY
8. Farrow NA, Muhandiram R, Singer AU, Pascal SM, Kay CM, Gish G, Shoelson SE, Pawson T, Forman-Kay JD, Kay LE (1994) Backbone dynamics of a free and phosphopeptide-complexed Src homology 2 domain studied by 15N NMR relaxation. Biochemistry 33(19):5984–6003
9. Delaglio F, Grzesiek S, Vuister GW, Zhu G, Pfeifer J, Bax A (1995) NMRPipe: a multidimensional spectral processing system based on UNIX pipes. J Biomol NMR 6(3):277–293
10. Vranken WF, Boucher W, Stevens TJ, Fogh RH, Pajon A, Llinas M, Ulrich EL, Markley JL, Ionides J, Laue ED (2005) The CCPN data model for NMR spectroscopy: development of a software pipeline. Proteins 59(4):687–696. doi:10.1002/prot.20449

Chapter 15

NMR Chemical Shift Mapping of SH2 Peptide Interactions

Marissa A. McKercher and Deborah S. Wuttke

Abstract

Heteronuclear single quantum coherence (HSQC) nuclear magnetic resonance (NMR) experiments offer a rapid and high resolution approach to gaining binding and conformational insights into a protein–peptide interaction. By tracking ^{1}H and ^{15}N chemical shift changes over the course of a peptide titration into isotopically labeled protein, amide NH pairs of amino acids whose chemical environment changes upon peptide binding can be identified. When mapped onto a structure of the protein, this approach can identify the peptide-binding interface or regions undergoing conformation changes within a protein upon ligand binding. Monitoring NMR chemical shift changes can also serve as a screening technique to identify novel interaction partners for a protein or to determine the binding affinity of a weak protein–peptide interaction. Here, we describe the application of NMR chemical shift mapping to the study of peptide binding to the C-terminal SH2 domain of PLCγ1.

Key words Nuclear magnetic resonance (NMR), SH2 domain, Peptide, Chemical shift mapping (CSM), Phosphotyrosine, PLCγ1

1 Introduction

SH2 domains are a classical signaling domain involved in many diverse signaling pathways, including the regulation of receptor tyrosine kinase pathways. These ubiquitous domains recognize peptide sequences containing a phosphorylated tyrosine in order to regulate downstream signaling events through the promotion of a desired enzymatic activity or by behaving as an adaptor to bridge protein-protein interactions [1–3]. SH2 domains are about 100 amino acids long, making them prime targets for study by nuclear magnetic resonance (NMR), which is well suited for the characterization of small proteins due to their faster tumbling rates and longer signal relaxation times, which confer experimentally desirable narrow linewidths [4]. NMR has many applications to the study of SH2 domains, the most universal being the determination of an SH2 domain's three-dimensional structure, frequently in complex with its ligand, to reveal the details of the binding interface. X-Ray crystallography is a complementary approach to obtaining this

Kazuya Machida and Bernard A. Liu (eds.), *SH2 Domains: Methods and Protocols*, Methods in Molecular Biology, vol. 1555, DOI 10.1007/978-1-4939-6762-9_15, © Springer Science+Business Media LLC 2017

high-resolution structural information, although crystallographic studies require the ability to grow high-quality crystals, a process that can be unpredictable. Solution NMR has the advantage of revealing structural features as they exist at ambient temperature in solution. NMR can further be used to study protein dynamics, from small vibrational motions of atoms to large conformational rearrangements of protein domains, yielding insight into many cellular processes including protein folding, ligand binding, and thermostability [5].

Although the high-resolution structure of an SH2 domain/peptide complex is invaluable to understanding the intermolecular interaction, it is often useful to be able to obtain basic structural insights more quickly than the months required to solve a complete structure. NMR chemical shift mapping (CSM) is a powerful strategy that can be completed over the course of just a few days, and is consequently an attractive technique for the rapid analysis of SH2 domains with various peptide ligands. If resonance assignments are available, this approach can be used to identify the most likely ligand-binding interface of a protein, or identify regions of the protein (e.g., flexible loops) that may undergo significant structural rearrangements in response to the ligand. Even in the absence of resonance assignments, it can additionally be used to obtain the binding dissociation constant (K_d) for systems within an appropriate (μM–mM) affinity regime [6].

CSM is performed by collecting a series of ^{1}H-^{15}N-heteronuclear single quantum coherence (^{1}H-^{15}N-HSQC) NMR spectra of an isotopically labeled SH2 domain protein as a function of added unlabeled peptide. The ^{1}H-^{15}N-HSQC experiment gives rise to a single crosspeak at the ^{1}H and ^{15}N chemical shifts of each amide in the backbone of the protein (as well as some ^{15}N-containing side-chains) [7]. The chemical shift of the proton and nitrogen is exquisitely sensitive to its chemical environment, and consequently is usually distinctive for the amide proton–nitrogen pair of every amino acid in the protein. Upon the addition of peptide, some amides, such as those buried within the protein, will experience little to no change and their crosspeaks will remain unperturbed. In contrast, a subset of crosspeaks will move as a consequence of being in a new chemical environment induced by peptide binding (Fig. 1a). This is often the result of a direct interaction with the peptide, but can also be due to secondary effects representing structural changes, such as the rearrangement of a flexible loop. If the resonance assignments of the protein are available, the amide proton–nitrogen pairs of the amino acids that experienced chemical shift changes can then be mapped onto an available structure of the protein, consequently illustrating the amino acids of the SH2 domain that are most likely in contact with the peptide at the protein–peptide interface or are flexible (Fig. 2).

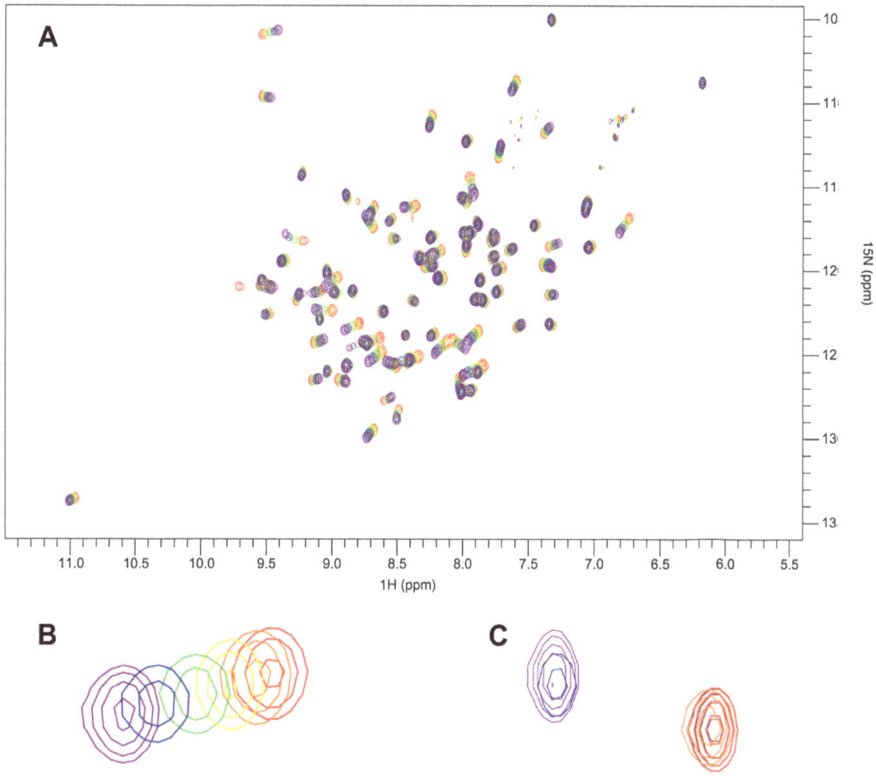

Fig. 1 NMR peak walking upon peptide titration into PLCγ1-SH2. (**a**) Spectra overlay of ^{1}H-^{15}N-HSQC data for 700 μM free PLCγ1-SH2 (*purple*) and PLCγ1-SH2 upon peptide titration (175 μM increments of a phosphopeptide, reaching a final saturated concentration of 875 μM, in *red*). (**b**) An example of an amide proton–nitrogen pair in fast exchange, which produces a characteristic peak "walking" as peptide is titrated into the system. (**c**) An example of an amide proton–nitrogen pair in slow exchange, in which the resonance line of the free protein decreases in intensity while the resonance line of the protein–peptide complex grows in intensity as peptide is titrated into the system

The most detailed structural insights from CSM are obtained by mapping the chemical shift changes onto an available structure. SH2 domains and their complexes are typically well-behaved systems, and have consequently been found to be amenable to structural studies by both NMR and X-ray crystallography. Of the approximately 120 SH2 domains identified in humans, over 50 of these have been structurally characterized on their own or in complex with peptide ligands or other scaffolding proteins, yielding over 200 total SH2 domain structures now available in the Protein Data Bank (PDB) [8, 9]. With this rich collection of structural data, NMR chemical shift mapping is readily feasible for many of the known SH2 domains. Due to their well-behaved nature, additional SH2 domains can also be studied by CSM, although a structure of the protein will first need to be obtained. Alternatively, homology modeling using a server such as SWISS-MODEL can be

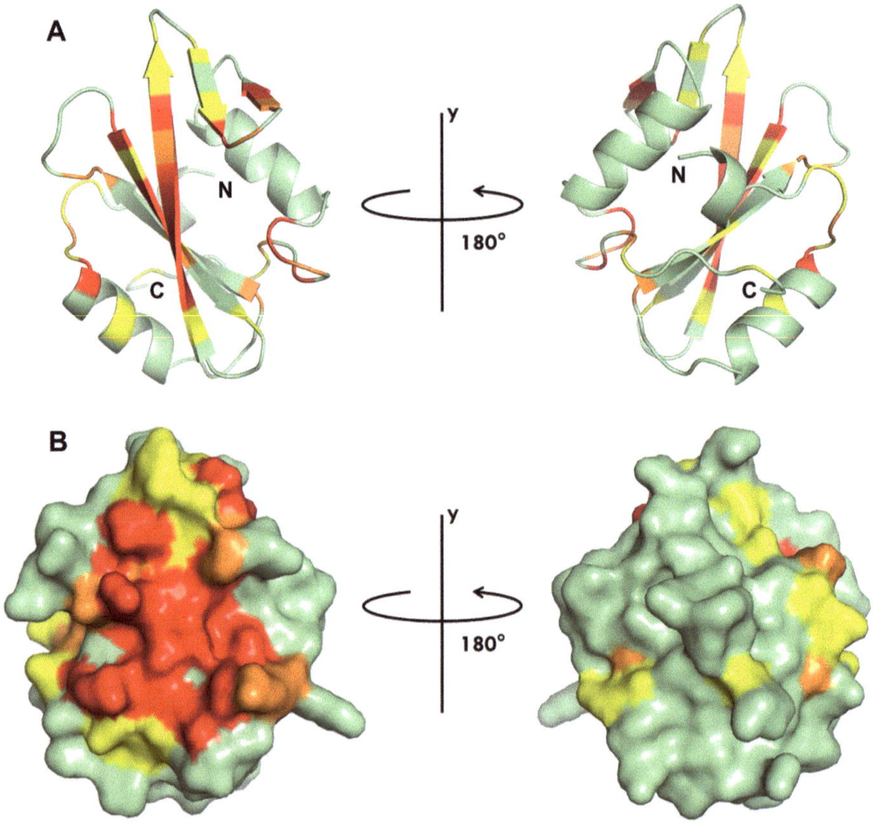

Fig. 2 Chemical shift mapping upon the titration of a phosphopeptide into PLCγ1-SH2. (**a**) Chemical shift perturbations were mapped onto the tertiary structure of PLCγ1-SH2 (PDB: 4K45) and colored as follows: $\Delta\delta < 0.05$ ppm (*green*), $0.05 < \Delta\delta < 0.08$ ppm (*yellow*), $0.08 < \Delta\delta < 0.1$ ppm (*orange*), and 0.1 ppm $< \Delta\delta$ (*red*). (**b**) The same chemical shift perturbations mapped onto the surface representation of PLCγ1-SH2, illustrating the peptide-binding interface

used to create a model of the protein using the structure of another closely related SH2 domain as a template [10]. Even in the absence of a structure, tracking the ^{1}H and ^{15}N chemical shift changes upon peptide addition has several additional applications. For instance, this strategy can be used to rapidly screen for novel peptides or small molecules that can bind to a protein (e.g., for drug design), or to estimate the K_d of a protein–ligand interaction, neither of which require explicit structural knowledge [6, 11].

Before performing CSM, backbone resonance assignments (i.e., assigning each crosspeak to the specific amino acid that gave rise to it) must be made for the free protein. If the protein has been studied previously by NMR, backbone assignments may already be available through the Biological Magnetic Resonance Bank (BMRB) repository [12]. By perfectly matching the conditions of the experiment (including the protein construct, temperature, pH, and additives) to those previously used, assignments can typically be transferred from the BMRB to the newly collected spectrum with ease and high confidence. In contrast, if backbone resonance assignments do not already exist for the protein of interest, or if the available assignments can't be reliably transferred, additional three-dimensional experiments on ^{15}N-^{13}C-isotopically labeled protein will need to be performed [13]. Common multidimensional experiments for completing assignments include the HNCA, HN(CO)CA, HNCACB, CBCA(CO)NH, HNCO, and the HN(CA)CO experiments [14–19]. Even if the BMRB contains assignments for the target protein, it is often still advisable to obtain the additional three-dimensional experiments in order to confirm the preexisting assignments in-house, as the observed crosspeaks can change unpredictably as a function of the construct and solution conditions. Backbone resonance assignments must also be completed for the peptide-bound protein, a process that is typically much more rapid than that of the free protein, though the approach, and consequently the efficiency of the process, depends on the chemical exchange regime of the protein.

When studying a protein–peptide interaction by CSM, it is necessary to consider the system's chemical exchange regime, as this has a significant impact on the appearance and behavior of the resonance lines observed for the exchanging species [20, 21]. To determine the chemical exchange regime, the following kinetic scheme is considered:

$$\text{Protein} + \text{Peptide} \underset{k_{\text{off}}}{\overset{k_{\text{on}}}{\rightleftharpoons}} \text{Complex} \quad (1)$$

where k_{on} and k_{off} are the kinetic rate constants for the forward and reverse binding reactions, respectively. An apparent exchange rate (k_{ex}) can further be defined as:

$$k_{ex} = k_{on} + k_{off} \quad (2)$$

Additionally, the difference between the resonance frequencies of a nuclear spin in the free protein versus peptide-bound environment is given by:

$$\Delta\omega = \left|\omega_{\text{Protein}} - \omega_{\text{Complex}}\right| \quad (3)$$

By comparing the k_{ex} and $\Delta\omega$ from Eqs. 2 and 3, three exchange regimes can be distinguished as follows:

Slow exchange: $k_{ex} \ll \Delta\omega$.

Intermediate exchange: $k_{ex} \approx \Delta\omega$.

Fast exchange: $k_{ex} \gg \Delta\omega$.

The most straightforward systems to study by chemical shift mapping are protein–peptide systems in the fast exchange regime.

Binding systems with weaker interactions (i.e., those with a larger dissociation constant, K_d, where $K_d = k_{off}/k_{on}$ at equilibrium) frequently (though not always) exhibit a faster k_{off}, and will thus have a greater k_{ex} in accordance with Eq. 2. Thus, weak affinity interactions are frequently fast exchange systems. For a fast exchange system, in which the k_{ex} of the protein–peptide interaction is much larger than $\Delta\omega$ (the chemical shift difference between the free and bound states), only a single resonance line is observed, representing the population-weighted average of the chemical shifts of the two interconverting states [20]. As peptide is titrated into the system, the population-weighted average will move towards the chemical shifts of the peptide-saturated state. If titrated through the transition from free to fully bound, the peak will appear to "walk" from the free to the bound state (Fig. 1b).

The ability to track the chemical shifts of an amide ^{1}H–^{15}N pair through the peptide titration greatly simplifies the process of assigning the backbone resonances of the peptide-saturated protein. This is because the assignments can be transferred between the free and bound states with higher confidence, without the need for independent assignment experiments. SH2 domains typically bind ligands with mid nanomolar to mid micromolar dissociation constants with diffusion-controlled on-rates, which frequently places them in the fast exchange regime [22, 23]. As such, backbone amides for many peptide-bound SH2 domains can be quickly assigned. This greatly hastens the process of CSM, which makes NMR an excellent approach for the rapid screening of whether, and how, many ligands or drugs may interact with an SH2 domain of interest.

In contrast to the population-weighted average resonance line observed for fast exchange systems, if the protein–peptide system is instead in the slow exchange regime, two resonance lines are observed (one for the free protein and one for the protein–peptide complex) since the spins exist in each environment long enough to establish a resonance frequency of their own (recall $k_{ex} \gg \Delta\omega$). The addition of peptide causes a decrease in intensity of the resonance lines for amino acids in the free state of the protein with a corresponding appearance of resonance lines with new chemical shifts for each amino acid in the peptide-saturated protein (Fig. 1c). Transferring assignments between the free and peptide-bound states of the protein is consequently more ambiguous for systems in the slow exchange regime, as the newly appearing crosspeak of an amino acid in the protein–peptide complex does not always correspond to the most proximal diminishing crosspeak from an amino acid in the free state of the protein. Additional three-dimensional experiments will likely be necessary to complete the backbone resonance assignments for the peptide-bound protein.

Due to the reliance of the exchange regime on both k_{ex} and the difference in chemical shift, it is not uncommon for a mixture of the resonance line behavior for fast and slow systems to be observed, either for different binding peptides or even within the same complex. Some amide NH pairs can behave entirely as if in the fast or slow exchange regime, while others will both "walk" and fade/reappear. If the vast majority of amide NH pairs are in the fast and intermediate exchange regimes, it may be possible to transfer all backbone chemical shift assignments from the free protein, depending on the extent of the intermediate exchange. Otherwise, three-dimensional experiments on the peptide-bound protein will need to be performed.

An additional advantage to working with a system in the fast exchange regime is the ability to determine the binding affinity of the protein–peptide interaction [6, 24]. For interactions with high micromolar to millimolar binding dissociation constants (K_d), the concentration of protein used for data collection (often low to mid micromolar, depending on the sample's signal-to-noise) is sufficiently below the K_d to allow for an accurate binding analysis. The degree to which the crosspeak has moved ($\Delta\delta_{obs}$) relative to the largest possible chemical shift change of the peptide-saturated protein ($\Delta\delta_{max}$) is an indicator of the fraction of total protein ($[P]_0$) that has been bound by the peptide ligand ([PL]). This is represented by the relationship:

$$\frac{\Delta\delta_{obs}}{\Delta\delta_{max}} = \frac{[\mathrm{PL}]}{[\mathrm{P}]_0} \tag{4}$$

The K_d of the interaction is further given by:

$$K_d = \frac{([P]_0 - [\mathrm{PL}])([L]_0 - [\mathrm{PL}])}{[\mathrm{PL}]} \tag{5}$$

where $[L]_0$ is the total peptide ligand added. By combining Eqs. 4 and 5, and solving the resulting quadratic equation, it can be shown that:

$$\Delta\delta_{obs} = \Delta\delta_{max} \frac{(K_d + [L]_0 + [P]_0) - \sqrt{(K_d + [L]_0 + [P]_0)^2 - (4[P]_0[L]_0)}}{2[P]_0} \tag{6}$$

Equation 6 can be fit to the experimentally obtained data as a plot of $\Delta\delta_{obs}$ versus $[P]_0$, using a program capable of nonlinear least-squares data fitting, such as the Solver add-in available in Microsoft Excel [25]. The data fitting will calculate the unknown K_d and $\Delta\delta_{max}$ values, thus illustrating the application of ^{1}H-^{15}N-HSQC titration experiments to the determination of an interaction's binding affinity.

NMR is a powerful approach by which structural insights, binding affinities, and protein dynamics can simultaneously be investigated. In this chapter, we discuss the study of peptide binding to SH2 domains through the use of NMR chemical shift mapping. We employ the binding of peptide ligands to the C-terminal SH2 domain of phospholipase C-γ1 (PLCγ1-SH2) as a model system. The processes of cloning and purifying PLCγ1-SH2 are described, followed by a description of the collection and analysis of ^{1}H-^{15}N-HSQC peptide titration data.

2 Materials

2.1 Subcloning of 6×-His-SH2 Construct

1. pET15b (*see* **Note 1**).
2. cDNA for SH2 domain of interest (*see* **Note 2**).
3. Custom oligonucleotide primers (*see* **Note 3**).
4. Pfu DNA polymerase (*see* **Note 4**).
5. PCR instrument or equivalent.
6. 1× TAE buffer: 40 mM Tris-acetate, 1 mM ethylenediamine tetraacetic acid, disodium salt dihydrate (EDTA).
7. 1 % agarose gel: 1 % agarose, 0.5 μg/mL ethidium bromide, in TAE buffer.
8. 6× DNA dye: 0.25 % bromophenol blue, 0.25 % xylene cyanol, 30 % glycerol.
9. Power source.
10. Ultraviolet (UV) imaging system, or equivalent.
11. Gel extraction kit.
12. Appropriate restriction enzymes (e.g., NdeI and BamHI) and corresponding restriction enzyme buffer.
13. Quick T4 DNA Ligase.
14. Quick T4 DNA Ligase Reaction Buffer: 200 mM Tris–HCl, 100 mM magnesium chloride ($MgCl_2$), 250 μg/ml acetylated bovine serum albumin (BSA), pH 7.6.
15. Competent cloning cells (e.g., DH5α or XL1-Blue).
16. Luria–Bertani (LB)-ampicillin agar plates: LB agar with 100 μg/mL ampicillin.
17. LB-ampicillin media: 100 μg/mL ampicillin in LB.
18. Plasmid DNA Mini Kit.

2.2 Purification of Soluble SH2 Protein

1. Expression cells (e.g., BL21-DE3).
2. 1000× metal mix: 46 mM boric acid, 102 mM calcium chloride dihydrate, 190 μM cobalt chloride hexahydrate, 780 μM copper sulfate pentahydrate, 1.02 M magnesium chloride

hexahydrate, 1 mM manganese chloride, 3 μM sodium molybdate, 1.7 mM zinc chloride.

3. 100× MEM vitamin solution.
4. Minimal media (MM): 24.7 mM sodium phosphate dibasic (Na_2HPO_4), 22 mM potassium phosphate monobasic (KH_2PO_4), 25.7 mM sodium chloride (NaCl), 11.2 mM isotopically labeled ammonium sulfate ($(^{15}NH_4)_2SO_4$) in 1 L of distilled deionized water. Autoclave this solution. After autoclaving, add 2 g glucose, 10 mL of the 100× vitamin mix, 1 mL of the 1000× metal mix, and 1 μM iron chloride ($FeCl_3$) to each 1 L of media.
5. MM-ampicillin: 100 μg/mL ampicillin in MM.
6. UV-Visible spectrophotometer.
7. Isopropyl-β-D-thiogalactoside (IPTG).
8. Centrifuge capable of at least 25,000 ×*g* force.
9. Purification buffer: 100 mM potassium phosphate, pH 7.4, 200 mM NaCl, 1 mM dithiothreitol (DTT), and 25 mM imidazole.
10. Protease inhibitor cocktail tablet.
11. Ni-NTA agarose.
12. Chromatography column, such as a Kimble-Chase Kontes™ FlexColumn™ gravity flow column with a 10 mm diameter and 40 mL capacity.
13. Wash buffer 1 (WB1): 100 mM potassium phosphate, pH 7.4, 1 M NaCl, 1 mM DTT, 25 mM imidazole.
14. Wash buffer 2 (WB2): 100 mM potassium phosphate, pH 7.4, 200 mM NaCl, 1 mM DTT, 60 mM imidazole.
15. Elution buffer: 100 mM potassium phosphate, pH 7.4, 200 mM NaCl, 1 mM DTT, 350 mM imidazole.
16. Bovine alpha-thrombin.
17. Size exclusion chromatography (SEC) column (HiLoad Superdex 75, or equivalent).
18. Fast protein liquid chromatography (FPLC) system.
19. NMR Buffer (NMRB): 100 mM sodium phosphate, pH 6.4, 500 μM DTT.
20. Concentrator with an appropriate molecular weight cut-off.

2.3 Nuclear Magnetic Resonance

1. Purified SH2 protein.
2. Deuterium oxide (D_2O).
3. Trimethylsilyl propanoic acid (TSP) or alternate referencing compound [26–28].

4. Purified peptide ligand. We typically use 11-mer phosphopeptides (containing a phosphorylated tyrosine at amino acid position 5), which have been purified to >95% purity by reversed-phase HPLC chromatography following standard Fmoc solid-phase peptide synthesis.
5. 5 mm Shigemi thin-wall sample tube, which is magnetic susceptibility matched to D_2O.
6. NMR spectrometer (all experiments herein were performed with an Oxford 600 MHz magnet equipped with an Agilent console and HCN PFG cold probe).
7. Spectrometer control software (e.g., VnmrJ).
8. Data processing programs (e.g., NMRPipe [29]).
9. NMR data visualization program (e.g., Ccp Nmr Analysis [30]).
10. Molecular visualization program (e.g., PyMOL [31]).

3 Methods

3.1 Subcloning of 6×-His-SH2 Construct

1. Amplify the SH2 domain DNA out of the cDNA template by PCR, using Pfu polymerase and the custom oligonucleotide primers (*see* **Notes 1–4**).
2. Add DNA dye to a 1× concentration to PCR product, and run sample on a 1% agarose gel at 80 V for 1 h to isolate the PCR product.
3. Use a UV gel imager to locate the PCR products on the gel. Excise the PCR product from the gel with a sterile razor blade.
4. Purify the excised sample using a gel purification kit.
5. Digest the SH2 DNA insert as well as the destination pET15b vector using the appropriate restriction enzymes. We used BamHI and NdeI.
6. Run samples on a second agarose gel and gel extract them, as done above.
7. Ligate the SH2 DNA insert into pET15b by mixing approximately 100 ng of the digested pET15b vector and 0.2 pmol of the DNA insert with 1 μL of Quick T4 DNA Ligase in 20 μL total volume of Quick T4 DNA Ligase Reaction Buffer. Incubate at room temperature (25 °C) for 5 min.
8. Transform the ligation product into a competent bacterial cell line, plate onto an LB-agar plate containing the appropriate selection antibiotic (for pET15b, this is ampicillin), and incubate at 37 °C overnight (*see* **Note 5**).
9. Inoculate 5 mL of LB-ampicillin (or appropriate selection antibiotic) with a colony of the cloning cells containing the SH2 DNA in the pET15b vector.

10. Shake at 225 rpm overnight at 37 °C to grow culture to saturation.
11. Purify DNA from the bacterial cells using a Plasmid DNA Mini Kit.
12. Confirm the success of the cloning by sequencing the DNA (*see* **Note 6**).

3.2 Purification of Soluble SH2 Protein

1. Transform the SH2-pET15b vector into a bacterial expression cell line.
2. Inoculate a small volume of MM-ampicillin (or appropriate selection antibiotic) with a colony of the SH2-pET15b vector-containing expression cells.
3. Shake this starter culture at 225 rpm overnight at 37 °C.
4. Inoculate 10 mL of the overnight bacterial culture into the desired number of 1 L MM-ampicillin (or appropriate selection antibiotic) baffled flasks for large scale growth.
5. Shake at 37 °C for several hours, monitoring the optical density (OD) at 600 nm approximately every 15–20 min after the initial 2 h.
6. When the OD_{600} reaches 0.6–1.0, remove flasks from shaker and place on ice for 20 min.
7. Add IPTG to 0.5 mM and shake at 20 °C overnight.
8. Pellet the bacterial cells by centrifugation at 3000 × *g* for 10 min. Store bacterial pellet in a −20 °C freezer for future use, or proceed immediately to the next step.
9. Resuspend the bacterial pellet from a 2 L growth in 50 mL of purification buffer with one protease-inhibitor tablet.
10. Sonicate on ice to lyse the cells (e.g., 15 s pulses at amplitude 8.0 followed by 45 s off; cycle repeated eight times for a total pulse time of 2 min).
11. Centrifuge lysate at 25,000 × *g* for 30 min to separate the soluble protein from cellular debris.
12. Promptly remove supernatant to another container to avoid resuspension of pellet.
13. Add the supernatant to a 2 mL nickel column that has been pre-equilibrated with purification buffer, and allow protein to move through the column by gravity flow (*see* **Note 7**).
14. Wash the column with 5 mL of WB1.
15. Wash the column with 5 mL of WB2 (*see* **Note 8**).
16. Elute SH2 protein off the column using 15–20 mL elution buffer (*see* **Note 9**).
17. Cleave off the 6×-His tag from the SH2 protein by adding 1–10 units of thrombin per mg of protein. Allow the cleavage reaction to proceed at 4 °C overnight.

18. Further purify the protein by size exclusion fast protein liquid chromatography (SEC-FPLC), using buffer NMRB. Alterations to this buffer may be considered if a different pH, ionic strength, or additional additives are necessary (*see* **Notes 10–13**).
19. Concentrate the protein-containing fractions as necessary using a concentrator with a membrane of an appropriate molecular weight cutoff.

3.3 Nuclear Magnetic Resonance

1. Prepare NMR sample for the isotopically labeled free SH2 protein by combining the protein with D_2O and a reference compound (e.g., TSP) in a Shigemi tube to a total volume of 350 μL. We used final concentrations of 700 μM protein, 10% D_2O, and 150 μM TSP (*see* **Notes 14–18**).
2. Collect a ^{1}H-^{15}N-HSQC spectrum of the free SH2 protein (*see* **Notes 19–22**). The spectrum should contain approximately the same number of crosspeaks as amino acids in the protein. Furthermore, the crosspeaks should be well-dispersed with uniform peak widths and intensities (*see* **Note 23**).
3. Confirm that all backbone chemical shift assignments have been made for the free protein (*see* **Note 24**).
4. Resuspend peptide ligand in buffer NMRB (*see* **Notes 25–26**).
5. Add peptide into the NMR sample such that a 1:0.25 protein to peptide ratio is achieved, with minimum dilution of the NMR sample if possible (*see* **Note 27**).
6. Collect a ^{1}H-^{15}N-HSQC spectrum, using identical spectral widths and collecting the same number of points in each dimension as was used for the free protein.
7. Repeat **steps 5–6**, adding additional 0.25 molar ratio increments of peptide until the SH2 protein has been completely saturated with peptide (*see* **Note 28**).
8. Process and reference data for all experiments identically using standard approaches [29, 32], and open data in a visualization program such as Ccp Nmr Analysis [30] (*see* **Note 29**).
9. Bring all spectra to similar contour levels and color them as desired.
10. If the system is in the fast exchange regime, chemical shift assignments may be transferred from the free protein to the peptide-saturated protein by tracking the chemical shift "walking" as the peptide concentration is increased (Fig. 1b). For a system in the slow exchange regime (Fig. 1c), additional three-dimensional experiments will need to be performed in order to complete the backbone chemical shift assignments for the protein–peptide complex (*see* **Note 30**).
11. For each amino acid in the SH2-protein, calculate the difference (in ppm) between the chemical shift for the amide proton of

each residue in the free protein versus in the peptide-saturated protein for both the hydrogen ($\Delta\delta_H$) and nitrogen ($\Delta\delta_N$) dimensions.

12. Compile the $\Delta\delta_H$ and $\Delta\delta_N$ values into a total observed chemical shift change ($\Delta\delta_{obs}$) for each amino acid using the following equation [33]:

$$\Delta\delta_{obs} = \sqrt{(\Delta\delta_H)^2 + (0.17 \times \Delta\delta_N)^2} \quad (7)$$

13. Determine appropriate $\Delta\delta_{obs}$ ranges for which chemical shift changes will be considered negligible, small, medium, or large (*see* **Note 31**).
14. Load a structure of the SH2-protein into PyMOL [31], or an alternative molecular visualization program, and color each amino acid in the protein in accordance with its classification as having a negligible, small, medium, or large chemical shift perturbation (Fig. 2) (*see* **Note 32**).
15. Analyze the resulting color-coded structure for a likely peptide-binding interface, or for possible conformational differences between multiple ligand-bound states.
16. If the system is in fast exchange and estimated to be in the appropriate affinity regime (a high micromolar to millimolar binding dissociation constant), the binding affinity can be further derived by fitting the chemical shift titration data to Eq. 6, as described in the introduction (*see* **Note 33**).

4 Notes

1. The pET15b vector contains an expression region transcribed by T7 RNA polymerase, and contains the DNA sequence encoding a 6× N-terminal histidine (6×-His) tag followed by a thrombin protease recognition site and three restriction enzyme cloning sites (those of NdeI, XhoI, and BamHI). The 6×-His tag will provide the SH2 protein with affinity for Ni-NTA agarose, and can be later removed from the protein after purification by the addition of thrombin protease. Similarly designed pET vectors or alternative proteolytic cleavage sites can alternatively be used.
2. cDNA encoding the SH2 protein of interest can often be requested from research labs who have previously used the construct. It can also be obtained commercially through companies such as I.M.A.G.E. Consortium, Addgene, or DNASU [34–36].
3. Primers were designed to isolate the DNA encoding the SH2 domain of interest, as well as to add a 3′ stop codon and to insert 3′ BamHI and 5′ NdeI restriction enzyme recognition sites.

4. Several polymerases are commercially available for use, including Pfu, Taq, and Vent. While all would be acceptable for use, Pfu polymerase has the highest fidelity and was consequently chosen to minimize the possibility of error introduction [37].
5. Including control reactions during the ligation and transformation steps is valuable for troubleshooting if the subcloning is unsuccessful. Positive control DNA is often included in ligation kits to test the ligase activity. If unavailable, one option is to include a ligation reaction in which only one restriction enzyme was used on the parental DNA plasmid. The ability of the parental plasmid to self-ligate will confirm the ligase activity. Another common control is to perform the ligation reaction and transformation on the digested parental plasmid, but in the absence of the insert DNA encoding the SH2 domain. This will illustrate the density of background transformation colonies that can be expected from undigested or re-ligated parental plasmid. One final positive control is to transform the undigested parental plasmid to verify the transformation protocol efficiency, antibiotic resistance, and cell competency.
6. Sequencing of all final constructs is essential. DNA can be sent directly to companies including Genewiz, Quintara, and ACGT Inc, for sequencing.
7. To optimize protein retention, the protein can be flowed over the column a second time, or alternatively mixed with the resin for several minutes before flow-through. Be careful, however, not to let too much time pass, as there are many proteins in the solution that may precipitate out over time and clog the column.
8. 60 mM imidazole is used in WB2 to elute unwanted proteins that may be weakly interacting with the nickel column. However, depending on the behavior of the individual protein, this may be a high enough concentration to begin eluting the SH2 protein of interest as well. If this occurs, consider using a gradient of imidazole concentrations to better optimize the concentration that will elute the desired protein. Design a more appropriate WB2 with an alternate concentration of imidazole accordingly. Likewise, the 350 mM imidazole used in the elution buffer can be better optimized for a specific protein, although this concentration will be suitably high for most proteins.
9. The 15–20 mL of elution buffer can be added as several smaller (e.g., 5 mL) increments. Only the fractions containing a significant amount of protein need to be pooled and further purified. At this point, samples of each step (e.g., lysate, supernatant, column flow-through, washes, elutions, remaining nickel beads) can be run on a 17%-SDS gel to confirm the quality of the purification, to determine which elution fractions are desired, and to troubleshoot as necessary.

10. The pH of an NMR sample is typically between pH 5–7 [38]. The amide proton exchange is both acid- and base-catalyzed, with a minimum rate of proton exchange typically occurring around a pH of 3.0 [39]. This results in an increase in the rate of the exchange as the pH is increased or decreased away from 3.0. At a high enough pH, the proton will exchange at a rate faster than the duration of the NMR pulse sequence, leading to a reduction in sensitivity or even a complete loss of detection of the amide proton signal. However, it is also desirable to perform experiments under conditions close to the physiological environment so that any structural data collected best reflects a functionally relevant state of the protein [21]. It may be advisable to collect NMR data at an ideal physiological pH, but to be prepared to collect data at a lower pH if sensitivity requires further optimization.
11. It is best to avoid buffers containing non-labile protons, as their signals can obscure those of the protein in the hydrogen dimension. Phosphate buffer is consequently one of the most prevalent choices, although a phosphate buffer may weaken the affinity of a phosphopeptide to the target SH2 domain. We recommend testing whether a phosphate buffer significantly impacts binding before using it for NMR studies. Additionally, sometimes alternate buffers are required in order to maintain consistency with other experiments or to increase the sensitivity of the protein signal via the use of low-conductivity buffers [40]. If it is necessary to use a buffer containing non-labile protons, such as Tris or acetate, consider purchasing the deuterated form of the buffer if the protons are not adequately suppressed in the spectrum.
12. Additional buffer additives should also be minimized if possible, including the chelating agent EDTA, protease inhibitors, and reducing reagents such as DTT. However, if these reagents are necessary for protein stability, including them will likely have little consequence on the detected protein signal for heteronuclear NMR experiments, such as the ^{1}H-^{15}N-HSQC performed here.
13. Avoid using high salt concentrations (>200 mM), especially when collecting data with a cold probe. Salt can generate ion currents during the radiofrequency pulses, resulting in a decline in the sensitivity of the NMR experiment as the ionic strength is increased, as well as a rise in the temperature of the sample [41]. Furthermore, high salt content can make probe tuning more challenging, will reduce the homogeneity of the applied radiofrequency pulses, and will increase the 90° pulse width (pw) required for excitation [41]. If a high salt concentration is crucial for the desired experiment, consider using a salt-tolerant probe and/or a more narrow Shigemi sample tube (e.g., 3 mm instead of 5 mm).

14. The signal-to-noise can be increased through several approaches, including increasing the number of scans or the concentration of the sample. For PLCγ1-SH2, we use a final concentration of 700 μM of protein to collect data from a single HSQC experiment in approximately 70 min (when nt = 16 and ni = 80), allowing us to collect a full set of peptide titration data in a single day. We increase the protein concentration to 1.0 mM if we anticipate performing additional three-dimensional NMR experiments (e.g., for backbone assignments) on the sample. Lower concentrations can also be used, especially if protein solubility is limiting. However, the signal-to-noise ratio improves as a result of the number of times the NMR signal is averaged [42]. Thus, reducing the protein concentration will require adjustments to several experimental parameters (such as the nt, ni, and sw) to achieve a spectrum with equivalent signal-to-noise. For SH2 domains, we typically do not raise the ni above approximately 256, as the corresponding longer delay time will only contribute noise as the ni is further increased.
15. A 5 mm thin-wall NMR Shigemi tube requires a minimum of 250 μL of protein. We use 350 μL samples in order to ease the process of shimming. If the protein and peptide are not limited in quantity, consider using regular 5 mm NMR tubes, which are much cheaper and less fragile, but require more sample volume (around 500 μL or more).
16. The addition of D_2O will dilute the buffer in addition to the protein, which is not problematic for most samples if properly planned for. However, if a certain ionic strength or other additive concentration is required for protein solubility, this can cause protein precipitation when neat D_2O is added. One solution to this is to lyophilize 1 mL of the NMR buffer and resuspend the lyophilized powder in an equivalent 1 mL of D_2O. This new sample of deuterated buffer can now be used to create the final NMR sample without risk of changing the ionic environment of the protein. Alternatively, the initial concentration of the buffer components can be increased, such that the addition of D_2O dilutes the buffer components to the desired final concentrations.
17. We use 10% D_2O in our protein's buffer so that even after dilution by peptide, the final solution will still contain sufficient deuterium to serve as the lock solvent for the spectrometer. If a lower initial concentration of D_2O needs to be used (e.g., 5% D_2O), consider adding an equivalent amount of D_2O to the peptide, so that the deuterium is not further diluted upon peptide titration.
18. A referencing compound is included so that all chemical shifts are calculated relative to the same standard and compared to a

centralized database, such as the Biological Magnetic Resonance Bank [12]. This establishes consistency between multiple HSQC experiments, which allows for an accurate comparison of chemical shifts for a peak of interest between different data sets. Here, we use TSP, although other referencing compounds can also be considered [26–28].

19. We utilized settings loaded with the Biopack gNhsqc.c pulse sequence available with the VnmrJ software. Calibrate all necessary parameters using procedures typically used for the system in use. The following parameters were used to perform a ^{1}H-^{15}N-HSQC on PLCγ1-SH2 using a Varian 600 MHz magnet with a Varian HCN PFG cold probe: sweep width (sw) = 11990.4 Hz, sweep width of indirect dimension (sw1) = 2200 Hz, steady state scans (ss) = 64, number of transients (nt) = 16, number of increments (ni) = 80, number of points (np) = 2048, gain = 32, relaxation delay (d1) = 1.5, pulse width (pw) = 9.0 ± 0.5, transmitter power (tpwr) = 53, pulse width for nitrogen dimension (pwN) = 36.0, power level for the nitrogen pulse (pwNlvl) = 60, carrier offset for nitrogen dimension (dof2) = 1337 Hz, ^{15}N decoupling power (dpwr2) = 37, and a wurst40 decoupling pattern for ^{15}N.

20. Spectral folding in the indirect dimensions occurs when one or more resonance lines of amide protons occur at frequencies outside of the range covered by the defined sweep width. Their peaks will be "folded", or reflected, over the edge of the spectrum, and will consequently appear at a mirrored location in the spectrum exactly as far into the spectrum as it was outside of the sweep width boundary. If spectral folding is observed, either tune it to minimize peak overlap and maximize digital resolution or increase the sweep width.

21. Optimize the length of the ^{1}H-^{15}N-HSQC experiment to find the fastest possible collection time in which peaks have sufficient signal-to-noise and can still be resolved from neighboring peaks. We used a number of transients (nt) of 16 and a number of increments (ni) of 80, leading to a data collection time of approximately 70 min. If sufficient sensitivity and/or peak resolution is not achieved, yet protein concentration cannot be raised due to solubility limits, consider increasing nt and ni (for an HSQC experiment, they should remain multiples of the experimental phase cycle), which will improve the spectrum at the expense of adding time to the data collection. The band-Selective Optimized Flip-Angle Short-Transient heteronuclear multiple quantum coherence (SOFAST-HMQC) experiment can also be considered to minimize the required data collection time [43, 44].

22. We perform binding reactions and NMR experiments at 298 K (25 °C). Lower temperatures may be considered for systems in

which the protein and/or peptide has poor solubility. If the protein is stable at even higher temperatures (e.g., 300–320 K), this can improve the resolution of the spectrum due to a decrease in the correlation time [42]. Furthermore, the K_d of a protein–peptide interaction can be temperature dependent. In some cases, the temperature can be altered to change the chemical exchange regime of the system, potentially moving it to fast exchange if the system is otherwise in intermediate or slow exchange.

23. A poor quality ^{1}H-^{15}N-HSQC spectrum can frequently be attributed to protein degradation, misfolded protein, aggregation, and/or sample impurity. Ensure that the free protein is monomeric at the experimental conditions, and has been properly folded and purified, producing a high quality HSQC spectrum before proceeding with the peptide titration.
24. If backbone chemical shifts have not yet been assigned for the SH2-protein of interest, additional three-dimensional experiments will need to be performed [13]. Express and purify the ^{1}H-^{15}N-^{13}C-protein exactly as done previously, but use uniform ^{13}C-glucose in the minimal media instead of unlabeled glucose. The predominant multidimensional experiments that we perform are the HNCACB and the CBCA(CO)NH experiments. Alternative HNCA and HN(CO)CA experiments offer greater signal-to-noise, although they lack information pertaining to the beta carbons, making the assignment process more challenging with these two experiments alone. If time permits, the four of these experiments in conjunction provide an excellent dataset for the assignment of backbone chemical shifts [14–17]. If the alpha carbon chemical shifts contain a lot of overlap, another popular approach is to perform HNCO and HN(CA)CO experiments, which instead observe carbonyl chemical shifts [18, 19].
25. The 11-mer phosphopeptide is obtained as a lyophilized powder following standard Fmoc solid-phase peptide synthesis and purification to >95% purity by reversed-phase HPLC chromatography [45, 46]. We prepare a peptide solution of ≥35 μL at a concentration of 8.75 mM. When a total of 35 μL of 8.75 mM peptide has been added to the 350 μL of 700 μM PLCγ1-SH2 after the final titration point, this will yield the desired 1.25-fold molar excess of peptide. For a protein–peptide interaction with a binding stoichiometry of 1:1, this should be sufficient to fully saturate the protein, assuming the concentration is sufficiently above the K_d of the interaction.
26. For peptides that do not absorb strongly at 280 nm due to minimal tryptophan or tyrosine residues, the concentration can instead be resolved by tracking the peptide absorbance at 205 nm [47, 48].

27. We inject 7.0 μL at a time of 8.75 mM peptide into the 350 μL solution of 700 μM PLCγ1-SH2. 7.0 μL injections contain one quarter (0.25) the number of moles of peptide as there are moles of protein present in the solution.
28. We perform five titration points of 7.0 μL each, yielding a final protein to peptide molar ratio of 1:1.25. A sixth titration point (bringing the molar excess of peptide to 1.5) produced identical chemical shift locations and intensities as the fifth titration point, allowing us to use the 1.25-fold excess peptide spectrum as the final spectrum. However, if changes are still observed, continue adding peptide until two consecutive additions yield identical ^{1}H-^{15}N-HSQC spectra, indicating full saturation has been achieved.
29. Details regarding NMR data processing are beyond the scope of this chapter. We assume the reader has a basic knowledge of, or access to knowledge of, NMR data acquisition and processing, although there are informative resources available elsewhere if needed [29, 32].
30. Chemical shift "walking" is characteristic of fast exchange systems (Fig. 1b). If, instead, the system is in slow exchange, separate resonance lines are observed for both the free and peptide bound conformations. As peptide is added to the solution, the intensity of the crosspeak for an amino acid in the free protein will decrease as a corresponding crosspeak appears and becomes more intense for the amino acid in the peptide-bound protein (Fig. 1c). This situation makes the transfer of assignments more difficult, and additional three-dimensional experiments (detailed in **Note 24**) may be necessary to complete the backbone assignments for the protein–peptide complex.
31. There is no convention for displaying chemical shift changes, and ranges should instead be optimized for each system. For PLCγ1-SH2, we used the following ranges: $\Delta\delta < 0.05$ ppm (negligible), $0.05 < \Delta\delta < 0.08$ ppm (small), $0.08 < \Delta\delta < 0.1$ ppm (medium), and 0.1 ppm $< \Delta\delta$ (large). Choose a low threshold (here, 0.05 ppm) that is above the distance moved by peaks with negligible change to ensure that noise is not interpreted as a significant change.
32. If a structure of the SH2-protein of interest is not available, homology modeling using a server such as SWISS-MODEL may be used to construct a model structure of the protein using its amino acid sequence and a template structure of a closely related SH2 domain [10].
33. We recommend making several adjustments to the data collection if it is to be used for the determination of the K_d. More data points will provide the most accurate data fitting, especially for the initial portion of the peptide titration in which the binding

curve is steepest. Additionally, it is valuable to collect data when there is a large excess of peptide, which will approach the $\Delta\delta_{max}$, thus better fitting the later portion of the binding curve. Typically, we collect 12 titration points spanning from the free state of the protein to a molar excess of five-fold peptide (specifically, the molar ratios of peptide we use are: 0, 0.10, 0.25, 0.40, 0.55, 0.75, 1, 1.25, 1.50, 2, 3, and 5). If sufficient signal can still be obtained, we also recommend lowering the concentration of total protein (e.g., down to 100 or 200 μM) to ensure that the concentration is below the expected K_d.

Acknowledgments

We would like to thank our collaborators Dr. Zhongping Tan and Dr. Xiaoyang Guan for the phosphopeptide synthesis, as well as Dr. Julie Forman-Kay for kindly providing us with a PLCγ1-pET11d construct. We would also like to thank Sabrina Hunt and Kathryn Wall for their generous feedback. This work was supported by NSF grant MCB1121842 (to D.S.W.).

References

1. Pawson T, Gish GD, Nash P (2001) SH2 domains, interaction modules and cellular wiring. Trends Cell Biol 11:504–511
2. Yaffe MB (2002) Phosphotyrosine-binding domains in signal transduction. Nat Rev Mol Cell Biol 3:177–186
3. Tinti M, Kiemer L, Costa S et al (2013) The SH2 domain interaction landscape. Cell Rep 3:1293–1305
4. Yu H (1999) Extending the size limit of protein nuclear magnetic resonance. Proc Natl Acad Sci U S A 96:332–334
5. Kleckner IR, Foster MP (2011) An introduction to NMR-based approaches for measuring protein dynamics. Biochim Biophys Acta 1814:942–968
6. Fielding L (2007) NMR methods for the determination of protein-ligand dissociation constants. Prog Nucl Magn Reson Spectrosc 51:219–242
7. Rule GS, Hitchens TK (2006) Two dimensional heteronuclear J-correlated spectroscopy. In: Kaptein R (ed) Fundamentals of protein nmr spectroscopy. Springer, Dordrecht, pp 197–211
8. Berman HM, Westbrook J, Feng Z et al (2000) The protein data bank. Nucleic Acids Res 28:235–242
9. Liu, B. (2010) SH2 Domain Structures. https://sites.google.com/site/sh2domain/sh2-structures. Accessed 15 April 2015.
10. Arnold K, Bordoli L, Kopp J et al (2006) The SWISS-MODEL workspace: a web-based environment for protein structure homology modelling. Bioinformatics 22:195–201
11. Shuker SB, Hajduk PJ, Meadows RP et al (1996) Discovering high-affinity ligands for proteins: SAR by NMR. Science 274:1531–1534
12. Ulrich EL, Akutsu H, Doreleijers JF et al (2008) BioMagResBank. Nucleic Acids Res 36:402–408
13. Rule GS, Hitchens TK (2006) Resonance assignments: heteronuclear methods. In: Kaptein R (ed) Fundamentals of protein NMR spectroscopy. Springer, Dordrecht, pp 277–312
14. Wittekind M, Mueller L (1993) HNCACB, a high-sensitivity 3D NMR experiment to correlate amide-proton and nitrogen resonances with the alpha- and beta-carbon resonances in proteins. J Magn Reson Ser B 101:201–205
15. Yamazaki T, Lee W, Arrowsmith CH et al (1994) A suite of triple resonance NMR experiments for the backbone assignment of ^{15}N, ^{13}C, ^{2}H labeled proteins with high sensitivity. J Am Chem Soc 116:11655–11666
16. Grzesiek S, Bax A (1992) Correlating backbone amide and side chain resonances in larger proteins by multiple relayed triple resonance NMR. J Am Chem Soc 114:6291–6293

17. Yamazaki T, Lee W, Revington M et al (1994) An HNCA pulse scheme for the backbone assignment of ^{15}N,^{13}C,^{2}H-labeled proteins: Application to a 37-kDa *Trp* repressor-DNA complex. J Am Chem Soc 116:6464–6465
18. Clubb RT, Thanabal V, Wagner G (1992) A constant-time three-dimensional triple-resonance pulse scheme to correlate intraresidue $^{1}H^{N}$, ^{15}N, and $^{13}C'$ chemical shifts in ^{15}N-^{13}C-labelled proteins. J Magn Reson 97:213–217
19. Ikura M, Kay LE, Bax A (1990) A novel approach for sequential assignment of ^{1}H, ^{13}C, and ^{15}N spectra of proteins: heteronuclear triple-resonance three-dimensional NMR spectroscopy. Application to calmodulin. Biochemistry 29:4659–4667
20. Rule GS, Hitchens TK (2006) Exchange processes. In: Kaptein R (ed) Fundamentals of protein NMR spectroscopy. Springer, Dordrecht, pp 403–430
21. Cavanagh J, Fairbrother WJ, Palmer AG III et al (2007) Experimental aspects of NMR spectroscopy. In: Protein NMR spectroscopy, 2nd edn. Elsevier Inc., Burlington, VT, pp 234–235
22. Songyang Z, Shoelson SE, Chaudhuri M et al (1993) SH2 domains recognize specific phosphopeptide sequences. Cell 72:767–776
23. Ladbury JE, Lemmon MA, Zhou M et al (1995) Measurement of the binding of tyrosyl phosphopeptides to SH2 domains: a reappraisal. Proc Natl Acad Sci U S A 92:3199–3203
24. Lian, L., Roberts, G. C. K. (1993) Effects of chemical exchange on NMR spectra. In: Roberts, G. C. K (ed) NMR of Macromolecules: A Practical Approach. Oxford University Press Inc., New York, pp. 153-182.
25. Kemmer G, Keller S (2010) Nonlinear least-squares data fitting in Excel spreadsheets. Nat Protoc 5:267–281
26. Harris RK, Becker ED, Cabral De Menezes SM et al (2001) NMR nomenclature. Nuclear spin properties and conventions for chemical shifts. Pure Appl Chem 73:1795–1818
27. Harris RK, Becker ED, Cabral De Menezes SM et al (2008) Further conventions for NMR shielding and chemical shifts. Pure Appl Chem 80:59–84
28. Wishart DS, Bigam CG, Yao J et al (1995) ^{1}H, ^{13}C and ^{15}N chemical shift referencing in biomolecular NMR. J Biomol NMR 6:135–140
29. Delaglio F, Grzesiek S, Vuister GW et al (1995) NMRPipe: a multidimensional spectral processing system based on UNIX pipes. J Biomol NMR 6:277–293
30. Vranken WF, Boucher W, Stevens TJ et al (2005) The CCPN data model for NMR spectroscopy: development of a software pipeline. Proteins Struct Funct Genet 59:687–696
31. Schrodinger LLC (2010) The PyMOL Molecular Graphics System, Version 1.5.0.4
32. Rule GS, Hitchens TK (2006) Introduction to signal processing. In: Kaptein R (ed) Fundamentals of protein NMR spectroscopy. Springer, Dordrecht, pp 65–88
33. Farmer BT II, Constantine KL, Goldfarb V et al (1996) Localizing the $NADP^{+}$ binding site on the MurB enzyme by NMR. Nat Struct Biol 3:995–997
34. Lennon G, Auffray C, Polymeropoulos M et al (1996) The I.M.A.G.E. consortium: an integrated molecular analysis of genomes and their expression. Genomics 33:151–152
35. Herscovitch M, Perkins E, Baltus A et al (2012) Addgene provides an open forum for plasmid sharing. Nat Biotechnol 30:316–317
36. Seiler CY, Park JG, Sharma A et al (2014) DNASU plasmid and PSI: biology-materials repositories: resources to accelerate biological research. Nucleic Acids Res 42:1253–1260
37. Cline J, Braman JC, Hogrefe HH (1996) PCR fidelity of *Pfu* DNA polymerase and other thermostable DNA polymerases. Nucleic Acids Res 24:3546–3551
38. Teng Q (2013) NMR sample preparation. In: Structural biology: practical NMR applications, 2nd edn. Springer, New York, NY, pp 103–116
39. Englander SW, Kallenbach NR (1984) Hydrogen exchange and structural dynamics of proteins and nucleic acids. Q Rev Biophys 16:521–655
40. Kelly AE, Ou HD, Withers R et al (2002) Low-conductivity buffers for high-sensitivity NMR measurements. J Am Chem Soc 124:12013–12019
41. Rule GS, Hitchens TK (2006) Practical aspects of N-dimensional data acquisition and processing. In: Kaptein R (ed) Fundamentals of protein NMR spectroscopy. Springer, Dordrecht, pp 313–315
42. Primrose, W. U. (1993) Sample preparation. In: Roberts, G. C. K (ed) NMR of Macromolecules: A Practical Approach. Oxford University Press Inc., New York, pp. 7-33.
43. Schanda P, Brutscher B (2005) Very fast two-dimensional NMR spectroscopy for real-time investigation of dynamic events in proteins on the time scale of seconds. J Am Chem Soc 127:8014–8015
44. Schanda P, Kupce E, Brutscher B (2005) SOFAST-HMQC experiments for recording two-dimensional heteronuclear correlation spectra of proteins within a few seconds. J Biomol NMR 33:199–211

45. Merrifield RB (1963) Solid phase peptide synthesis. I. The synthesis of a tetrapeptide. J Am Chem Soc 85:2149–2154
46. Attard TJ, O'Brien-Simpson N, Reynolds EC (2007) Synthesis of phosphopeptides in the Fmoc mode. Int J Pept Res Ther 13: 447–468
47. Anthis NJ, Clore GM (2013) Sequence-specific determination of protein and peptide concentrations by absorbance at 205 nm. Protein Sci 22:851–858
48. Goldfarb AR, Saidel LJ, Mosovich E (1951) The ultraviolet absorption spectra of proteins. J Biol Chem 193:397–404

Chapter 16

Calorimetric Measurement of SH2 Domain Ligand Affinities

Marissa A. McKercher and Deborah S. Wuttke

Abstract

Isothermal titration calorimetry (ITC) has emerged as a leading approach in the characterization of protein/ligand interactions. This technique measures the heat change of a system upon binding of a ligand to a biomolecule, and thereby requires no immobilization, intrinsic fluorescence, or labeling of any kind of either species. If properly designed, a single experiment can not only measure the binding affinity, but also determine additional binding and thermodynamic parameters, including the enthalpy, entropy, and the stoichiometry of the interaction. Here, we describe the protocol for the collection of calorimetric data for the binding of peptides to SH2 protein domains.

Key words Isothermal titration calorimetry, ITC, SH2 domain, Peptide, Binding affinity, Thermodynamics, Enthalpy, Entropy

1 Introduction

SH2 domain-containing proteins are key components of cellular signaling pathways, and can promote a variety of diverse downstream events through the interaction with phosphotyrosine-containing peptide ligands [1–3]. Characterizing the thermodynamics underlying the formation of these protein/peptide complexes can help elucidate the driving forces that promote the biomolecular interaction. In addition to the binding affinity (K_d) and standard Gibbs free energy change ($\Delta G°$), isothermal titration calorimetry (ITC) can also measure the binding stoichiometry (n), enthalpy change ($\Delta H°$), and entropy change ($\Delta S°$) allowing for a much richer understanding of the energetics driving a binding reaction [4–8]. ITC can be used to measure a wide range of binding affinities, spanning from approximately 1 nM to 100 μM [5]. Furthermore, ITC requires no fluorophore tag or solid support, and is a true equilibrium technique (i.e., does not require separation of the free and bound states), providing advantages relative to alternative popular methods such as fluorescence polarization and surface plasmon resonance.

Kazuya Machida and Bernard A. Liu (eds.), *SH2 Domains: Methods and Protocols*, Methods in Molecular Biology, vol. 1555, DOI 10.1007/978-1-4939-6762-9_16, © Springer Science+Business Media LLC 2017

Isothermal titration calorimeters measure a solution's heat change upon protein/ligand binding. This change in heat can be used to measure binding parameters as long as there is a nonzero change in enthalpy. Many instruments are available to make calorimetric measurements, including temperature change and heat conduction instruments, although the most common is the power compensation instrument [9]. In this approach, constant power is supplied to a reference cell heater to hold the reference water or buffer solution at a constant temperature. Separate variable power is supplied to the protein-containing sample cell to maintain the sample at the same temperature as the reference cell. As peptide is incrementally titrated into the protein solution, heat is released or absorbed as a result of the binding event. The sample cell power source will accordingly adjust its power supply to accommodate the change in temperature. The change in power (microcalories per second) applied to the sample cell heater is the raw signal that serves as the initial output data (Fig. 1a). Using knowledge of the concentration of peptide and the volume per injection, the raw data can be integrated to produce a binding curve relating the heat released (or more rarely, absorbed) per mole of peptide titrant (kcal/mol) to the molar ratio of protein to peptide (Fig. 1b). From this binding curve, the values of the binding constant (K, where $K = 1/K_d$), n, and $\Delta H°$ can be directly determined.

For a macromolecule approximated to have a single identical ligand-binding site ($n = 1$) per unit, a nonlinear fit can be used to determine the best approximations for K, $\Delta H°$, and the actual n. Three starting relationships must be understood to rationalize this nonlinear fit [10]. First, the binding constant is given by:

$$K = \frac{\theta}{(1-\theta)[L]} \quad (1)$$

where θ is the fractional saturation (fraction of macromolecule sites occupied by ligand), and $[L]$ is the concentration of free ligand. $[L]$ is given by:

$$[L] = [L_T] - n\theta[M_T] \quad (2)$$

where $[L_T]$ and $[M_T]$ are the total concentrations of ligand and macromolecule (protein), respectively. Lastly, the total heat content (Q) of the solution in the sample cell (with a volume V_0) is:

$$Q = n\theta[M_T](\Delta H)V_0 \quad (3)$$

Combining these relationships and solving the resulting quadratic equation from combining Eqs. (1, 2) then yields the following expression:

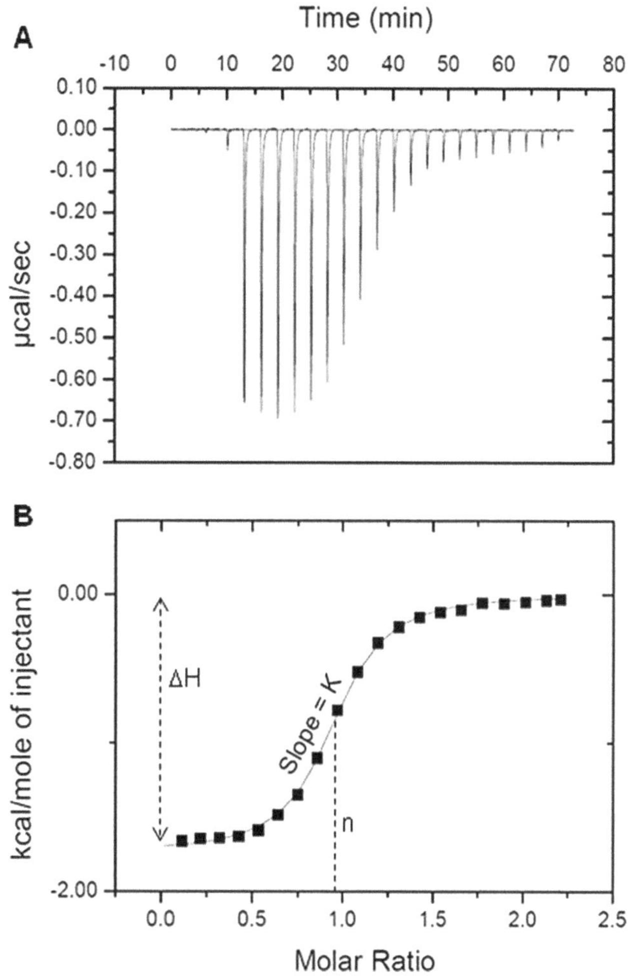

Fig. 1 ITC data collection and manipulation. (**a**) Raw heat signal upon the titration of 2 mM of the phosphopeptide ADND-pY-IIPLPD into 200 μM PLCγ1-SH2. After an initial 10 min delay followed by a dummy injection of 0.2 μL of peptide, 19 peptide injections of 2 μL each and one final injection of 1.3 μL were made at 3 min intervals into 203 μL of protein. (**b**) Heat signals from the raw data can be integrated, producing a binding curve from which the binding constant (*K*), stoichiometry (*n*), and change in enthalpy (ΔH) for the interaction can be derived. In this example, $n = 0.923$ sites, $K = 144{,}000\ M^{-1}$, and $\Delta H = -1.74$ kcal/mol. Figure was created using Origin® software

$$Q = \frac{n[M_T](\Delta H)V_0}{2}\left[1 + \frac{[L_T]}{n[M_T]} + \frac{1}{nK[M_T]} - \sqrt{\left(1 + \frac{[L_T]}{n[M_T]} + \frac{1}{nK[M_T]}\right)^2 - \frac{4[L_T]}{n[M_T]}}\right] \quad (4)$$

The final parameter that must be computed for the fit of the graphical data is the change in heat content (ΔQ) for each injection (i) upon the completion of the previous injection ($i-1$). The data fitting must further take into account the displacement of some volume of solution out of the cell with each titration of ligand (where ΔVi is the injection volume, or the volume by which the total solution volume increases with each titration). The displaced volume is approximated to contribute about 50% as much of a heat effect as the solution that remained in the sample cell [10]. The correction for these factors produces the following relationship, which identifies the heat released from the *i*th injection (ΔQi):

$$\Delta Q_i = Q_i + \frac{\Delta V_i}{V_0}\left[\frac{Q_i + Q_{i-1}}{2}\right] - Q_{i-1} \quad (5)$$

To fit the data, the values of K, n, and ΔH° are approximated, producing a hypothetical ΔQi that is then compared to the actual experimentally measured heat per injection. The parameters are repeatedly adjusted until minimal error has been achieved, and additional iterations provide no further improvement to the fit [10]. ΔG° and ΔS° values can then be calculated from the standard thermodynamic relationships:

$$\Delta G^\circ = -\mathrm{RT}\ln K \quad (6)$$

$$\Delta G^\circ = \Delta H^\circ - T\Delta S^\circ \quad (7)$$

where R is the universal gas constant, and T is the temperature in Kelvin [11].

The binding constants of protein/ligand interactions are often exclusively compared by researchers when evaluating the efficacy of potential drug targets or when determining whether a small molecule may be a physiological binding partner for a corresponding protein. However, the binding affinities of two different interactions may appear similar, when the comprising enthalpic and entropic contributions of the interactions may actually be very different [8]. For example, as seen in Eq. (7), a large enough ΔS can overcome an endothermic (positive ΔH) binding event to allow the system to achieve a similar free energy to that of an exothermic reaction with a smaller entropic contribution. Accordingly, it is valuable to dissect these thermodynamic contributions to better understand the likely response of a protein to different small molecule ligands [6]. The enthalpy of binding reflects the changes in non-covalent bond energies between the free protein and free ligand in buffer and the protein/ligand complex after binding [5]. During a binding event, there may be a reduction in the hydrogen bonds and van der Waals contacts between each species and the buffer, with a corresponding increase in the hydrogen bonds, van der Waals contacts, and electrostatic interactions between the protein and ligand. The ΔH reflects the overall sum of these changes. The primary contribution to the ΔS of binding is the change in solvation energy, which is indicative of the reordering of solvent molecules as the solvent-accessible surface area of protein changes upon ligand binding. Additional entropic contributions arise from changes in the conformational freedom of the protein and ligand backbone and side-chains [5, 7, 8]. Taken together, an understanding of the thermodynamics underlying a protein/ligand interaction can provide a much deeper understanding of the driving forces behind the interaction than can knowledge of the binding constant alone.

Despite its many advantages, one main drawback of ITC is the requirement of a fairly large quantity of sample, typically on the order of 0.05–1.5 mg of protein and around twofold more moles of peptide needed per run [4]. However, the unique ability to determine all of the binding parameters in a single experiment using an unmodified system has upheld ITC as an ideal technique for the study of biomolecular interactions in well-behaved systems.

In this chapter, we discuss the application of ITC to the study of peptide binding to SH2 domains. We employ the binding of a canonical phosphotyrosine-containing peptide (ADND-pY-IIPLPD) to the C-terminal SH2 domain of phospholipase C-γ1 (PLCγ1-SH2) as a model system. The processes of cloning and purifying PLCγ1-SH2 are described, followed by a description of the collection and analysis of calorimetric data.

2 Materials

2.1 Subcloning of 6×-His-SH2 Construct

1. pET15b (*see* **Note 1**).
2. cDNA for SH2 domain of interest (*see* **Note 2**).
3. Custom oligonucleotide primers (*see* **Note 3**).
4. Pfu DNA polymerase (*see* **Note 4**).
5. PCR instrument or equivalent.
6. 1× TAE buffer: 40 mM Tris-acetate, 1 mM ethylenediamine tetraacetic acid, disodium salt dihydrate (EDTA).
7. 1% agarose gel: 1% agarose, 0.5 μg/mL ethidium bromide, in TAE buffer.
8. 6× DNA dye: 0.25% bromophenol blue, 0.25% xylene cyanol, 30% glycerol.
9. Power source.
10. Ultraviolet (UV) imaging system, or equivalent.
11. Gel extraction kit.
12. Appropriate restriction enzymes (e.g., Nde1 and BamH1) and corresponding restriction enzyme buffer.
13. Quick T4 DNA Ligase.

14. Quick T4 DNA Ligase Reaction Buffer: 200 mM Tris–HCl, 100 mM magnesium chloride ($MgCl_2$), 250 μg/ml acetylated bovine serum albumin (BSA), pH 7.6.
15. Competent cloning cells (e.g., DH5α or XL1-Blue).
16. Luria-Bertani (LB)-ampicillin agar plates: LB agar with 100 μg/mL ampicillin.
17. LB-ampicillin media: 100 μg/mL ampicillin in LB.
18. Plasmid DNA Mini Kit.

2.2 Purification of Soluble SH2 Protein

1. Expression cells (e.g., BL21-DE3).
2. UV-Visible spectrophotometer.
3. Isopropyl-β-D-thiogalactoside (IPTG).
4. Centrifuge capable of at least 25,000 × *g* force.
5. Purification buffer: 100 mM potassium phosphate, pH 7.4, 200 mM sodium chloride (NaCl), 1 mM dithiothreitol (DTT), and 25 mM imidazole.
6. Protease inhibitor cocktail tablet.
7. Ni-NTA agarose.
8. Chromatography column, such as a Kimble-Chase Kontes™ FlexColumn™ gravity flow column with a 10 mm diameter and 40 mL capacity.
9. Wash buffer 1 (WB1): 100 mM potassium phosphate, pH 7.4, 1 M NaCl, 1 mM DTT, 25 mM imidazole.
10. Wash buffer 2 (WB2): 100 mM potassium phosphate, pH 7.4, 200 mM NaCl, 1 mM DTT, 60 mM imidazole.
11. Elution buffer: 100 mM potassium phosphate, pH 7.4, 200 mM NaCl, 1 mM DTT, 350 mM imidazole.
12. Bovine alpha-thrombin.
13. Size exclusion chromatography (SEC) column (HiLoad Superdex 75, or equivalent).
14. Fast protein liquid chromatography (FPLC) system.
15. ITC Buffer (ITCB): 100 mM potassium phosphate, pH 7.4, 200 mM NaCl.
16. Concentrator with an appropriate molecular weight cutoff.

2.3 Isothermal Titration Calorimetry

1. Purified SH2 protein.
2. Purified peptide ligand. The peptide used here is the 11-mer ADND-pY-IIPLPD, which has been purified to >95% purity by reversed-phase HPLC chromatography following standard Fmoc solid-phase peptide synthesis.
3. Isothermal titration calorimeter (MicroCal iTC200, or equivalent).
4. Data analysis software (e.g., Origin®).

3 Methods

3.1 Subcloning of 6×-His-SH2 Construct

1. Amplify the SH2 domain DNA out of the cDNA template by PCR, using Pfu polymerase and the custom oligonucleotide primers (*see* **Notes 1–4**).
2. Add DNA dye to a 1× concentration to PCR product, and run sample on a 1% agarose gel with standards at 80 V for 1 h to isolate the PCR product.
3. Use a UV gel imager to locate the PCR products on the gel. Excise the PCR product from the gel with a sterile razor blade.
4. Purify the excised sample using a gel purification kit.
5. Digest the SH2 DNA insert as well as the destination pET15b vector using the appropriate restriction enzymes. We used Ndel and BamH1.
6. Run samples on a second agarose gel and gel extract them, as done above.
7. Ligate the SH2 DNA insert into pET15b by mixing approximately 100 ng of the digested pET15b vector and 0.2 pmol of the DNA insert with 1 μL of Quick T4 DNA Ligase in 20 μL total volume of Quick T4 DNA Ligase Reaction Buffer. Incubate at room temperature (25 °C) for 5 min.
8. Transform the ligation product into a competent bacterial cell line, plate onto an LB-agar plate containing the appropriate selection antibiotic (for pET15b, this is ampicillin), and incubate at 37 °C overnight (*see* **Note 5**).
9. Inoculate 5 mL of LB-ampicillin (or appropriate selection antibiotic) with a colony of the cloning cells containing the SH2 DNA in the pET15b vector.
10. Shake at 225 rpm overnight at 37 °C to grow culture to saturation.
11. Purify DNA from the bacterial cells using a Plasmid DNA Mini Kit.
12. Confirm the success of the cloning by sequencing the DNA (*see* **Note 6**).

3.2 Purification of Soluble SH2 Protein

1. Transform the SH2-pET15b vector into a bacterial expression cell line.
2. Inoculate a small volume of LB-ampicillin (or appropriate selection antibiotic) with a colony of the SH2-pET15b vector-containing expression cells.
3. Shake this starter culture at 225 rpm overnight at 37 °C.
4. Inoculate 10 mL of the overnight bacterial culture into the desired number of 1 L LB-ampicillin (or appropriate selection antibiotic) baffled flasks for large-scale growth.

5. Shake at 37 °C for several hours, monitoring the optical density (OD) at 600 nm approximately every 15–20 min after the initial 2 h.
6. When the OD_{600} reaches 0.6–1.0, remove flasks from the shaker and place on ice for 20 min.
7. Add IPTG to 0.5 mM and shake at 20 °C overnight.
8. Pellet the bacterial cells by centrifugation at 3000 × *g* for 10 min. Store bacterial pellet in a −20 °C freezer for future use, or proceed immediately to the next step.
9. Resuspend the bacterial pellet from a 2 L growth in 50 mL of purification buffer with one protease-inhibitor tablet.
10. Sonicate on ice to lyse the cells (e.g., 15 s pulses at amplitude 8.0 followed by 45 s off; cycle repeated eight times for a total pulse time of 2 min).
11. Centrifuge lysate at 25,000 × *g* for 30 min to separate the soluble protein from cellular debris.
12. Promptly remove supernatant to another container to avoid resuspension of pellet.
13. Add the supernatant to a 2 mL nickel column that has been pre-equilibrated with purification buffer, and allow protein to move through the column by gravity flow (*see* **Note 7**).
14. Wash the column with 5 mL of WB1.
15. Wash the column with 5 mL of WB2 (*see* **Note 8**).
16. Elute SH2 protein off the column using 15–20 mL elution buffer (*see* **Note 9**).
17. If desired, cleave off the 6×-His tag from the SH2 protein by adding 1–10 units of thrombin per 1 mg of protein. Allow the cleavage reaction to proceed at 4 °C overnight.
18. Further purify the protein by size exclusion fast protein liquid chromatography (SEC-FPLC), using buffer ITCB.
19. Concentrate the protein-containing fractions as necessary using a concentrator with a membrane of an appropriate molecular weight cutoff.

3.3 Isothermal Titration Calorimetry

1. Concentrate or dilute the SH2 protein to an appropriate concentration (*see* **Note 10**). For the MicroCal iTC200, prepare at least 300 μL of protein per run (*see* **Note 11**).
2. Resuspend peptide ligand in buffer ITCB (*see* **Notes 12** and **13**) to obtain at least 50 μL per run at an appropriate concentration (*see* **Notes 14–17**).
3. Create an appropriate ITC method on your instrument by entering relevant parameters including the temperature, precise concentrations of protein and peptide, number and frequency

of injections, the initial time delay, and the stirring speed (*see* **Notes 18–21**).

4. Make sure that the sample cell and syringe have been scrupulously cleaned since their previous use. Follow all manufacturer's instructions for cleaning and routine maintenance.
5. After all cleaning is complete, remove any leftover cleaning solution from the cell, and use the Hamilton syringe to manually rinse the cell several times with deionized water.
6. Attach the syringe fill port adaptor into the syringe (*see* **Note 22**).
7. Perform at least three consecutive cell and syringe wash methods (which should be pre-programmed in the instrument) to eliminate any remaining cleaning solution or other contaminants from the system.
8. Perform a water into water run, in which both the cell and syringe are filled with deionized water, to ensure that there is no spurious heat signal remaining from previous contaminations or other instrumental issues.
9. Remove any excess water from the sample cell.
10. Replace the water in the buffer container with buffer ITCB.
11. Perform a cell and syringe wash to equilibrate the system with buffer ITCB.
12. Remove any excess buffer from the sample cell.
13. Fill sample cell with the SH2 protein (*see* **Note 23**).
14. Load the syringe with the peptide solution.
15. Remove the syringe fill port adaptor from the syringe, and gently insert the syringe down into the sample cell.
16. Begin data collection.
17. Repeat data collection for control experiments. This includes peptide into buffer, buffer into protein, and buffer into buffer experiments (*see* **Note 24**). An additional protein into peptide control experiment is also recommended if protein solubility permits (*see* **Note 25**).
18. Open the results in a data software program, such as Origin®.
19. Subtract heat signal from the control experiments (*see* **Note 26**).
20. Perform a model fitting for the data [9]. For SH2/peptide systems, restrain the system to one set of binding sites. Perform 100 or more iterations of the data fitting until the fit is finalized (that is, additional iterations have no impact on the fit) (*see* **Note 27**).
21. The dataset will now include the binding curve of the integrated raw data, as well as the K, n, and $\Delta H°$ values. Additional parameters, $\Delta G°$ and $\Delta S°$, can be further calculated (*see* **Notes 28** and **29**).

4 Notes

1. The pET15b vector contains an expression region transcribed by T7 RNA polymerase, and contains the DNA sequence encoding a 6× N-terminal histidine (6×-His) tag followed by a thrombin protease recognition site and three restriction enzyme cloning sites (those of Nde1, Xho1, and BamH1). The 6×-His tag will provide the SH2 protein with affinity for Ni-NTA agarose, and can be later removed from the protein after purification by the addition of thrombin protease. Similarly designed pET vectors or alternative proteolytic cleavage sites can alternatively be used.
2. cDNA encoding the SH2 protein of interest can often be requested from research labs that have previously used the construct. It can also be obtained commercially through companies such as IMAGE Consortium, Addgene, or DNASU.
3. Primers were designed to isolate the DNA encoding the SH2 domain of interest, as well as to add a 3′ stop codon and to insert 3′ BamH1 and 5′ Nde1 restriction enzyme recognition sites.
4. Several polymerases are commercially available for use, including Pfu, Taq, and Vent. While all would be acceptable for use, Pfu polymerase has the highest fidelity and was consequently chosen to minimize the possibility of misincorporation [12].
5. Including control reactions during the ligation and transformation steps is valuable for troubleshooting if the subcloning is unsuccessful. Positive control DNA is often included in ligation kits to test the ligase activity. If unavailable, one option is to include a ligation reaction in which only one restriction enzyme was used on the parental DNA plasmid. The ability of the parental plasmid to self-ligate will confirm the ligase activity. Another common control is to perform the ligation reaction and transformation on the digested parental plasmid, but in the absence of the insert DNA encoding the SH2 domain. This will illustrate the density of background transformation colonies that can be expected from un-digested or re-ligated parental plasmid. One final positive control is to transform the un-digested parental plasmid to verify the transformation protocol efficiency, antibiotic resistance, and cell competency.
6. Sequencing of all final constructs is essential. DNA can be sent directly to companies including Genewiz, ACGT Inc, and Quintara for sequencing.
7. To optimize protein retention, the protein can be flowed over the column a second time, or alternatively mixed with the resin for several minutes before flow-through. Be careful,

however, not to let too much time pass, as there are many proteins in the solution that may precipitate out over time and clog the column.

8. 60 mM imidazole is used in WB2 to elute unwanted proteins that may be weakly interacting with the nickel column. However, this may be a high enough concentration to begin eluting the SH2 protein of interest as well. If this occurs, consider using a gradient of imidazole concentrations to better optimize the concentration that will elute the desired protein. Select a more appropriate WB2 with an alternate concentration of imidazole accordingly. Likewise, the 350 mM imidazole used in the elution buffer can be better optimized for a specific protein, although this concentration will be suitably high for most proteins.
9. The 15–20 mL of elution buffer can be added as several smaller (e.g., 5 mL) increments. Only the fractions containing a significant amount of protein need to be pooled and further purified. At this point, samples of each step (e.g., lysate, supernatant, column flow-through, washes, elutions, remaining nickel beads) can be run on a 17%-SDS gel to confirm the quality of the purification, to determine which elution fractions are desired, and to troubleshoot as necessary.
10. The best protein concentration to use for the ITC experiment will vary based on the expected equilibrium dissociation constant (K_d) for the reaction. A balance needs to be achieved between producing an amount of heat that is detectable, but not too large to be measured, and achieving an appropriate thermogram curvature from which binding and thermodynamic variables can be accurately determined. A unitless constant termed the *c*-value (c) can be used to determine a concentration range ideal for the protein of interest. A *c*-value between 10 and 100 is recommended, and can be determined with the equation:

$$c = \frac{[M]n}{K_d}$$

where $[M]$ is the concentration of the macromolecule (the SH2 protein), and n is the binding stoichiometry (the number of peptide binding sites on the protein) [11, 13]. Thus, for a protein/peptide interaction for which $n = 1$, $[M]$ should be 10–100 times larger than the K_d.

11. While 203 μL of protein will fill the sample cell of the MicroCal iTC200, we recommend preparing at least 300 μL of protein to ensure that no air bubbles are introduced as the protein is loaded into the cell. The sample cell is loaded using a 1.0 mL Hamilton syringe with a plastic sleeve, which

serves to protect the calorimeter from scratches as the needle is inserted and removed.

12. The phosphopeptide is obtained as a lyophilized powder following standard Fmoc solid-phase peptide synthesis and purification to >95% purity by reversed-phase HPLC chromatography [14, 15].
13. It is critical that the protein and peptide are in identical buffers, as buffer mismatch leads to a non-negligible heat of mixing, which can interfere with the heat signal from protein/peptide binding. If the exact batch of buffer ITCB that was used to purify the SH2 protein is not available, dialyze or further concentrate the protein, and use the dialysate or the concentrator flow-through to resuspend the peptide. Many standard buffers and buffer additives (including Tris buffer and DTT) should also be avoided or minimized since the heat evolved by these compounds can be substantial, which can overwhelm the heat signal from the protein/peptide binding [16–18].
14. For a protein/ligand interaction with 1:1 stoichiometry ($n = 1$), a ligand concentration that is 10- to 20-fold higher than $[M]$ is ideal. For PLCγ1-SH2, which binds peptides with low-micromolar affinity in phosphate buffer, we chose a peptide concentration of 2 mM and a protein concentration of 200 μM. The choice of buffer (here, phosphate) may alter the affinity of a phosphopeptide to the target SH2 domain, and alternate protein and peptide concentrations should be chosen accordingly if a different buffer system is used.
15. For peptides that do not absorb strongly at 280 nm due to minimal tryptophan or tyrosine residues, the concentration can instead be resolved by tracking the peptide absorbance at 205 nm [19, 20].
16. Our ligand syringe is calibrated to hold approximately 40 μL of peptide. Make sure that the method used (i.e., the number and volume of injections) does not exceed this amount. A typical method we use is as follows: one 0.2 μL injection, nineteen 2 μL injections, and one final 1.3 μL injection, summing to a total of 39.5 μL of peptide used. The first injection is typically error-prone and consequently removed from the final data set. As such, only a very small 0.2 μL volume is used to minimize the loss of peptide. The final injection is reduced to 1.3 μL to ensure that the total volume of peptide used (39.5 μL) remains sufficiently below 40 μL.
17. Each ITC run will use approximately 40 μL of peptide. However, to avoid introducing air into the syringe needle, we find that at least an extra 10 μL of peptide should be present, for a total volume of 50 μL or greater.

18. Standard binding reactions are performed at 25 °C. Alternative temperatures may be considered for systems in which the protein and/or peptide has poor solubility.
19. Although the specifics may differ based on the capabilities of the calorimeter used, a typical experiment is composed of approximately 20 injections of 2 μL of peptide (detailed in **Note 16**) every 3 min into a solution of approximately 200 μL of protein.
20. An initial time delay of 1–10 min can help eliminate interference from air bubbles by allowing the bubbles time to rise out of the sample before peptide injections begin.
21. Stirring speed is typically set to the default 1000 rpm for a titration syringe with a traditional straight paddle. Newer titration syringe models have twisted paddles, for which a stirring speed of 750 rpm is recommended. An increase in speed may aid in data collection of more viscous solutions, but may simultaneously increase the baseline noise level.
22. The titration syringe is very delicate, and great care should be taken not to overtighten the fill port adaptor, which can easily break the glass body of the titration syringe. Additionally, carefully avoid bending the syringe, as even minor bends invisible to the human eye can lead to strong baseline noise and poor data quality.
23. The presence of air bubbles can introduce spurious heat signals. As such, be careful that no bubbles have been introduced upon adding SH2 protein into the cell.
24. Mixing or dilution effects for the biomolecules and buffer can produce heat signals that can artificially raise the heat signals observed during a protein/peptide titration experiment. To account for these effects, three standard control experiments must be performed: peptide into buffer, buffer into protein, and buffer into buffer experiments [5, 9]. In all of these experiments, it is critical that the ITC method (volumes, frequency of injections, etc.) remains identical to that used for the actual peptide into protein experiments. For the peptide into buffer control, peptide is loaded into the syringe as normal; however, the sample cell contains only buffer. The second control experiment is the reverse, in which buffer in the syringe is titrated into protein in the sample cell. These first two controls will identify any heat effects from dilution of the peptide and protein. The buffer into buffer control experiment serves as a final instrument blank to account for any heat of mixing or other artifacts. Large or spurious heat signals in the control experiments are commonly due to buffer mismatch, impurities in the protein and/or peptide solution, a dirty calorimeter sample cell and/or syringe, or the presence of air bubbles in the cell or

syringe. Clean the calorimeter or further optimize the sample purity to ensure that the control experiments produce small heat signals before proceeding with the protein into peptide data collection.

25. Alternatively, if the solubility of the protein permits, SH2 protein can be titrated into a solution of peptide. Fill the sample cell with peptide, using a concentration that will achieve a c-value between 10 and 100. The SH2 protein titrant will accordingly need a concentration 10- to 20-fold greater than that of the peptide. This titration reversal experiment serves as a control, as it should yield identical binding affinity, thermodynamic, and stoichiometry values as those found for the peptide into protein experiment.
26. To correct the raw heat (Q) signal for heats of mixing or dilution from the three control experiments, the following equation is used [9]:

$$Q_{\text{corrected}} = Q_{\text{measured}} - Q_{\text{dil, peptide}} - Q_{\text{dil,protein}} - Q_{\text{blank}} \quad (8)$$

27. At this time, reevaluate whether the concentration of protein and peptide used are appropriate based on the determined binding affinity. Collect additional experimental data accordingly. Recall that ITC is best at measuring binding dissociation constants between approximately 1 nM and 100 μM. If the K_d for the interaction of interest appears to be outside this range, an alternate binding technique may be needed to better assess the affinity.
28. $\Delta G°$ and $\Delta S°$ values can be derived from the initial data using Eqs. (6, 7), described in the introduction.
29. In this chapter, we use the "naught" version of thermodynamic parameters (e.g., $\Delta G°$), though it should be noted that the system is not, in fact, under standard conditions. Specifically, our experiments are performed in a buffered solution at pH 7.4.

Acknowledgments

We would like to thank Dr. Annette Erbse for her generous feedback and technical support. We would also like to thank our collaborators Dr. Zhongping Tan and Dr. Xiaoyang Guan for the phosphopeptide synthesis, as well as Dr. Julie Forman-Kay for kindly providing us with a PLCγ1-pET11d construct. This work was supported by NSF grant MCB1121842 (to D.S.W.).

References

1. Pawson T, Gish GD, Nash P (2001) SH2 domains, interaction modules and cellular wiring. Trends Cell Biol 11:504–511
2. Yaffe MB (2002) Phosphotyrosine-binding domains in signal transduction. Nat Rev Mol Cell Biol 3:177–186
3. Tinti M, Kiemer L, Costa S et al (2013) The SH2 domain interaction landscape. Cell Rep 3:1293–1305
4. Brown RK, Brandts JM, O'Brien R et al (2009) ITC-derived binding constants using microgram quantities of protein. In: Cooper MA (ed) Label-free biosensors: techniques and applications. Cambridge University Press, New York, NY, pp 223–250
5. Perozzo R, Folkers G, Scapozza L (2004) Thermodynamics of protein-ligand interactions: history, presence, and future aspects. J Recept Signal Transduct Res 24:1–52
6. Chaires JB (2008) Calorimetry and thermodynamics in drug design. Annu Rev Biophys 37:135–151
7. Doyle L (1997) Characterization of binding interactions by isothermal titration calorimetry. Curr Opin Biotechnol 8:31–35
8. Leavitt S, Freire E (2001) Direct measurement of protein binding energetics by isothermal titration calorimetry. Curr Opin Struct Biol 11:560–566
9. Freyer MW, Lewis EA (2008) Isothermal titration calorimetry: experimental design, data analysis, and probing macromolecule/ligand binding and kinetic interactions. Methods Cell Biol 84:79–113
10. MicroCal LLC (2004) ITC data analysis in Origin®: tutorial guide, version 7.0, pp 105–107
11. Wiseman T, Williston S, Brandts JF et al (1989) Rapid measurement of binding constants and heats of binding using a new titration calorimeter. Anal Biochem 179:131–137
12. Cline J, Braman JC, Hogrefe HH (1996) PCR fidelity of Pfu DNA polymerase and other thermostable DNA polymerases. Nucleic Acids Res 24:3546–3551
13. Turnbull WB, Daranas AH (2003) On the value of c: can low affinity systems be studied by isothermal titration calorimetry? J Am Chem Soc 125:14859–12866
14. Merrifield RB (1963) Solid phase peptide synthesis. I. The synthesis of a tetrapeptide. J Am Chem Soc 85:2149–2154
15. Attard TJ, O'Brien-Simpson N, Reynolds EC (2007) Synthesis of phosphopeptides in the Fmoc mode. Int J Pept Res Ther 13:447–468
16. Fukada H, Takahashi K (1998) Enthalpy and heat capacity changes for the proton dissociation of various buffer components in 0.1 M potassium chloride. Proteins Struct Funct Genet 33:159–166
17. Goldberg RN, Kishore N, Lennen RM (2002) Thermodynamic quantities for the ionization reaction of buffers. J Phys Chem Ref Data 31:231–370
18. MicroCal LLC (2004) VP-DSC MicroCalorimeter: user's manual, pp 88–90
19. Anthis NJ, Clore GM (2013) Sequence-specific determination of protein and peptide concentrations by absorbance at 205 nm. Protein Sci 22:851–858
20. Goldfarb AR, Saidel LJ, Mosovich E (1951) The ultraviolet absorption spectra of proteins. J Biol Chem 193:397–404

Chapter 17

Binding Assays Using Recombinant SH2 Domains: Far-Western, Pull-Down, and Fluorescence Polarization

Kazuya Machida and Bernard A. Liu

Abstract

Recognition of phosphotyrosine-containing sequences by SH2 domains confers specificity in tyrosine kinase pathways. By assessing interactions between isolated SH2 domains and their binding proteins, it is possible to gain insight into otherwise inaccessible complex cellular systems. Far-Western, pull-down, and fluorescence polarization (FP) have been frequently used for characterization of phosphotyrosine signaling. Here, we outline standard protocols for these established assays using recombinant SH2 domain, emphasizing the importance of appropriate sample preparation and assay controls.

Key words SH2 domain, SH2 binding assay, Far-Western, Pull-down, Fluorescence polarization, Pervanadate treatment

1 Introduction

There is great interest in studying protein–protein interactions mediated by Src Homology 2 (SH2) domains and their recognition sites because of their role in conferring specificity in tyrosine kinase pathways. The ability to study isolated SH2 domains and their binding sites using in vitro assays has afforded us a powerful means by which to understand otherwise inaccessible signaling modules within a complex cellular system [1]. Various SH2 domain binding assays have been developed to assess interactions between SH2 domains and their binding proteins [2–4].

Far-Western and pull-down are widely adopted methods for studying SH2 binding sites (≈ phosphotyrosine sequences) on cellular proteins. In far-Western (or overlay assay), protein samples are separated by gel electrophoresis, immobilized on a membrane, and probed with a labeled SH2 domain [5, 6](Fig. 1). In this assay, an SH2 probe can directly bind to a pTyr-containing short amino acid sequence in either its native or denatured structure. On the other hand, pull-down using an SH2 domain bait may isolate

Kazuya Machida and Bernard A. Liu (eds.), *SH2 Domains: Methods and Protocols*, Methods in Molecular Biology, vol. 1555,
DOI 10.1007/978-1-4939-6762-9_17, © Springer Science+Business Media LLC 2017

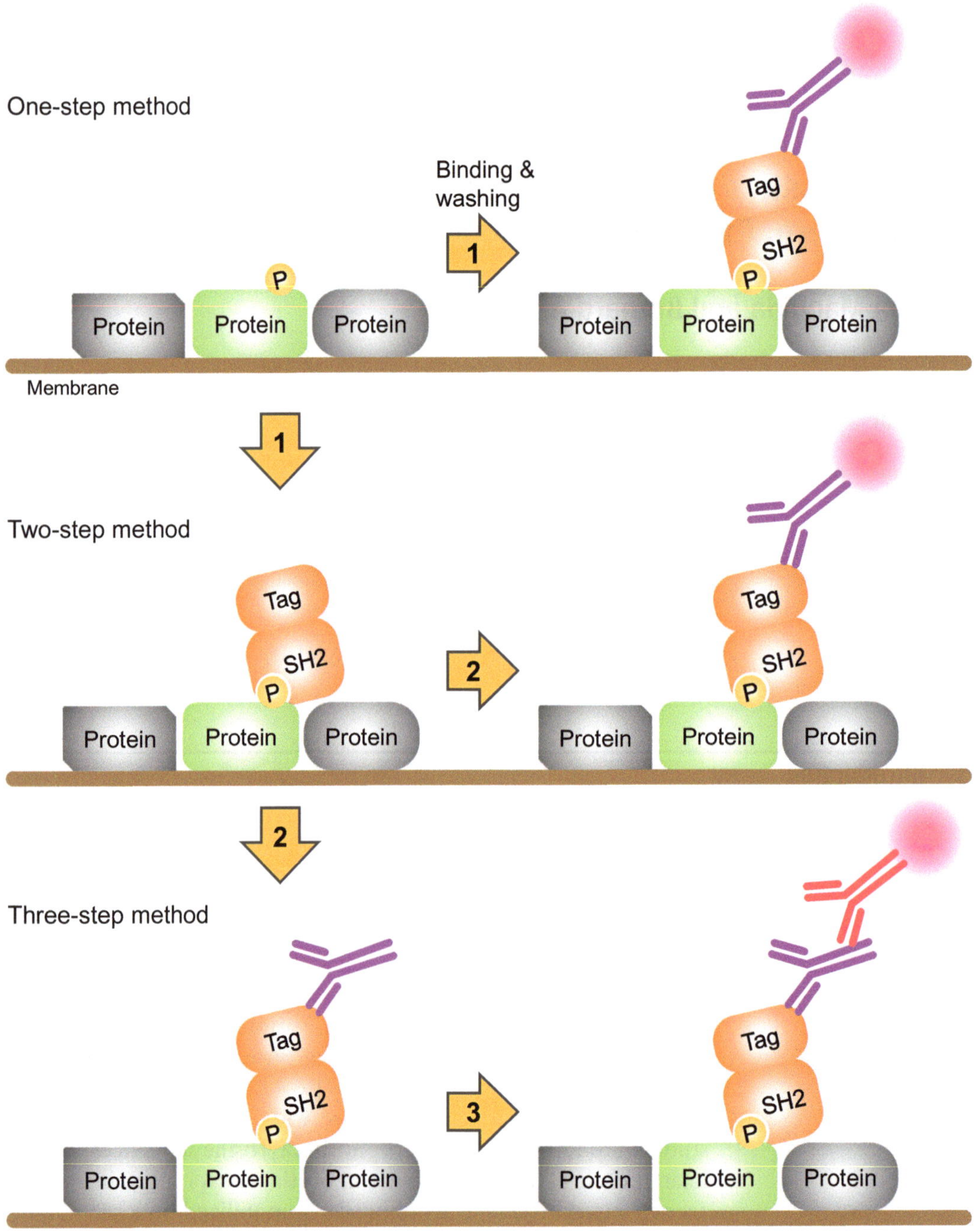

Fig. 1 Far-Western methods. Proteins are separated on an SDS-PAGE gel and transferred to a membrane for Western blotting. The tagged-SH2 domains are pre-conjugated with a dye reagent and then incubated with the membrane (one-step method). Alternatively, unlabeled tagged-SH2 domain is incubated, washed, and then sequentially incubated with anti-tag and dye-labeled antibodies (two and three-step methods). In far-Western, direct interactions between the SH2 probe and proteins containing SH2 binding sites can be detected

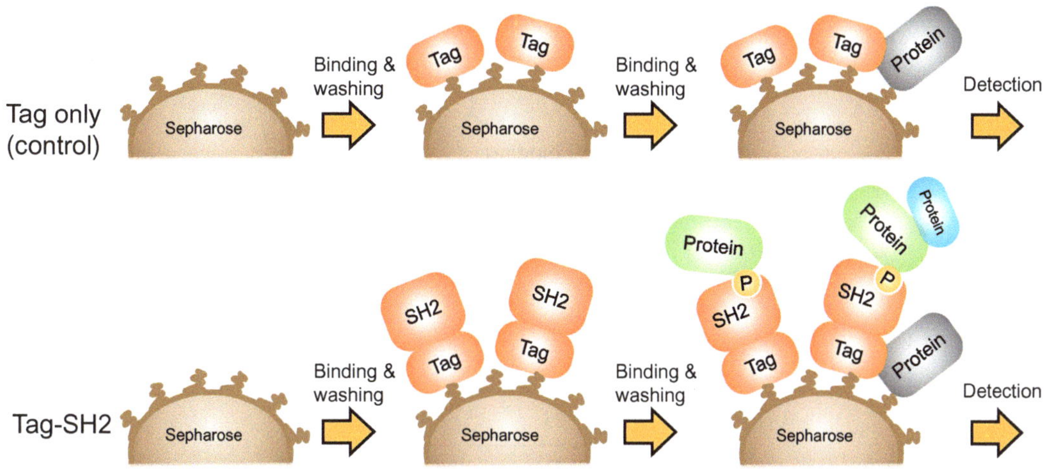

Fig. 2 Pull-down assay. The SH2 domain with a fusion tag (e.g., GST) is pre-conjugated with Sepharose beads and then incubated with cell lysates containing proteins with SH2 binding sites. The SH2-beads conjugate can interact with not only direct SH2 binding proteins (indicated in *green*) but also indirect binding proteins (indicated in *blue*) and tag binding proteins (indicated in *gray*). Thus it is important to perform a pull-down with a tag only control to assess the background binding

proteins that bind both directly and indirectly to the SH2 domain, similar to immunoprecipitation [7](Fig. 2).

To measure the binding affinities between the SH2 binding and phosphotyrosine ligand, surface plasmon resonance (SPR) and fluorescence polarization (FP) have become conventional methods. FP, also known as fluorescence anisotropy, is in-solution binding assay that measures the changes in polarized light during molecular binding [8, 9]: a rapidly rotating fluorescently labeled short peptide has a large depolarization (i.e., low FP). Upon SH2 binding, the complex will move slower, resulting in a reduced depolarization (i.e., high FP, Fig. 3).

In the following sections, we describe representative methods for sample preparation including controls (subheadings 3.1–3.8), far-Western (3.9), pull-down (3.10), and FP (3.11).

2 Materials

2.1 Lysate Preparation

1. Phosphate Buffered Saline (PBS).
2. 50 mM sodium orthovanadate: Dissolve powdered sodium orthovanadate to ~100 mM, adjust with NaOH to pH of 10 (solution will turn bright yellow); boil in microwave until colorless, then stir until cooled to room temperature; repeat this process until pH stays at 10.0 after boiling; adjust final concentration to 50 mM with water, filter, and store at room temperature.

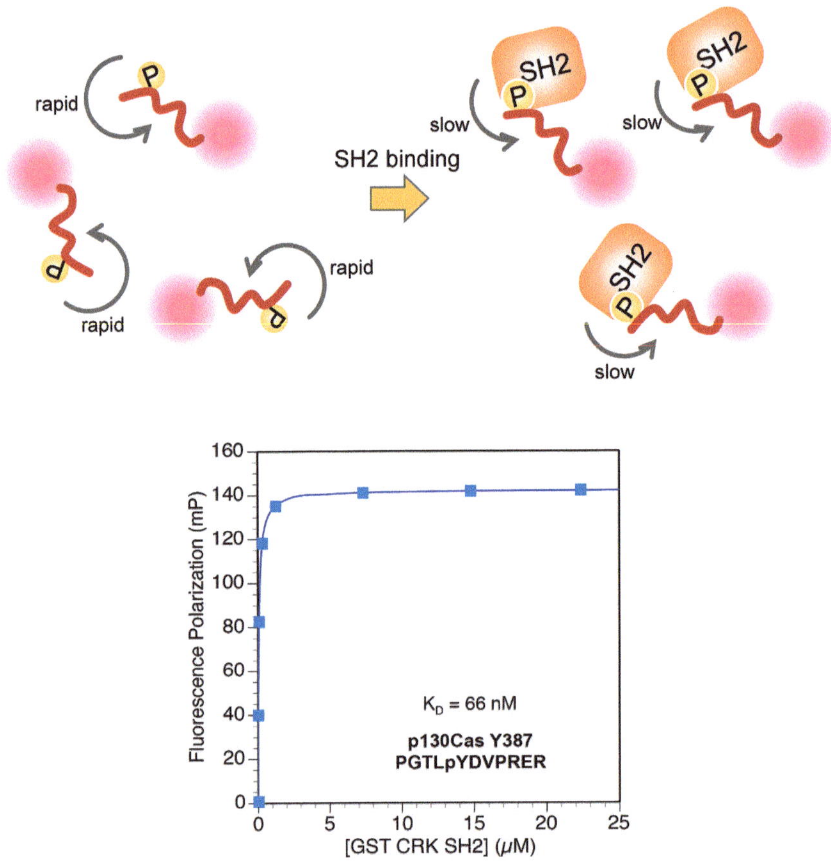

Fig. 3 Fluorescence Polarization. The rapidly rotating fluorescently labeled phosphopeptide produces a low FP value (left). Upon SH2 binding, the complex moves slower, resulting in an increased FP value (right). A saturation binding curve of the GST-tagged CRK SH2 domain with an FITC-labeled peptide from p130Cas Y387 (lower panel)

3. PBS-vanadate: PBS with 1 mM sodium orthovanadate.
4. Kinase lysis buffer (KLB): 25 mM Tris–HCl pH 7.4, 150 mM NaCl, 5 mM ethylene diamine tetraacetic acid (EDTA), 1% Triton X-100, 10% glycerol, 0.1% sodium pyrophosphate, 10 mM β-glycerophosphate, 10 mM sodium fluoride (NaF), 1 mM sodium orthovanadate. Store at 4 °C. Add 1% aprotinin (3 TIU/mL) and 1 mM phenylmethylsulfonyl fluoride (PMSF) just before use. (*see* **Note 1**).
5. Radioimmunoprecipitation assay (RIPA) buffer: 25 mM Tris–HCl pH 7.4, 150 mM NaCl, 5 mM EDTA, 0.1% sodium dodecyl sulfate (SDS), 1% Triton X-100, 0.5% sodium deoxycholate, 1 mM sodium orthovanadate. Store at 4 °C. Add 1% aprotinin (3 TIU/mL) and 1 mM phenylmethylsulfonyl fluoride (PMSF) just before use (*see* **Note 2**).
6. Denaturing lysis buffer: 10 mM Tris–HCl pH 7.4, 1% SDS, 1 mM sodium orthovanadate. Store at room temperature (*see* **Note 3**).

7. 5X SDS sample buffer: 300 mM Tris–HCl pH 6.8, 10% SDS, 50% glycerol, 25% β-mercaptoethanol, 0.1 mg/ml bromophenol blue. Store aliquots at –20 °C.
8. BCA protein assay kit (Thermo Fisher Scientific).
9. Laboratory ice pan.
10. Cell scraper.
11. –80 °C freezer.
12. Sonicator with microtip probe (e.g., Branson Sonifier 450 or equivalent).
13. CryoPrep CP02 pulverizer (Covaris) (Items 13–18 are required for tissue sample only).
14. TT1 tissue tubes, adaptors, tubes, and caps 12×24 mm (Covaris).
15. Focused-ultrasonicator system and software (Covaris).
16. Liquid nitrogen with a small container.
17. Steel hammer.
18. Cotton gauze.

2.2 Positive Control

1. Cells, cell culture supplies, and incubator.
2. Pervanadate solution: Mix 16 μL of 30% (w/w) hydrogen peroxide and 100 μL of 50 mM sodium orthovanadate; incubate at room temperature for 30 min (not stable, needs to be freshly prepared before use).
3. PBS.
4. KLB.
5. 5X SDS sample buffer.
6. BCA protein assay kit.
7. Sodium dodecyl sulfate–polyacrylamide gel electrophoresis (SDS-PAGE) system.
8. Anti-phosphotyrosine antibody, e.g., PY100.
9. 15-mL and 50-mL conical tubes.
10. –80 °C freezer.

2.3 Negative Control

1. Cells, cell culture supplies, and incubator.
2. PBS.
3. KLB without vanadate: 25 mM Tris–HCl pH 7.4, 150 mM NaCl, 5 mM EDTA, 1% Triton X-100, 10% glycerol. Store at 4 °C. Add 1% aprotinin (3 TIU/mL) and 1 mM PMSF just before use.
4. Dithiothreitol (DTT): Dissolve in water. Store 1 M aliquots at –20 °C.

5. Recombinant GST-PTP1B: Purify GST-tagged PTP-1B protein using bacterial protein purification system (*see* Chapter 8). pGEX-PTP1B is available from Addgene (Plasmid #8602).
6. Glutathione (GSH) Sepharose 4B.
7. Spin column (e.g., Spin-X Plastic Centrifuge Tube Filter).
8. 5X SDS sample buffer.
9. BCA protein assay kit.
10. SDS-PAGE system.
11. Anti-phosphotyrosine antibody, e.g., PY100.
12. 15-mL and 50-mL conical tubes.
13. –80 °C freezer.

2.4 Far-Western

1. Tagged SH2 domain and control tag only proteins: e.g., purified GST-tagged SH2 domain and GST tag proteins diluted to 0.1 μg/μL in 10% glycerol–PBS. Store aliquots at –80 °C (*see* Chapter 8).
2. SDS-PAGE system. (*see* **Note 4**).
3. Nitrocellulose membrane (0.2 μm pore size).
4. Transfer buffer: 10 mM 3-(cyclohexylamino)-1-propanesulfonic acid (CAPS), pH 11.0, 20% methanol, kept at 4 °C.
5. Tris buffered saline–Tween 20 (TBST): 25 mM Tris–HCl, pH 8.0, 150 mM NaCl, and 0.05% (v/v) Tween 20.
6. Blocking solution: 5–10% fat-free milk in TBST.
7. Odyssey Blocking Buffer [PBS] (LI-COR Biotechnology).
8. Labeling reagents for chemiluminescent detection: e.g., anti-GST-HRP conjugate (GST-tagged SH2 domain); anti-poly His-HRP conjugate (poly His-tagged SH2 domain); or streptavidin–HRP conjugate (biotin-tagged SH2 domain) (*see* **Note 5**).
9. Labeling reagent for near-infrared (IR) fluorescent detection: e.g., anti-GST-IR dye conjugate (GST-tagged SH2 domain); anti-poly His-IR dye conjugate (poly His-tagged SH2 domain); or streptavidin–IR dye conjugate (biotin-tagged SH2 domain) (*see* **Note 6**).
10. Chemiluminescence kit for HRP-based detection (e.g., Western Lightning Plus-ECL, PerkinElmer).
11. Chemiluminescence imager and software, e.g., Carestream Image Station and Carestream MI.
12. Near-IR imager and software, e.g., Odyssey Imager and Odyssey 3.0 software (LI-COR Biotechnology).
13. Acidic stripping buffer: 100 mM glycine-HCl pH 2.0.

14. SDS-ME stripping buffer: 2 % SDS, 100 mM β-mercaptoethanol, 62.5 mM Tris–HCl pH 6.8.
15. NewBlot stripping buffer (LI-COR Biotechnology).

2.5 Pull-Down

1. Cell lysates in KLB or RIPA solutions (*see* **Note** 7).
2. GST and GST-SH2 expressing bacterial clones or purified GST tag and GST-tagged SH2 domain proteins (*see* Chapter 8). Purified GST-SH2 proteins can also be obtained from commercial sources, e.g., BioCat GmbH (https://goo.gl/pSK4r4).
3. LB-ampicillin: LB broth with 50 μg/mL ampicillin.
4. Bacterial shaker.
5. Isopropyl β-D-1-thiogalactopyranoside (IPTG). Dissolve in water to 10 mM and store aliquots.
6. Bacteria Triton X-100 lysis buffer (BXB): PBS, 100 mM EDTA, 1 % Triton X-100, 10 % glycerol; add PMSF to 1 mM, aprotinin solution to 1 % v/v, and DTT to 1 mM just before use.
7. Sonicator with microtip probe.
8. 5X SDS sample buffer.
9. GSH Sepharose 4B.
10. Protein concentration measurement: Bradford assay, BCA assay, UV spectrometer.
11. Optional materials for post-assay confirmation (items 10–13). SDS-PAGE system with 12 % mini gels.
12. CBB gel staining solution: 40 % methanol, 10 % acetic acid, 0.25 % Coomassie Brilliant Blue R-250.
13. Destaining solution: 20 % methanol, 10 % acetic acid.
14. Anti-GST antibody and secondary antibody for Western blot.

2.6 Fluorescence Polarization

1. Recombinant GST or His-tagged SH2 domains (*see* Chapter 8).
2. Fluorescein isothiocyanate (FITC) labeled phosphotyrosine peptide (*see* **Note 8**).
3. Fluorescein (FITC, Sigma)
4. Dimethyl sulfoxide (DMSO).
5. 1 M DTT
6. Bovine serum albumin (BSA; 2 mg/mL concentration).
7. PBS.
8. FP buffer: PBS containing 100 μg/mL BSA, 1 mM DTT.
9. Black 96-well or 384-well flat or round bottom plates.
10. Aluminum foil plate seal.

11. EnVision Multilabel Plate Reader (PerkinElmer) or equivalent plate readers capable of fluorescent polarization (FITC: Ex/Em = 488/520 nm).
12. Centrifuge with a 96/384-well plate adaptor.
13. GraphPad Prism.

3 Methods

3.1 Cell Lysis Conditions

For far-Western, lysates can be prepared in either native and denaturing conditions in the same way as for Western blots. For pull-down, the choice of lysis buffers may affect the result, and therefore preliminary experiments may be needed to identify the optimal lysis buffer (*see* **Note 1 and 2**). KLB is usually appropriate for preparation of both far-Western and pull-down samples. If the proteins of interest are difficult to solubilize with KLB, such as membrane and nuclear proteins, RIPA buffer may work. If the target proteins are highly insoluble or unstable, direct lysis with denaturing lysis buffer may help (see subheading 2.1). It is important to add sodium orthovanadate to the lysis buffers to avoid dephosphorylation of samples.

3.2 Lysis of Adherent Cells

1. Culture adherent cells until subconfluent (*see* **Note 9**).
2. Place culture plates on ice (*see* **Note 10**).
3. Wash the cells twice with ice-cold PBS-vanadate.
4. Tilt the plates to 45° on ice and aspirate the supernatant completely.
5. Add 100–200 μL lysis buffer for a 6-cm dish and 250–500 μL for a 10-cm dish.
6. Tilt plates so that the lysis buffer covers the entire cell culture surface.
7. Allow cells to lyse on ice for 10 min.
8. Collect lysates and transfer to labeled microcentrifuge tubes.
9. Spin at 16,000 × *g* for 1 min in a microcentrifuge at 4 °C.
10. Sonicate the lysate on ice three times for 2–5 s each at a low power. Try to avoid foaming and cool tubes on ice between cycles.
11. Spin at 16,000 × *g* for 10 min at 4 °C.
12. Transfer supernatants into new labeled microcentrifuge tubes on ice.
13. Determine lysate concentration by BCA protein assay and store the aliquots at –80 °C (*see* **Note 11**).

3.3 Lysis of Suspension Cells

1. Culture suspension cells until subconfluent (*see* **Note 9**).
2. Transfer cells to 15-mL or 50-mL centrifuge tubes and spin again at 500 × *g* for 1 min at room temperature.

3. Aspirate media, add 10 mL ice cold PBS-vanadate, vortex briefly, and spin at 500 × *g* for 5 min at 4 °C.
4. Decant the PBS-vanadate and aspirate the excess supernatant.
5. Add lysis buffer and pipet up and down to mix on ice: 250 μL for a 25-cm^2 original culture flask and 600 μL for a 75-cm^2 flask.
6. Allow cells to lyse on ice for 10 min.
7. Transfer lysates to labeled microcentrifuge tubes.
8. Spin at 16,000 × *g* for 1 min in a microcentrifuge at 4 °C.
9. Sonicate the lysate on ice three times for 2–5 s each at a low power. Try to avoid foaming and cool tubes on ice between cycles.
10. Spin at 16,000 × *g* for 10 min at 4 °C.
11. Transfer supernatants to new labeled microcentrifuge tubes on ice.
12. Determine lysate concentration by BCA protein assay and freeze aliquots at −80 °C (*see* **Note 11**).

3.4 Lysis of Tissue Specimens

1. Place frozen tissue vials on dry ice.
2. For a large frozen tissue sample, use hammer and cotton gauze to break the sample into appropriate size fragments (*see* **Note 12 and 13**). Avoid thawing of tissue by addition of liquid nitrogen as needed. This step is not necessary if the whole piece is to be pulverized.
3. Set up the CryoPrep pulverizer system.
4. Insert the tissue sample into the center portion of the TT1 bag.
5. Attach adaptor and glass tube loosely to the TT1 bag (*see* **Note 14**).
6. Submerge the TT1 bag in liquid nitrogen to flash-freeze the tissue sample.
7. Place the TT1 with attached tube into the Cryoprep holder.
8. Pulverize the sample at a medium strength setting.
9. Remove the TT1 with tube and repeat steps 6–8.
10. Remove the TT1 and submerge it in liquid nitrogen.
11. Transfer the pulverized tissue to the glass tube by turning the bag upside down.
12. Tap the bag to transfer any additional tissue. Avoid thawing by submerging the TT1 bag in liquid nitrogen as needed.
13. Remove the bag and put a cap on the 12 × 24 mm glass tube.
14. Keep the tube on dry ice.
15. Repeat this process for the rest of samples.
16. Set up the focused-ultrasonicator system following the manufacturer's instructions (*see* **Note 15**).

17. Adjust the ultrasonicator settings using the provided software. Our setting is: 80% frequency and intensity; pulverization 30 sec; rest 30 sec; 2-cycles; water temperature at 6 °C.
18. Remove a glass tube for the dry ice (step 12) and add 500–750 μL of ice-cold RIPA buffer containing proteinase inhibitor and sodium orthovanadate.
19. Place the tube into the holder and insert the holder into the homogenizer.
20. Start homogenization using the software.
21. After the second cycle, place the tube on ice.
22. Repeat this process for the rest of samples.
23. Transfer lysates to labeled microcentrifuge tubes.
24. Spin at 16,000 × *g* for 10 min at 4 °C.
25. Transfer supernatants to new labeled microcentrifuge tubes on ice.
26. Determine lysate concentration by BCA protein assay and store the aliquots at −80 °C (*see* **Note 11**).

3.5 Positive Control

To ensure phosphotyrosine-dependent binding of SH2 domains, we recommend to include positive (highly tyrosine phosphorylated) and negative (unphosphorylated) control samples in each experiment. Pervanadate is a potent inhibitor of protein tyrosine phosphatases that irreversibly oxidizes their active sites. Thus, pervanadate treatment can be a convenient method for obtaining lysates with a maximum level of tyrosine phosphorylation which can be used as a positive control. To sufficiently diversify SH2 binding sites in the positive control, we recommend combining multiple cell lines of different tissue origins (Fig. 4) (*see* **Note 16**). Methods for preparing a positive control are described in subheadings 3.5 and 3.6.

3.6 Selection of Cell Lines

1. Culture five or more different cell lines with various tissue origins in 60-mm size plates or flasks and allow cells to grow until confluence.
2. Add pervanadate solution directly to culture media (final concentration at 50 μM) and incubate for 30 min at 37 °C.
3. Wash the cells twice with ice-cold PBS (*see* **Note 17**).
4. Add 200 μL KLB solution.
5. Tilt plates so that the lysis buffer covers the entire cell culture surface.
6. Allow cells to lyse on ice for 5–10 min.
7. Scrape cells and collect lysates in labeled microcentrifuge tubes.
8. Spin down cell lysates at 16,000 × g for 10 min in a microcentrifuge at 4 °C.

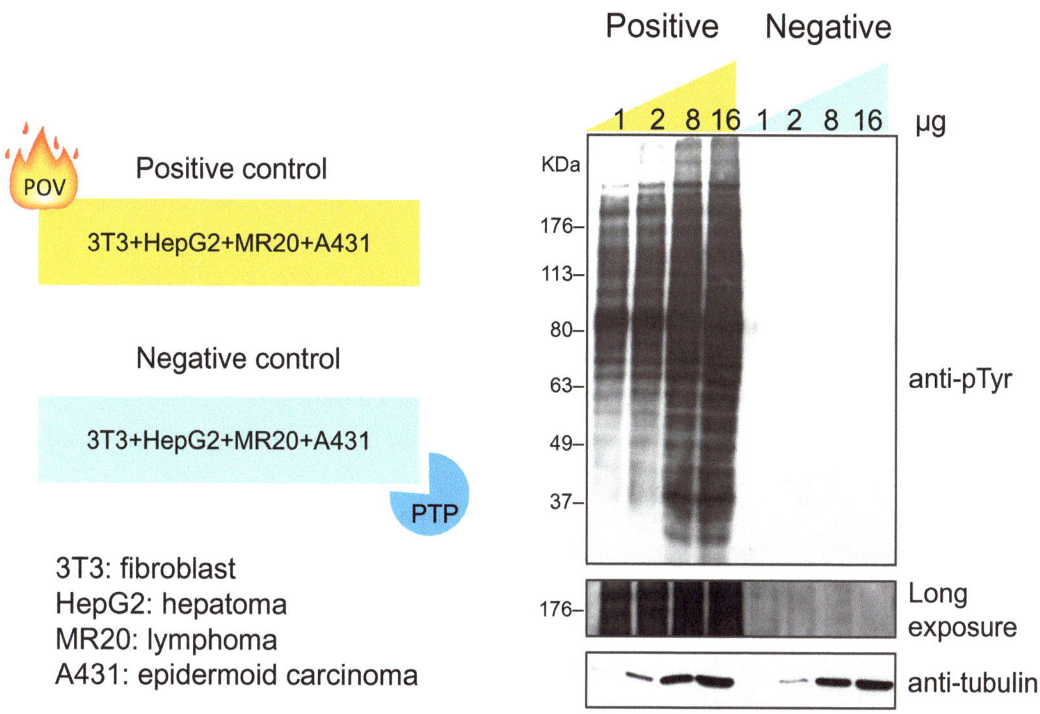

Fig. 4 Positive and negative controls for far-Western. Cell lysates from different tissue origins were pooled to cover a broad spectrum of phosphotyrosine sites. To prepare a positive control, cells were treated with pervanadate, lysed in KLB solution, and an equal amount of cleared lysates were combined. The negative control was prepared similarly but in the absence of vanadate and additional tyrosine phosphatase (PTP) treatment. A broad range of tyrosine phosphorylated proteins are detectable with the positive control on the pTyr blot. In contrast, virtually no signal (or a trace of signal with long exposure) is detectable with the negative control

9. Transfer supernatants into new labeled microcentrifuge tubes on ice.
10. Measure the protein concentration by BCA protein assay.
11. Run 10 µg samples on 10% SDS-PAGE gels and transfer to nitrocellulose membrane.
12. Perform Western blot with an anti-phosphotyrosine antibody.
13. Select several cell lines with distinctive band patterns on the pTyr blot (*see* **Note 18**).
14. Store aliquots of lysates at –80 °C.

3.7 Large-Scale Preparation

It is valuable to have a large amount of positive (and negative) control lysates for SH2 binding assays. By using the same validated controls in every experiment, the quality of assays can always be assured. In addition, the control can be used as a standard when quantitatively comparing multiple experiments. Skip this section if the small-scale product from 3.6 is sufficient for your study.

1. Grow each cell line in eight 10-cm size plates or flasks until confluent (*see* **Note 19**).
2. Add pervanadate solution directly to the culture media (final concentration at 50 μM) and incubate for 30 min at 37 °C (*see* **Note 20**).
3. Wash the cells twice with ice-cold PBS. For suspension cells, follow the method in subheading 3.3 as appropriate.
4. Add 500 μL KLB solution to the plate or cell pellet.
5. Tilt plates or vortex the tube so that the lysis buffer covers the entire cell culture surface.
6. Allow cells to lyse on ice for 10 min.
7. Collect lysates in four labeled microcentrifuge tubes (~1 ml per tube).
8. Spin at 16,000 ×*g* for 1 min in a microcentrifuge at 4 °C.
9. Sonicate the lysate on ice three times for 2–5 s each at a low power. Try to avoid foaming and cool tubes on ice between cycles.
10. Spin at 16,000 ×*g* for 10 min at 4 °C.
11. Combine supernatants into a single 10-ml tube on ice.
12. Measure the protein concentration by BCA protein assay.
13. Transfer 10 mg of lysate to labeled microcentrifuge tube. Store this 10 mg aliquot and remaining lysate at –80 °C.
14. Repeat steps 1–13 for additional cell lines selected in subheading 3.6.
15. Thaw all the 10 mg pervanadate-treated cell lysates on ice.
16. Combine the 10 mg lysates into a 15-mL or 50-mL tube on ice.
17. Measure the protein concentration by BCA assay and dilute the lysate mix to 5 μg/μL with KLB.
18. Aliquot the lysate into multiple tubes, e.g., 100 μg, 200 μg, 500 μg, or 1 mg per tube, and store at –80 °C.

3.8 Negative Control

Negative control lysates are prepared using a lysis buffer without sodium vanadate. This procedure alone is sufficient to dephosphorylate the majority of phosphotyrosines in the lysate (due to high endogenous phosphatase activity). Additionally, tyrosine phosphatase treatment can be performed to remove residual tyrosine phosphorylation.

1. Repeat steps in subheading 3.7 with the following modifications: 1) Do not perform step 2 (pervanadate treatment); 2) Use KLB without vanadate for cell lysis (steps 4 and 15).
2. Optional: Perform tyrosine phosphatase treatment, steps 3–12. Otherwise proceed to step 13. (*see* **Note 21**).

3. Incubate 1 mg of purified GST-PTP1B protein and 400 μL of a 50% slurry of GSH Sepharose at for 1 h at room temperature.
4. Wash three times with KLB without vanadate.
5. Add fresh PMSF (1 mM), aprotinin (1%), and DTT (2%) to the combined lysate immediately before PTP treatment.
6. Add bead-bound GST-PTP1B to the lysate and incubate at room temperature for 30 min with gentle agitation.
7. Spin the reaction tube at 1,000 × *g* for 1 min.
8. Transfer the supernatants to new tubes and remove residual beads using a spin column or gravity flow.
9. Add 1 mL of fresh GSH Sepharose beads to the flow-through and incubate at 4 °C for 2 h to remove any remaining GST-PTP1B protein.
10. Spin the reaction tube at 1,000 × *g* for 1 min.
11. Transfer the supernatants to new tubes and remove residual beads using a gravity flow or spin column.
12. Measure the protein concentration by BCA protein assay.
13. Aliquot the lysate into multiple tubes, e.g., 100 μg, 200 μg, 500 μg, or 1 mg per tube, and store at −80 °C.
14. Evaluate the tyrosine phosphorylation state of positive and negative control lysates by pTyr blots (Fig. 4).

3.9 Far-Western

Protein membranes are prepared in the same way as for Western blot. Protein samples are separated by SDS-PAGE and transferred to a nitrocellulose or polyvinylidene fluoride (PVDF) membrane. The membrane is blocked, incubated with a labeled SH2 domain probe, washed, and bound probes are visualized (steps 4–10, one-step probe binding) (see Fig. 1 and Fig. 5A and B). Alternatively, membranes can be sequentially incubated with a tagged-SH2 domain and anti-tag antibody conjugated with dye/peroxidase (steps 11–16, two-step probe binding)(*see* **Note 22**).

1. Run samples with positive and negative controls on SDS-polyacrylamide gels.
2. Transfer proteins to nitrocellulose or PVDF membranes following a general western blotting protocol (*see* **Note 23**).
3. Block membranes in blocking solution for about 1 h at room temperature or at 4 °C overnight (*see* **Note 24**).
4. One-step binding (steps 4–10): Thaw the stored recombinant SH2 protein and labeling reagents on ice.
5. Mix the SH2 domain and labeling reagent (*see* **Note 25**).

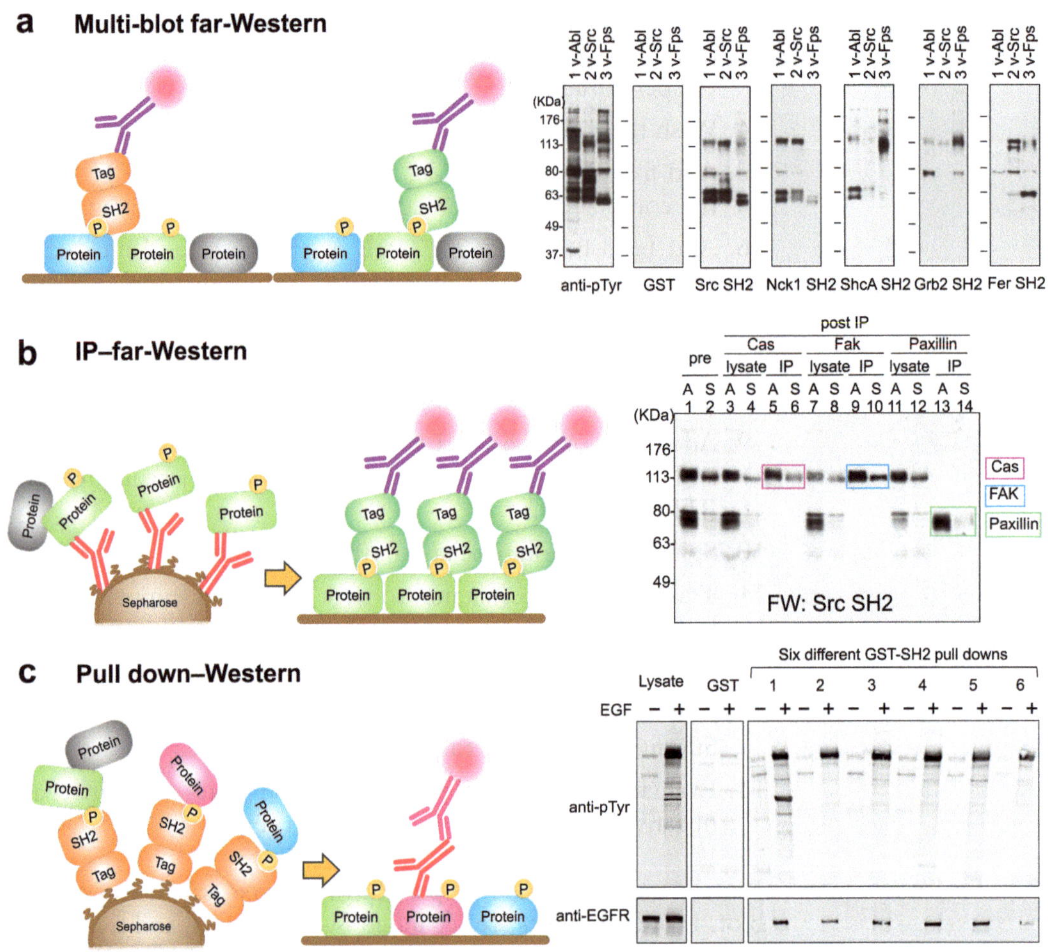

Fig. 5 Applications of far-Western and pull-down. (**a**) In the multi-blot far-Western, replicated membranes are differently probed with SH2 domains for a comparison purpose. The global tyrosine phosphorylation state is reflected by SH2 binding patterns. Right panel shows distinct SH2 profiles of oncogenic kinases expressing 3 T3 cells [10]. (**b**) Immunoprecipitation (IP) and far-Western are combined to identify SH2 binding proteins. Right panel shows identification of adhesion-dependent Src SH2 binding proteins [10]. A and S denote adherent and suspended cells, respectively. (**c**) Similarly, SH2 binding to specific proteins can be assessed using pull-down and Western blot. Right panel shows that binding of six different SH2 domains to phosphorylated EGFR from EGF-stimulated cells

Example 1: GST tag, chemiluminescent detection.

GST-SH2	0.1 μg/μL	50 μL
anti-GST-HRP	0.5 μg/μL	5 μL
Total		55 μL

Example 2: Biotin tag, chemiluminescent detection.

Biotin-SH2	0.1 μg/μL	50 μL
Streptavidin-HRP	0.5 μg/μL	1 μL
Total		51 μL

Example 3: GST tag, Near-IR detection.

GST-SH2	0.1 μg/μL	50 μL
Anti-GST-IR dye	1 μg/μL	1 μL
Total		51 μL

6. Incubate the mix at 4 °C for at least 1 h (*see* **Note 26**).
7. Dilute labeled probe to the optimal concentration with blocking buffer.
   Suggested dilution (*see* **Note 27**).

Labeled SH2	0.1 μg/μL	20 μL
Blocking solution		1 mL
Total	2 μg/mL (~50 nM)	1.02 mL

8. Incubate the blocked membrane with diluted SH2 domain in a sealed nylon bag with occasional agitation at room temperature for 2 h or at 4 °C for overnight (*see* **Note 28**).
9. Wash the membrane with a sufficient volume of TBST for 20 min with buffer changes every five minutes (*see* **Note 29**).
10. Proceed to step 17 for detection.
11. Two-Step binding (steps 11–16): Thaw the stored SH2 protein and labeling reagents on ice.
12. Dilute the SH2 protein to the optimal concentration with blocking buffer and incubate with the blocked membrane.
    Suggested dilution (*see* **Note 27**).

SH2	0.1 μg/μL	20 μL
Blocking solution		1 mL
Total	2 μg/mL (~50 nM)	1.02 mL

13. Incubate the blocked membrane with the diluted SH2 domain in a sealed nylon bag with occasional agitation at room temperature for 2 h or at 4 °C for overnight (*see* **Note 28**).
14. Wash the membrane with a sufficient volume of TBST for 15 min with buffer changes every five minutes (*see* **Note 29**).
15. Incubate the washed membrane with a diluted dye-labeled reagent in a sealed nylon bag with occasional agitation for 1.5 h.
    Example 1: GST tag, chemiluminescent detection.

anti-GST-HRP	0.5 μg/μL	5 μL
Odyssey Blocking Buffer		2 mL
Total		2.05 mL

Example 2: Biotin tag, chemiluminescent detection.

Streptavidin-HRP	0.5 μg/μL	1 μL
Odyssey Blocking Buffer		2 mL
Total		2.01 mL

Example 3: GST tag, near-IR detection.

Anti-GST-IR dye	1 μg/μL	1 μL
Odyssey Blocking Buffer		2 mL
Total		2.01 mL

16. Wash the membrane with a sufficient volume of TBST for 20 min with buffer changes every five minutes.
17. Detection: Perform chemiluminescent or near-IR detection according to the manufacturer's instructions (*see* **Note 30**).
18. When the exposure is done, remove the membrane from the imager and keep it in TBST solution. Perform membrane stripping immediately or store the membrane with plastic wrap at –20 °C. Do not store the membrane at room temperature for a long time.
19. Open the imager software and export the blot image for data analysis.
20. Optional: stripping and reprobing. If necessary, stripping and reprobing is possible, though the signal-to-noise ratio can be lower than when using a fresh membrane.
21. Immerse the membrane in TBST. If the membrane is frozen with plastic wrap, place it in TBST and remove the plastic when thawed.
22. Rinse the membrane twice with TBST.
23. Strip the membrane with acidic stripping buffer at room temperature for 20 min at ~60 rpm agitation using an orbital or rocking shaker.
24. Wash the membrane in a large volume of TBST at room temperature for 45 min with frequent buffer changes (*see* **Note 31**).
25. Proceed to blocking and reprobing.

3.10 Pull-Down

Pull-down assay is performed using purified SH2 domain, typically tagged with GST, and bound proteins are analyzed by immunoblot [10](Fig. 5), mass spectrometry [11], or functional assays [12]. Since contamination of nonspecific binding proteins can confound the result, it is important to include appropriate controls in each assay. In the steps below, a method for pull-down using bacterial expression clones for GST-SH2 is described. If purified GST-SH2 and GST control are already available, start from step 18 (*see* **Note 32**).

1. Inoculate 2 ml LB-ampicillin with colonies from LB plates or frozen bacteria stocks of GST-SH2 and control GST and grow at 37° overnight in a 200 rpm shaker.
2. The following day, inoculate 0.4 mL of the overnight culture in 1.6 mL fresh pre-warmed LB-ampicillin with.
3. Shake tubes at 37 °C for 1–2 h.
4. Add IPTG to 0.1 mM and shake at 37 °C for 1 h to induce protein expression.
5. Transfer the bacteria to microcentrifuge tubes and spin at 10,000 × g at 4 °C for 3 min in a microcentrifuge.
6. Remove the supernatant and resuspend the bacterial pellet in 0.2–0.4 mL BXB.
7. Vortex to resuspend, then sonicate briefly (e.g., 2–5 s at relatively low power) on ice to lyse cells. Try to avoid foaming.
8. Remove 10 μL of total lysate, add 2.5 μL of 5X sample buffer, boil for 5 min and store at −20 °C for future use (whole cell fraction)(*see* **Note 33**).
9. Spin the rest of the total lysate for 5 min in a microcentrifuge at 10,000 × *g* at 4 °C, and transfer the supernatant to a new tube.
10. Take 10 μL of the cleared supernatant lysate, add 2.5 μL of 5X sample buffer, boil for 5 min and store at −20 °C for future use (soluble fraction).
11. Add 20 μL of a 50% slurry of GSH Sepharose 4B.
12. Rotate at 4 °C for 1 h.
13. Spin briefly and wash beads 4 times with 1 mL cold BXB.
14. Remove the supernatant leaving ~200 μL in the tube, and vortex briefly.
15. Transfer 10 μL to a new tube, add 2.5 μL of 5X sample buffer, boil for 5 min, and store at−20 °C for future use (GSH bead bound fraction).
16. Spin the original tube again and remove the supernatant.
17. This sample is ready for pull-down (GST-SH2 bound beads). Proceed to step 23.
18. If the purified SH2 protein is already available, take 5–10 μg of the purified GST-SH2 protein and mix with 20 μL of a 50% slurry of glutathione Sepharose 4B. Repeat this step with GST protein as a control.
19. Rotate at 4 °C for 1 h.
20. Spin briefly and wash beads two times with 1 mL cold BXB.
21. Spin down again and remove the supernatant.

22. Optional: quickly measure the concentration of bead-bound proteins by protein concentration assay or absorbance at 280 nm (*see* **Note 34**).
23. Keep the GST-SH2 bound beads on ice until the lysate is ready.
24. Dilute fresh or stored lysates to 1 μg/μL in lysis buffer containing 1 mM sodium orthovanadate and proteinase inhibitor and keep on ice (*see* **Note 35**).
25. Take 10 μL of the lysate, add 2.5 μL of 5X sample buffer, boil for 5 min and store at in −20 °C for future use (pre-pull-down lysate).
26. Add lysate to the tubes with GST-SH2 and control GST bound beads (step 17 or 22), mix by tapping, and incubate at 4 °C for 1 h with gentle agitation.
27. Spin the tubes at 2,500 × *g* for 30 sec at 4 °C.
28. Take 10 μL of the supernatant, add 2.5 μL of 5X sample buffer, boil for 5 min and store at in −20 °C for future use (post-pull-down lysate).
29. Remove the supernatant and add 1 mL of cold KLB.
30. Spin briefly and wash beads four times with 1 mL of cold KLB.
31. Remove the supernatant leaving ~20 μL in the tube, and vortex briefly.
32. Add 2.5 μL of 5X sample buffer and boil for 5 min (pull-down fraction).
33. Proceed to downstream detection such as gel staining, immunoblotting (Fig. 5C) or mass spectrometry (*see* Chapters 22 and 23) (*see* **Note 36**).
34. Optional: Check the loading of GST proteins by blotting with anti-GST.
35. Alternative conditions. If the control GST gives a high background, consider blocking the GSH Sepharose beads with filtered BSA (0.1 %–3 %) prior to incubation with lysate. For issues caused by nucleic acid contamination to the protein preparations, micrococcal nuclease treatment is suggested [13].

3.11 Fluorescence Polarization

The steps below are used to determine the binding affinity (K_D) using an equilibrium saturation binding assay between an SH2 domain and a labeled phosphotyrosine peptide. The FP protocol described below is using a 96-well plate format but can be adjusted to a 384-well format. Check your plate reader whether it can accommodate these plate formats.

1. Dissolve the FITC-labeled peptide in DMSO and generate a 10 μM stock. Keep peptide stocks away from light. Freeze stocks at −20 °C for long-term storage.

2. Dilute the FITC-labeled phosphotyrosine peptide in FP buffer ranging from 0.005 nM to 2 nM (*see* **Note 37**).
3. Load 100 μL of the diluted FITC-peptide onto a 96-well black well plate.
4. Insert plate into the Envision Multilabel Plate Reader and read the plate using the FP 488 nm /520 nm (excitation/emission) settings.
5. Determine the optimal concentration of FITC-labeled peptide for the assay and prepare a 2X solution where the final concentration is the optimal concentration (*see* **Note 38**).
6. Generate a 2X SH2 domains stock in FP buffer (10–15 dilutions ranging from 2 nM to 100 μM) (*see* **Note 39**).
7. Mix in the 96-well black-well plate 50 μL the 2X SH2 domains and 50 μL of the 2X peptide solution.
8. Pipette up and down to mix the samples together (*see* **Note 40**).
9. Include a well containing the 50 μL 2X peptide solution and 50 μL FP buffer as a negative or background control (*see* **Note 41**).
10. Gently shake the plate for 15 min at ~22 °C using an orbital shaker or the orbital shaking mode in the plate reader. Keep plate covered with aluminum foil plate seal.
11. Centrifuge at 300 x*g* for 2 min at ~22 °C.
12. Remove aluminum foil.
13. Read the plate on the EnVision Multilabel Plate Reader (PerkinElmer) or equivalent
14. Export data in an Excel file or tab-delimited text format
15. Calculate the equilibrium dissociation constant (K_D) and V_{max} using GraphPad Prism with the Binding Saturation (One-site) fit (*see* **Note 42**)(Fig. 3).

4 Notes

1. As KLB is nonionic detergent-based, proteins are retained in their native state and protein–protein interactions will likely be preserved. KLB is suitable for preparation of lysates for both far-Western and pull-down.
2. RIPA buffer contains the denaturing detergent SDS and thus can solubilize more proteins than KLB but results in dissociation of some protein complexes. Thus cellular protein complexes involving SH2 binding proteins may not be recoverable by pull-down with RIPA lysis. However, RIPA buffer is suitable for far-Western.

3. Insoluble proteins using KLB and RIPA buffers can be solubilized by the denaturing lysis buffer. However, denatured lysates may not be suitable for pull-down.
4. Chose appropriate acrylamide percentage for optimal separation of the target bands/proteins. Precast gel systems are helpful to minimize gel-to-gel variability (e.g., NuPAGE gel system).
5. We use anti-GST-HRP conjugate (Sigma A7340) for GST-tagged SH2 proteins and HisProbe-HRP (15165, Thermo scientific) for poly His-tagged SH2 proteins.
6. Chemiluminescent detection has higher detection sensitivity than fluorescent detection, but we usually observe sufficient signal and better image resolution (shaper bands) with near-IR detection. In addition, a multi-color channel fluorescent imager can allow for dual probe detection (e.g., SH2 and antibody).
7. The amount of lysate required for SH2 pull-down should be determined empirically. Typically, 100–200 μg of growth factor stimulated cell lysate and 1–2 mg of unstimulated or tumor specimens are sufficient. For unknown samples, prepare 500 μg lysate per assay.
8. Most peptides are generated with the FITC dye at the N-terminus with additional amino acids (2-5 amino acids) between the FITC tag and the phosphotyrosine. Avoid using long peptides with more than 5 amino acids between the FITC tag and the phosphotyrosine as the FITC dye may be able to rotate independent of SH2 domain binding. Other dyes can be coupled to the peptide such as 5-TAMRA (5-carboxytetramethyrhodamine) but be sure to have the correct filters.
9. Usually, far-Western and pull-down require 25 μg and 500 μg per probe (SH2 domain), respectively. For example, a subconfluent 10-cm culture dish, equivalent to ~1 mg protein, would be sufficient for 40 far-Western and 2 pull-down assays.
10. For lysis with the denaturing lysis buffer, perform steps 3–8 at room temperature.
11. Ideally the lysate concentration is 2 μg/μL or higher. If protein concentration is too low, the sample can be concentrated by ultrafiltration or precipitation by cold ethanol.
12. Tissue specimens contain varying degree of cell and non-cell materials typically, 0.5–1 mg protein may be extracted from a 2–5 mm^3 tissue fragment.
13. Tumor samples that have been embedded in frozen OCT compound or DMSO can interfere with the SDS-PAGE. This should be removed prior to pulverization. For samples in DMSO, quickly rinse the frozen sample with ice cold PBS-vanadate. For OCT-embedded samples, attempt to remove as

much OCT as possible using forceps and scalpels on a surface chilled with dry ice.

14. Do not attach the tube too tight as it can result in bag rupture during pulverization.
15. Turn on the water bath cooling system 1 h before use.
16. We use a cocktail of lysates consisting of fibroblast, hepatoma, B-cell lymphoma, and epithelial type cell lines. Most SH2 domains we tested using this control gave readily detectable signal in our far-Western analysis [10], but other cocktails may also be considered.
17. For suspension cells, transfer the medium to a conical tube and spin for 1 min at 1,000 × *g* and wash the cell pellet.
18. For evaluation of cell lines, we also employed far-Western blots with representative SH2 domains [10]. Gene expression databases can also be informative, e.g., eCellLineNavigator (http://medicalgenomics.org).
19. Preparation of 10 mg of lysate per cell line is intended. The yield of protein is variable for individual cell lines. Change the number of plates/flasks accordingly.
20. A higher dose (100–200 μM) and longer incubation time (~1 h) of pervanadate may help to increase the pTyr level, though it may cause detachment of adherent cells due to cell toxicity. Collect detached cells by centrifugation for 1 min at 1,000 × *g*.
21. Phosphatase treatment at room temperature could result in degradation of proteins during the reaction by endogenous proteases, and contamination of a small amount of GST-PTP protein to the lysate. Thus, we suggest checking the residual phosphorylation by pTyr blot prior to PTP treatment.
22. The SH2 domain(unlabeled), anti-tag antibody (unlabeled), and dye/peroxidase conjugated-secondary antibody can also be incubated sequentially (three-step binding, Fig. 1)[14]. Generally, we observed a stronger far-Western signal with GST-tagged SH2 domains than His-tagged SH2 domains with equivalent probe concentrations, likely due to GST-GST dimers.
23. Usually 10–50 μg of lysate per lane is optimal for far-western using GST-SH2 domains. In our hands, nitrocellulose gives lower background signal than PVDF membrane. Transferred membranes can be stored at −20 °C for later use: rinse with TBST once and wrap in plastic wrap. Care should be taken to avoid damaging the membrane, which causes undesirable background signal.
24. The type of blocking buffer, such as nonfat milk, bovine serum albumin (BSA), and ovalbumin may significantly affect the sig-

nal and noise. For SH2-far-Western, we routinely use 5% nonfat milk TBST buffer for blocking and probe incubation, and increase it up to 10% if the background is high.

25. The optimal condition for labeling should be empirically determined.
26. Purification of the labeled probe is not needed. Labeled probe can be stored at 4 °C up to a month in the dark.
27. Usually ~50 nM of GST-SH2 domain solution gives sufficient signal without GST background. If signal is weak, probe concentration can be increased up to 150 nM, although we recommend preparing a control blot with the same concentration of GST only probe to assess background binding.
28. Do not allow the membrane to dry at any point during the experiment.
29. In our hands, washing at 40–50 rpm using an orbital shaker is appropriate. To reduce nonspecific background, we recommend washing for an additional 5–10 min without increasing the velocity.
30. For chemiluminescent detection, perform the procedure promptly to avoid signal loss. For image scanning, place the wet membrane on the scanner plane and cover the membrane with plastic wrap or silicon mat. Use a roller to remove any excess solution from the membrane surface (retained solution can cause a blurred image and nonspecific background).
31. If retained signal is present after stripping, a harsher strip with SDS-ME stripping buffer is available at the expense of potential loss of signal. Incubate the membrane in SDS-ME buffer at 50 °C for 30 min under vigorous agitation, followed by extensive washing for 60 min in TBST with frequent buffer change. If needed, NewBlot stripping buffer is suited to remove Near-IR dye secondary antibodies.
32. We routinely use pre-verified stock of GST fusion proteins for pull-down.
33. These samples are useful for a post assay evaluation of GST-SH2 proteins (e.g., expression, solubility, purity, and assay loading).
34. Using the concentration values, the amount of GST proteins in different tubes can be adjusted. If it is too low, check the expression and solubility by running the purification fractions (step 8, 10, 15) to a 12% SDS gel stained with Coomassie blue. GST and GST-SH2 should appear as 26 KDa and ~40 KDa bands, respectively.
35. The amount of lysate required for SH2 pull-down should be determined empirically. Typically, 100–200 μg of growth factor stimulated cell lysate and 1–2 mg of unstimulated or

tumor specimens are sufficient. For unknown samples, load 500 μg lysate per assay.

36. As tyrosine phosphorylated SH2 binding proteins are often of low abundance, the bead-bound proteins may not be detectable with conventional Coomassie staining. Consider more sensitive dye and phosphorylation-specific gel staining, e.g., Pro-Q (https://goo.gl/PPxJOe) and Phos-tag (http://goo.gl/g1lEu0), or pTyr-immunoblot.
37. The FITC-labeled peptide should be measured to determine the polarization (mP) signal on the EnVision reader. Fluorescein can also be used as control to calibrate or modify the amount of peptide to add to the assay to obtain a proper mP signal. The mP signal should be between 0 and 100. If the mP signal is negative the peptide concentration may be too high and if the mP is too high reduce the concentration of the peptide.
38. A good starting point is 0.5 nM of peptide but this can vary depending on the polarization signal from the plate reader. Check if the mP value is between 0 and 100 as in step 5.
39. To obtain saturation in the assay estimate the approximate K_D to determine range in concentration of SH2 domain protein. For example, an estimated K_D of 100 nM may require at 100-fold range from 0 to 10 μM.
40. Pipette gently while avoid creating bubbles.
41. Fluorescein-only or GST-only can be used a negative controls. Also an FITC labeled version of the non-phosphorylated peptide can be used as a control for phospho-dependent binding.
42. Subtract the mP values from the sample containing the FITC-peptide with no SH2 domain added.

Acknowledgments

We thank Joshua Jadwin for his assistance with editing this chapter, and Bruce Mayer for his continuous encouragement and support. This study was partly supported by grant CA1154966 from the National Institutes of Health and Quest for CURES (QFC) grant from the Leukemia and Lymphoma Society (to K.M.).

References

1. Nash PD (2012) Why modules matter. FEBS letters 586 (17):2572-2574. doi:10.1016/j.febslet.2012.04.049
2. Jadwin JA, Ogiue-Ikeda M, Machida K (2012) The application of modular protein domains in proteomics. FEBS letters 586 (17):2586-2596. doi:10.1016/j.febslet.2012.04.019
3. Kaneko T, Joshi R, Feller SM, Li SS (2012) Phosphotyrosine recognition domains: the typical, the atypical and the versatile. Cell Commun

Signal 10 (1):32. doi:10.1186/1478-811x-10-32

4. Liu BA, Engelmann BW, Nash PD (2012) High-throughput analysis of peptide-binding modules. Proteomics 12 (10):1527-1546. doi:10.1002/pmic.201100599
5. Nollau P, Mayer BJ (2001) Profiling the global tyrosine phosphorylation state by Src homology 2 domain binding. Proceedings of the National Academy of Sciences of the United States of America 98 (24):13531-13536. doi:10.1073/pnas.241215998
6. Jadwin JA, Mayer BJ, Machida K (2015) Detection and quantification of protein-protein interactions by far-western blotting. Methods in molecular biology (Clifton, NJ) 1312:379-398. doi:10.1007/978-1-4939-2694-7_38
7. Blagoev B, Kratchmarova I, Ong SE, Nielsen M, Foster LJ, Mann M (2003) A proteomics strategy to elucidate functional protein-protein interactions applied to EGF signaling. Nature biotechnology 21 (3):315-318. doi:10.1038/nbt790
8. Hause RJ, Jr., Leung KK, Barkinge JL, Ciaccio MF, Chuu CP, Jones RB (2012) Comprehensive binary interaction mapping of SH2 domains via fluorescence polarization reveals novel functional diversification of ErbB receptors. PloS one 7 (9):e44471. doi:10.1371/journal.pone.0044471
9. Rossi AM, Taylor CW (2011) Analysis of protein-ligand interactions by fluorescence polarization. Nature protocols 6 (3):365-387. doi:10.1038/nprot.2011.305
10. Machida K, Thompson CM, Dierck K, Jablonowski K, Karkkainen S, Liu B, Zhang H, Nash PD, Newman DK, Nollau P, Pawson T, Renkema GH, Saksela K, Schiller MR, Shin DG, Mayer BJ (2007) High-throughput phosphotyrosine profiling using SH2 domains. Molecular cell 26 (6):899-915. doi:10.1016/j.molcel.2007.05.031
11. Schweigel H, Geiger J, Beck F, Buhs S, Gerull H, Walter U, Sickmann A, Nollau P (2013) Deciphering of ADP-induced, phosphotyrosine-dependent signaling networks in human platelets by Src-homology 2 region (SH2)-profiling. Proteomics 13 (6):1016-1027. doi:10.1002/pmic.201200353
12. Tian J, Cai T, Yuan Z, Wang H, Liu L, Haas M, Maksimova E, Huang XY, Xie ZJ (2006) Binding of Src to Na+/K+-ATPase forms a functional signaling complex. Molecular biology of the cell 17 (1):317-326. doi:10.1091/mbc.E05-08-0735
13. Nguyen TN, Goodrich JA (2006) Protein-protein interaction assays: eliminating false positive interactions. Nature methods 3 (2):135-139. doi:10.1038/nmeth0206-135
14. Gao Z, Poon HY, Li L, Li X, Palmesino E, Glubrecht DD, Colwill K, Dutta I, Kania A, Pawson T, Godbout R (2012) Splice-mediated motif switching regulates disabled-1 phosphorylation and SH2 domain interactions. Molecular and cellular biology 32 (14):2794-2808. doi:10.1128/mcb.00570-12

Chapter 18

In-Solution SH2 Domain Binding Assay Based on Proximity Ligation

Kazuya Machida

Abstract

Protein–protein interactions mediated by SH2 domains confer specificity in tyrosine kinase pathways. Traditional assays for assessing interactions between an SH2 domain and its interacting protein such as far-Western and pull-down are inherently low throughput. We developed SH2-PLA, an in-solution SH2 domain binding assay, that takes advantage of the speed and sensitivity of proximity ligation and real-time PCR. SH2-PLA allows for rapid assessment of SH2 domain binding to a target protein using only a few microliters of cell lysate, thereby making it an attractive new tool to study tyrosine kinase signaling.

Key words SH2 domain, In-solution binding assay, Proximity ligation, Real-time PCR, Grb2, EGFR

1 Introduction

There is great interest in studying phosphotyrosine dependent protein–protein interactions in tyrosine kinase pathways that play a critical role in many aspects of cellular function. Various SH2 domain binding assays have been developed to study interactions between SH2 domains and their binding proteins in cell lysate [1–3]. Far-Western blotting and pull-down assays are representative of traditional methods to assess interactions between purified SH2 domains and cellular proteins. However, these methods are inherently low throughput and require gel electrophoresis and membrane blotting. Further, identification of SH2 binding proteins or verification of a specific protein–SH2 interaction often requires additional analyses such as immunoprecipitation and mass spectrometry [4–6].

In an effort to implement a convenient SH2 binding assay, we have developed SH2-PLA, an in-solution binding assay, that takes advantage of the speed and sensitivity of proximity ligation assay (PLA) and real-time PCR [7, 8]. This assay is designed to detect interactions between tagged SH2 domains (or other modular domains) and a target protein containing SH2 binding sites in cell lysate.

Kazuya Machida and Bernard A. Liu (eds.), *SH2 Domains: Methods and Protocols*, Methods in Molecular Biology, vol. 1555, DOI 10.1007/978-1-4939-6762-9_18, © Springer Science+Business Media LLC 2017

Notable features of SH2-PLA over traditional methods are low sample requirement (i.e., few microliters of lysate) and rapid analysis (i.e., 6–7 h assay runtime) without requiring SDS-PAGE and protein transfer. Using A431 and COS1 cells, we confirmed that this assay provides results equivalent to traditional far-Western blotting for the detection of SH2–EGFR interaction. Conveniently, SH2-PLA assay can allow users to rapidly validate SH2 binding to a protein of interest using a small amount of cell lysate, without requiring a labor and time-intensive workflow.

In SH2-PLA, binding of an SH2 and its target protein is recognized by a pair of antibodies conjugated with oligonucleotides (Fig. 1). One antibody binds to the SH2 domain via tag moiety (e.g., GST) and another antibody to the target protein (e.g., epidermal growth factor receptor, EGFR). The distal ends of the oligo tags can be ligated in the presence of a complementary oligo connector when the antibody binding sites are in proximity (<30–40 nm), and the ligation can be quantified by real-time PCR [9, 10]. This ligation and subsequent amplification are an indicator of interaction of the SH2 domain and the target protein. A typical SH2-PLA assay is performed using growth factor stimulated and unstimulated cells; as the stimulated cells contain a greater number of tyrosine phosphorylated target proteins that serve as SH2 binding sites. In this system, the increase in binding sites leads to a greater number of ligation events which is reflected by a reduction

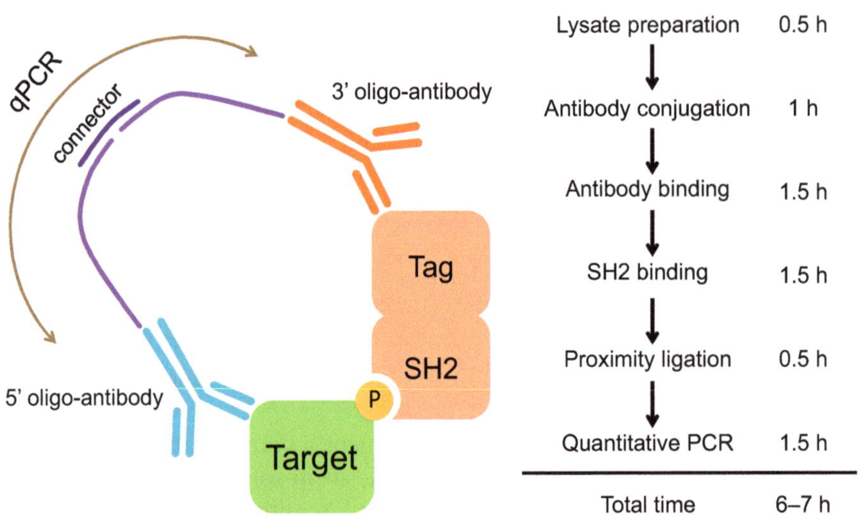

Fig. 1 SH2-PLA strategy. To detect interaction between a tagged SH2 domain protein and target protein in solution, a pair of oligonucleotide-tagged antibodies recognizing the tag and target protein are mixed together and allowed to form a complex as depicted above. The oligonucleotide portion of the antibodies is then ligated in the presence of a connector oligo (supplied in reaction buffer) and subjected to real-time PCR analysis. The whole process including lysate preparation and antibody conjugation takes only 6–7 h and repetitive experiments using stored probes are even faster

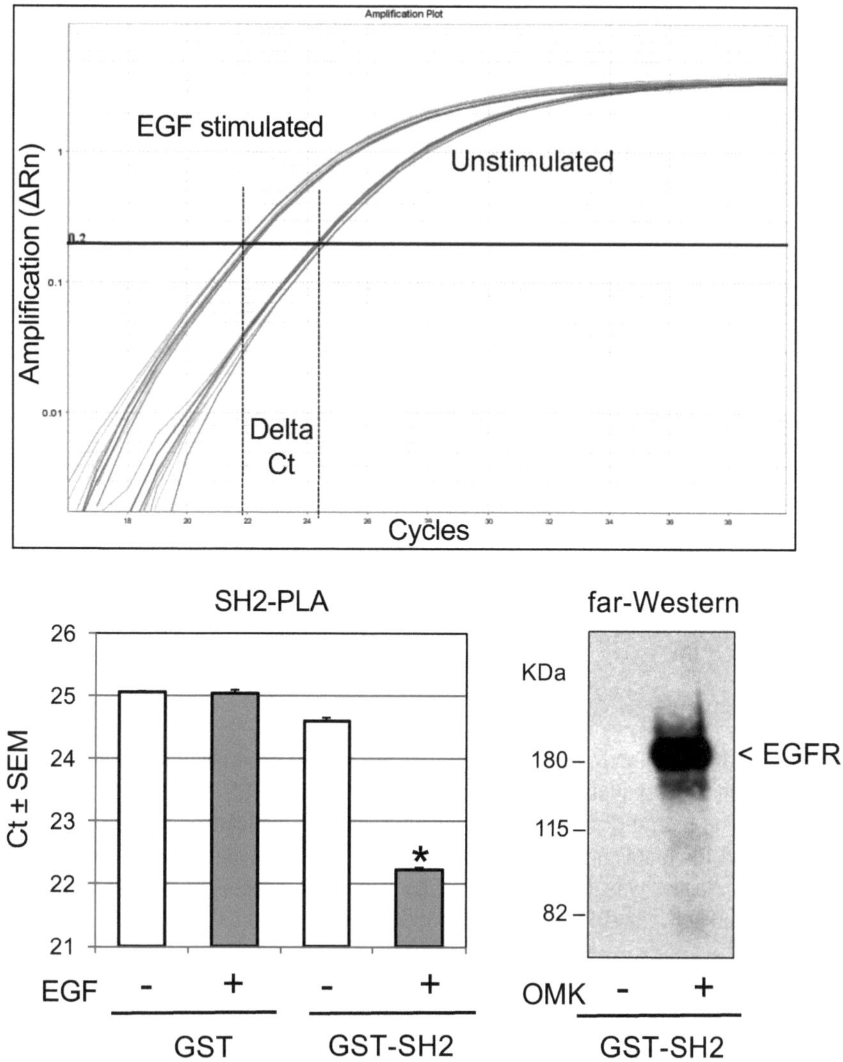

Fig. 2 SH2–EGFR interactions determined by SH2-PLA and far-Western. SH2-PLA comparing EGF stimulated and unstimulated cells was performed using a GST-SH2 probe. *Upper panel*, raw amplification plot shows reduced Ct values in stimulated cells relative to unstimulated cells (delta Ct = ~2.2). *Lower left panel*, Ct values for GST control and GST-SH2 probes. The reduced in Ct value is EGF and SH2 domain dependent (*asterisk*), as the control GST probe shows no change with EGF stimulation. *Lower right panel*, a similar result was obtained using far-Western blotting with the same sample and SH2 domain (SH2-bound EGFR band is indicated)

in real-time PCR Ct value (Fig. 2). This change in Ct values correlated well with quantified far-Western data using the same lysates and SH2 domain probe [8].

In this chapter, I describe an SH2-PLA protocol that has been developed using TaqMan Protein Assays Open Kit (Applied Biosystems) [11–13]. For a new experiment, antibodies for each target protein must first be validated. Subheadings 3.2–3.7 describe preparation and validation of an oligo-conjugated anti-GST

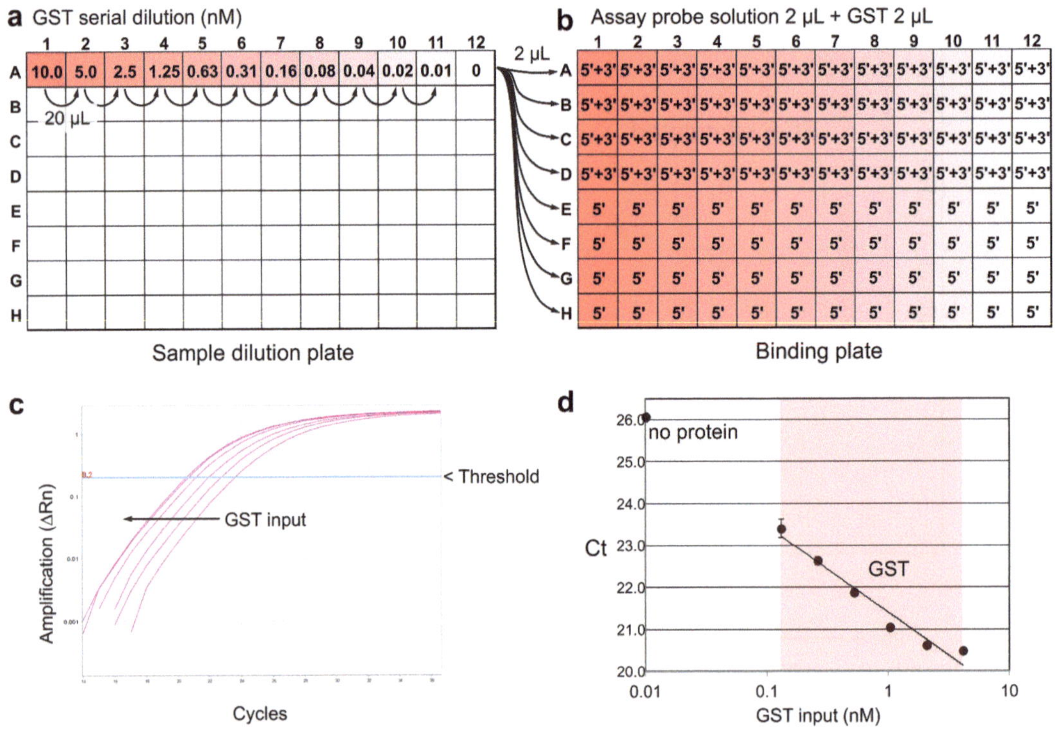

Fig. 3 Validation of GST tag antibody. (**a**, **b**) Purified GST tag protein is serially diluted in the sample dilution plate and transferred in 2-μL aliquots to the binding plate. The diluted GST protein is incubated with assay probe solution containing 5′ and 3′ Prox-oligo conjugated anti-GST antibody (*rows A–D*) or only the 5′ GST antibody (*rows E–H*). (**c**) The reaction mixture was subjected to proximity ligation. Real-time PCR amplification of ligated 5′ and 3′ Prox-oligos was attained in a GST concentration-dependent manner. (**d**) A linear real-time PCR signal was obtained between 0.13 and 4.17 nM of GST protein input

antibody (Fig. 3). If an antibody shows linear signal increases (reduced Ct values) in response to an increasing amount of GST tag input, the antibody is suitable for SH2-PLA. Similarly, an oligo-conjugated anti-EGFR antibody is validated using EGFR expressing cell lysate. When the optimal concentration range for the antibody pair are found, the assay can be tested using biological samples such as EGF stimulated and unstimulated cells (Subheadings 3.9–3.13). If the appropriate concentrations of antibodies are known in advance, SH2-PLA optimization can be performed immediately (from Subheadings 3.9). Common applications of SH2-PLA are briefly covered in Subheadings 3.14–3.16.

2 Materials

The SH2-PLA protocol described here is intended to detect SH2 domain binding sites on EGFR using a recombinant GST-SH2 domain protein. Use antibodies and cells specific to your study, e.g., anti-His tag (*see* **Note 1**).

2.1 Lysate Preparation

1. EGFR expressing A431 cells (ATCC) or appropriate cell line expressing proteins of interest.
2. Dulbecco's modified Eagle's medium (DMEM) with 10% FBS, 1% penicillin and streptomycin.
3. Starvation media: DMEM with 0.1% FBS, 1% penicillin and streptomycin.
4. 35 or 60 mm cell culture dish.
5. Cell culture incubator.
6. Epidermal growth factor (EGF), human recombinant (Millipore).
7. Phosphate buffered saline (PBS) 1×.
8. Sample lysis buffer (Protein Quant Sample Lysis Kit, Applied Biosystems) (*see* **Note 2**). Add PMSF to 1 mM, Aprotinin solution to 1% (Sigma A-6279) and Na-orthovanadate to 0.5 mM before use.
9. Laboratory ice pan.
10. Cell scraper.
11. BCA protein assay.
12. –80 °C freezer.

2.2 Antibody Conjugation

1. TaqMan Protein Assays Open Kit (*see* **Note 3**).

   Streptavidin-linked oligos.

   5′ Prox-oligo, 200 nM (50 μL).

   3′ Prox-oligo, 200 nM (50 μL).

   Antibody dilution buffer (1 mL).

   Assay probe storage buffer (1 mL).
2. Biotinylated antibodies (*see* **Note 4**).

   Biotinylated rabbit anti-GST-tag polyclonal antibody (A00202, Genscript).

   Biotinylated goat anti-EGFR polyclonal antibody (BAF231, R&D Systems).
3. Molecular biology grade water.

2.3 Antibody Binding

1. Purified GST tag and GST-tagged SH2 domain proteins diluted to 0.1 μg/μL in 10% glycerol–PBS. Store aliquots at –80 °C (*see* Chapter 8).
2. EGFR expressing cell lysate, e.g., A431. Adjust the concentration to ~1 μg/μL in lysate dilution buffer.
3. TaqMan Protein Assays Open Kit.

   Assay probe dilution buffer (7 mL).

   Lysate dilution buffer (25 mL).

4. 96-well PCR plates compatible with your thermal cycler and real-time PCR unit.
5. Adhesive sealing film for 96-well PCR plate.
6. Plastic paddle or silicone compression mat for 96-well PCR plate.
7. Centrifuge with a 96-well plate adaptor.
8. Thermal cycler.

2.4 Proximity Ligation

1. TaqMan Protein Assays Core Reagents Base Kit (Applied Biosystems).

   DNA Ligase, 500× (1 × 10 μL).

   Ligase dilution buffer, 1× (2 × 1.5 mL).

   Ligation reaction buffer, 20× (1 × 0.7 mL).

   PBS, 1×, pH 7.4 (1 × 1 mL).

   Protease, 100× (1 × 10 μL).
2. 96-well PCR plates.
3. Adhesive sealing film for 96-well PCR plate.
4. Plastic paddle or silicone compression mat for 96-well PCR plate.
5. Centrifuge with a 96-well plate adaptor.
6. Thermal cycler.

2.5 Real-Time PCR

1. TaqMan Protein Assays Core Reagents Base Kit (Applied Biosystems).

   Universal PCR Assay, 20× (1 × 120 μL).
2. TaqMan Protein Expression Fast Master Mix (Applied Biosystems).
3. 96-well PCR plates.
4. Optical adhesive cover, e.g., MicroAmp Optical Adhesive Film.
5. Plastic paddle or silicone compression mat for 96-well PCR plate.
6. Centrifuge with a 96-well plate adaptor.
7. Real-time PCR system such as StepOnePlus and 7900HT Fast systems.

2.6 Data Analysis

1. Real-time PCR system software, e.g., StepOnePlus Software v2.3.
2. ProteinAssist Software (Applied Biosystems) (*see* **Note 3**).

3 Methods

3.1 Lysate Preparation

Lysates should be prepared using a lysis buffer suitable for proximity ligation (*see* **Note 2**). For assay optimization, a pair of growth factor stimulated/unstimulated cell lysates containing an SH2 binding protein is recommended. Higher concentration lysates previously prepared with other buffers (>2 μg/mL) can be used for optimization by diluting with lysate dilution buffer.

1. Grow cells in 35 or 60 mm dish until semi-confluent.
2. Replace media with starvation media and incubate at 37 °C for 16 h.
3. Stimulate cells with 25–50 ng/mL EGF or other growth factor for 10 min.
4. Place the culture plate and sample lysis buffer containing protease and phosphatase inhibitors on ice.
5. Wash the cells twice with ice-cold 1× PBS.
6. Add sample lysis buffer. 50–100 μL for a 35 mm dish and 100–200 μL for a 60 mm dish.
7. Tilt plates so that the lysis buffer covers the entire cell culture surface.
8. Allow cells to lyse on ice for 5–10 min.
9. Scrape cells and collect lysates in labeled microcentrifuge tubes.
10. Spin down cell lysates at 16,000 × *g* for 10 min in a microcentrifuge at 4 °C.
11. Transfer supernatants into new labeled microcentrifuge tubes on ice.
12. Determine lysate protein concentration using the BCA protein assay, adjust concentration to 1 μg/mL, and store 50 μL aliquots at −80 °C.

3.2 Antibody Conjugation

1. Dilute biotinylated antibodies with antibody dilution buffer to 200 nM (~30 μg/mL) in 20 μL.
2. Set up the following mixtures in separate tubes.

   5 μL of 200 nM biotinylated anti-GST antibody + 5 μL of 200 nM 5′ Prox-oligo.

   5 μL of 200 nM biotinylated anti-GST antibody + 5 μL of 200 nM 3′ Prox-oligo.
3. Mix gently by pipetting up and down 4–6 times and spin the liquid to the bottom.
4. Incubate at room temperature for 60 min to allow conjugation of the Prox-oligo and antibody.
5. Add 90 μL assay probe storage buffer.

6. Mix gently by pipetting up and down 4–6 times and spin the liquid to the bottom.
7. Incubate at room temperature for 20 min.
8. Immediately use created 3′ and 5′ Prox-oligo anti-GST (10 nM) or store at –20 °C (*see* **Note 5**).

3.3 Antibody Binding

1. Thaw purified GST tag protein on ice.
2. Adjust concentration of the GST tag to 0.5 μg/mL (~20 nM) with lysate dilution buffer to a volume of 100 μL.
3. Put a 96-well PCR plate on ice (sample dilution plate).
4. In the 96-well PCR plate, pipette 20 μL of lysate dilution buffer into wells A1 through A12.
5. Transfer 20 μL of diluted GST tag to well A and mix thoroughly by pipetting without forming bubbles (*see* **Note 6**).
6. Serially dilute the solution by transferring 20 μL from A1 to A11 as in Fig. 3a, mixing thoroughly after each transfer. Leave only buffer in A12 well (no protein control).
7. Thaw probe dilution buffer on ice.
8. Prepare the 5′ + 3′ assay probe solution as below.

Assay probe dilution buffer	108 μL
5′ Prox-oligo anti-GST (Subheading 3.2 **step 8**)	6 μL
3′ Prox-oligo anti-GST (Subheading 3.2 **step 8**)	6 μL
Total	120 μL

9. Briefly vortex the tube to mix the solution, then centrifuge the tube at 500 × *g* for 30 s.
10. Prepare the 5′ assay probe solution as below. This solution is the negative control.

Probe dilution buffer	114 μL
5′ Prox-oligo anti-GST (Subheading 3.2 **step 8**)	6 μL
Total	120 μL

11. Briefly vortex the tube to mix the solution, then centrifuge the tube at 500 × *g* for 30 s.
12. Put a 96-well plate on ice (binding plate).
13. Pipette 2 μL of the 5′ + 3′ assay probe solution (**step 8**) into each well of rows A–D of the binding plate.
14. Pipette 2 μL of the 5′ assay probe solution (**step 10**) into each well of rows E–H of the binding plate.
15. Pipette 2 μL of the diluted GST tag protein from the sample dilution plate (**step 6**) into corresponding columns on the

binding plate as shown in Fig. 3b. For example, A1 of the sample dilution plate to A1–H1 of the binding plate.

16. Seal the plate with using an adhesive film.
17. Briefly spin the plate at 500 × *g* for 30 s to remove any bubbles.
18. Press the film to the plate using a plastic paddle or silicone compression mat to insure a complete seal.
19. Incubate the sealed plate at 37 °C for 60 min using a thermal cycler with a heated cover.

3.4 Proximity Ligation

1. Thaw TaqMan Protein Assays Core Reagents (Subheading 2.4) on ice.
2. Dilute the DNA ligase with ligase dilution buffer and keep on ice.

DNA ligase, 500	0.1 μL
Ligase dilution buffer	49.9 μL
Total	50 μL

3. Prepare the ligation solution. Multiply the volume below by the number of reactions/wells (*see* **Note 7**).

Ligation reaction buffer, 20×	5 μL
H_2O	90.9 μL
Diluted ligase (**step 2**)	0.1 μL
Total per reaction	96 μL

4. Remove the binding plate from the thermal cycler, remove the sealing film, and put the plate on ice.
5. Pipette 96 μL of the ligation solution into each well of the binding plate. Pipette up and down once to mix.
6. Seal the plate with using an adhesive film.
7. Briefly centrifuge the plate.
8. Insure the seal by pressing the film onto the plate.
9. Incubate the sealed plate at 37 °C for 10 min and keep at 4 °C using a thermal cycler with a heated cover.

3.5 Protease Reaction

1. Thaw protease and PBS on ice.
2. Dilute the protease with PBS.

Protease, 100×	1.5 μL
PBS	148.5 μL
Total	150 μL

3. Remove the ligation reaction plate from the thermal cycler, remove the sealing film, and put the plate on ice.
4. Pipette 2 μL of the protease solution into each well of the binding plate.
5. Seal the plate with an adhesive film.
6. Incubate the sealed plate at 37 °C for 10 min, then at 95 °C for 5 min, and keep at 4 °C using a thermal cycler with a heated cover (*see* **Note 8**).

3.6 Real-Time PCR

1. Put a 96-well PCR plate (compatible to your real-time PCR unit), TaqMan Protein Expression Fast Master Mix, and Universal PCR Assay on ice.
2. Prepare the PCR solution. Multiply the volume below by the number of reactions/wells.

TaqMan protein expression fast master mix, 2×	10 μL
Universal PCR Assay, 20×	1 μL
Total per reaction	11 μL

3. Pipette 11 μL of the PCR solution into each well of the 96-well PCR plate.
4. Remove the protease reaction plate from the thermal cycler, remove the sealing film, and put the plate on ice.
5. Pipette 9 μL of protease-treated ligation product into the corresponding wells of the 96-well PCR plate. Pipette up and down once.
6. Seal the plate with an optical adhesive cover (*see* **Note 9**).
7. Briefly centrifuge the plate.
8. Load the 96-well PCR plate into your real-time PCR instrument.
9. Run real-time PCR using the parameter values shown below (*see* **Note 10**).

Sample volume	20 μL
Reporter	FAM dye
Quencher	Nonfluorescent

Thermal cycling conditions

Hold	95 °C	20 s
Cycle	95 °C 60 °C	1 s 20 s
40 Cycles		

3.7 Data Analysis

1. Open the software equipped with the PCR system and view the amplification plots (Fig. 3c).
2. Set the threshold to 0.2 with automatic baseline and plot the Ct value against the GST input (Fig. 3d).
3. Determine the linearity between GST input (nM) and signal (Ct value). The GST concentration in the linear range will be used for SH2-PLA.

3.8 Additional Antibody Validation (Optional)

Perform validation of antibodies for your target protein, e.g., EGFR. Repeat Subheadings 3.2–3.7 considering the following points.

Probe: a polyclonal antibody, ideally raised against a full-length or near-full-length protein (>100 aa).

Target: purified recombinant protein (10 pM to 10 nM) or target expressing cell lysate (0.1–100 μg/mL).

As mentioned above, validation of each antibody is not always required but can be informative for optimizing SH2-PLA experiments.

3.9 SH2-PLA Optimization

Optimization of SH2-PLA is performed using a pair of the validated Prox-oligo antibodies, e.g., anti-GST and anti-EGFR antibodies as well as positive and negative controls. First, the anti-GST and anti-EGFR antibodies are incubated with GST-SH2 and the lysate, respectively (Subheadings 3.10–3.11), then the antibody-bound GST-SH2 probe is added to serially diluted lysate to allow SH2–target protein binding (Subheading 3.12), and finally, the antibody–protein complex is detected by real-time PCR (Subheading 3.13) (*see* **Note 11**). This experiment is useful to find optimal parameters for assay performance. A suggestive assay plate setup is shown in Fig. 4b (*see* **Note 12**).

3.10 Pre-binding Antibody to GST-SH2

1. Thaw stored GST/GST-SH2 proteins (0.1 μg/μL) and 3′ Prox-oligo anti-GST (10 nM, Subheading 3.2 **step 8**) on ice.
2. Dilute GST proteins and 3′ Prox-oligo anti-GST with Assay probe dilution buffer and mix as below.

GST or GST-SH2	0.2 μg/mL (5–10 nM, 500× dil.)	50 μL
3′ Prox-oligo anti-GST	1 nM (10× dil.)	50 μL
Total		100 μL

3. Incubate at 4 °C for 1.5–2 h.
4. Proceed to Subheading 3.11 during the incubation.

3.11 Pre-binding Antibody to EGFR

1. Thaw EGF stimulated/unstimulated lysates (1 μg/μL) and 5′ Prox-oligo anti-EGFR (10 nM) on ice.
2. Dilute 5′ Prox-oligo anti-EGFR to 1 nM (10× dil.) with lysate dilution buffer to a volume of 250 μL on ice.
3. Dilute EGF stimulated/unstimulated lysates to 125 ng/μL (8× dil.) with lysate dilution buffer to a volume of 16 μL on ice.
4. Put a 96-well PCR plate on ice (sample dilution plate).
5. In the 96-well plate, pipette 10 μL of lysate dilution buffer into wells A1 through A12 and E1 through E12 (Fig. 4a).
6. Transfer 10 μL of the unstimulated lysate solution to well A1 and mix thoroughly by pipetting without forming bubbles.

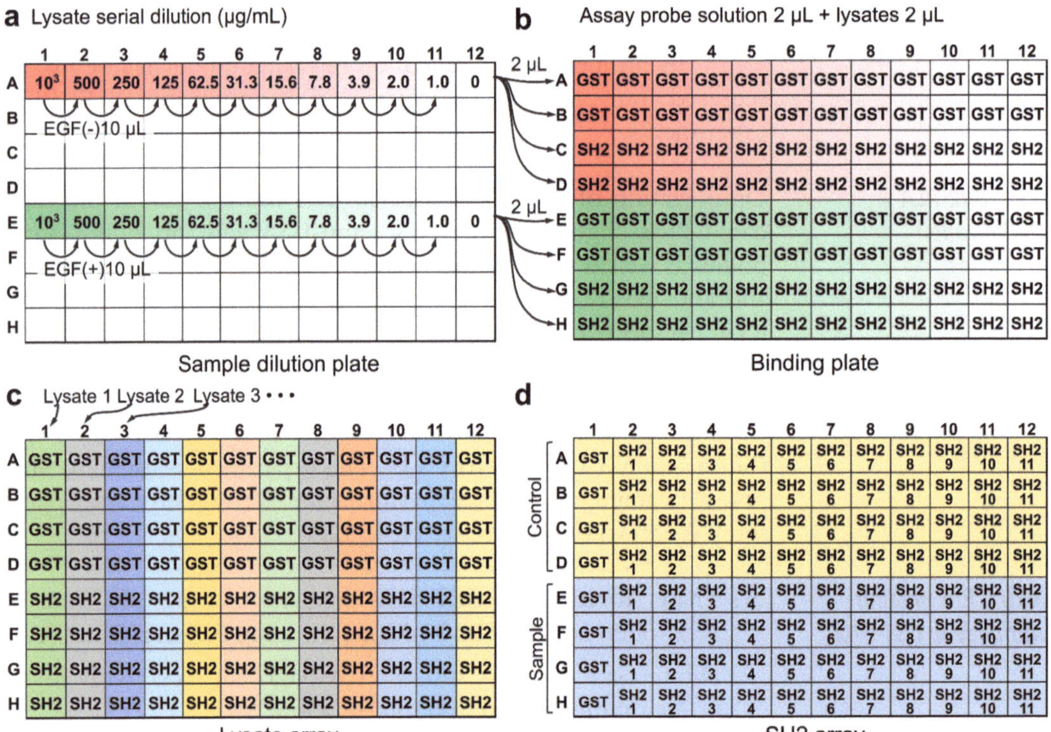

Fig. 4 Optimization and applications of SH2-PLA. (**a**, **b**) EGF stimulated and unstimulated cell lysates are serially diluted in the sample dilution plate and transferred in 2-μL aliquots to the binding plate, followed by proximity ligation and real-time PCR. Ct values for *rows A/B* and *E/F* represent GST background signal. Delta Ct values ($Ct_{unstimulated} - Ct_{EGF\text{-}stimulated}$) for *rows C/D* and *G/H* represent SH2–target interaction and can be used to determine the linear range. These optimization experiments should be performed prior to the analysis of unknown samples. (**c**) Suggested plate setup for an SH2-PLA based lysate array experiment. Lysate samples *1*, *2*, *3*, … are incubated with a GST-SH2 probe (*rows A–D*) or GST as control (*rows E–H*) for real-time PCR analysis. Positive signal indicates the presence of SH2 binding site on the target protein in the lysates. (**d**) Suggested plate setup for an SH2 domain array experiment. Sample and control lysates, e.g., growth factor stimulated/unstimulated, are incubated with different GST-SH2 domain proteins (SH2 1–11). This assay is suitable to identify SH2 domains that bind to the target in a stimulation or treatment dependent manner

7. Serially dilute the solution by transferring 10 µL from A1 to A11. Mix each transfer thoroughly. Leave the A12 well buffer only (no protein control).
8. Transfer 10 µL of the stimulated lysate solution to well E1 and mix thoroughly by pipetting without forming bubbles.
9. Serially dilute the solution by transferring 10 µL from E1 to E11. Mix each transfer thoroughly. Leave the E12 well buffer only (no protein control).
10. Add 10 µL of 5′ Prox-oligo anti-EGFR (1 nM from **step 2**) to A1–A11 and E1–E11.
11. Seal the plate with an adhesive film.
12. Briefly centrifuge the plate.
13. Insure the seal by pressing the film on the plate using a plastic paddle or silicone compression mat.
14. Incubate the sealed plate at 4 °C for 1.5 h using a thermal cycler with a heated cover.

3.12 SH2 Binding

1. Remove the PCR plate from the thermal cycler, remove the sealing film, and put the plate on ice.
2. Place a new PCR plate on ice (binding plate).
3. Pipette 2 µL from wells in row A of the sample dilution plate to corresponding wells in rows A–D of the binding plate as shown in Fig. 4b. For example, A1 of the sample dilution plate to A1–D1 of the binding plate.
4. Pipette 2 µL from wells in row E of the dilution plate to corresponding wells in rows E–H of the binding plate as shown.
5. Add 2 µL of 5′ Prox-oligo anti-GST + GST solution (Subheading 3.10) to wells of rows A, B, E, and F of the binding plate.
6. Add 2 µL of 5′ Prox-oligo anti-GST + GST-SH2 solution (Subheading 3.10) to wells of rows C, D, G, and H of the binding plate.
7. Seal the plate with an adhesive film.
8. Briefly centrifuge the plate.
9. Insure the seal by pressing the film on the plate using a plastic paddle or silicone compression mat.
10. Incubate the sealed plate at 4 °C for 1.5 h using a thermal cycler with a heated cover.

3.13 Proximity Ligation and Real-Time PCR

Perform proximity ligation, protease reaction, real-time PCR, and data analysis as described in Subheadings 3.4–3.7. If the signal from the positive control (stimulated cell lysate + GST-SH2 probe) is reproducibly greater than the negative controls (unstimulated cell lysate + GST probe, unstimulated cell lysate + GST-SH2 probe, or stimulated cell lysate + GST probe), the 5′ and 3′ antibody pair is suitable for SH2-PLA. If not, repeat the assay with modified conditions (*see* **Note 13**). Once the assay is optimized, several types of SH2-PLA experiments can be performed: (1) Fold change estimation; (2) Lysate array screening; and (3) SH2 domain array screening.

3.14 Fold Change Estimation

Using the serial dilution experiment (e.g., Fig. 4a, b), SH2 binding for stimulated and unstimulated samples can be quantitatively assessed. The conventional real-time PCR $\Delta\Delta$Ct method is not easily applicable to the analysis of antibody-based proximity ligation assay data due to unequal PCR efficiencies [11]. The fold change estimate can be performed using ProteinAssist (Applied Biosystems), a software tool developed for TaqMan Protein Expression Assays based on the ΔCt squared method [8] (*see* **Note 14**).

3.15 Lysate Array

Similar to the reverse phase protein array, the presence or absence of SH2 binding sites on a target protein in a battery of lysates can be investigated using SH2-PLA. An array of concentration-adjusted lysates are loaded into the binding plate and incubated with a pair of pre-optimized antibody/SH2 probes. This application would be suitable for screening human malignancies for SH2 binding sites. A suggested plate setup is presented in Fig. 4c.

3.16 SH2 Array

Akin to an SH2 domain-based microarray, SH2-PLA. is suitable to identify SH2 domains that have high affinity binding sites on the target protein. Sample and control sets such as agonist-stimulated/unstimulated and inhibitor-treated/untreated are prepared and the difference in SH2 binding is determined. SH2 domains with enhanced signal (reduced Ct value) indicate that they are likely involved in the signaling pathway (if the SH2-containing protein is expressed in cell). A suggested plate setup is presented in Fig. 4d.

4 Notes

1. Both monoclonal and polyclonal antibodies are applicable for SH2-PLA but only polyclonal antibodies are applicable to the antibody verification (*see* Subheading 3.8). Make sure the SH2 domain and antibody recognition sites on the target protein do not overlap. For example, the EGFR antibody in this protocol recognizes the extracellular portion of EGFR without interfering with the intracellular SH2 domain binding sites.
2. Compatible lysis buffers include but are not limited to cell lysis reagent and sample lysis buffer (Applied Biosystems). Cell lysis

reagent (part of Protein Expression Sample Preparation Kit) containing a carrier protein is suitable for lysing a limited number of cells, e.g., single colony. Once lysate is prepared with cell lysis reagent, protein concentration cannot be estimated. Thus the sample concentration has to be adjusted by cell counting, e.g., 500 cells/μL prior to lysis. Sample lysis buffer (part of Protein Quant Sample Lysis Kit) which does not contain the carrier protein is suitable for preparation of high concentration samples. Protein concentration assays, e.g., BCA protein assay, can be performed after cell lysis. Although not tested, other commercial or in-house lysis buffers, should also be compatible with proximity ligation. A component compatibility table for lysis buffers is available in the TaqMan Protein Assays Sample Prep and Assay Protocol (https://goo.gl/vzDmA2) and TaqMan Protein Assays Sample Prep Quick Reference card (http://goo.gl/2rrLbG).

3. Instructions for the TaqMan Protein Assays listed below are helpful for performing antibody conjugation, proximity ligation, real-time PCR, and trouble shooting. Basically, the original TaqMan Protein Assays is designed for a protein expression assay, while SH2-PLA is intended to detect SH2–target protein binding. Thus, some procedures such as the probe binding step are unique.

   TaqMan Protein Assays Probe Development Protocol (http://goo.gl/854IpX).

   TaqMan Protein Assays Chemistry Guide (https://goo.gl/3txkj2).

   TaqMan Protein Expression Assay Protocol (http://goo.gl/osx0kt).ProteinAssist Software Getting Started Guide (http://goo.gl/hfdDvB).

4. Obtain appropriate pre-biotinylated antibodies for your study or perform biotinylation using a labeling reagent e.g., EZ-Link Sulfo-NHS-LC-Biotin (Pierce) and Biotin-XX Microscale Protein Labeling Kit (Molecular Probes). Free biotin in the antibody solution can be the source of background. *See* http://goo.gl/854IpX for a rigorous test for free biotin and other considerations.

5. The conjugated antibody can be stored at –20 °C for up to 6 months.

6. Since the assay is performed in very low volumes, assay precision and reproducibility highly depend on precise pipetting. Use calibrated pipettes and make sure that the pipette tips are well seated. Try to avoid creating air bubbles when mixing the reagents.

7. Prepare >20 % excess volume to provide for any loss that occurs during reagent transfers.

8. The plate can be stored at −20 °C until needed for real-time PCR.
9. Make sure to use an adhesive cover compatible with your real-time PCR system.
10. These parameter values are included in the "Fast" preset of StepOne Plus and 7900HT real-time PCR units (Applied Biosystems).
11. In this method (Fig. 1), antibodies are premixed with target proteins prior to SH2 binding. However, GST-SH2 domains and the lysate can be incubated first, prior to antibody binding [8].
12. In this experiment, wells containing EGF stimulated A431 cell lysates and the probe pair (5′ Prox-oligo anti-GST + GST-SH2) should results a positive signal (positive control). PCR amplification in wells containing GST tag probe is considered background amplification (*see* Fig. 2).
13. In our experience, lysate and probe concentrations are important for control of signal and noise. Try lower concentration first to see if there is any improvement in the delta Ct value ($Ct_{negative\ control} - Ct_{positive\ control}$). Biotinylation of antibodies may also affect the result (*see* **Note 4**). For other considerations about TaqMan Protein Assays, refer to "TaqMan Protein Assays: best practices" in TaqMan Protein Assays Chemistry Guide (https://goo.gl/3txkj2).
14. *See* http://goo.gl/hfdDvB for manual of ProteinAssist Software.

Acknowledgments

I thank Christopher Thompson and Lee Bloom for their efforts in assay development, Elana Swartzman for helpful discussions, Joshua Jadwin for assistance with editing this chapter, and Bruce Mayer for his continuous encouragement and support. This study was partly supported by grant CA1154966 from the National Institutes of Health and Quest for CURES (QFC) grant from the Leukemia and Lymphoma Society (to K.M.).

References

1. Jadwin JA, Ogiue-Ikeda M, Machida K (2012) The application of modular protein domains in proteomics. FEBS Lett 586(17):2586–2596. doi:10.1016/j.febslet.2012.04.019
2. Kaneko T, Joshi R, Feller SM, Li SS (2012) Phosphotyrosine recognition domains: the typical, the atypical and the versatile. Cell Commun Signal 10(1):32. doi:10.1186/1478-811x-10-32
3. Liu BA, Engelmann BW, Nash PD (2012) High-throughput analysis of peptide-binding modules. Proteomics 12(10):1527–1546. doi:10.1002/pmic.201100599
4. Blagoev B, Kratchmarova I, Ong SE, Nielsen M, Foster LJ, Mann M (2003) A proteomics strategy to elucidate functional protein-protein interactions applied to EGF signaling. Nat

Biotechnol 21(3):315–318. doi:10.1038/nbt790

5. Schweigel H, Geiger J, Beck F, Buhs S, Gerull H, Walter U, Sickmann A, Nollau P (2013) Deciphering of ADP-induced, phosphotyrosine-dependent signaling networks in human platelets by Src-homology 2 region (SH2)-profiling. Proteomics 13(6):1016–1027. doi:10.1002/pmic.201200353
6. Meyer K, Selbach M (2015) Quantitative affinity purification mass spectrometry: a versatile technology to study protein-protein interactions. Front Genet 6:237. doi:10.3389/fgene.2015.00237
7. Machida K, Khenkhar M, Nollau P (2012) Deciphering phosphotyrosine-dependent signaling networks in cancer by SH2 profiling. Genes Cancer 3(5-6):353–361. doi:10.1177/1947601912459048
8. Thompson CM, Bloom LR, Ogiue-Ikeda M, Machida K (2015) SH2-PLA: a sensitive in-solution approach for quantification of modular domain binding by proximity ligation and real-time PCR. BMC Biotechnol 15:60. doi:10.1186/s12896-015-0169-1
9. Fredriksson S, Gullberg M, Jarvius J, Olsson C, Pietras K, Gustafsdottir SM, Ostman A, Landegren U (2002) Protein detection using proximity-dependent DNA ligation assays. Nat Biotechnol 20(5):473–477. doi:10.1038/nbt0502-473
10. Soderberg O, Gullberg M, Jarvius M, Ridderstrale K, Leuchowius KJ, Jarvius J, Wester K, Hydbring P, Bahram F, Larsson LG, Landegren U (2006) Direct observation of individual endogenous protein complexes in situ by proximity ligation. Nat Methods 3(12):995–1000. doi:10.1038/nmeth947
11. Ruff DW, Shannon ME, Chen S-M, Fredriksson S, Landegren U (2013) Proximity ligation assay for protein quantification. PCR Technol 2013:393–407
12. Swartzman E, Shannon M, Lieu P, Chen SM, Mooney C, Wei E, Kuykendall J, Tan R, Settineri T, Egry L, Ruff D (2010) Expanding applications of protein analysis using proximity ligation and qPCR. Methods (San Diego, CA) 50(4):S23–S26. doi:10.1016/j.ymeth.2010.01.024
13. Ruff D, MacArthur C, Tran H, Bergseid J, Tian J, Shannon M, Chen SM, Fontes A, Laurent L, Swartzman E, Taliana A, Rao M, Lieu PT (2012) Applications of quantitative polymerase chain reaction protein assays during reprogramming. Stem Cells Dev 21(4):530–538. doi:10.1089/scd.2011.0032

Part V

High-Throughput Screening and Proteomics

Chapter 19

Alpha-Based Multiplexed Assay for Identifying SH2 Domain Antagonists

Akira Asai and Kazuyuki Takakuma

Abstract

Constitutive activation of STAT3/5b frequently occurs in various human malignancies. STAT3/5b activation involves dimerization via intermolecular pTyr-SH2 binding; therefore, antagonizing this interaction is a feasible approach to inhibit STAT3/5b activation for cancer therapy. We have developed a multiplexed assay to assess STAT3- and STAT5b-SH2 binding in a single well by combining AlphaLISA and AlphaScreen beads. In this chapter, we describe application of the method for the purpose of identifying new STAT3 and STAT5b antagonists.

Key words STAT5b, STAT3, Src homology 2 (SH2), High-throughput screening (HTS), Multiplexed assay, Alpha assay

1 Introduction

High-throughput screening (HTS) is one of the most important tools to generate new lead compounds in modern-day drug discovery research. Hence, there is a growing need for the development of new HTS systems with fast, cost-effective, and robust performance. Multiplexing, where multiple targets or readouts are monitored in the same experiment, is a particularly useful technique for HTS. Drug screening in multiplexed formats has been reported for enzymes, nuclear receptors, and GPCRs thus far [1–5]. However, there are no examples of multiplexed assay systems for identifying inhibitors of SH2-containing disease-related proteins, such as signal transducer and activator of transcription (STAT) family transcription factors. Among the STAT family transcription factors, STAT3 and STAT5b are thought to be important therapeutic targets for anticancer chemotherapy [6, 7]. Diminishing the formation of STAT dimers by targeting their SH2 domains is one of the most investigated strategies (*see* Chapter 1, Introduction).

Kazuya Machida and Bernard A. Liu (eds.), *SH2 Domains: Methods and Protocols*, Methods in Molecular Biology, vol. 1555, DOI 10.1007/978-1-4939-6762-9_19, © Springer Science+Business Media LLC 2017

For the purpose of identifying new SH2 antagonists for STAT3 and STAT5b, we have developed a multiplexed assay to assess STAT3- and STAT5b-SH2 binding simultaneously by combining AlphaLISA and AlphaScreen beads, which produce different colors, in a single-well assay as illustrated in Fig. 1 [8]. In AlphaLISA or AlphaScreen assays, two kinds of beads, the donor and the acceptor beads, are used. When these beads are in proximity, the acceptor beads excited by the singlet oxygen that is released from the excited donor beads emit the luminescence. The AlphaLISA and the AlphaScreen acceptor beads emit the different peak wavelength of luminescence, around 615 and 570 nm, respectively. These assay systems are often used to assess the interaction between two molecules such as protein-protein interactions [5, 9, 10]. We conducted an HTS campaign of our in-house chemical libraries using this multiplex assay, and successfully identified a cell permeable small molecule that preferentially antagonizes STAT3-SH2 binding as previously reported [8]. Thus, this multiplexed assay can serve as powerful tool for SH2-related drug discovery and basic protein chemistry research.

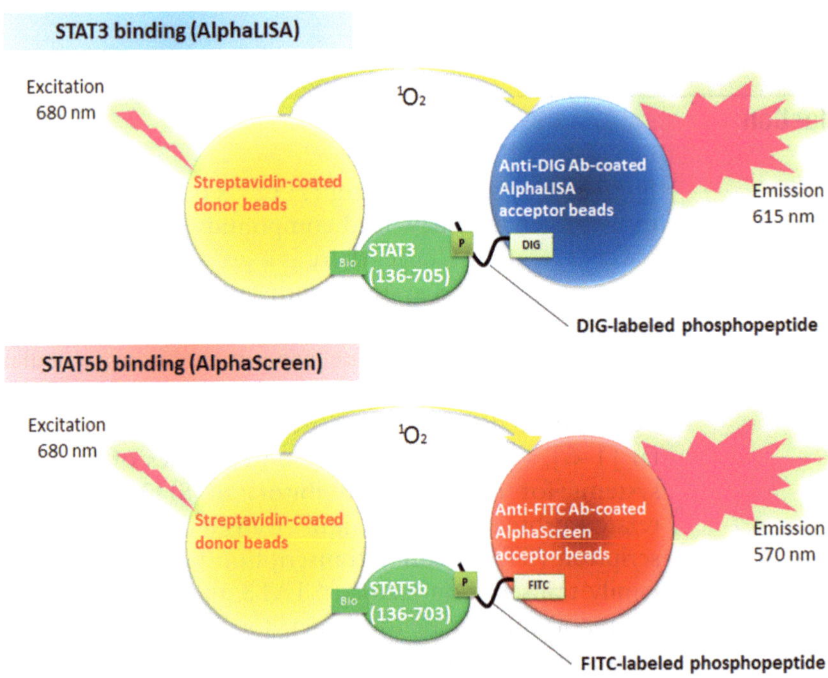

Fig. 1 Schematic diagram of the multiplexed Alpha assay for STAT3- and STAT5b-SH2 binding [8]. The STAT3 binding is detected with the DIG-labeled peptide and AlphaLISA beads, and the STAT5b binding is detected with the FITC-labeled peptide and AlphaScreen beads. By measuring the two different signals from the same well, selective STAT3- or STAT5b-SH2 binding inhibitors can be simultaneously screened

2 Materials

Prepare all solutions with Milli-Q water (prepared with Milli-Q® Water Purification System, Merck Millipore) and the analytical grade reagents. Prepare and store all reagents at room temperature unless described otherwise.

1. Biotinylated STAT3(136–705) and STAT5b(136–703) proteins (*see* Chapter 9).
2. Synthetic peptide DIG-C6-GpYLPQTV: digoxigenin (DIG)-labeled high affinity ligand for the STAT3-SH2 domain.
3. Synthetic peptide FITC-C6-GpYLVLDKW: FITC-labeled high affinity ligand for the STAT5b-SH2 domain.
4. Synthetic peptide Ac-GpYLPQTV-NH2.
5. Synthetic peptide Ac-GYLPQTV-NH2.
6. Synthetic peptide Ac-GpYLVLDKW-NH2.
7. Synthetic peptide Ac-GYLVLDKW-NH2.
8. Assay buffer: 10 mM HEPES-NaOH (pH 7.4), 50 mM NaCl, 1 mM EDTA, 0.1 % (w/v) NP-40, and 0.001 % (w/v) BSA (*see* **Note 1**).
9. 96-well microplate (*see* **Note 2**).
10. AlphaScreen Streptavidin Donor beads (PerkinElmer).
11. Anti-FITC AlphScreen Acceptor Beads (PerkinElmer).
12. Anti-DIG AlphaLISA Acceptor Beads (PerkinElmer).
13. EnVison Xcite multilabel plate reader (PerkinElmer Life Sciences) or other AlphaScreen compatible reader.

3 Methods

In this section, a stepwise protocol for the Stat inhibitor screening is described. Carry out all procedures at room temperature unless described otherwise. Preparation of required Stat proteins is described in Chapter 9.

1. Dilute the biotinylated STAT3(136–705) and STAT5b(136–703) proteins, labeled ligand peptides DIG-C6-GpYLPQTV and FITC-C6-GpYLVLDKW, donor and acceptor beads, and the inhibitors to be tested with the assay buffer.
2. Mix 7.5 μL of assay buffer, 5 μL of 500 nM STAT3(136–705) protein, 5 μL of 100 nM STAT5b(136–703) protein, and 2.5 μL of the test compound, and incubate for 30 min at 25 °C.

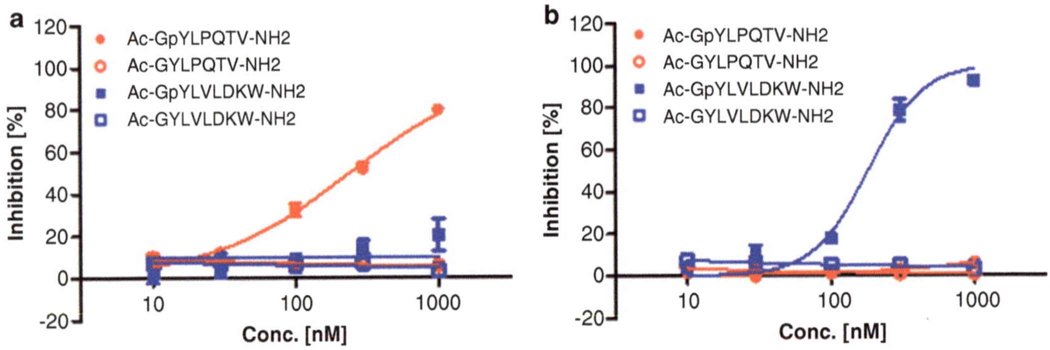

Fig. 2 Inhibitory effects of the non-labeled peptides on STAT3- and STAT5b-SH2 binding [8]. (**a**) Dose-dependent inhibition of STAT3 binding in the multiplexed assay. (**b**) Dose-dependent inhibition of STAT5b binding in the multiplexed assay. Phosphorylated or non-phosphorylated unlabeled peptides were used as competitors. Each data point represents the mean from three replicates, and the error bars represent the standard deviation from the mean

3. Add 2.5 μL of 10 nM DIG-C6-GpYLPQTV and 2.5 μL of 10 nM FITC-C6-GpYLVLDKW, and incubate for 90 min at 25 °C (*see* **Note 3**).
4. Mix 12.5 μL of 0.040 μg/μL Streptavidin-coated donor beads, 6.25 μL of 0.040 μg/μL anti-DIG AlphaLISA acceptor beads, and 6.25 μL of 0.040 μg/μL anti-FITC AlphaScreen acceptor beads, and incubate for 90 minutes at room temperature under dark conditions (*see* **Note 4**).
5. Measure the AlphaLISA signal at around 615 nm and the AlphaScreen signal between 520 and 620 nm by 680 nm excitation with EnVision Xcite multilabel plate reader (*see* **Note 5**).
6. Represent the "bound" and "free" signals as 0 and 100%, respectively, and calculate the inhibition percent of the inhibitors for AlphaLISA signal (STAT3 binding) and the AlphaScreen signal (STAT5b binding) (*see* **Note 6**).
7. Plot % inhibition of each concentration of test compounds with Graph pad prism 4.0 (*see* **Note 7**) (Fig. 2).
8. Calculate the Z′ factor by using the following equation for both AlphaLISA and the AlphaScreen signals. Z′ = 1 − (3 × Standard deviation_bound + 3 × Standard deviation _free)/(Signal_bound-Signal_free) (*see* **Note 8**) (Table 1) [11].

4 Notes

1. Simple method of preparing assay buffer: weigh 0.24 g of HEPES, 0.29 g of NaCl, 29 mg of EDTA, 0.1 mL of NP-40, and 0.013 mL of 7.5% BSA, and dissolve in about 80 mL of water. Adjust the pH to 7.35–7.45 with 5 or 1 N NaOH solution, and then adjust the total assay buffer to 100 mL with water.

Table 1
Determination of the Z′ factors by the multiplexed assay [8]

	Z′ values			
STST3 binding	0.79	0.83	0.89	0.69
STST5b binding	0.66	0.64	0.73	0.69

The Z′ factors acquired from four independent experiments are shown. All values are greater than 0.5 demonstrating a high accuracy of the multiplexed assay

2. We often use the 96-well half area microplate to reduce the total assay volume and cost of the beads.
3. The assay volume is 25 μL consisting of a test compound, two different biotinylated proteins, and the labeled peptides. The final concentrations of the biotinylated STAT3(136–705) and STAT5b(136–703) proteins are 100 and 20 nM, respectively. The final concentrations of DIG-C6-GpYLPQTV and FITC-C6-GpYLVLDKW are both 1.0 nM. The 25 μL reaction mixture contains 2.5 μL of a test compound dissolved in 10% DMSO solution yielding a final concentration of 1% DMSO. Concentrations of the biotinylated proteins and the labeled peptides need to be optimized by performing a saturation assay. DMSO concentration up to 1% should be checked not to affect the signal intensity in this assay.
4. Prior to running the assay, prepare streptavidin-coated donor beads, AlphaLISA acceptor beads and anti-FITC AlphaScreen acceptor beads to 0.040 μg/μL. These beads suspensions are preferably stored under dark conditions because of their sensitivity to light. They can be stored at room temperature prior to running the assay, but it is most preferable to be prepared just prior to use. The final weights of AlphaScreen Streptavidin Donor beads, Anti-FITC AlphScreen Acceptor Beads, and Anti-DIG AlphLISA Acceptor Beads are 0.50 μg/well, 0.25 μg/well, and 0.25 μg/well, respectively.
5. The AlphaLISA and AlphaScreen signals indicate the STAT3 and STAT5b bindings, respectively. The AlphaLISA and the AlphaScreen signals can be separately measured by using a dysprosium 572 nm and a europium 615 nm narrow bandwidth filter. We measured them with the EnVision Xcite multilabel plate reader, but any other comparable instrument can be used.
6. The "bound" represents the signal without the addition of inhibitors or non-labeled peptides. The "free" indicates the signal upon addition of the 10 μM of non-labeled phosphorylated peptide. The "bound" and "free" represent 0% inhibition and 100% inhibition, respectively.

7. We calculated the mean from three replicates of % inhibition in each concentration. When the IC_{50} values of the inhibitors need to be calculated, we often calculate them by Graph pad prism 4.0.
8. Z′ factor means the accuracy by calculating with the signal amplitude and the data variation of "bound" and "free." It is generally considered that the value over 0.5 is suitable for high-throughput screening. To have a good Z′ factor, the beads suspension has to be well mixed by pipetting prior to use for the purpose of lowering the data variation.

Acknowledgment

This work was supported by National Institute of Biomedical Innovation (NIBO) grant number 06–02, JSPS KAKENHI Grant Number 23510281/26430166, and Pharma Valley Center (PVC) Shizuoka, Japan.

References

1. Grover GS, Turner BA, Parker CN, Meier J, Lala DS, Lee PH (2003) Multiplexing nuclear receptors for agonist identification in a cell-based reporter gene high-throughput screen. J Biomol Screen 8:239–246
2. Liu K, Southall N, Titus SA, Inglese J, Eskay RL, Shinn P, Austin CP, Heilig MA, Zheng W (2010) A multiplex calcium assay for identification of GPCR agonists and antagonists. Assay Drug Dev Technol 8:367–379
3. Surviladze Z, Young SM, Sklar LA (2012) High-throughput flow cytometry bead-based multiplex assay for identification of Rho GTPase inhibitors. Methods Mol Biol 827: 253–270
4. Pötz O, Schneiderhan-Marra N, Henzler T, Herget T, Joos TO (2012) Receptor tyrosine kinase inhibitor profiling using bead-based multiplex sandwich immunoassays. Methods Mol Biol 795:191–202
5. Arcand M, Roby P, Bosse R, Lipari F, Padros J, Beaudet L, Marcil A, Dahan S (2010) Single-well monitoring of protein-protein interaction and phosphorylation-dephosphorylation events. Biochemistry 49:3213–3215
6. Turkson J, Jove R (2000) STAT proteins: novel molecular target for cancer drug discovery. Oncogene 19:6613–6626
7. Yu H, Jove R (2001) The STATs of cancer--new molecular targets come of age. Nat Rev Cancer 4:97–105
8. Takakuma K, Ogo N, Uehara Y, Takahashi S, Miyoshi N, Asai A (2013) Novel multiplexed assay for identifying SH2 domain antagonists of STAT family proteins. PLoS One 8:71646. doi:10.1371/journal.pone.0071646
9. Oikawa T, Unno Y, Matsuno K, Sawada J, Ogo N, Tanaka K, Asai A (2010) Identification of a small-molecule inhibitor of the interaction between Survivin and Smac/DIABLO. Biochem Biophys Res Commun 393: 253–258
10. Bielefeld-Sevigny M (2009) AlphaLISA immunoassay platform- the "no-wash" high-throughput alternative to ELISA. Assay Drug Dev Technol 7:90–92
11. Zhang JH, Chung TD, Oldenburg KR (1999) A simple statistical parameter for use in evaluation and validation of high throughput screening assays. J Biomol Screen 4:67–73

Chapter 20

Characterizing SH2 Domain Specificity and Network Interactions Using SPOT Peptide Arrays

Bernard A. Liu

Abstract

Src Homology 2 (SH2) domains are protein interaction modules that recognize and bind tyrosine phosphorylated ligands. Their ability to distinguish binding to over thousands of potential phosphotyrosine (pTyr) ligands within the cell is critical for the fidelity of receptor tyrosine kinase (RTK) signaling. Within humans there are over a hundred SH2 domains with more than several thousand potential ligands across many cell types and cell states. Therefore, defining the specificity of individual SH2 domains is critical for predicting and identifying their physiological ligands. Here, in this chapter, I describe the broad use of SPOT peptide arrays for examining SH2 domain specificity. An orientated peptide array library (OPAL) approach can uncover both favorable and non-favorable residues, thus providing an in-depth analysis to SH2 specificity. Moreover, I discuss the application of SPOT arrays for paneling SH2 ligand binding with physiological peptides.

Key words Src Homology 2 (SH2), Phosphotyrosine (pTyr), Specificity, SPOT peptide arrays, Oriented Peptide Array Library (OPAL)

1 Introduction

The Src Homology 2 (SH2) domain was initially discovered by Tony Pawson and colleagues through the observation that an ~100 amino acid sequence in the v-Fps/Fes oncoprotein was necessary for cellular transformation [1]. This region was subsequently named the SH2 domain due to its protein sequence homology to a corresponding region in the Src and Abl family of cytoplasmic tyrosine kinases. Since that time 121 SH2 domains were identified in 111 human proteins with diverse cellular functions, thereby mediating distinct signaling events upon RTK activation [2]. SH2 domains achieve fidelity in signal transduction through recognizing a phosphorylated tyrosine residue within the context of a specific amino acid sequence [3], often termed its specificity. The specificity allows SH2 containing proteins to discriminate binding

Kazuya Machida and Bernard A. Liu (eds.), *SH2 Domains: Methods and Protocols*, Methods in Molecular Biology, vol. 1555, DOI 10.1007/978-1-4939-6762-9_20, © Springer Science+Business Media LLC 2017

to specific activated RTKs and other tyrosine phosphorylated signaling effectors [1, 4–6].

The SH2 domain structure shows a high degree of structural homology consisting of a core fold of three or four β strands, which make up an antiparallel β sheet, flanked by two α helices. A positively charged binding pocket utilizes a critical arginine (Arg) residue in the β2 strand (within a very highly conserved FLVR motif; [7, 8]) to bind the phosphotyrosine (pTyr) of target ligands. The standard mode of binding depends upon the residues surrounding the critical Arg, which directly determine the specificity of amino acids from position +1 to +6 (C-terminal of the pTyr) [3, 9, 10]. Therefore, the pTyr binding pocket of an SH2 domain is adjacent to the region that dictates binding specificity and determines whether or not the signaling effector is recruited to a given RTK or scaffold protein. This in turn determines which pathway(s) is subsequently activated downstream of the receptor (reviewed in [3, 9–11]). For example, the SH2 domain of Grb2 preferentially binds pY-X-N-X (where X represents any of the natural 20 amino acids) motifs present among its interaction partners. While SH2 domain studies reveal mostly preferential motifs, recent studies reveal that the sequence context and disfavorable amino acids at the specific positions, termed nonpermissive factors, play an important role in mediating SH2-phosphopeptide interactions [12, 13]. Identifying both permissive and nonpermissive factors are essential for informed predictions on SH2 domain specificity.

Over the past decade, multiple high-throughput approaches have been applied toward analyzing SH2-peptide interactions [14]. Proteomic techniques such as protein-microarrays, peptide arrays, phage display peptide libraries and mass-spectrometry have shown to be powerful tools for studying SH2 binding and specificity. The concept of high-density peptide arrays of synthetic peptides using the SPOT method, pioneered by Ronald Frank and Mario Geysen, is a powerful tool for mapping domain-mediated protein-peptide interactions at the proteome level [15, 16]. During SPOT synthesis droplets of activated FMOC-amino acids are spotted onto the planar surface of a porous activated cellulose membrane effectively generating an open reactor for synthesis of cellulose-bound peptides. Peptides are coupled with the C-terminus immobilized to the solid support (e.g., nitrocellulose) with a free N-terminus. Automation of the technique made it accessible enough to be widely used [17] (Fig. 1). One of the benefits of SPOT peptide synthesis is that it circumvents any limitation as to the types and modifications of amino acids that may be employed. As with other forms of solid phase peptide synthesis (SPPS), SPOT synthesis allows the use of unnatural amino acids as well as modified amino acids as building blocks in the synthesis. This is highly advantageous for the study of interactions that depend upon post-translational modification of the peptide ligand such a protein tyrosine phosphorylation.

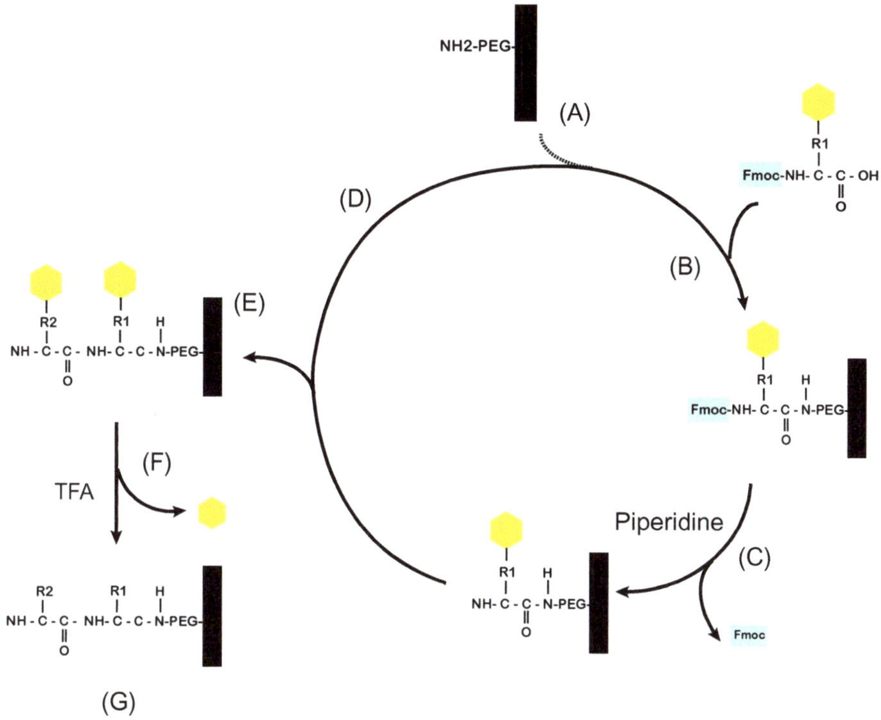

Fig. 1 SPOT synthesis on a nitrocellulose membrane. (**a**) Cellulose membranes containing a free amino-term PEG linker provide a platform for coupling amino acids. (**b**) Amino acids with an activated COOH terminus react to generate a peptide bond on the cellulose membrane. (**c**) The protecting FMOC group is removed using piperidine. (**d**) The addition of additional amino acids is repeated as the peptide is built from C- to N- till the last amino acid is coupled (**e**). (**f**) The final step of synthesis is the removal of the side-chain protective group (*yellow hexagon*) using TFA giving rise to the final peptide product (**g**)

Synthesis of hundreds or even thousands of individual peptides on a single cellulose membrane can yield large sets of binary interaction data. Protein interaction domain studies can easily detect peptide binding using large sets of purified recombinant protein interaction domain incubated individually with the SPOT membranes and subsequently detected using antibodies against specific expression tags (e.g., GST, His) or by using radiolabeled or fluorescently labeled protein. Another advantage of this technique is the ability to observe weak binding or nonbinding peptides and thus generate data not only about the peptides that bind, but also those that do not. In some cases, knowing the certain peptides that do not bind to a protein interaction domain may be as informative as knowing ones that do.

Designing arrays of defined peptides generally requires some degree of a priori knowledge of the binding profile for the domain of interest. Oriented peptide arrays library (OPAL) synthesis using the SPOT approach does not require an understanding of binding preference but generally work best when at least one critical determinant for binding is fixed. This library approach fixes the pY position at 0 and alters the amino acids at one particular position surrounding the pY (e.g., pY+1, pY+2, or pY+3) while using a randomized mixture of amino acids at all other positions [18]. OPAL is commonly used to identify favorable residues at the positions surrounding the pY that can be used to generate position weight matrixes (PWM) or position-specific scoring matrixes (PSSM) or general specificity profiles. This type of data set provides the framework for generating predictive algorithms such as Scansite and SMALI (*see* Chapter 5). The OPAL approach appears has been successful with domains where a major factor is fixed such as pTyr for SH2 domains or PxxP for SH3 domains [18–20]. The relatively high affinity of SH2 domains for pTyr peptides means that library arrays need only fix the pTyr and other positions may be varied, whereas domains such as SH3 and PDZ domains that bind at one magnitude lower affinity generally require at least two fixed residues. Thus, SPOT peptide arrays provide a semiquantitative method for studying SH2 domain interaction networks and for defining specificity.

This chapter will discuss two applications of SPOT peptide arrays to identify and profile the specificity of SH2 domains. One is to characterize the specificity of an SH2 domain using an OPAL-based method and the other is to identify a network of interactions among a large set of physiological (tyrosine sites on known proteins) peptides. Both applications require nearly identical setups and protocols with the main difference in designing the peptide array layout (Subheading 3.1) and the data analysis (Subheading 3.5).

2 Materials

2.1 SPOT Peptide Synthesis (See Note 1)

1. MultiPep high-throughput peptide synthesizer (INTAVIS).
2. Allen key.
3. Acid-hardened cellulose membrane.
4. 2.0 mL screw cap microcentrifuge tubes.
5. F-moc amino acids.
6. FMOC-phosphotyrosine.
7. *N*-Methylpyrrolidone (NMP).
8. Dimethylformamide (DMF).
9. Dichloromethane (DCM).
10. Capping Solution: 2% Acetic Anhydride (Ac_2O) in DMF.
11. Deprotection solution: 20% (v/v) Piperidine in NMP.
12. *N*,*N*-Diisopropylethylamine (DIPEA).
13. HOBt buffer: 0.75 M 1-Hydroxybenzotriazole anhydrous (HOBt) dissolved in NMP.

14. *N,N'*-Diisopropylcarbodiimide (DIC).
15. Triisopropylsilane (TIPS).
16. Trifluoroacetic acid (TFA).
17. Methanol (MeOH).
18. 95 % Ethanol (EtOH).
19. HPLC grade water.

2.2 Membrane Side Chain Deprotection

1. Triisopropylsilane (TIPS).
2. Trifluoroacetic acid (TFA).
3. Methanol (MeOH).
4. 95 % Ethanol (EtOH).
5. HPLC grade water.
6. Chemically resistant container (*see* **Note 2**).
7. Sidechain deprotection cocktail: 80 % TFA, 12 % DCM, 4 % H_2O, 4 % TIPS. Add in this order. Add water slowly and ensure adequate mixing.
8. Ultra Violet (UV) Lamp (*see* **Note 3**).
9. Pencil.

2.3 Confirming Synthesis of Peptides on the Membrane

1. Bromophenol Blue (BPB).
2. BPB Indicator Solution: 10 mg/mL BPB in DMF, store at room temp (*see* **Note 4**).
3. 95 % Ethanol.
4. Nonfat powered milk.
5. TBS: 50 mM Tris–Cl, pH 7.6; 150 mM NaCl.
6. Blocking Buffer: 5 % nonfat powered milk in TBS-T.
7. Anti-phosphotyrosine mouse monoclonal antibody (4G10 clone) (*see* **Note 5**).
8. TBS-T: TBS with 0.5 % Tween-20.
9. Anti-Mouse-IR-800 or Anti-Mouse-HRP (Horse Radish Peroxidase) (*see* **Note 6**).

2.4 Far Western SPOT Detection Using SH2 Domains

1. Recombinant GST-tagged SH2 domains (*see* Chapter 8).
2. Dithiothreitol (DTT): 1 M stock.
3. Nonfat powered milk.
4. TBS: 50 mM Tris–Cl, pH 7.6; 150 mM NaCl.
5. TBS-T: TBS with 0.5 % Tween-20.
6. Anti-GST antibody (rabbit antibody).
7. Anti-Rabbit-IR-680 or Anti-Rabbit-HRP (*see* **Note 6**).
8. Enhanced Chemiluminescence (ECL).

9. Orbital table-top shaker or a plate rocker.
10. Odyssey LI-COR Imaging System.
11. Optional (for films): Film Developer or a Chemiluminescence imaging machine.

2.5 Data Analysis

1. LI-COR Odyssey Image Studio Software.
2. Optional (for films): BioRad Quantity One Software (http://www.bio-rad.com/en-us/product/quantity-one-1-d-analysis-software).
3. Web-based programs such as WebLogo (http://weblogo.berkeley.edu).

3 Methods

3.1 SPOT Peptide Synthesis

Prior to setting up a SPOT experiment determine the best array design to address your question. While both approaches may provide insight into the specificity of an SH2 domain, it is worth noting the limitations and challenges to each array design. If the interest is in obtaining a detailed understanding of SH2 specificity, OPAL may be your best option which can define favorable amino acids at positions surrounding the pTyr [19] (Fig. 2a). In addition, OPAL can be redesigned to fixed both pTyr and a second favorable position (for example, an array for Grb2 can be fixed with x-pY-x-N-x-x) to identify nonfavorable residues that abolish binding termed nonpermissive factors [12]. It should be noted that there are several SH2 domains where identifying the specificity using OPAL may not reveal strong preferences at any one position due to contextual specificity or weak affinity. For these SH2 domains paneling multiple known pTyr peptides may reveal binding specificity [12, 21]. To grasp the pTyr network for a particular SH2 domain, one may be interested in compiling a list of known or potential pTyr sites on proteins (e.g., insulin, FGF, or EGF signaling pathway) to screen for potential SH2 domain binding sites (Fig. 2b). Physiological peptides can be synthesized onto an array and then probed using recombinant SH2 domains to semiquantitatively identify pTyr sites on these signaling proteins. For those who purchased commercially synthesized SPOT arrays, proceed to Subheading 3.3 or 3.4.

1. *Setting up a SPOT run.* Turn On the MultiPep peptide synthesizer and the laptop computer.
2. Open the MultiPep application on the computer and click on the "Edit Sequence" Tab.
3. Determine the type of peptide array you wish to generate. If you plan to generate an OPAL array proceed to **step 2**. If generating an array of physiological peptides proceed to **step 3**. Both types of arrays can be generated on a single membrane, if so proceed to **step 2** (*see* **Note 7**).

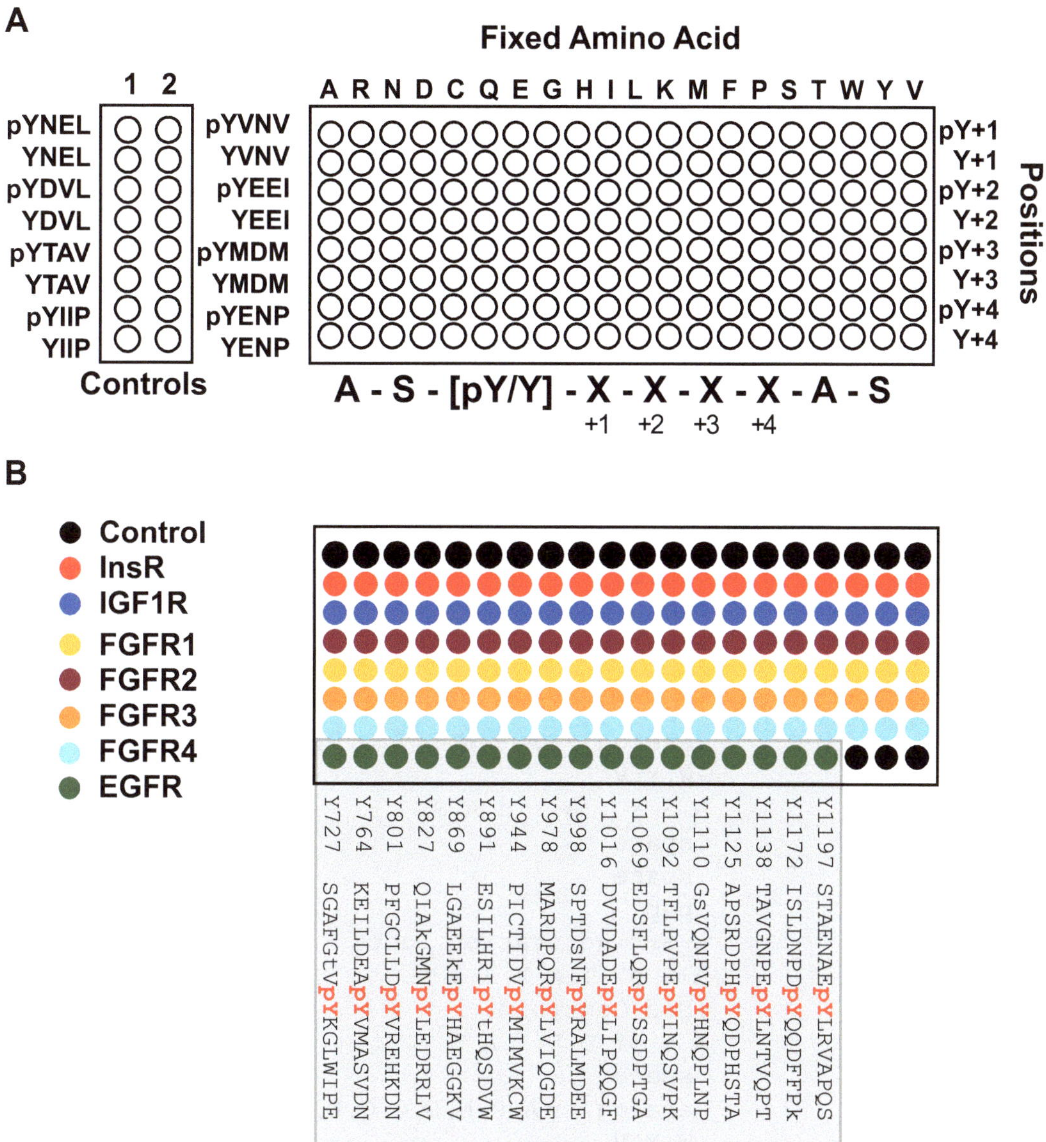

Fig. 2 Layout of peptide array approaches. (**a**) Oriented Peptide Array Library include phosphotyrosine (pY) and non-phosphotyrosine (Y) peptides where at each position represents one of the natural 20 amino acids along the *x*-axis. Control peptides are displayed on the left side of the array. (**b**) Array of physiological phosphotyrosine peptides. Each spot represents a distinct physiological ligand from a specific receptor tyrosine kinase (e.g., *InsR* insulin receptor, *IGFR1* insulin growth factor receptor 1, *FGFR* fibroblast growth factor receptor, *EGFR* epidermal growth factor receptor)

4. *Computing an OPAL Library.* The OPAL library that is encoded below is meant for identifying favorable amino acids at various positions surrounding the Tyr or pTyr. Note that the command lines listed below include positive control sequences for various SH2 domains. These control peptides can be excluded or changed accordingly. In this library includes both

pTyr and non-phosphorylated tyrosine as a control. The array is setup up where at an * is used to represent one of the natural 20 amino acids that will be printed as a single spot across a line on the array (*see* **Note 8**).

```
;SH2 library
;ITAM positive control
.pep
.space
A S pY N E L N A S
;Grb2 positive control
A S pY V N V A S
.space
.analog,*=A,R,N,D,C,Q,E,G,H,I,L,K,M,F,P,S,T
,W,Y,V
A S pY * a r n A S
.space
.space
.pep
A S Y N E L N A S
A S Y V N V A S
.space
.analog,*=A,R,N,D,C,Q,E,G,H,I,L,K,M,F,P,S,T
,W,Y,V
A S Y * a r n A S
.space
.space
.pep
; Src activation loop
A S pY T A V A S
;Src Control
A S pY E E I A S
.space
.analog,*=A,R,N,D,C,Q,E,G,H,I,L,K,M,F,P,S,T
,W,Y,V
A S pY a * r n A S
.space
.space
.pep
A S Y T A V A S
A S Y E E I A S
.space
.analog,*=A,R,N,D,C,Q,E,G,H,I,L,K,M,F,P,S,T
,W,Y,V
A S Y a * r n A S
.space
.space
.pep
;Shp-2 control
```

```
A S pY T A V Q P A S
;SHF control
A S pY I D L D A S
.space
.analog,*=A,R,N,D,C,Q,E,G,H,I,L,K,M,F,P,S,T
,W,Y,V
A S pY a r * n A S
.space
.space
.pep
A S Y T A V Q P A S
A S Y I D L D A S
.space
.analog,*=A,R,N,D,C,Q,E,G,H,I,L,K,M,F,P,S,T
,W,Y,V
A S Y a r * n A S
.space
.space
.pep
```

5. *Computing a Library of Phosphotyrosine Peptides.* Generate a list of phosphotyrosine peptides (*see* **Note 9**). To create a quantitative approach using CPC microarrays please *see* Chapter 21. Be sure to generate internal controls such as duplicate peptides on a single array to ensure peptide synthesis is consistent across the array. Also include non-phosphotyrosine peptides as negative controls.
6. Produce amino acids working stock solutions. Dissolve 0.5 mmol of each FMOC Amino Acid in 1 mL of HOBt buffer. Bring the solution up to 1.5 mL, yielding a ~0.333 M amino acid and 0.5 M HOBt stock solution in NMP. Aliquot inactivated stock solutions (500 μl per 2 mL screw cap tube) and stored at −80 °C until further use.
7. Navigate to the "Show Report" page within the MultiPep software. Print the report and note of the volume of each amino acid required (this is calculated from the phosphopeptide list and protocol design) for each set of couplings, separated by pause points. Thaw enough amino acid stock solution for all required couplings for a synthesis run. For labile or specialty moieties, weigh out the required amount and store at room temperature in a vacuum desiccator protected from light until ready for use.
8. Prepare a 1.1 M DIC solution in NMP by adding 800 μl DIC to 4.2 mL of NMP in a 15 mL polypropylene tube. This activating solution is good for 1 day and this volume is enough (generally) for one cycle (*see* **Note 10**).
9. Combine ¼ parts activating solution with ¾ parts amino acid stock solution such that the final molar ratio of amino acid/HOBt/DIC = 1/1.5/1.1. Pre-activate for 30 min. Spin down in a tabletop centrifuge at 10,000 × *g* for 5 min at room temp.

10. Place the activated amino acid solutions in the correct positions of the rack according to the location designed on the Multipep software. Be sure to remove caps.
11. Prepare the capping and deprotection solutions. Prepare enough capping and deprotection solution for the first round of synthesis, ensuring enough dead volume for proper aspiration by the needle.
12. Fill solvent wash bottles with DMF and EtOH for membrane washes for the first cycle.
13. Remove the acid-hardened cellulose membrane from the packaging and swell the membrane in water for at least half an hour.
14. At the MultiPep machine, find the printing tray. Remove the screws of the top metal (rectangular) plate using the Allen Key and safely remove the top portion of the metal plate.
15. Place the wet membrane on the porous plate with each of the four corners contacting the edges of the screws.
16. Place the top metal plate back onto the printing tray where it covers the edges of the membrane. Cross-tighten the plate evenly.
17. Go to "Manual" and turn Extraction Pump: ON. Wash the membrane with ethanol (3×) and extract the solvent. Dry the last wash for 10 min. Inspect the pump tubing, as no droplets should be seen.
18. Dry the membrane completely before the start of synthesis and then turn off the extraction pump.
19. Go to "Run Synthesis" and click task 1 start. This step will take approximately 8–16 h to complete the first half of the synthesis. Recommend running this overnight.
20. The run will pause at the designated pause point. Perform amino acid preactivation, prepare fresh capping and deprotection solution, and refill solvent wash containers for second round of synthesis. This step will take approximately 8–16 h to complete the second half of the synthesis. Recommend running this overnight.
21. The run will pause after coupling of the N-terminal amino acid. Prepare fresh deprotection and capping solution, and ensure there is enough washing solvent for the end of the run. Finish the run (~1–2 h).
22. The synthesis is now complete. Ensure that the membrane is completely dry, and wash a few more times with EtOH if necessary. If not moving directly onto deprotection and membrane dissolution, place the membrane in a plastic sheet and heat seal the borders. Synthesized membranes can be stored for months at −20 °C.

3.2 Membrane Side Chain Deprotection

After the peptide assembly is complete, it is necessary to remove all side chain protecting groups from the peptides (Fig. 1f). This must be performed under a fume hood as trifluoroacetic acid is very volatile and harmful!

1. Make sure the membrane is completely dry. Using a UV lamp and pencil, trace a grid on the membrane outlining in rows and columns each peptide spot on the membrane.
2. Place blots in the chemically resistant plastic container (with lid).
3. Assemble the sidechain deprotection cocktail in a 50 mL polypropylene tube or corrosion resistant tray with lid just prior to use.
4. Add sidechain deprotection cocktail to the tray containing the membrane. Incubate for 2 h (no shaking, but move around to keep blots covered).
5. Remove the sidechain deprotection cocktail.
6. Wash membranes six times with DCM (Place waste in a beaker, TFA reacts violently with DMF).
7. Wash membranes three times with 95 % EtOH.
8. Wash membranes one time with 100 % MeOH.
9. If not using the membrane right away, air dry membrane and store membranes sealed in a desiccator till future use.

3.3 Confirming Synthesis of Peptides on the Membrane

To confirm whether peptides were properly synthesized on the array there are several methods to determine the quality and uniformity of the SPOTs. One approach is using BPB to detect the free amines on the synthesized peptides (**steps 3–6**). This can be visualized and scanned on a LI-COR Odyssey machine (*see* **Note 11**). An alternative approach to measure synthesis of pTyr peptides is through the use of phosphotyrosine specific antibodies (**steps 7–16**) (*see* **Notes 12** and **13**).

1. Line a grid on the blot under the UV light using pencil.
2. Cut the blot accordingly and remove any excess and unnecessary pieces.
3. Incubate the membrane with 20 mL of BPB indicator solution.
4. Wash membrane 3× with EtOH or MeOH.
5. Dry the membrane.
6. Scan the membrane on the LI-COR imaging system (*see* **Note 11**).
7. Test the synthesis of phosphotyrosine peptides on the SPOT array.
8. Wash the blot in 100 % EtOH for 5 min.
9. Wash the blot in 50 % EtOH for 5 min.
10. Wash the blot in TBS for 5 min.
11. Block in a solution of 5 % nonfat milk in TBS overnight at 4 °C.
12. Add a mouse monoclonal phosphotyrosine specific antibody to 5 % nonfat milk in TBS and incubate for 1 h at room temperature on an orbital shaker.
13. Wash membranes 3× in TBS-T for 5–10 min.

14. Add a secondary anti-mouse IR-800 antibody to 5% nonfat milk in TBS-T. Incubate for 1 h at room temperature on an orbital shaker.
15. Wash membranes 3× in TBS-T for 5–10 min.
16. Scan membranes using the LICOR Odyssey scanner using the IR-800 channel (*see* **Note 6**).

3.4 Far Western SPOT Detection Using SH2 Domains

Detection of SH2 binding to SPOT peptide arrays requires recombinant SH2 protein, please read Chapter 8 for protocols on recombinant expression of SH2 domains. Recombinant protein can usually be expressed as a tagged fusion protein, usually His-tag, GST-tag or biotin. In this protocol, GST-tagged SH2 domains will be used in the Far-western approach for detecting binding to specific pTyr peptides (Fig. 3). For additional approaches *see* **Note 14**.

1. Line a grid on the blot under the UV light using pencil.
2. Cut the blot accordingly and remove any excess and unnecessary pieces.

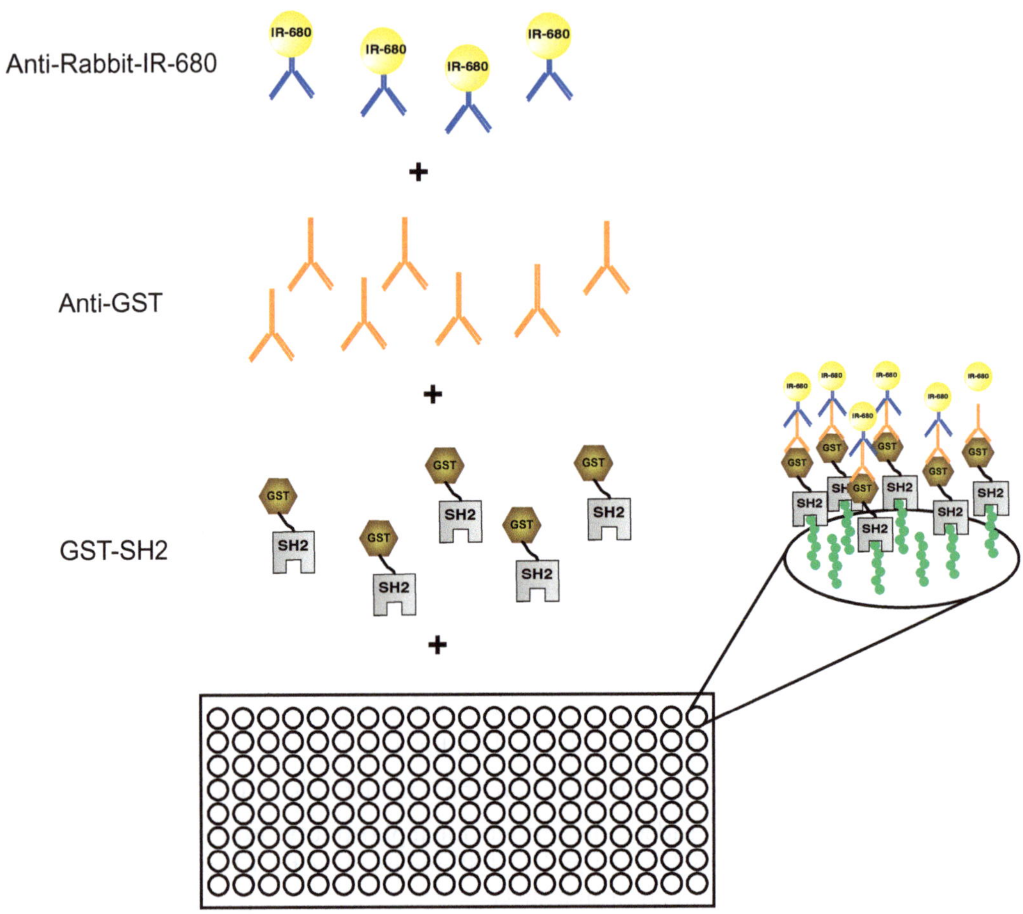

Fig. 3 Detection of recombinant SH2 domains by far-Western Blotting. SPOT peptide arrays are incubated with recombinant SH2 domains. Detection of bound SH2 domains is through Western using anti-GST followed by a secondary antibody that recognizes the primary anti-GST antibody

3. Wash the blot in 100% EtOH for 5 min.
4. Wash the blot in 50% EtOH for 5 min.
5. Wash the blot in TBS for 5 min.
6. Block in a solution of 5% nonfat milk in TBS overnight at 4 °C.
7. Optional: Wash the blocking solution off the blot with TBS.
8. Probe with a GST-SH2 recombinant protein using 5 μg of SH2 in 5% nonfat milk in TBS for 1 h (*see* **Note 15**).
9. Wash the blot with TBST 3× for 5–10 min.
10. Probe with 1° anti-GST at 1:1000 in 5 mL of 5 milk/TBS for 1 h rocking at room temp.
11. Wash with TBST 3× for 5–10 min.
12. Probe with 2° goat anti-Rabbit-IR-680 at 1:10,000 in 5 mL 5% milk/TBS. *See* **Note 16** if using HRP for detection.
13. Wash with TBS-T 3× for 5–10 min.
14. Scan membranes using the LI-COR Odyssey imaging system using the 700 IR channel according to the manufacturer's instructions (*see* **Note 17**).

3.5 Data Analysis

1. After scanning your membranes use the Licor ImageStudio software program to measure the relative intensity of each SPOT (*see* **Notes 18** and **19**).
2. Orientated peptide array libraries can be visualized by capturing a snapshot of the scanned membrane (Fig. 4).
3. Using SPOT arrays, one can visualize their data represented as a position-specific-scoring matrix (PSSM) (Fig. 5). If using an array of physiological phosphotyrosine peptides, one way to set a threshold is through taking the average of all the SPOT intensities of the array (Fig. 5a). Using a cut-off of three times (3×) the mean, select peptides that exceed this threshold. For example, in the case of Grb2, 11 peptides exceeded the >3× mean threshold (Fig. 5b). Web-based programs such as WebLogo (http://weblogo.berkeley.edu) allow you to input your peptide sequences in order to visualize your SPOTs specificity data.

4 Notes

1. SPOT peptide arrays can be purchased commercially. Contact the commercial vendor to discuss the format and layout of your peptide array.
2. Make sure the container is able to fit the full size cellulose membrane.
3. Use UV protective eyewear when using the UV lamp.

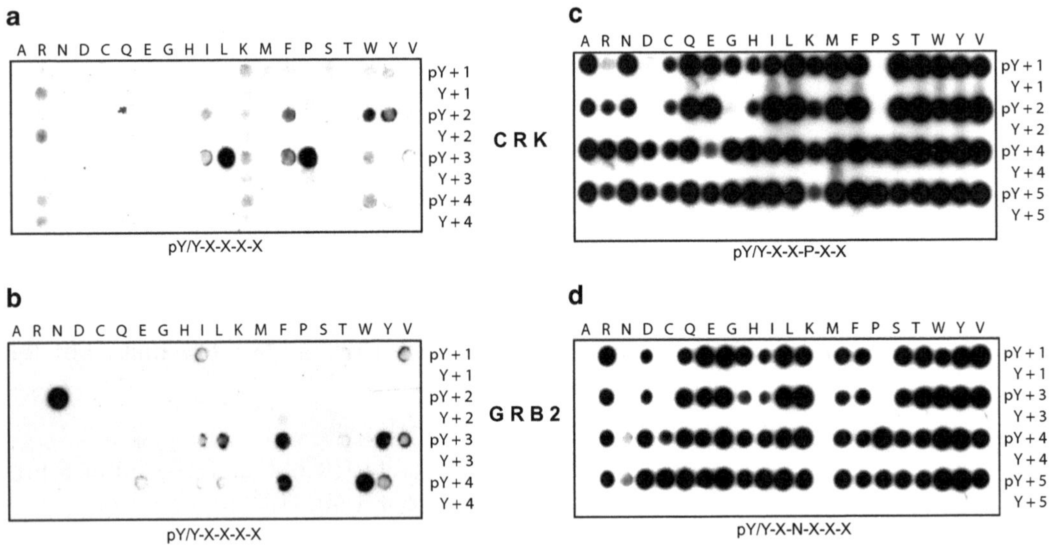

Fig. 4 Specificity of SH2 domains using OPAL. Orientated peptide array libraries display preferential amino acids for the CRK (**a**) and Grb2 (**b**) SH2 domains. For example, Leu and Pro are preferred for CRK at pY + 3 while an Asn is preferred at pY + 2 for Grb2. Nonpermissive factors were identified using OPAL fixed with pY-x-x-P-x-x for Crk SH2 domain (**c**) and with pY-x-N-x-x for Grb2 SH2 domain (**d**). Prolines (P) are considered nonpermissive for Crk and Grb2 at pY + 1 and pY + 2

4. The BPB indicator solution should be discarded if the color has turned to green.
5. Phosphotyrosine antibodies have their own inherent specificity so the intensities do not correlate with the concentration of peptide on the membrane. We recommend using at least two phosphotyrosine antibodies from various suppliers to confirm and validate the synthesis is complete.
6. IR detection is compatible with the LI-COR Odyssey imaging system. Alternatively, you can use HRP conjugated secondary antibodies, which can be measured using film or a chemiluminescence imaging system.
7. Be sure to configure and design the array appropriately if attempting to generate OPAL and physiological peptides onto a single membrane.
8. The * represents fixation of each individual spot to each of the separate amino acids listed accordingly (the order can be rearranged based on the users preference). Cysteine or Tryptophan can be removed to reduce background binding on the arrays.
9. Phosphotyrosine peptide sequences can be collected from various databases including PhosphoSite or PhosphoELM. For

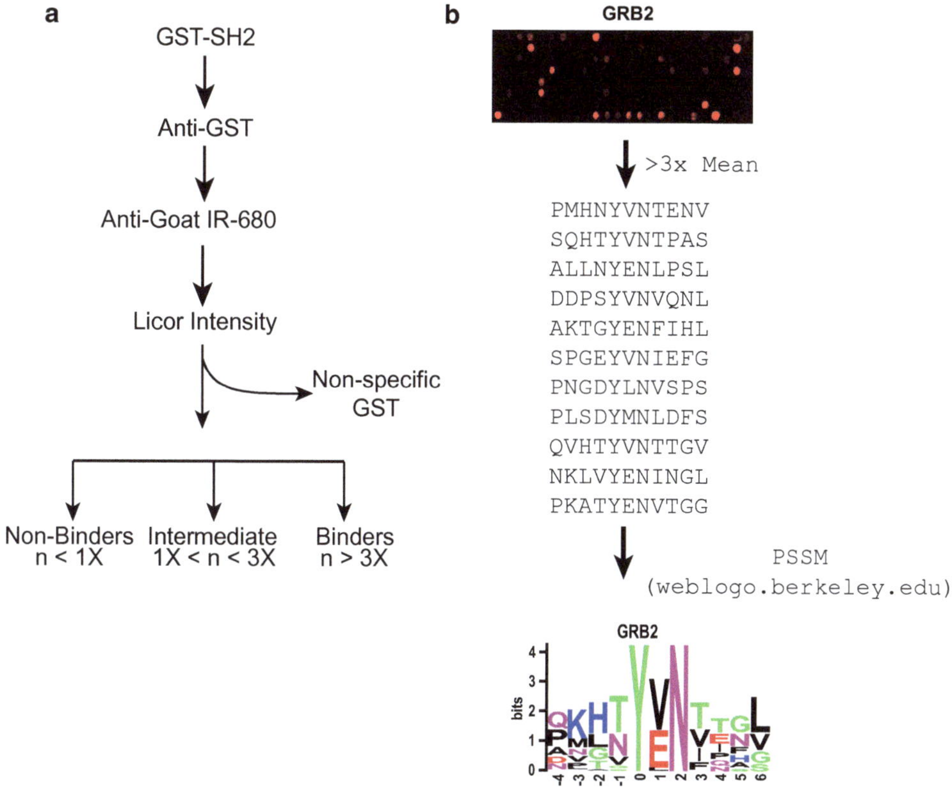

Fig. 5 Analysis of SH2 phosphotyrosine networks and specificity using physiological peptide arrays. (**a**) Setting a threshold for positive binders. After probing and detecting binding using the Odyssey LICOR system, peptides are separated into three categories based on their intensity. Peptide >3× mean are determined as binders, between 1× and 3× are intermediate binders and <1× are non-binders. Prior to analysis background peptides (e.g., GST binding) should be removed from the analysis. (**b**) Specificity of the Grb2 SH2 domain using an array of physiological phosphotyrosine peptides is represented as a PSSM logo

this instance, it is recommended that the length of peptide to be consistent across the array. Be sure the position of the tyrosine is centered in the middle of your peptide with amino acid sequences that include up to the +4 position to capture the binding potential of an SH2 domain. Once a list has been assembled, exchange any Cys to Ser or Ala to eliminate potential disulfide interactions.

10. DIC can readily break down to form crystals of urea. Be sure to check for crystals and to use fresh bottle of DIC.
11. BPB intensities do not necessarily correlate with the amount of peptide at each SPOT. The side chains and protecting groups of other amino acids can strongly influence the staining intensity. It can offer a qualitative sense of uniform synthesis across the array.

12. Since peptides on the peptide array do not undergo purification, the yield and efficiency of peptide synthesis on a solid-support is not uniform and may be difficult to assess.
13. After the peptide array is used for phosphotyrosine antibody detection, it is not recommended that this array be reused. Stripping of the membrane can be done but this is not recommended.
14. Other methods that can be used to detect recombinant SH2 domains include radiolabeling of proteins, biotinylation, or labeling with a fluorescent marker. When designing your assay take into consideration the tag of desire and ways to determine background noise. For instance, using GST-SH2 domains may require GST alone as a negative control to identify peptides that may be sticky or bind nonspecifically.
15. The addition of a reducing reagent such as 1 mM DTT may prevent GST aggregation and nonspecific oxidation induced binding between the SH2 domain and peptides on the array. Be cautious as some SH2 domains contain cysteine(s) within the structure as this may influence proper protein folding. Concentrations of GST-SH2 domains may vary based on the affinity of some SH2 domains to pTyr ligands. Adjust the concentration of GST-SH2 domains (e.g., higher concentrations for low affinity binding SH2 domains) if necessary.
16. An alternative is using anti-rabbit-HRP. Add ECL enough to cover blot if using HRP for detection. Expose blot on film or a chemiluminescence imaging machine if using ECL.
17. If using film scan the film (be sure to note the orientation) using a scanner and save the image as a JPEG file. Download the BioRad Quantity One software (or other western blot quantification software) to measure the intensity of each SPOT.
18. Scanning using the LI-COR Odyssey can be set using the Grid function. The diameter of each circle can be changed so each SPOT is measured accurately.
19. Regardless of the application, SPOT peptide arrays are generally a semiquantitative method and specific interactions require verification or quantification by other methods [22]. Synthesis failure or low yield of a peptide could thus result in false-negative results.

Acknowledgments

I would like to thank Piers Nash, Tony Pawson, Brett Engelmann, and the members of the Nash and Pawson laboratory for helpful discussions. This work was supported by the Canadian Institutes of Health Postdoctoral Fellowship, Bernice Goldblatt Fellowship, Abbott Laboratories Graduate Fellowship, and the Cancer Research Foundation.

References

1. Sadowski I, Stone JC, Pawson T (1986) A noncatalytic domain conserved among cytoplasmic protein-tyrosine kinases modifies the kinase function and transforming activity of Fujinami sarcoma virus P130gag-fps. Mol Cell Biol 6(12):4396–4408
2. Liu BA, Jablonowski K, Raina M, Arce M, Pawson T, Nash PD (2006) The human and mouse complement of SH2 domain proteins-establishing the boundaries of phosphotyrosine signaling. Mol Cell 22(6):851–868. doi:10.1016/j.molcel.2006.06.001
3. Songyang Z, Shoelson SE, Chaudhuri M, Gish G, Pawson T, Haser WG, King F, Roberts T, Ratnofsky S, Lechleider RJ et al (1993) SH2 domains recognize specific phosphopeptide sequences. Cell 72(5):767–778, doi:0092-8674(93)90404-E [pii]
4. Blaikie P, Immanuel D, Wu J, Li N, Yajnik V, Margolis B (1994) A region in Shc distinct from the SH2 domain can bind tyrosine-phosphorylated growth factor receptors. J Biol Chem 269(51):32031–32034
5. Kavanaugh WM, Williams LT (1994) An alternative to SH2 domains for binding tyrosine-phosphorylated proteins. Science 266(5192):1862–1865
6. Anderson D, Koch CA, Grey L, Ellis C, Moran MF, Pawson T (1990) Binding of SH2 domains of phospholipase C gamma 1, GAP, and Src to activated growth factor receptors. Science 250(4983):979–982
7. Johnson LN, Noble ME, Owen DJ (1996) Active and inactive protein kinases: structural basis for regulation. Cell 85(2):149–158
8. Hidaka M, Homma Y, Takenawa T (1991) Highly conserved eight amino acid sequence in SH2 is important for recognition of phosphotyrosine site. Biochem Biophys Res Commun 180(3):1490–1497
9. Pawson T, Gish GD, Nash P (2001) SH2 domains, interaction modules and cellular wiring. Trends Cell Biol 11(12):504–511, doi:S0962-8924(01)02154-7 [pii]
10. Pawson T (1995) Protein modules and signalling networks. Nature 373(6515):573–580. doi:10.1038/373573a0
11. Wagner MJ, Stacey MM, Liu BA, Pawson T (2013) Molecular mechanisms of SH2- and PTB-domain-containing proteins in receptor tyrosine kinase signaling. Cold Spring Harb Perspect Biol 5(12):a008987. doi:10.1101/cshperspect.a008987
12. Liu BA, Jablonowski K, Shah EE, Engelmann BW, Jones RB, Nash PD (2010) SH2 domains recognize contextual peptide sequence information to determine selectivity. Mol Cell Proteomics 9(11):2391–2404. doi:10.1074/mcp.M110.001586
13. Liu BA, Engelmann BW, Nash PD (2012) The language of SH2 domain interactions defines phosphotyrosine-mediated signal transduction. FEBS Lett 586(17):2597–2605. doi:10.1016/j.febslet.2012.04.054
14. Liu BA, Engelmann BW, Nash PD (2012) High-throughput analysis of peptide-binding modules. Proteomics 12(10):1527–1546. doi:10.1002/pmic.201100599
15. Frank R, Heikens W, Heisterberg-Moutsis G, Blocker H (1983) A new general approach for the simultaneous chemical synthesis of large numbers of oligonucleotides: segmental solid supports. Nucleic Acids Res 11(13):4365–4377
16. Geysen HM, Meloen RH, Barteling SJ (1984) Use of peptide synthesis to probe viral antigens for epitopes to a resolution of a single amino acid. Proc Natl Acad Sci U S A 81(13):3998–4002
17. Frank R (2002) The SPOT-synthesis technique. Synthetic peptide arrays on membrane supports—principles and applications. J Immunol Methods 267(1):13–26, doi:S002217590200 1370 [pii]
18. Rodriguez M, Li SS, Harper JW, Songyang Z (2004) An oriented peptide array library (OPAL) strategy to study protein-protein interactions. J Biol Chem 279(10):8802–8807. doi:10.1074/jbc.M311886200, M311886200 [pii]
19. Huang H, Li L, Wu C, Schibli D, Colwill K, Ma S, Li C, Roy P, Ho K, Songyang Z, Pawson T, Gao Y, Li SS (2008) Defining the specificity space of the human SRC homology 2 domain. Mol Cell Proteomics 7(4):768–784. doi:10.1074/mcp.M700312-MCP200
20. Jia CY, Nie J, Wu C, Li C, Li SS (2005) Novel Src homology 3 domain-binding motifs identified from proteomic screen of a Pro-rich region. Mol Cell Proteomics 4(8):1155–1166. doi:10.1074/mcp.M500108-MCP200, M500108-MCP200 [pii]
21. Liu BA, Engelmann BW, Jablonowski K, Higginbotham K, Stergachis AB, Nash PD (2012) SRC Homology 2 Domain Binding Sites in Insulin, IGF-1 and FGF receptor mediated signaling networks reveal an extensive potential interactome. Cell Commun Signal 10(1):27. doi:10.1186/1478-811X-10-27
22. Weiser AA, Or-Guil M, Tapia V, Leichsenring A, Schuchhardt J, Frommel C, Volkmer-Engert R (2005) SPOT synthesis: reliability of array-based measurement of peptide binding affinity. Anal Biochem 342(2):300–311. doi:10.1016/j.ab.2005.04.033, doi:S0003-2697(05)00329-5 [pii]

Chapter 21

High-Throughput Quantification of SH2 Domain–Phosphopeptide Interactions with Cellulose–Peptide Conjugate Microarrays

Brett W. Engelmann

Abstract

The Src Homology 2 (SH2) domain family primarily recognizes phosphorylated tyrosine (pY) containing peptide motifs. The relative affinity preferences among competing SH2 domains for phosphopeptide ligands define "specificity space," and underpins many functional pY mediated interactions within signaling networks. The degree of promiscuity exhibited and the dynamic range of affinities supported by individual domains or phosphopeptides is best resolved by a carefully executed and controlled quantitative high-throughput experiment. Here, I describe the fabrication and application of a cellulose–peptide conjugate microarray (CPCMA) platform to the quantitative analysis of SH2 domain specificity space. Included herein are instructions for optimal experimental design with special attention paid to common sources of systematic error, phosphopeptide SPOT synthesis, microarray fabrication, analyte titrations, data capture, and analysis.

Key words Peptide microarray, SH2 domain, Specificity, Quantitative, High-throughput screening, Context specificity, Proteomics, Interactome, Affinity, Specificity space, Phosphotyrosine peptides

1 Introduction

Phosphotyrosine (pY) mediated cellular signaling is achieved via a dynamic interplay between protein tyrosine kinases, protein tyrosine phosphatases, and proteins containing pY recognition domains, predominantly the SH2 domain [1, 2]. This three component system is a uniquely metazoan adaptation of prominent importance in many major cellular signal transduction pathways [3]. At the biochemical level, it is common for kinase target proteins (or the kinases themselves in the case of receptor tyrosine kinases) to contain multiple in vivo confirmed pY modification sites [4]. Considering that the SH2 domain family consists of over 120 different domains dedicated to the recognition of pY [5] and many SH2 domain containing proteins are expressed concurrently [3, 6], the combination of possible in vivo pY-SH2 domain interactions within just a *single*

Kazuya Machida and Bernard A. Liu (eds.), *SH2 Domains: Methods and Protocols*, Methods in Molecular Biology, vol. 1555, DOI 10.1007/978-1-4939-6762-9_21, © Springer Science+Business Media LLC 2017

multiply phosphorylated protein can be great [7]. The particular SH2-pY complex formed can have profound consequences in vivo, as evidenced in oncogenic [8], pathogenic [9], and even synthetic [10] contexts. Like many other domains, the SH2 domain recognizes a "linear motif" [11] encompassing the central pY moiety and surrounding residues. These surrounding residues dictate the affinity of the interaction between a given SH2 domain and a pY modification site, and can be used to establish preferred "core" motifs that cluster SH2 domains by subfamily [12]. Interestingly, there is a high degree of binding overlap between SH2 domains for physiological ligands [6], and this overlap can be modulated by subtle changes within a "core" affinity defining motif [13]. However, the extent of promiscuity and interaction overlap between competing SH2 domains and phosphopeptides remains poorly resolved biophysically. Understanding the relative preferences between competing domains for phosphorylation sites—specificity space—improves our biochemical understanding of signaling networks by providing greater biophysical resolution than is achievable from in vivo interaction mapping approaches such as affinity purification mass spectrometry and binary in vitro methods (Fig. 1). The pursuit

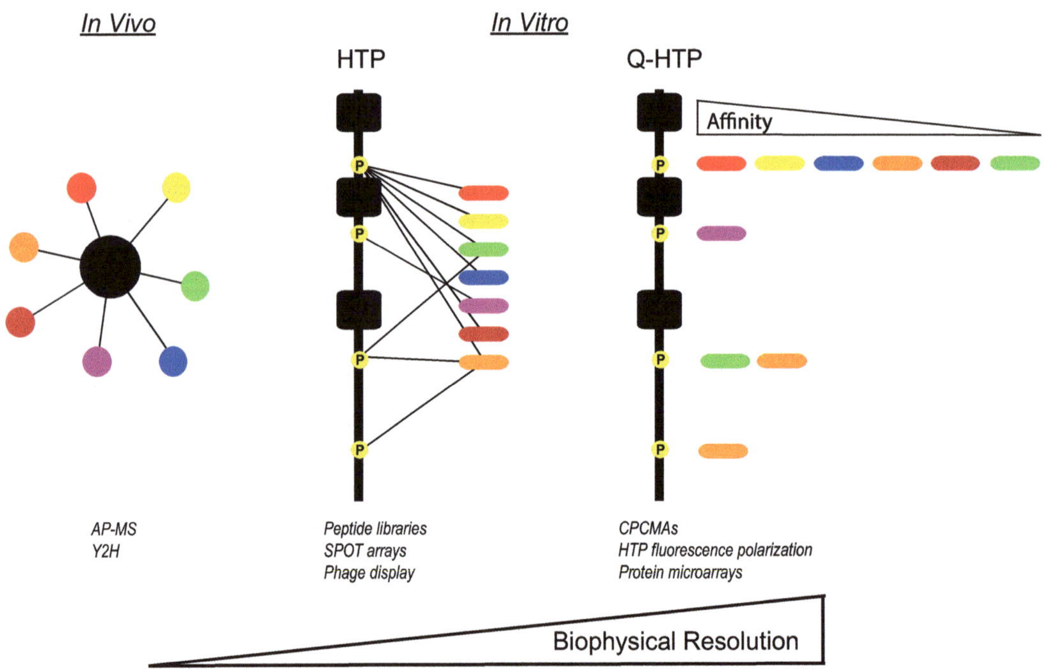

Fig. 1 The biophysical resolution of SH2 domain–phosphopeptide interactions improves with increasingly quantitative in vitro approaches. From left to right, the information recovered from HTP experiments in order of increasing biophysical resolution. Examples of approaches that fit into each category are given. In vivo approaches provide a list of functionally relevant interacting proteins, binary in vitro approaches provide a list of interacting proteins with binding site resolution, while quantitative in vitro approaches provide a ranked list of interacting proteins with binding site resolution. A multi-phosphorylated hub protein is shown in *black*, with sites of tyrosine phosphorylation represented by *yellow circles* with the letter "P" circumscribed. *Colored circles* and *ovals* represent SH2 domains. HTP, high throughput; Q-HTP, quantitative high throughput

of this high resolution binding site information motivates the development of in vitro high-throughput quantitative interaction mapping efforts [14].

Recent technical developments have enabled the economic miniaturization of the SPOT synthesis [15] technique to produce hundreds of microarrays from a single synthesis run, referred to as spotting compound conjugates (SC^2) [16]. The adsorbed material is a conjugate of inert, low-fouling cellulose and a peptide ligand. The "gentle" printing process employed is advantageous because it does not require chemistries that alter peptide activity and provides a three-dimensional scaffolding that increases the amount of ligand per unit area relative to monolayers [16–18]. In addition, recent interfacial kinetic experiments on microarrays have highlighted the important role of controlling for mass action in order to achieve accurate estimates of affinity [19, 20]. I have integrated these findings to develop a carefully controlled, sensitive, and reproducible platform to estimate interaction affinities in high throughput [21].

Here, I describe in detail each step in my approach for the development and application of our CPCMA platform (Fig. 2). Included in this protocol are important peptide library, array design, and image capture considerations related to the precision and accuracy of the final affinity estimates. Multiple control experiments derived from these considerations are also provided. These experiments address multiple types of "nonspecific" binding and systematic sources of experimental error, both of which are common in high-throughput contexts (Fig. 2b) [22–25]. Ultimately, the binding isotherm produced by a titration of 12 concentrations of glutathione S-transferase (GST) fused SH2 domains is fitted to a mass action derived model (Fig. 2c, Eq. 1). Assuming signals have reached equilibrium and conditions such that for each SH2 domain i at every applied concentration $j = 1,2,\ldots 12, [SH2]_j - [SH2:CPC] \approx [SH2]_j$, for each qualifying cellulose–peptide conjugate (CPC) $n = 1,2,\ldots$, the mass action derived function θ is fitted:

$$\theta_n([SH2_i]) = \frac{SI_{i,n}^{\max} \cdot [SH2_i]_j}{[SH2_i]_j + K_{D_{i,n}}} \quad (1)$$

where, $SIi_{,}nmax$ and the apparent dissociation constant for each SH2-CPC combination, $K_{D_{i,n}}$ are free parameters.

There are many excellent reviews and protocols available for phosphopeptide synthesis [26–29], SPOT synthesis [30–36], and a few excellent protocols available for the SC^2 approach [37, 38], commercialized by Intavis, AG and marketed as Celluspots™. The reader is strongly encouraged to consult these excellent protocols and reviews for more detailed information and troubleshooting on this aspect of the protocol.

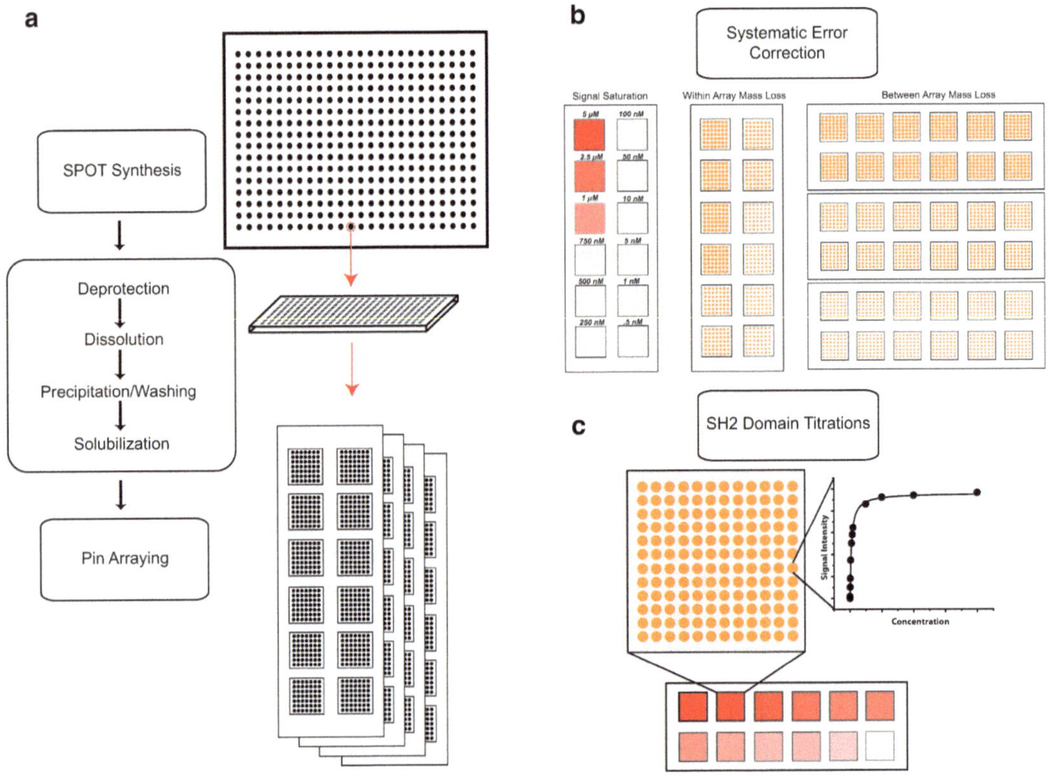

Fig. 2 Overview and schematic representation CPCMA fabrication and application. (**a**) CPCMA fabrication. 11-mer phosphopeptides are synthesized on SPOT macroarrays via Fmoc (Solid Phase Peptide Synthesis, SPPS). The CPC spots are manually removed by hole punch (*dashed red circle* and *arrow*) and placed into microplates where they are simultaneously deprotected and acid-solubilized. Following precipitation, washing and re-solubilization (*red arrow*) they are pin arrayed onto glass substrates with addressable sub-arrays. Each sub-array serves as an independent reaction vessel for each bioassay. (**b**) The indicated sources of systematic error are addressed by the experimental control approaches outlined herein and explained in detail elsewhere [21]. (**c**) Application of CPCMAs via SH2 domain titration experiments is depicted. Each CPC-SH2 domain titration is fitted to Eq. (1)

2 Materials

Organic solvents should be peptide synthesis grade or higher (*see* **Note 1**). It is recommended to use HPLC grade water for all synthesis steps. It is also beneficial to have a cylinder of dry Argon (Ar) with inline desiccant/filtration available for displacing air within chemical containers and for drying arrays after incubations.

2.1 SPOT Synthesis

1. MultiPep high-throughput peptide synthesizer (Intavis #32.000).
2. Cellulose membrane (e.g., Intavis #32.105).
3. F-moc amino acids.
4. Phosphotyrosine (Fmoc-Tyr[PO(OBzl)OH]-OH).

5. N-Methylpyrrolidone (NMP).
6. Dimethylformamide (DMF).
7. Dichloromethane (DCM).
8. Capping solution: 2% (v/v) acetic anhydride (Ac_2O) in DMF.
9. Deprotection solution: 20% (v/v) piperidine in NMP.
10. N,N-diisopropylethylamine (DIPEA).
11. 1-hydroxybenzotriazole anhydrous (HOBt).
12. N,N′-diisopropylcarbodiimide (DIC).
13. Triisopropylsilane (TIPS).
14. Trifluoroacetic acid (TFA).
15. Methanol (MeOH).
16. Ethanol 95% (EtoH).
17. Bromophenol blue (BPB).
18. Parafilm®.

2.2 Membrane Deprotection, Dissolution, and Solubilization

1. UV lamp (various manufacturers).
2. Hole-puncher (various manufacturers).
3. Tweezers (various manufacturers).
4. 96-deep well plates with removable wells.
5. Capstrips.
6. Sidechain deprotection cocktail (by volume): 80% TFA, 12% DCM, 4% H_2O, 4% TIPS. Add in this order. Add water slowly and ensure adequate mixing.
7. Dissolving solution (by volume): 93.5% TFA, 2.5% trifluoromethanesulfonic acid (TFMSA), 2.5% TIPS, 1.5% H_2O. Add in this order. Ensure proper mixing (*see* **Note 2**).
8. *Tert*-butyl methyl ether (MTBE).
9. DMSO, extra dry.
10. Compact liquid handling robot.
11. Repeater pipette.
12. Explosion proof centrifuge (various manufacturers).
13. 384-well microplate
14. Teflon® bottle.

2.3 CPCMA Fabrication

1. Pin arraying device (e.g., Digilab Omnigrid 100).
2. Microarraying pins.
3. Sonication bath (various manufacturers).
4. Detergent for sonication cleaning
5. Amine functionalized slides.
6. Canister of compressed air.

2.4 Array Incubations

1. ArrayMix incubation device (Gel Company, #AM24-4).
2. ArraySlide high-volume incubation chamber(s) (Gel Company, #AHS 16-4-HV).
3. PBS: 137 mM NaCl, 2.7 mM KCl, 10 mM Na_2HPO_4, 1.8 mM KH_2PO_4, pH 7.4. For a 10× solution, add 80 g NaCl, 2 g KCl, 14.4 g Na_2HPO_4, and 2.4 g KH_2PO_4 to 800 mL in a 1 L graduated cylinder. Bring volume to 1 L. Dilute to 1× and adjust pH if necessary before use.
4. PBS-T: add Tween 20 to PBS at .1% v/v.
5. PBS-TM: add Dry milk to PBS-T at 2% w/v.
6. PBS-TMD: add dithiothreitol (DTT) to PBS-TM to 1 mM.
7. Slide staining dish and tray (e.g., Ted Pella, Inc. #21054).
8. Aluminum coated sealing film.
9. Access to a cold room or deli fridge with an electrical outlet.
10. Cellulose binding domain—protein L fusion (CBD) (Sigma #59898) diluted 1:1000 in PBS-TM.
11. Monoclonal anti-GST primary antibody diluted 1:1000 in PBS-TM
12. Human Igg diluted 1:1000 in PBS-TM.
13. AF633 conjugated anti-mouse secondary antibody diluted 1:2000 in PBS-TM.
14. AF633 conjugated anti-human secondary antibody diluted 1:2000 in PBS-TM.
15. Highly pure (>95% as judged by SDS-PAGE) SH2 domains produced as GST-fusion proteins.
16. Highly pure (>95% as judged by SDS-PAGE) GST.

2.5 Scanning and Feature Identification

1. Microarray scanner capable of high resolution (<20 μm) scans (e.g., Genepix 4000B, Molecular Devices).
2. Image analysis software for spot finding (e.g., Genepix Pro 7, Molecular Devices).

2.6 Data Analysis

A personal computer with the R statistical computing environment (http://www.R-project.org/) installed is recommended.

3 Methods

CPCMA manufacture requires the use of dangerous solvents and acids. Corrosive acids should be handled with care and stored in a secondary container. Handling of chemicals, including automated handling steps should be performed in a fume hood. Appropriate personal protective equipment should be worn throughout the procedure, and butyl gloves should be worn when handling TFMSA. Organic, caustic, and biohazard waste should be disposed of in accordance with local regulations.

3.1 SPOT Synthesis

1. Produce stock solutions of amino acids. Dissolve 0.5 mmol of each Fmoc Amino Acid in 1 mL of 0.75 M (HOBt/NMP). The solution is then brought up to 1.5 mL, yielding a ~0.333 M amino acid and 0.5 M HOBt stock solution in NMP (*see* **Note 3**). These inactivated stock solutions are aliquoted at 500 μL and stored at −80 °C until used (*see* **Note 4**). After thawing, these stocks solutions are good for ~1 month at −20 °C.
2. Design phosphopeptide panel and SPOT synthesis protocol. The choice of phosphopeptide is experiment specific, as is the length of the peptide synthesized, however for this protocol we assume 11-mer (11 amino acid) peptides are synthesized (*see* **Note 5**). Input the phosphopeptide and control sequences using the sequence editor within the Spotter.exe software suite. Ensure the method is appropriate and uses the proper synthesis scale, peptide number, and pause points (*see* **Note 6**).
3. Navigate to "Show Report" page within the spotter software suite. Make note of the volume of each amino acid required (this is calculated from the phosphopeptide list and protocol design) for each set of couplings, separated by pause points. Thaw enough amino acid stock solution for all required couplings for a synthesis run. For labile or specialty moieties, weigh out the required amount and store at room temperature in a vacuum desiccator protected from light until ready for use.
4. Prepare a 1.1 M DIC solution in NMP by adding 800 μL DIC to 4.2 mL of NMP in a 15 mL polypropylene tube (*see* **Note 7**). This activating solution is good for 1 day and this volume is enough (generally) for one cycle.
5. Combine ¼ parts activating solution with ¾ parts amino acid stock solution such that the final molar ratio of amino acid/HOBt/DIC = 1/1.5/1.1. Preactivate for 30 min. Spin down in tabletop centrifuge at 10 K rcf for 5 min at room temp. Place the preactivated amino acid solutions loosely in the correct positions of the rack in the Multipep.
6. Prepare the capping and deprotection solutions. Prepare enough capping and deprotection solution for first round of synthesis, ensuring enough dead volume for proper aspiration by the needle (*see* **Note 8**).
7. Fill solvent wash bottles with enough DMF and EtOH for membrane washes for the first cycle.
8. Pre-swell the cellulose membrane in nanopure water for at least 15-min. Place the wet membrane, using forceps, dead centered and wrinkle/air-free on the plate. Cross-tighten the

plate evenly. Go to “Manual” and turn Extraction Pump: ON. Wash the membrane with ethanol (3×) and extract the solvent. Dry the last wash for 10-min. Inspect the pump tubing—no droplets should be seen. Dry the membrane completely before the start of synthesis. Turn off extraction pump.

9. Go to “Run Synthesis” and click task 1 start (*see* **Note 9**).
10. The run will pause at the designated pause point. Perform amino acid preactivation (see **step 5**), prepare fresh capping and deprotection solution (see **step 6**), and refill solvent wash containers for second round of synthesis.
11. The run will pause after coupling of the N-terminal amino acid. Prepare fresh deprotection and capping solution, and ensure there is enough washing solvent for the end of the run. Finish the run.
12. The synthesis is now complete. Ensure that the membrane is completely dry, and wash a few more times with EtOH if necessary. If not moving directly on to deprotection and membrane dissolution, place the membrane in a plastic sheath and heat seal the borders (e.g., Thomas Scientific #1303R98). Synthesized membranes can be stored for months at −20 °C in a desiccator cabinet.

3.2 Membrane Deprotection, Dissolution, and Solubilization

1. Ensure that membrane is completely dry. Using UV lamp and pencil, trace a grid on the membrane outlining in rows and columns each peptide spot on the membrane (*see* **Note 10**). Using a grid as a guide, hole-punch each CPC disk into the corresponding well in the deep-well plates (*see* **Note 11**). Punch-remove some cellulose disks with no peptide synthesized to serve as controls in downstream experimentation.
2. Make up sidechain deprotection cocktail in 50 mL polypropylene tube or corrosion-resistant Erlenmeyer flask.
3. Using a repeater pipette, add 150 μL of sidechain deprotection cocktail to each well using a repeater pipette (*see* **Note 12**). Incubate capped for 1.5 h. If spots turn yellow to red, add more TIPS to another aliquot of deprotection solution and add to the affected wells.
4. Remove cleavage solution from each well.
5. Make up dissolving solution in a Teflon® bottle.
6. Add 250 μL of dissolving solution to each well using a repeater pipette (*see* **Note 12**). Incubate capped overnight with light orbital agitation. Chill MTBE at −20 °C overnight.
7. The next day confirm that all disks have been dissolved. Check a representative sample for dissolution by pipetting 10 μL up and down using a gel loading tip. There should be no clogging of the tip.

8. Once peptides are completely dissolved, add 750 μL of *ice cold* MTBE to each well to begin precipitation. Cap all wells and gently vortex plates. Let plates cool down at −20 or −80 °C for 1 h or leave overnight if desired.
9. Ensure plates are covered and spin at 1500 × *g*, 4 °C for 10 min (*see* **Note 13**).
10. Remove 500 μL from each tube, being careful to remove volume from the meniscus (*see* **Note 14**).
11. Wash three more times as above with 500 μL ice cold MTBE. Vortex plate after every addition of MTBEMTBE.
12. After final wash, manually decant each well and allow gel-like pellets to dry, uncovered for 10 min (*see* **Note 15**).
13. After allowing pellets to dry, add 100 μL of extra DMSO to each well. Vortex each plate and verify solubilization of CPC pellets.
14. Gently spin down plates at 500 × *g* at room temperature. Cap then Flash or flush freeze plates and store in desiccator cabinet at −20 °C until ready for use (*see* **Note 16**).

3.3 Microarray Fabrication

1. Transfer CPCs to 384 microplate for printing. Aspirate an appropriate volume of dry DMSO immediately before transfer. Add 2.5 μL of DMSO to each microplate well, followed by 2.5 μL of each CPC. Add 5 μL of DMSO to multiple wells of a separate 384-well plate. This will be used to "prime" the pin heads before array manufacture. After addition of CPCs cover the plate, gently vortex and then spin down at 500 × *g* for 2 min.
2. Program your print run for SH2 domain titration arrays (Fig. 3a, *see* **Note 17**).
3. Prepare the arrayer for printing. Ensure arraying pallet is clear of debris by spraying surface with compressed air. Fill all wash basins with H_2O and ensure sonication bath is filled to the appropriate level for pin cleaning during the run. Insert printing pin(s) into printhead (*see* **Note 18**).
4. Using the microplate containing only DMSO, prime the pins by printing a few arrays using low cost slides (these can be reused). Ensure that liquid is being deposited on the slide, and deposited volume is consistent across the desired number of repeated depositions before returning to the microplate for more volume. Ensure that wash and drying steps are efficient and that the pins return to "home" position after the print run.
5. Load the arraying palette with the desired number of slides to be printed and run the printing protocol (*see* **Note 19**). Ensure the successful printing of the first few CPCs.
6. After the print run has completed, remove each slide from the pallet and place the printed arrays in a slide rack (*see* **Note 20**). Place rack in desiccator cabinet at 4 °C until ready for later use. Slides can be stored for up to 6 months before use.

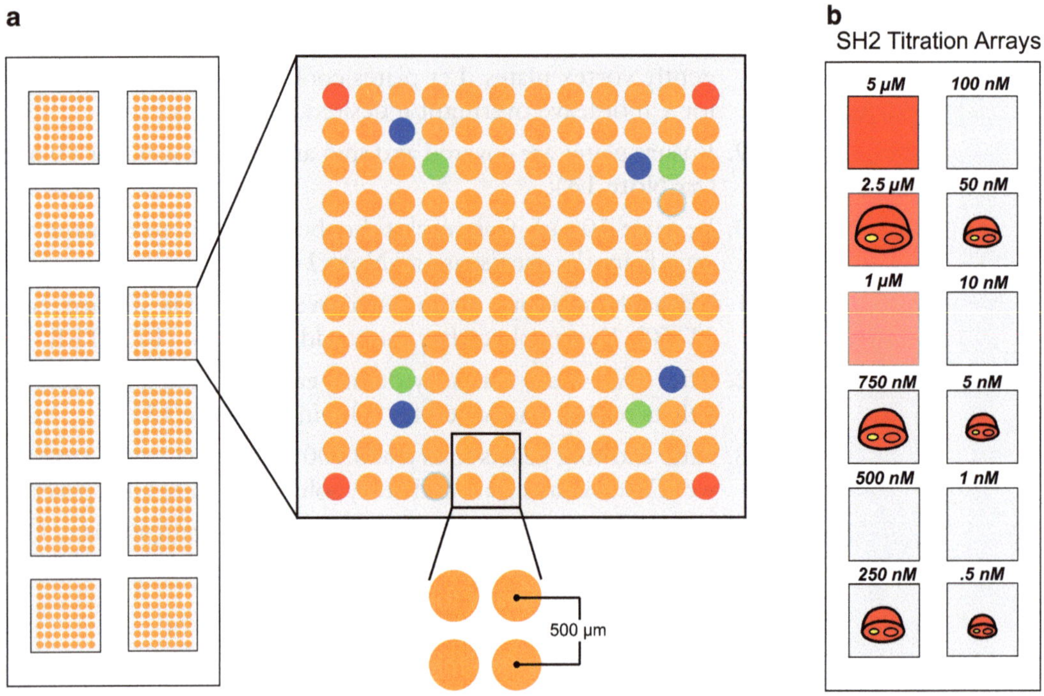

Fig. 3 CPCMA layout and SH2 domain titration design. (**a**) CPCMA microarray with inset of individual well with 12 × 12 array of CPCs. *Red*, *blue*, and *green circles* represent cellulose alone as well as replicated CPC controls for precision assessment. (**b**) SH2 domain titration design with applied concentrations indicated

3.4 Array Control Experiments

3.4.1 Assessment of Background Binding

Nonspecific background is assessed by applying GST and anti-GST antibody. All incubations are performed with the ArrayMix™ and ArraySlide™ apparatus. All incubations are performed at 300 rpm and at room temperature unless otherwise indicated. It is recommended to use filtered solutions for microarray incubations. Liquid removal during wash steps is best achieved by forcibly inverting the ArraySlide™ and blotting the surface dry with an absorbent material such as a paper towel or lint-free wipe. A repeater pipette greatly facilitates liquid handling steps.

1. Remove the slide storage box from the refrigerator or freezer and allow it to equilibrate to room temperature before opening. Remove four slides and load into the four slots on the ArraySlide™ device (*see* **Note 21**). Place well/gaskets cover over the arrays and lightly tighten the screws.
2. Add 500 μL H_2O to each well. Wash the arrays for 10 min. Repeat once.
3. Add 500 μL PBS to each well. Wash the arrays for 5 min. Repeat once.
4. Add 500 μL PBS-T to each well. Wash the arrays for 5 min. Repeat once.

5. Add 250 μL PBS-TMD to each well of the array. Cover the plate with aluminum adhesive film to prevent evaporation. Block the arrays for 1.5 h.
6. Add 250 μL PBS-T to each well. Wash the arrays for 5 min. Repeat twice.
7. If assessing detection antibody background binding proceed directly to **step 9**. For GST nonspecific binding assessment, add 100 μL of 1 μM GST in PBS-TMD to each well of the array (*see* **Note 21**). Cover the plate with aluminum adhesive film to prevent evaporation. Incubate primary analyte for 20 h at 4 °C (*see* **Note 22**).
8. Add 250 μL PBS-T to each well. Wash the arrays for 5 min. Repeat twice.
9. Add 100 μL of Anti-GST to each well of the array. Cover the plate with aluminum adhesive film to prevent evaporation. Incubate primary analyte for 5 h at room temp.
10. Add 250 μL PBS-T to each well. Wash the arrays for 5 min. Repeat twice.
11. Add 100 μL of Anti-GST to each well of the array. Cover the plate with aluminum adhesive film to prevent evaporation. Incubate primary antibody for 5 h at room temp.
12. Add 250 μL PBS-T to each well. Wash the arrays for 5 min. Repeat twice.
13. Add 100 μL of AF633 conjugated anti-mouse secondary antibody to each well of the array. Cover the plate with aluminum adhesive film to prevent evaporation. Incubate secondary antibody for 2.5 h at room temp.
14. Add 250 μL PBS-T to each well. Wash the arrays for 5 min. Repeat twice.
15. Add 500 μL PBS to each well. Wash the arrays for 5 min. Repeat once.
16. Add 500 μL H_2O to each well. Wash the arrays for 5 min. Repeat once.
17. Remove slides from incubation device and place into slide staining tray. Gently immerse the slides into a water bath and cover with aluminum foil. Gently rock the slides for 5 min. Remove the slides one at a time from the rack and remove water droplets using a stream of dry Ar (*see* **Note 23**).
18. Place slides in a slide box and store in a desiccator at 4 °C until scanned. Slides should be scanned within 1 week of arraying.

3.4.2 Assessment of Mass Loss Within Array

Mass loss *within* an array (a source of systematic error) can be assessed by applying the CBD analyte, which has affinity towards the cellulose matrix (*see* **Note 24**). At least three replicate arrays should be used for this purpose, printed in the same format used in the SH2 domain titration experiments. This set of control experiments should be performed as outlined in Subheading 3.3, but with the following changes:

1. The analyte used is CBD.
2. The primary detection antibody is replaced with Human Igg.
3. The secondary detection antibody is AF633 conjugated anti-human secondary antibody.

3.5 SH2 Domain Titrations

The primary experiment is a titration of GST-SH2 domains applied across replicate arrays from 0.5 nM to 5 μM (Fig. 3b). A common protein stock should be used for replicate titration arrays (*see* **Note 25**). SH2 domain titration experiments should be performed as outlined in Subheading 3.3, but with the following changes:

1. The analytes used are GST fusions of SH2 domains, diluted from stocks to the appropriate concentrations in PBS-TMD. The concentration of SH2 domain added to each well on the array is outlined in Fig. 3b.

3.6 Array Scanning and Feature Identification

1. Remove arrayed CPCMAs from desiccator and allow box to equilibrate to room temperature before opening.
2. Turn on scanner and allow laser to warm up (10 min).
3. Start the GenePix Pro software. Insert slide and scan at maximum laser power and resolution. Scan the slide at least three times. Each scan is done with a different gain or photomultiplier tube setting. Start with a photomultiplier tube voltage such that no pixels are saturated. Additional scans can be performed in steps of 100 V (*see* **Note 26**).
4. Identify the features within each block on the array with static circles. Create a 2 by 6 grid with 12 blocks. Within each block create a 12 by 12 array of 144, 300 μm circular features (*see* **Note 27**). Find the features within each block of the array. Score the array and export the 635 channel results to a text file. Repeat these steps using the same exact features for each image of the array.
5. Within each block, align 20, 300 μm circular features along the periphery of each side of the well (*see* **Note 28**). Score the array and export the 635 channel results to a text file. Repeat these steps using the same exact features for each image of the array.
6. Identify the features within each block on the array with adaptive circles (*see* **Note 29**). Create a 2 by 6 grid with 12 blocks. Within each block create a 12 by 12 array of 144, 200 μm circular features. Find the features within each block of the array, allowing the circle to expand or contract depending on the size

of the feature. Score the array and export the 635 channel results to a text file. Repeat these steps using the same exact features for each image of the array.

3.7 Data Analysis

A more in depth description of the data analysis pipeline, including formal definitions of the statistical models used to correct for scanner bias and systematic errors and a detailed information based classification scheme for interactions and peptides can be found in a recent report [21].

1. Using the R package "aroma light," combine the multiple fluorophore estimates for a single feature on each array into a single, bias-free fluorophore estimate.
2. Using a linear model and the *rlm* function within the R package "MASS," subtract the between-array effect (estimated from replicate arrays, caused by mass loss between arrays) and within-array effect (estimated by CBD experiments, caused by mass loss within an array).
3. Subtract the background signal (average value per well) from each feature. For SH2-CPC interactions that pass the signal to noise threshold (use the adaptive circle data for this), fit the titration derived signals from each replicate array to Eq. (1) (*see* **Note 30**).

4 Notes

1. It is generally recommended that all solvents, especially those used for membrane dissolution and resolubilization be kept away from water. To keep solvents as pure as possible, displace the air from the void of open containers after using them with dry Ar, and seal containers with Parafilm®. Hygroscopic solvents, such as TFA should be used within 6 months to prevent inconsistencies between runs. Especially hygroscopic materials such as DIC should be stored in a desiccator cabinet or vacuum desiccator.
2. TFMSA is extremely caustic! Use butyl gloves when opening container and handling neat acid. Always store in secondary container. Yellow color may be observed when cocktail is produced. This color will disappear upon addition to tubes.
3. NMP used for making AA stocks should be treated with molecular sieves to guard against amine and H_2O contamination. The solvent should be stored under Ar and the cap Parafilm™ wrapped. Before using, it should be tested for free amines with bromophenol blue (BPB) testing solution (1 % w/v in DMF). Add 10 μL BPB testing solution to 1 mL NMP. If the solution turns yellow then proceed. If not discard NMP and dry a new batch under sieves for a sufficient period.

4. Some amino acids, such as Arg and pTyr are not stable for long term storage and should be made up immediately before use. Oxidizable amino acids such as Cys and Met should also be made up immediately before use. However I strongly recommend replacing Cys with Ser and Met with Norleucine (Nle) to avoid such issues [39].
5. We have found that an 11-amino acid phosphopeptide with the pTyr in the fifth position relative to the N-terminus is capable of capturing the selectivity preferences of SH2 domains, which are generally defined within 1–5 residues C-terminal to the central pY and 1–3 residues N-terminal to the central pY. In addition, this length is compatible with good performance in biophysical assays such as fluorescence polarization (FP), an important consideration if such an assay is to be used for orthogonal corroboration or quantitative benchmarking. It is strongly advised to include a small panel of non-phosphorylated analog peptides and cellulose alone (no synthesized peptide) spots to serve as controls. It is also strongly advised to synthesize a set of duplicate phosphopeptides known to interact with a subset of the domains of interest. These phosphopeptides will eventually be printed in the microarray format at nonadjacent positions of the print block in order to assess and potentially control for spatial contributions to signal intensity (Fig. 3a).
6. The repetitive couplings of SPOT synthesis along with best practices have been covered in detail previously (see the introduction for SPOT synthesis references). We have found the most natural pause points for phosphopeptide synthesis to be just before the addition of the pY residue, and after the addition of the N terminal amino acid. This way a freshly prepared pY stock and capping solution is used for pY coupling and acetylation, respectively.
7. DIC is extremely hygroscopic and readily breaks down to form urea. If crystals are seen discard solution and use a fresh container of DIC.
8. DMF and NMP solvents used here should be stored under Ar with sieves. They should be tested for amine contamination as outlined in **Note 3**. Ac_2O is very reactive with water and should be stored under Ar in aliquots with organic acids. DIPEA can be added to the capping solution at 2% (v/v). DIPEA is water and air sensitive much like DIC and should be handled appropriately.
9. Observe cycle 1 (at least up to the first wash, after the deprotection with piperidine solution). Observe if the solvent is extracted 5–10 s before the vacuum pump turns off. Make sure the vacuum release sensor is off 5–10 s before the preset time.

10. It is a good idea to label columns alphabetically and rows numerically with your pencil to provide a reference point during hole-punching. A spreadsheet linking row/column position from the membrane to row/column position within the 96-well plate is also strongly advised. In addition, you will eventually be transferring material to 384-well plates for microarray manufacture, so a third set of peptide coordinates should be indicated, ideally in the same spreadsheet for integrated sample handling and troubleshooting.
11. This step can be tedious, but provides the operator with more flexibility in terms of throughput and synthesis yield per run, which are inversely proportional and proportional to spot size, respectively. A larger or smaller hole-puncher can be used depending on the application. Alternatively, there are single spot size multi-hole punch devices (Intavis) available to facilitate this process.
12. If a pipette model not resistant to corrosion is used, it is recommended to replace the syringe after every plate. In order to prevent equipment damage, only positive displacement pipettes should be used for acid pipetting.
13. MTBE is highly volatile and flammable substance! An explosion proof centrifuge is recommended for spins with ether and no open flames should be in the vicinity.
14. A liquid handling robot greatly facilitates wash steps. Ensure volume is being removed at the meniscus in order to not perturb the partition. All handling steps should be performed in a chemical fume hood.
15. It is important to not wait too long before adding DMSO. Pellets allowed to dry for too long will become insoluble. However, excess ether will negatively impact printing. It is therefore advised to allow the wells access to ambient air for another 10 min after the addition of DMSO.
16. If multiple print runs are anticipated, now is the ideal time to aliquot sample into 384 or other formats. We commonly aliquot multiple 25 μL from each well into 3, 384 "source plates" for later use. Ensure plates are covered to prevent evaporation. Try to allow some extra time for residual MTBE to flash off before aliquoting (*see* **Note 15**). Again, a liquid handling robot greatly facilities this process.
17. In this configuration two columns of eight blocks are used. For amine coated slides with a split pin (SMP3) configuration the following (albeit rather conservative in respect to number of spots block) layout has produced consistent results. Within each block, 12 rows and 12 columns (12 × 12) with a spot-to-spot center distance of 500 μm are used (Fig. 3a). The 144 CPCs per block are to be centered in the middle of each 7 × 7 mm

gasket within the ArraySlide™ incubation device. For initial calibration, it is advised to use a colored dye or similar solution in combination with a microscope slide coated with an absorbent material, such as filter paper to outline the print boundaries and ensure proper alignment with the ArraySlide™ gaskets.

18. Be mindful to not handle pin tip directly, a magnetic pen-like tool can help with this. Always wear nitrile gloves. Ensure pins are periodically cleaned by sonication. Kits are available for this purpose.
19. It is highly recommended to print more arrays than you think you will need per batch because batch to batch print variability can be a significant source of systematic error. Be mindful that enough material is deposited in each well of the microplate. If large batches are to be printed, 10 μL of CPC/DMSO may be required. It is also prudent to be mindful of CPC precipitation across the print run. This effect can be observed as irregular spot morphology in downstream analysis. Prints can be performed more quickly by using multiple pins or by printing smaller batches of arrays.
20. It is a good idea to remove dust or any other potentially accumulated debris from the slides at this time with a dry source of inert gas, such as Ar. When placing the slides in the rack, place them in such a way that the print number and orientation is apparent.
21. For the assessment of background binding two replicate arrays for each assay are usually sufficient. Note that alternative array designs can be incorporated specifically for these control experiments, where all peptides from a library larger than 144 can be printed on the same array. These arrays can also be used to remove nonspecific binding peptides from future print batches. It is useful to separate fusion tag derived background binding from detection antibody derived background binding for future troubleshooting. For example if a "sticky" peptide is especially important biologically, one can alter the expression tag or the detection antibodies used given the binding signature. However, background binding resulting from the application of GST followed by the detection antibodies is sufficient to remove nonspecific interactions from downstream analysis.
22. The timing for incubation is a critical step. The mass action derived model that will be used to fit the binding isotherms assumes signals have reached equilibrium. The time until equilibrium is reached is dependent on a number of factors such as mass transport, surface type, protein diffusion coefficient, and temperature. Pilot arrays experiments should be performed to identify the optimal incubation times for a given experimental design.

23. This step helps prevent streaking on the array. When drying the arrays, take care to not spray droplets from one well into a neighboring well. Always "push" the droplets towards the periphery of the slide.
24. It is also possible to use a pan-specific antibody for this purpose. The advantage of this choice is that the antibody can verify successful synthesis of the peptide, at least up to the point of the pY residue. However, I have noticed that pan-specific antibodies themselves have their own specificities depending on the amino acid context surrounding the pY, oftentimes not recognizing successfully synthesized peptides. I have found a similar phenomenon to be at play with biotin-streptavidin (where biotin is added to the N-terminus of the peptide) mediated quality control. These problems limit the number of features that can be identified on an array, and therefore limit the statistical power of the resulting corrections. Therefore quality control—either for synthesis or to correct for systematic artifacts—using pan-specific antibodies or a biotin "tag" is not recommended. The CBD has an equal affinity to the common cellulose matrix regardless of the particular synthesized peptide. This makes the CBD ideal for the control of printing artifacts and a flexible quality control analyte for mass deposition in any conceivable cellulose conjugate microarray context.
25. It is recommended to use arrays that were printed in succession from the same batch to minimize differences in deposited mass across arrays. Three replicate arrays are recommended. These arrays should be processed at the same time with the same stock of SH2 protein. If orthogonal biophysical corroboration will be performed it is highly recommended that the same protein stock is used for this assay as well.
26. Because we use a broad range of SH2 domain concentrations, and given the heterogeneity in parameters such as protein background binding, microarray surface characteristics and peptide binding affinity it is not possible to scan all arrays across all SH2 domains at the same voltage *without* having saturated pixels. Moreover, each scan voltage imparts its own characteristic bias into the estimation of fluorophore content. Multiple scans allow for the correction of this bias, and the estimation of a high confidence estimate of fluorophore quantity on a common scale across replicates [40].
27. Ensure that the size of the circle used is large enough to circumscribe the largest feature on the array. This file will be used for quantification, and is robust to slight changes in morphology and/or signal maturation across the concentration gradient.
28. These features will be used for background subtraction. This is done to avoid inaccuracies in background call due to "cometing,"

a common artifact caused by residual analyte adsorbed to the background regions adjacent to high-signal spots of the array.

29. The large circle sometimes underestimates the signal to noise ratio for small features. This file will be processed in parallel and used solely for signal to noise calculations.

30. The K_D parameter should be fitted in logarithmic form so that the resulting error estimate is normally distributed, thereby facilitating downstream statistical analysis [41]. It is best to fit Eq. (1) after taking the logarithm of both sides of the equation. This transformation corrects for common nonuniform variance in the titration data [42].

References

1. Lim WA, Pawson T (2010) Phosphotyrosine signaling: evolving a new cellular communication system. Cell 142(5):661–667. doi:10.1016/j.cell.2010.08.023
2. Liu BA, Engelmann BW, Nash PD (2012) The language of SH2 domain interactions defines phosphotyrosine-mediated signal transduction. FEBS Lett 586(17):2597–2605. doi:10.1016/j.febslet.2012.04.054
3. Liu BA, Shah E, Jablonowski K, Stergachis A, Engelmann B, Nash PD (2011) The SH2 domain-containing proteins in 21 species establish the provenance and scope of phosphotyrosine signaling in eukaryotes. Sci Signal 4(202):ra83. doi:10.1126/scisignal.2002105
4. Hornbeck PV, Kornhauser JM, Tkachev S, Zhang B, Skrzypek E, Murray B, Latham V, Sullivan M (2012) PhosphoSitePlus: a comprehensive resource for investigating the structure and function of experimentally determined post-translational modifications in man and mouse. Nucleic Acids Res 40(Database issue):D261–D270. doi:10.1093/nar/gkr1122
5. Liu BA, Jablonowski K, Raina M, Arce M, Pawson T, Nash PD (2006) The human and mouse complement of SH2 domain proteins-establishing the boundaries of phosphotyrosine signaling. Mol Cell 22(6):851–868. doi:10.1016/j.molcel.2006.06.001
6. Liu BA, Engelmann BW, Jablonowski K, Higginbotham K, Stergachis AB, Nash PD (2012) SRC homology 2 domain binding sites in insulin, IGF-1 and FGF receptor mediated signaling networks reveal an extensive potential interactome. Cell Commun Signal 10(1):27. doi:10.1186/1478-811X-10-27
7. Hlavacek WS, Faeder JR, Blinov ML, Perelson AS, Goldstein B (2003) The complexity of complexes in signal transduction. Biotechnol Bioeng 84(7):783–794
8. Pawson T, Warner N (2007) Oncogenic re-wiring of cellular signaling pathways. Oncogene 26(9):1268–1275. doi:10.1038/sj.onc.1210255
9. Gruenheid S, DeVinney R, Bladt F, Goosney D, Gelkop S, Gish GD, Pawson T, Finlay BB (2001) Enteropathogenic E. coli Tir binds Nck to initiate actin pedestal formation in host cells. Nat Cell Biol 3(9):856–859. doi:10.1038/ncb0901-856, ncb0901-856 [pii]
10. Howard PL, Chia MC, Del Rizzo S, Liu FF, Pawson T (2003) Redirecting tyrosine kinase signaling to an apoptotic caspase pathway through chimeric adaptor proteins. Proc Natl Acad Sci U S A 100(20):11267–11272. doi:10.1073/pnas.1934711100, 1934711100 [pii]
11. Puntervoll P, Linding R, Gemund C, Chabanis-Davidson S, Mattingsdal M, Cameron S, Martin DM, Ausiello G, Brannetti B, Costantini A, Ferre F, Maselli V, Via A, Cesareni G, Diella F, Superti-Furga G, Wyrwicz L, Ramu C, McGuigan C, Gudavalli R, Letunic I, Bork P, Rychlewski L, Kuster B, Helmer-Citterich M, Hunter WN, Aasland R, Gibson TJ (2003) ELM server: a new resource for investigating short functional sites in modular eukaryotic proteins. Nucleic Acids Res 31(13):3625–3630
12. Songyang Z, Cantley LC (1995) SH2 domain specificity determination using oriented phosphopeptide library. Methods Enzymol 254:523–535
13. Liu BA, Jablonowski K, Shah EE, Engelmann BW, Jones RB, Nash PD (2010) SH2 domains recognize contextual peptide sequence information to determine selectivity. Mol Cell

Proteomics 9(11):2391–2404. doi:10.1074/mcp.M110.001586, doi:M110.001586 [pii]

14. Liu BA, Engelmann BW, Nash PD (2012) High-throughput analysis of peptide-binding modules. Proteomics 12(10):1527–1546. doi:10.1002/pmic.201100599
15. Frank R (1992) Spot-synthesis: an easy technique for the positionally addressable, parallel chemical synthesis on a membrane support. Tetrahedron 48(42):9217–9232. doi:10.1016/S0040-4020(01)85612-X
16. Dikmans A, Beutling U, Schmeisser E, Thiele S, Frank R (2006) SC2: a novel process for manufacturing multipurpose high-density chemical microarrays. QSAR Comb Sci 25(11):1069–1080. doi:10.1002/qsar.200640130
17. Zubtsov DA, Savvateeva EN, Rubina AY, Pan'kov SV, Konovalova EV, Moiseeva OV, Chechetkin VR, Zasedatelev AS (2007) Comparison of surface and hydrogel-based protein microchips. Anal Biochem 368(2):205–213. doi:10.1016/j.ab.2007.04.040
18. Mori T, Yamanouchi G, Han X, Inoue Y, Shigaki S, Yamaji T, Sonoda T, Yasui K, Hayashi H, Niidome T, Katayama Y (2009) Signal-to-noise ratio improvement of peptide microarrays by using hyperbranched-polymer materials. J Appl Phys 105(10):102020
19. Kusnezow W, Syagailo YV, Ruffer S, Baudenstiel N, Gauer C, Hoheisel JD, Wild D, Goychuk I (2006) Optimal design of microarray immunoassays to compensate for kinetic limitations: theory and experiment. Mol Cell Proteomics 5(9):1681–1696. doi:10.1074/mcp.T500035-MCP200
20. Kusnezow W, Syagailo YV, Ruffer S, Klenin K, Sebald W, Hoheisel JD, Gauer C, Goychuk I (2006) Kinetics of antigen binding to antibody microspots: strong limitation by mass transport to the surface. Proteomics 6(3):794–803. doi:10.1002/pmic.200500149
21. Engelmann BW, Kim Y, Wang M, Peters B, Rock RS, Nash PD (2014) The development and application of a quantitative peptide microarray based approach to protein interaction domain specificity space. Mol Cell Proteomics 13(12):3647–3662. doi:10.1074/mcp.O114.038695
22. Yang YH, Dudoit S, Luu P, Lin DM, Peng V, Ngai J, Speed TP (2002) Normalization for cDNA microarray data: a robust composite method addressing single and multiple slide systematic variation. Nucleic Acids Res 30(4):e15
23. Quackenbush J (2002) Microarray data normalization and transformation. Nat Genet 32(Suppl):496–501. doi:10.1038/ng1032
24. Smyth GK, Speed T (2003) Normalization of cDNA microarray data. Methods 31(4):265–273
25. Sboner A, Karpikov A, Chen G, Smith M, Mattoon D, Freeman-Cook L, Schweitzer B, Gerstein MB (2009) Robust-linear-model normalization to reduce technical variability in functional protein microarrays. J Proteome Res 8(12):5451–5464. doi:10.1021/pr900412k
26. Coin I, Beyermann M, Bienert M (2007) Solid-phase peptide synthesis: from standard procedures to the synthesis of difficult sequences. Nat Protoc 2(12):3247–3256. doi:10.1038/nprot.2007.454
27. Toth GK, Kele Z, Varadi G (2007) Phosphopeptides—chemical synthesis, analysis, outlook and limitations. Curr Org Chem 11(5):409–426. doi:10.2174/138527207780059295
28. Attard T, O'Brien-Simpson N, Reynolds E (2007) Synthesis of phosphopeptides in the Fmoc mode. Int J Pept Res Ther 13(4):447–468. doi:10.1007/s10989-007-9107-y
29. McMurray JS, Coleman DR, Wang W, Campbell ML (2001) The synthesis of phosphopeptides. Biopolymers 60(1):3–31. doi:10.1002/1097-0282(2001)60:1<3::aid-bip1001>3.0.co;2-l
30. Zander N, Gausepohl H (2002) Chemistry of Fmoc peptide synthesis on membranes. In: Koch J, Mahler M (eds) Peptide arrays on membrane supports. Springer lab manuals. Springer, Berlin, Heidelberg, pp 23–39. doi:10.1007/978-3-662-09229-3_2
31. Hilpert K, Winkler DF, Hancock RE (2007) Peptide arrays on cellulose support: SPOT synthesis, a time and cost efficient method for synthesis of large numbers of peptides in a parallel and addressable fashion. Nat Protoc 2(6):1333–1349. doi:10.1038/nprot.2007.160
32. Kramer A, Reineke U, Dong L, Hoffmann B, Hoffmuller U, Winkler D, Volkmer-Engert R, Schneider-Mergener J (1999) Spot synthesis: observations and optimizations. J Pept Res 54(4):319–327
33. Winkler DF, Campbell WD (2008) The spot technique: synthesis and screening of peptide macroarrays on cellulose membranes. Methods Mol Biol 494:47–70. doi:10.1007/978-1-59745-419-3_4
34. Volkmer R (2009) Synthesis and application of peptide arrays: quo vadis SPOT technology. Chembiochem 10(9):1431–1442. doi:10.1002/cbic.200900078
35. Gausepohl H, Behn C (2002) Automated synthesis of solid-phase bound peptides. In: Koch J, Mahler M (eds) Peptide arrays on membrane

supports. Springer lab manuals. Springer, Berlin, Heidelberg, pp 55–68. doi:10.1007/978-3-662-09229-3_4

36. Frank R (2002) The SPOT-synthesis technique. Synthetic peptide arrays on membrane supports—principles and applications. J Immunol Methods 267(1):13–26, doi:S0022175902001370 [pii]

37. Beutling U, Stading K, Stradal T, Frank R (2008) Large-scale analysis of protein-protein interactions using cellulose-bound peptide arrays. Adv Biochem Eng Biotechnol 110:115–152. doi:10.1007/10_2008_096

38. Winkler DF, Hilpert K, Brandt O, Hancock RE (2009) Synthesis of peptide arrays using SPOT-technology and the CelluSpots-method. Methods Mol Biol 570:157–174. doi:10.1007/978-1-60327-394-7_5

39. Roller PP, Otaka A, Nomizu M, Smyth MS, Barchi JJ Jr, Burke TR Jr, Case RD, Wolf G, Shoelson SE (1994) Norleucine as a replacement for methionine in phosphatase-resistant linear and cyclic peptides which bind to p85 SH2 domains. Bioorg Med Chem Lett 4(15):1879–1882. doi:10.1016/S0960-894X(01)80389-9

40. Bengtsson H, Jonsson G, Vallon-Christersson J (2004) Calibration and assessment of channel-specific biases in microarray data with extended dynamical range. BMC Bioinformatics 5:177. doi:10.1186/1471-2105-5-177

41. Motulsky H, Christopoulos A (2004) Fitting models to biological data using linear and nonlinear regression: a practical guide to curve fitting. Oxford University Press, New York

42. Findlay JW, Dillard RF (2007) Appropriate calibration curve fitting in ligand binding assays. AAPS J 9(2):E260–E267. doi:10.1208/aapsj0902029

Chapter 22

SH2 Domains as Affinity Reagents for Phosphotyrosine Protein Enrichment and Proteomic Analysis

Mi Ke, Bizhu Chu, Lin Lin, and Ruijun Tian

Abstract

Dynamic tyrosine phosphorylation is a key molecular modulation for many signal transduction events. Because of their low abundance and dynamic nature in cells, the detection and enrichment of phosphotyrosine proteins has long relied on specific antibodies, such as 4G10 and P-Tyr-100. Another well-established approach for phosphotyrosine proteins recognition and enrichment is by their specific binding domains, such as Src homology 2 (SH2) domains. In this chapter, we describe a typical analytical approach for purifying specific SH2 domains, enriching specific phosphotyrosine proteins from activated cells, mass spectrometry analysis, and related data analysis.

Key words Phosphotyrosine, SH2 domain, Affinity reagent, Signal transduction, Proteomics, LC-MS

1 Introduction

Tyrosine phosphorylation is a key molecular mechanism for regulating cellular signaling [1]. Commonly, cell surface receptors are activated as a result of ligand-induced clustering and tyrosine phosphorylation of the receptor itself. The activation signals are then transmitted through short tyrosine phosphorylated peptide motifs in cytoplasmic domains that engage cytoplasmic targets with specific protein-interaction modules such as the SH2 domain. These cytoplasmic targets include kinases and phosphatases for further phosphorylation and dephosphorylation of a broad spectrum of downstream signaling proteins, such as adaptor proteins for scaffolding protein complexes [2]. The comprehensive characterization of these tyrosine phosphorylation sites and related protein complexes is therefore critical for understanding specific signaling pathways and deciphering related diseases, such as cancer.

Mass spectrometry-based proteomics has been one of the major approaches for systematic characterization of tyrosine phosphorylation on a global scale [3]. Due to their relative low abundance and dynamic

Kazuya Machida and Bernard A. Liu (eds.), *SH2 Domains: Methods and Protocols*, Methods in Molecular Biology, vol. 1555, DOI 10.1007/978-1-4939-6762-9_22, © Springer Science+Business Media LLC 2017

nature, phosphotyrosine proteins usually need to be enriched for the following mass spectrometry analysis. The development of phosphotyrosine-specific antibodies, such as 4G10 and P-Tyr-100, has greatly facilitated such efforts. Rikova et al. performed the largest scale of tyrosine phosphorylation profiling in 41 non-small cell lung cancer (NSCLC) cell lines and 150 NSCLC tumors [4]. More than 4500 tyrosine phosphorylation sites have been successfully identified by mass spectrometry. In a typical procedure of tyrosine phosphoproteomic analysis, total proteins were first extracted from cell and tissue samples and then digested into peptides. Tyrosine phosphorylated peptides were specifically enriched by bead-conjugated P-Tyr-100 antibody followed by LC-MS analysis. Due to its high enrichment efficiency and specificity, this approach has been widely used for tyrosine phosphoproteomic analysis of various signaling pathways [5, 6].

Another approach for enriching and characterizing phosphotyrosine proteins is utilizing specific binding domains such as the SH2 domain. SH2 domains represent the largest family of phosphotyrosine-specific binding domains in the human proteome [1, 7–9]. It is composed by about 100 amino acid residues with a conserved tertiary structure composed of a central core of three to four beta-strands flanked by two alpha-helices. SH2-containing proteins dock onto the phosphorylated tyrosine residues of other proteins using their SH2 domains, usually with high affinity and in a tyrosine phosphorylation-dependent manner. To date, over 120 SH2 domains have been well characterized by various experiments and detailed bioinformatics analysis [8]. Reverse phase protein array (RPPA) is a well-established technology for high-throughput screening of SH2 domain-binding proteins across various samples. Machida et al. successfully cloned and purified more than 70 recombinant human SH2 domains with enough solubility and stability [10]. By overlaying these SH2 domain probes onto protein-lysate arrays, the authors systematically investigated their binding specificities and applied them for phosphotyrosine profiles of oncogene-transformed cell lines. Similarly, synthetic phosphopeptide library array according to predicated and mass spectrometry-identified phosphorylated peptide motif has been extensively applied for characterizing SH2 domain binding specificity and related protein interactome [11–13]. Forward phase arrays in which a library of SH2 domains is immobilized onto solid supports have also been applied for the screening of SH2 domain interactions [14, 15].

As compared with array-based approaches, mass spectrometry offers an unbiased profiling of SH2 domain-binding proteins and accurate characterization of related phosphorylation sites. Blagoev et al. first demonstrated that recombinant SH2 domain could function as an affinity probe with high binding specificity for proteomic analysis of specific signaling pathway [16]. By using growth factor receptor-bound protein 2 (GRB2) SH2 domain as affinity probe and stable isotope labeling by amino acids in cell culture (SILAC)-based quantitative proteomics, the authors successfully characterized the activation of epidermal growth factor receptor (EGFR) signaling and activation-dependent association of 28 signaling proteins. By using purified SH2 domain pulldown and western blot, Machida et al. showed that most of stably expressed SH2 domains could efficiently pull down phosphotyrosine proteins, demonstrating the broad application of SH2 domain-based affinity purification [10]. An inherent limitation of any in vitro affinity purification is the loss of interacting proteins due to their low-abundance and low-affinity. Uezu et al. developed an unnatural amino acid-modified SH2 domain that can covalently trap phosphotyrosine proteins in vivo with UV light [17]. By further introducing subcellular compartment orientation signal and characterizing by mass spectrometry, the authors successfully identified associated phosphotyrosine proteins at different subcellular localizations, including actin-targeted, mitochondria, and membrane.

In this protocol, we employ SH2 domains as an affinity reagent for enrichment of phosphotyrosine proteins from mammalian cells. We used pervanadate stimulation, which inhibits the activity of tyrosine phosphatases, as a mode of phosphotyrosine stimulation. The SH2 domain of GRB2 and its R86M mutant with loss of phosphotyrosine binding were used as an example [18]. Mass spectrometry-based proteomic approach was employed to identify the associated proteins. The workflow is depicted in Fig. 1.

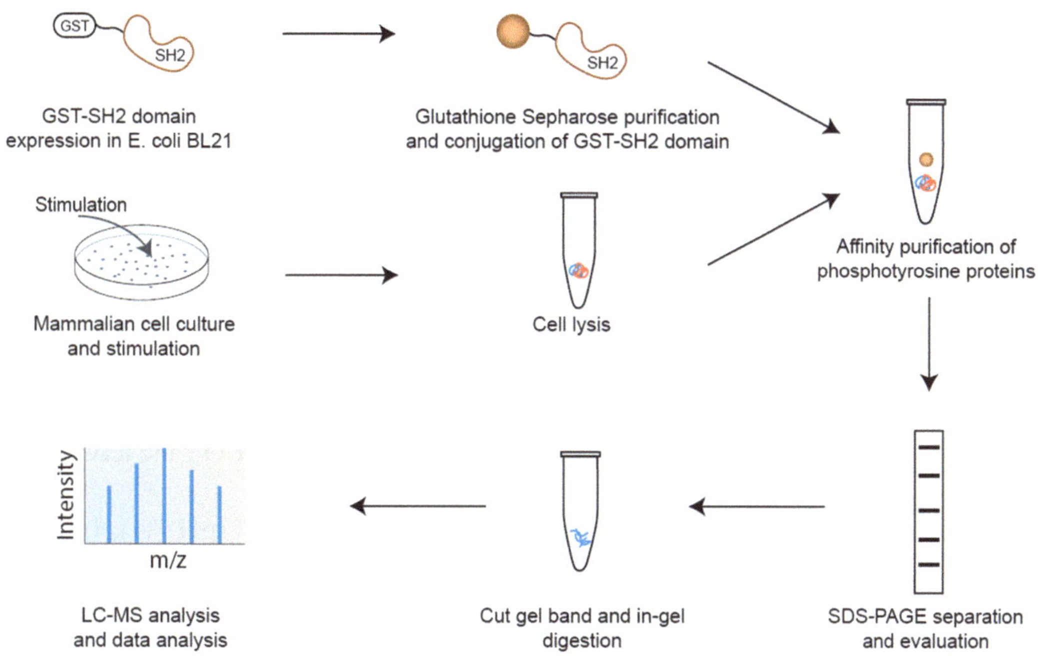

Fig. 1 A diagram of the workflow described in this protocol. GST tagged SH2 domain is expressed in *E. coli* BL21, and the expressed protein is purified and conjugated onto glutathione Sepharose beads. The cells are stimulated and lysed for affinity purification of phosphotyrosine proteins. After that, the enriched proteins are separated by SDS-PAGE. Gel bands are cut and digested by trypsin. The obtained peptide samples are analyzed by LC-MS followed by database searching

2 Materials

2.1 GST-GRB2 SH2 Expression and Conjugation to Glutathione Sepharose 4B

2.1.1 GRB2-SH2 Domain Expression

1. LB medium: 1 % (w/v) tryptone; 0.5 % (w/v) yeast extract; 1 % (w/v) sodium chloride, 1.5 % (w/v) agar. The medium was sterilized at 121 °C, 0.1 MPa, for 15 min.
2. Ampicillin solution: 50 μg/mL of ampicillin.
3. Isopropyl β-D-1-thiogalactopyranoside solution (IPTG): 0.8 M IPTG in water, filter with 0.22 μm water filter, store at 4 °C.
4. 1× Phosphate Buffered Saline (PBS): 0.14 M NaCl, 2.6 mM KCl, 0.01 M Na_2HPO_4, 1.7 mM KH_2PO_4, adjust pH to 7.4 with HCl.
5. Bacteria cell lysis buffer: 1 × PBS, 1 % (v/v) Triton X-100, 3.3 mM benzamidine, 1.5 mM dithiothreitol (DTT), 2 mM ethylene diamine tetraacetic acid (EDTA). Adjust to pH 8.0.
6. Sonicator: JY-96 II N, SCIENTZ.

2.1.2 Conjugation of GST-SH2 Domain to Glutathione Sepharose 4B

1. Glutathione Sepharose 4B.
2. Storage buffer: 40 mM HEPES (pH 7.5), 150 mM NaCl, 1 mM DTT, 1 mM EDTA, 0.01 % (w/v) NaN_3.
3. 5 × SDS protein loading buffer: 0.25 M Tris–HCl (pH 6.8), 10% (w/v) SDS, 50% (v/v) glycerol, 0.5 M DTT (or 5% (w/v) β-mercaptoethanol), 0.1% (w/v) bromophenol blue, store at 4 °C.
4. Gelcode: GelCode blue staining reagent, Thermo Scientific.
5. Protein Ladder: 26617, Thermo Scientific.
6. SDS-PAGE electrophoresis system, Bio-Rad.

2.2 Cell Stimulation and SH2 Domain Pulldown

2.2.1 Cell Culture and Pervanadate Stimulation

1. Culture medium: RPMI medium 1640, 10 % (v/v) fetal bovine serum, 1 % penicillin and streptomycin solution.
2. 75 cm^2 flask.
3. Jurkat T cells in the culture medium, at 37 °C and 5 % CO_2.
4. FBS-free medium: RPMI medium 1640 alone.
5. Pervanadate solution: Prepare 500 μL of 100 mM sodium orthovanadate, add 5 μL of 30 % (w/w) H_2O_2, and leave on ice for 10 min.
6. NP-40 cell lysis buffer: 50 mM Tris–HCl (pH 7.4), 100 mM NaCl, 1 % (v/v) Nonidet P-40 (NP-40), 1 mM sodium orthovanadate (Na_3VO_4).

2.2.2 SH2 Domain Pulldown

1. Phenyl phosphate solution: 100 mM of phenyl phosphate.
2. Protease inhibitors cocktails.

3. Phosphatase inhibitors cocktails.
4. Polyvinylidene difluoride (PVDF) membrane: pore size, 0.45 μm.
5. 4G10 antibody: anti-phosphotyrosine, clone 4G10.
6. Secondary antibody: horse radish peroxidase (HRP)-conjugated goat anti-mouse IgG.
7. ECL western blotting substrate: SuperSignal West Pico Chemiluminescent Substrate kit, Thermo Scientific.

2.3 Mass Spectrometry Sample Preparation and Analysis

2.3.1 In-Gel Digestion

1. Solution A: 50 mM ammonium bicarbonate (ABC, pH 7.8).
2. Solution B: 50% (v/v) acetonitrile (ACN, HPLC grade), 25 mM ABC.
3. Solution C: 10 mM DTT dissolved in 50 mM ABC.
4. Solution D: 100 mM iodoacetamide (IAA) dissolved in 50 mM ABC.
5. Trypsin solution: 0.02 μg/μL of trypsin dissolved in 50 mM ABC.
6. Solution E: 25 mM ABC.
7. Solution F: 5% (v/v) formic acid (FA), 50% (v/v) ACN.
8. SpeedVac: CentriVap® Concentrator, LABCONCO.

2.3.2 StageTip C18 Desalting

1. Methanol (MeOH, HPLC grade).
2. Solution G: 80% (v/v) ACN, 0.5% (v/v) acetic acid (HOAc).
3. Solution H: 1% (v/v) FA.

2.3.3 LC-MS Analysis

1. StageTip C18 is made in house with C18 membrane (3M).
2. Fused silica capillary, 75 μm ID (Polymicro).
3. ReproSil-Pur 120 C18-AQ, 3 μm (Dr. Maisch GmbH, Germany).
4. Mobile phase A: water (HPLC grade) with 0.1% (v/v) FA.
5. Mobile phase B: ACN with 0.1% (v/v) FA.
6. LC-MS analysis is performed on an Orbitrap Fusion Tribrid coupled to an EASY-nano-LC System (Thermo Scientific).
7. A home-made 75 μm × 15 cm capillary tip column packed with 3 μm C18 beads is used as a separation column.

2.3.4 Database Search and Statistical Evaluation

1. Sequest HT algorithm, Thermo Scientific.
2. Proteome Discoverer v1.4.0, Thermo Scientific.

3 Methods

3.1 GRB2-SH2 Domain Expression and Conjugation to Glutathione Sepharose 4B

Expression vectors for the GRB2-SH2 domain are expressed in *E. coli* BL21 upon IPTG induction. The SH2 domains are purified and conjugated to Glutathione beads. The process is confirmed by SDS-PAGE separation and Gelcode staining. The details are described below.

3.1.1 GRB2-SH2 Domain Expression

1. 1 μL (about 0.2 μg) of wild-type GRB2 SH2 domain and GRB2 SH2 domain with R86M mutation plasmids are transformed into BL21 cells, and the cells are plated on LB medium plate with 50 μg/mL ampicillin and incubated at 37 °C for 16 h (*see* **Note 1**).
2. Pick one colony and culture in 2 mL LB medium with 50 μg/mL ampicillin at 37 °C, in a bacteria shaker at 220 rpm, overnight.
3. Transfer 0.5 mL cultured cells to 50 mL fresh LB medium with ampicillin resistance and continue the culture at 37 °C, 220 rpm, for 2–3 h.
4. As OD_{600} of cultured cells reach to 0.6, add 1 mM of IPTG into the medium for induction and culture for another 1.5–3 h (*see* **Note 2**).
5. Collect the cells by centrifugation at 6000 × *g*, 30 min. Wash with 1 × PBS once and centrifuge again under the same condition.
6. Discard the supernatant and resuspend the cell pellet in 2.5 mL bacteria cell lysis buffer. Sonicate 20 s at a power of 400 W for four times with 60 s break on ice in between (*see* **Note 3**).
7. Centrifuge the mixture at 12,000 × *g* for 30 min at 4 °C, and leave the cell lysate on ice (*see* **Note 4**).

3.1.2 Conjugation of GST-SH2 Domain to Glutathione Sepharose 4B

1. Wash 40 μL of Glutathione Sepharose with 1 × PBS for three times.
2. Mix the GRB2-SH2 domain containing cell lysate (Subheading 3.1.1, **step** 7) and the Glutathione Sepharose 4B beads together, and gently rotate for 3 h at 4 °C.
3. Centrifuge at 500 × *g*, 1 min for collecting the beads.
4. Remove supernatant and wash the beads four times with bacteria cell lysis buffer by rotating the tube briefly (*see* **Note 5**).
5. Finally, resuspend the beads with 80 μL storage buffer and store at 4 °C.
6. Take 10 μL of obtained beads suspension, add 2.5 μL of 5 × SDS-PAGE loading buffer, and boil the sample for 5 min at 95 °C.
7. Check protein expression and purification by SDS-PAGE electrophoresis and Gelcode staining. as shown in Fig. 2.

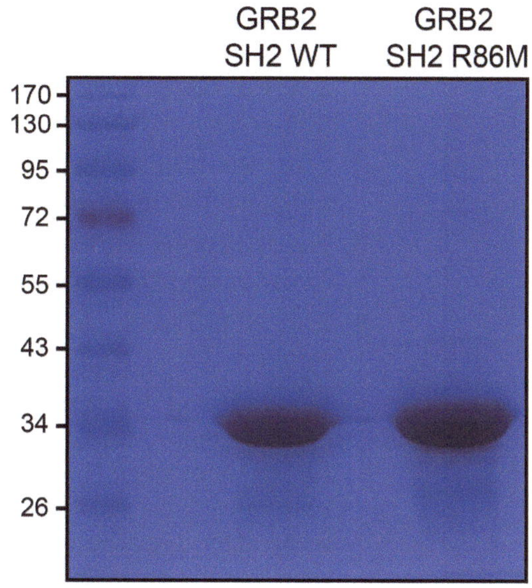

Fig. 2 Glutathione Sepharose 4B pulldown of GST-tagged wild type GRB2 SH2 domain (GRB2 SH2 WT) and GRB2 SH2 domain with R86M mutation (GRB2 SH2 R86M). As shown in the Gelcode stained SDS-PAGE gel, glutathione Sepharose 4B specifically recognized and affinity captured the GST tagged GRB2 SH2 proteins

3.2 Cell Stimulation and SH2 Domain Pulldown

Jurkat T cells with pervanadate stimulation are used as a model system to test the activity and efficiency of the purified GRB2 SH2 domains for enriching phosphotyrosine proteins. The process is confirmed by western blot for visualizing phosphotyrosine proteins. The details are described as below.

3.2.1 Cell Culture and Pervanadate Stimulation

1. Culture two 75 cm^2 flask of Jurkat T cells in the culture medium, at 37 °C and 5 % CO_2.
2. Collect the cultured cells by centrifuge and wash with 1 × PBS twice. Starve the cells with 20 mL FBS-free medium for 2 h at 37 °C (*see* **Note 6**).
3. Prepare 500 μL of pervanadate solution, and keep it on ice for 10 min before use (*see* **Note** 7).
4. For pervanadate stimulation, concentrate the cultured cells in 2 mL FBS-free medium and add 20 μL of pervanadate solution, and pipette up and down briefly. Leave the cells at 37 °C for 5 min.
5. Stop the stimulation by adding 4 mL of ice-cold 1 × PBS and collect the cells by centrifugation at 4 °C.
6. Add 2 mL of NP-40 cell lysis buffer, vortex briefly, and leave it on ice for 30 min to allow complete protein dissolving.
7. Centrifuge the lysate at 12,000 × *g*, 4 °C for 30 min (*see* **Note 4**).

3.2.2 SH2 Domain Pulldown

1. Mix the cell lysate with 10 μL of agarose beads conjugated with GRB2 SH2 domains, and rotate the mixture at 4 °C, for 2 h. For competitive incubation, add phenyl phosphate solution to the mixture (*see* **Note 8**).
2. Wash the beads with 1 mL NP-40 cell lysis buffer for three times, centrifuge at 400 × *g* for 1 min at 4 °C.
3. For denaturing elution, add 5 μL of 5 × protein loading buffer to the beads pellet, and boil the sample at 95 °C, for 5 min.
4. Run SDS-PAGE electrophoresis to separate the enriched proteins.
5. For verifying the enrichment efficiency of the conjugated SH2 domains, transfer the protein bands in the gel onto PVDF membrane. Incubate the membrane with 4G10 antibody and followed by incubation with the secondary antibody. Develop the membrane by developing solution to visualize the enriched phosphotyrosine proteins, as shown in Fig. 3.

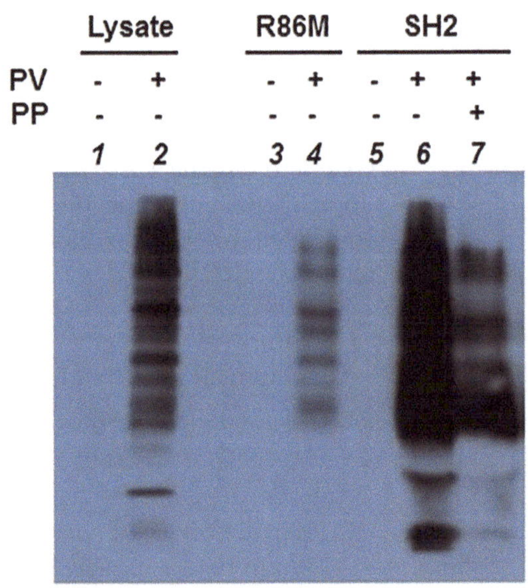

Fig. 3 Conjugated GRB2 SH2 pulldown of phosphotyrosine proteins. The overall tyrosine phosphorylation level is greatly stimulated by pervanadate treatment (*lanes 1* and *2*). GRB2-SH2 domain successfully affinity precipitated and enriched phosphotyrosine proteins (*lanes 5* and *6*) as compared with GRB2 domain with R86M mutation (*lanes 3* and *4*). The enriched phosphotyrosine proteins could be selectively competed by adding phenyl phosphate (*lane 7*). *PV* pervanadate, *PP* phenyl phosphate, *R86M* GRB2 SH2 domain with mutation at arginine 86 to methionine, *SH2* GRB2 SH2 domain

3.3 Mass Spectrometry Sample Preparation and Analysis

For identifying the associated proteins with the conjugated GRB2 SH2 domain, the SDS-PAGE gel from Subheading 3.2.2, **step 4** is processed for mass spectrometry analysis with the following steps.

3.3.1 In-Gel Digestion

1. Stain the SDS-PAGE gel with Gelcode kit.
2. Put the SDS-PAGE gel on a clean glass, and use a new scalpel to cut the gel band into small pieces with size around 1 mm^3. Clean the blade between bands to avoid contamination. Transfer the gel slices into properly labeled eppendorf tubes (*see* **Note 9**).
3. Wash the gel fragments by vortexing briefly in 50 μL of solution A. Centrifuge tubes briefly and discard the liquid.
4. Shrink the gel fragments in 50 μL of solution B for 15 min at room temperature. Fragments should become cloudy.
5. Briefly centrifuge and discard the liquid then dry the gel pieces by Speedvac without heating for 5–10 min.
6. Swell the gel fragments in 50 μL of solution C and incubate at 60 °C for 30 min. Then remove the tubes from hot bath and allow the sample to be cooled down to room temperature.
7. Discard the supernatant and add 30 μL of solution D to the gel pieces and incubate for 30 min in the dark at room temperature.
8. Remove the supernatant and add 100 μL of solution A to wash the gel pieces and discard the liquid.
9. Repeat **steps 4** and **5**.
10. Add 25 μL of trypsin solution to each sample and keep on ice for 10 min (*see* **Note 10**). Incubate the samples at 37 °C overnight. To keep the gel pieces with enough moisture, add an additional 10 μL of solution E.
11. Briefly centrifuge the samples and transfer the supernatant to clean and properly labeled tubes.
12. Wash the gel pieces with 50 μL of solution E and combine solution collected from the previous step.
13. Add 50 μL of solution F to gel pieces and vortex for 15 min at room temperature. Briefly centrifuge and collect supernatant to the solution of previous steps.
14. Speedvac the samples to dryness and store at −20 °C or continue for desalting.

3.3.2 StageTip C18 Desalting

1. Wash the StageTip with C18 membrane with 20 μL of MeOH, 2000 × *g* for 10 s.
2. Discard the liquid and wash the tip with 20 μL of solution G at 2000 × *g* for 10 s.

3. Wash the tip with 20 μL of solution H for twice.
4. Resuspend the sample with 100 μL of solution H, and load sample twice at 400×*g* for 5 min.
5. Wash the tip with 20 μL of solution H twice.
6. Elute the peptides with 40 μL of solution G.
7. Speedvac the sample till dryness and store at −20 °C for LC-MS analysis.

3.3.3 LC-MS Analysis

1. Dissolve the digested peptides in 10 μL of mobile phase A and load 4 μL of sample directly onto the separation column at a flow rate of 400 nL/min. After equilibration with mobile phase A, the sample is separated with the following gradient setting at 200 nL/min: 5 min for 5 % mobile phase B, 60 min for 5–35 % mobile phase B, 3 min for 35–80 % B, 5 min for 80 % B, 2 min for 80 –5 % B, and 15 min for 5 % B.
2. The nano source is operated at 2 kV, with no sheath gas flow, and with the ion transfer tube at 320 °C. Orbitrap Fusion is operated in positive mode, with a data-dependent "Top Speed" mode with 3 s cycle time. Full MS scan from 300 to 1500 m/z is acquired in Orbitrap at a resolution of 120,000 (at $m/z = 200$) with an automated gain control (AGC) target of 200,000, and a max injection time of 50 ms. Precursors were filtered according to charge state (2–6), and monoisotopic peak assignment. Previously interrogated precursors were excluded using a dynamic window (60 s ± 10 ppm). The MS/MS scan is performed in ion trap by collision induced dissociation (CID) with collision energy of 35 %. The max injection time and AGC target were set as 35 ms and 4000, respectively.

3.3.4 Database Search and Statistical Evaluation

1. Raw data is processed and searched using Sequest HT algorithm implemented in Proteome Discoverer v1.4.0 against human FASTA database with trypsin as digestion enzyme (*see* **Note 11**).
2. Sequest HT parameters used to search the MS data are: precursor tolerance, 5 ppm; fragment ion tolerance, 0.6 Da; two missed cleavages; static modification of carbamidomethyl on cysteine; variable modification of oxidation on methionine. Resulting peptide hits were filtered for maximum 1 % false discovery rate (FDR) using the Percolator algorithm.

4 Notes

1. R86M is a critical mutation in the GRB2 SH2 domain binding pocket that disrupts the specific interaction between SH2 domain and phosphotyrosine motif.
2. OD_{600} measurements and IPTG induction conditions are critical for proper protein expression.

3. In order to prevent protein degradation, the cell lysates should be strictly kept on ice.
4. To avoid protein degradation, do not keep lysates at this stage for a long period of time. Continue immediately to the next steps.
5. The affinity of the GST tag to the Glutathione Sepharose 4B is high. Wash the SH2 domain conjugated beads extensively and thoroughly to remove any contamination from the bacteria cell lysate.
6. Cell starvation is a necessary step for reducing the basal level of tyrosine phosphorylation.
7. Activated pervanadate needs to be prepared in fresh.
8. Phenyl phosphate carries with the same chemical structure as phosphotyrosine and can be used to compete for binding between the SH2 domain and tyrosine phosphorylated proteins.
9. In order to avoid keratin contamination, it is recommended to perform the whole in-gel digestion process in a clean hood and wear gloves during the whole process.
10. Make sure the gel pieces are swelled completely in the trypsin solution, if not, add more solution E.
11. Latest protein FASTA database can be downloaded from Uniprot website (http:// www.uniprot.org). The database search can also be performed with Mascot searching engine with the same protein database. Generally, the minimum Mascot score is 30.

Acknowledgment

This work was supported by South University of Science and Technology of China basic research fund to R.T. (FRG-SUSTC1501A-19) and National Natural Science Foundation of China (21575057).

References

1. Sadowski I, Stone JC, Pawson T (1986) A noncatalytic domain conserved among cytoplasmic protein-tyrosine kinases modifies the kinase function and transforming activity of Fujinami sarcoma virus P130gag-fps. Mol Cell Biol 6:4396–4408
2. Seet BT, Dikic I, Zhou MM, Pawson T (2006) Reading protein modifications with interaction domains. Nat Rev Mol Cell Biol 7:473–483
3. Bensimon A, Heck AJ, Aebersold R (2012) Mass spectrometry-based proteomics and network biology. Annu Rev Biochem 81:379–405
4. Rikova K, Guo A, Zeng Q, Possemato A, Yu J, Haack H et al (2007) Global survey of phosphotyrosine signaling identifies oncogenic kinases in lung cancer. Cell 131:1190–1203
5. Wolf-Yadlin A, Hautaniemi S, Lauffenburger DA, White FM (2007) Multiple reaction monitoring for robust quantitative proteomic analysis of cellular signaling networks. Proc Natl Acad Sci U S A 104:5860–5865
6. Jorgensen C, Sherman A, Chen GI, Pasculescu A, Poliakov A, Hsiung M et al (2009) Cell-specific information processing in segregating

populations of Eph receptor ephrin-expressing cells. Science 326:1502–1509
7. Pawson T, Gish GD, Nash P (2001) SH2 domains, interaction modules and cellular wiring. Trends Cell Biol 11:504–511
8. Liu BA, Jablonowski K, Raina M, Arce M, Pawson T, Nash PD (2006) The human and mouse complement of SH2 domain proteins-establishing the boundaries of phosphotyrosine signaling. Mol Cell 22:851–868
9. Liu BA, Shah E, Jablonowski K, Stergachis A, Engelmann B, Nash PD (2011) The SH2 domain-containing proteins in 21 species establish the provenance and scope of phosphotyrosine signaling in eukaryotes. Sci Signal 4:ra83
10. Machida K, Thompson CM, Dierck K, Jablonowski K, Karkkainen S, Liu B et al (2007) High-throughput phosphotyrosine profiling using SH2 domains. Mol Cell 26:899–915
11. Songyang Z, Shoelson SE, Chaudhuri M, Gish G, Pawson T, Haser WG et al (1993) SH2 domains recognize specific phosphopeptide sequences. Cell 72:767–778
12. Tinti M, Kiemer L, Costa S, Miller ML, Sacco F, Olsen JV et al (2013) The SH2 domain interaction landscape. Cell Rep 3:1293–1305
13. Engelmann BW, Kim Y, Wang M, Peters B, Rock RS, Nash PD (2014) The development and application of a quantitative peptide microarray based approach to protein interaction domain specificity space. Mol Cell Proteomics 13:3647–3662
14. Kaushansky A, Allen JE, Gordus A, Stiffler MA, Karp ES, Chang BH et al (2010) Quantifying protein-protein interactions in high throughput using protein domain microarrays. Nat Protoc 5:773–790
15. Liu H, Li L, Voss C, Wang F, Liu J, Li SS (2015) A comprehensive immunoreceptor phosphotyrosine-based signaling network revealed by reciprocal protein-peptide array screening. Mol Cell Proteomics 14:1846–1858
16. Blagoev B, Kratchmarova I, Ong SE, Nielsen M, Foster LJ, Mann M (2003) A proteomics strategy to elucidate functional protein-protein interactions applied to EGF signaling. Nat Biotechnol 21:315–318
17. Uezu A, Okada H, Murakoshi H, del Vescovo CD, Yasuda R, Diviani D et al (2012) Modified SH2 domain to phototrap and identify phosphotyrosine proteins from subcellular sites within cells. Proc Natl Acad Sci U S A 109:E2929–E2938
18. Bisson N, James DA, Ivosev G, Tate SA, Bonner R, Taylor L et al (2011) Selected reaction monitoring mass spectrometry reveals the dynamics of signaling through the GRB2 adaptor. Nat Biotechnol 29:653–658

Chapter 23

Identification of Tyrosine Phosphorylated Proteins by SH2 Domain Affinity Purification and Mass Spectrometry

Sophia Buhs, Helwe Gerull, and Peter Nollau

Abstract

Phosphotyrosine signaling plays a major role in the control of many important biological functions such as cell proliferation and apoptosis. Deciphering of phosphotyrosine-dependent signaling is therefore of great interest paving the way for the understanding of physiological and pathological processes of signal transduction. On the basis of the specific binding of SH2 domains to phosphotyrosine residues, we here present an experimental workflow for affinity purification and subsequent identification of tyrosine phosphorylated proteins by mass spectrometry. In combination with SH2 profiling, a broadly applicable platform for the characterization of phosphotyrosine profiles in cell extracts, our pull down strategy enables researchers by now to identify proteins in signaling cascades which are differentially phosphorylated and selectively recognized by distinct SH2 domains.

Key words SH2 domain, Signal transduction, Phosphotyrosine, Cancer, Protein-protein interaction, Pull down, Affinity purification, Mass spectrometry

1 Introduction

SH2 domains of which 120 different domains are encoded by the human genome [1], are classical modular binding domains playing a central role in phosphotyrosine-dependent signaling. Present in many signaling proteins and recognizing phosphotyrosine residues with high specificity, SH2 domains intracellularly sense, integrate, and process important biological signals triggered by the action of the many tyrosine kinases and phosphatases [1–3]. Given the large number of signaling proteins present in a cell, phosphotyrosine-dependent signaling is of high complexity. Understanding the organization and activation of these signaling networks is of great importance, particularly in diseases such as cancer in which aberrant phosphotyrosine signaling is a hallmark of the disease [2, 4]. To decipher these complex networks, we developed SH2 profiling, an analytical platform that utilizes the unique binding properties of SH2 domains for the characterization of the cellular state of

Kazuya Machida and Bernard A. Liu (eds.), *SH2 Domains: Methods and Protocols*, Methods in Molecular Biology, vol. 1555,
DOI 10.1007/978-1-4939-6762-9_23,

tyrosine phosphorylation [5]. Applying a panel of SH2 domains with distinct binding specificities as probes in Far Western blot analyses or reverse phase assays, our previous studies demonstrate that disease and cell type specific patterns of phosphotyrosine activity can be uncovered by SH2 profiling [4–6]. However, our analytical approaches were developed to primarily classify cells and tissues by their phosphotyrosine patterns without further information on the individual phosphoproteins contributing to pattern formation. Here, we describe a detailed experimental protocol for affinity purification and identification of tyrosine-phosphorylated proteins using SH2 domains as bait proteins [7]. For affinity purification, SH2 domains are expressed as GST fusion proteins in bacteria, monobiotinylated during bacterial expression and immobilized on streptavidin beads after purification (Fig. 1a). Subsequently, phosphoproteins are captured from whole cellular extracts by incubation with the SH2 domain coupled beads, enriched proteins are eluted, separated by SDS-PAGE and bands of interest visualized by silver staining are excised from the gel (Fig. 1b). After tryptic in-gel digestion, peptides are separated by nano liquid chromatography (LC) followed by electrospray ionization (ESI) and identification by mass spectrometry (MS).

2 Materials

Prepare all buffers with ultrapure, sterile water. All reagents need to be of analytical grade. When preparing samples for mass spectrometry, contamination of buffers, containers, surfaces, or instruments by environmental sources such as keratin from skin is a serious problem interfering with the quality of mass spectrometry results. To prevent contaminations, specific precautions need to be taken before experiments are started (*see* **Note 1**). For handling of hazardous substances and waste disposal follow the regulations of your institution and the recommendations of the manufacturer. Always wear protective clothing, safety gloves, and safety glasses when handling hazardous substances or solutions.

2.1 Pull Down Components

1. SH2 domains: express SH2 domains as GST-fusion proteins in AVB100 *E. coli* bacteria (Avidity, CO, USA) carrying a GST-tag at the N-terminus and a monobiotinylation site at the C-terminus [8]. Generally, we use SH2 domains cloned in the bacterial expression vector pGEX-6P (GE Healthcare, Little Chalfont, UK) with a PreScission protease cleavage site localized between the GST-tag and the SH2 domain for optional, enzymatic removal of the GST tag (Fig. 1a; see Subheading 3.4, **step 4**). Alternatively, other pGEX expression vectors can be used such as the pGEX-2T vector (GE Healthcare) with a thrombin cleavage site between the GST

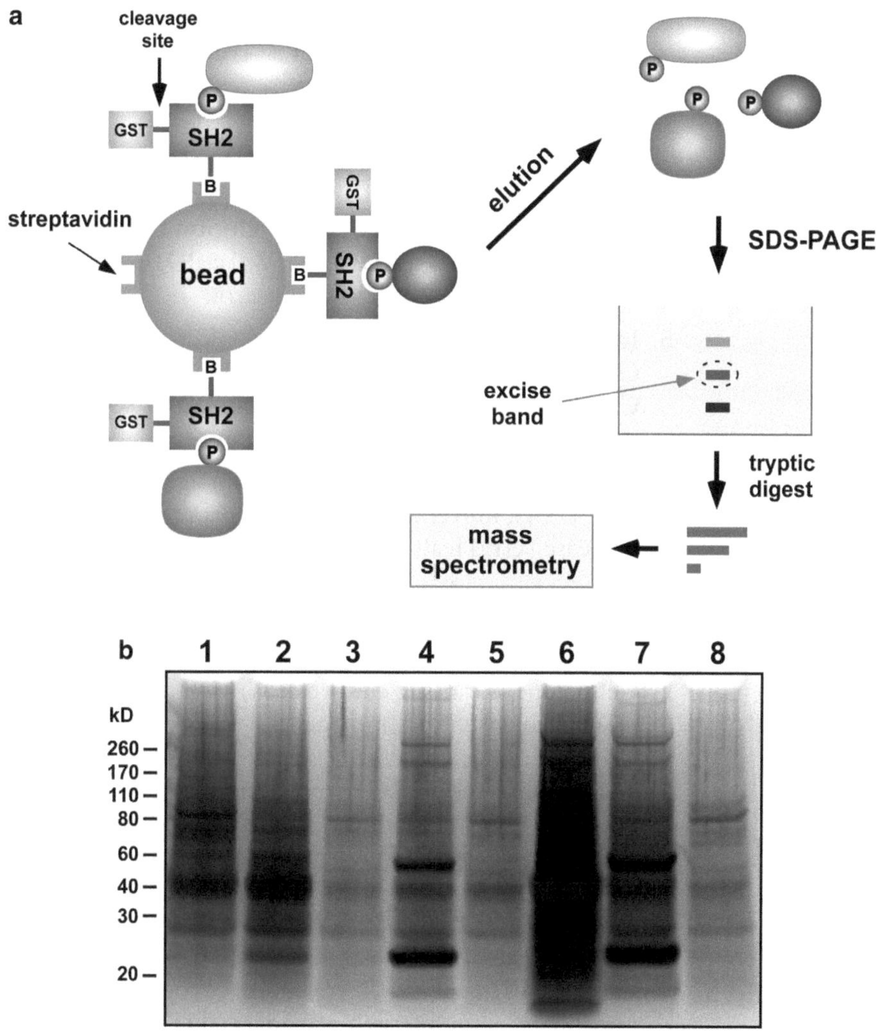

Fig. 1 Affinity purification and identification of tyrosine phosphorylated proteins by SH2 domains. (**a**) Schematic illustration of the experimental workflow. Biotinylated (**B**) SH2 domains are immobilized via streptavidin to agarose beads and used as bait proteins for affinity purification of tyrosine phosphorylated proteins (**P**) from cell extracts. After elution under alkaline conditions, phosphoproteins are separated by SDS-PAGE, visualized by silver staining and bands of interest are cut out from the gel for subsequent analysis by mass spectrometry. (**b**) Representative results of silver stained proteins affinity purified from extracts of different human leukemia samples (*lanes 1–8*) using the N-terminal SHP-2 SH2 domain for pull down

and the SH2 domain or the pAC4 vector (Avidity) with a GST-SH2 fusion construct inserted in the multiple cloning site.

2. Components for bacterial growth: Luria Broth Base, isopropyl-β-D-thiogalacto-pyranoside (IPTG), L-arabinose (Sigma, St. Louis, USA), 100 mg/mL Ampicillin, 25 mM D-biotin (AppliChem, Darmstadt, Germany; dissolved in water, adjusted to pH 7.0 with NaOH).

3. Bacteria lysis buffer: aprotinin solution (0.1 TIU/mL; Sigma), 1 mM PMSF, 1 mM DTT, 1% Triton X-100, and 100 mM EDTA (pH 8.0) in PBS.
4. SH2 domain protein purification: Glutathione (Sigma), Gluthathione (GSH)-Sepharose 4B (GE Healthcare), PBS, and glycerol. Column wash buffer: 50 mM Tris–HCl (pH 7.4), 150 mM EDTA (pH 8.0); GSH elution buffer: 20 mM glutathione, 100 mM Tris–HCl (pH 8.0); PD10-column equilibration buffer: PBS/20% (v/v) glycerol.
5. Instrumentation and labware: sonicator (Bandelin, Berlin, Germany), Econo-Pac columns (20 mL bed volume; BioRad, CA, USA) for affinity purification, PD10 sephadex columns (GE Healthcare) for buffer exchange.
6. Pull down: high capacity streptavidin agarose (Thermo Fisher, MA, USA), universal indicator pH paper, optional: PreScission protease (GE Healthcare); overhead shaker (4 °C).
7. Bradford assay: Bradford reagent, microtiter plates, microplate reader.

2.2 Cell Lysates

1. Cell scraper for tissue culture cells. Mortar and pestle for native tissue samples, liquid nitrogen.

2.3 SDS PAGE

1. 4–12% NuPAGE Bis-Tris protein gels (10-well, 1.0 mm gel thickness), protein ladder, 20× MES SDS running buffer, 4× LDS sample buffer, antioxidant, dithiothreitol (DTT), gel chamber (all from Thermo Fisher), gel loading tips.

2.4 Buffers

1. Components: Tris base, HCl, NaCl, Tween 20, EDTA, Glycerol, Triton X-100, Sodium pyrophosphate, ß-glycerol phosphate, sodium orthovanadate, DTT, PMSF, Aprotinin, sodium fluoride, hydrogen peroxide solution, triethylamine hydrochloride; skim milk.
2. 1 M Tris–HCl, pH 8: Weigh in 24.2 g Tris base and transfer to a graduated 250 mL cylinder. Add 180 mL ultrapure, sterile water and stir with a magnetic stir bar until Tris base is dissolved. Adjust to pH 8 with HCl (*see* **Note 2**) and fill up with sterile water to a final volume of 200 mL.
3. TBS-T: Weigh in 4.5 g NaCl and transfer to a 500 mL autoclaved glass bottle. Add 5 mL 1 M Tris–HCl, pH 8 and 450 mL ultrapure, sterile water. Stir solution until dissolved and add 250 μL Tween 20 (*see* **Note 3**). Fill up to 500 mL with ultrapure, sterile water (final concentration of Tween 20: 0.05%).
4. KLB lysis buffer: 25 mM Tris–HCl (pH 7.4), 150 mM NaCl, 5 mM EDTA (pH 8.0), 10% glycerol, 1% Triton X-100, 10 mM sodium pyrophosphate, 10 mM ß-glycerol phosphate, 1 mM sodium orthovanadate. Store stock solution at 4 °C in an auto-

claved glass bottle. Immediately before cell lysis, add the following components per mL of stock solution: 1 μL DTT (1 M stock solution), 10 μL PMSF (100 mM stock solution), 10 μL aprotinin solution, 10 μL sodium fluoride (1 M stock solution), and 2 μL of sodium pervanadate which needs to be freshly prepared by combining 16 μL H_2O_2 with 100 μL of sodium orthovanadate (50 mM sock solution) and incubated for 30 min at room temperature before adding to the lysis buffer.

5. Ice-cold PBS.
6. Blocking buffer: weigh in 5 g skim milk and transfer into a 50 mL graduated cylinder. Add 45 mL TBS-T and stir until skim milk is dissolved (*see* **Note 4**). Adjust to 50 mL with TBS-T. Cover cylinder with parafilm to avoid contaminations.
7. MES buffer: For 1× MES buffer dilute 50 mL of 20× MES buffer with 950 mL sterile water and store at room temperature in an autoclaved glass bottle.
8. 0.1 M triethylamine (pH 11.5): dissolve 0.7 g triethylamine hydrochloride in a sterile, polypropylene Falcon tube in 50 mL ultrapure, sterile water and store at 4 °C. Triethylamine hydrochloride is a strong base and should be handled with appropriate caution.
9. 3 N HCl: for preparing a 3 N HCl solution dilute one part of 37% HCl (12 N) with three parts of ultrapure, sterile water. Always prepare freshly, e.g., by adding 25 μL of 37% HCl to 75 μL ultrapure water in a 1.5 mL Eppendorf tube. Handle concentrated acids and bases with extreme caution and pipette under a fume hood.
10. Fixing solution: 50% ethanol, 10% acetic acid in ultrapure, autoclaved water; ethanol wash solution: 20% (v/v) ethanol/H_2O.
11. Glassware: 0.05, 0.1, and 0.25 L graduated cylinders; 0.5 and 1 L bottles.

2.5 Silver Staining

1. FireSilver Staining Kit (Proteome Factory, Berlin, Germany).
2. Sterile petri dish (diameter 10 cm) and scalpels.

3 Methods

The experimental workflow for the identification of proteins captured by SH2 domains starts with the preparation of a cell lysate under conditions preserving the phosphorylation state of the sample. Subsequently, biotinylated SH2 domains are immobilized to streptavidin-coated agarose beads. At this point of the workflow, the GST-tag that may affect the quality of mass spectrometry results by increasing the background of nonspecifically bound proteins, can be cleaved off from the SH2 domain by incubation with a protease. Immobilized SH2 domains are then incubated with cellular extracts, beads are extensively washed, and bound proteins are eluted under alkaline conditions. Subsequently, eluates of the pull down are separated by one-dimensional SDS-PAGE and proteins are stained by silver staining compatible with MS [9]. Bands of interest are cut from the gel, proteins are in-gel digested by trypsin, and peptides are subsequently identified by nano LC-ESI-MS/MS.

3.1 SH2 Domain Purification

1. Remove bacteria from frozen stocks with a sterile pipette tip, inoculate 40 mL of LB medium supplemented with 100 μg/mL ampicillin, and incubate under shaking overnight at 37 °C.
2. The next day, inoculate 360 mL of LB medium with the entire overnight culture and incubate in the presence of 100 μg/mL ampicillin at 37 °C until OD (595 nm) reaches 0.5–0.6.
3. Add 40 μL of 1 M IPTG to the culture. After 15 min add 1.2 g L-arabinose and 800 μL D-biotin solution (25 mM).
4. Incubate under shaking at 37 °C for 3 h.
5. Centrifuge bacteria at 5000 × *g*, 4 °C for 10 min and resuspend the bacteria pellet in 8 mL of ice-cold bacteria lysis buffer.
6. Sonicate two times for 2.5 min each (5 cycles, 40% power) on ice.
7. Clear lysate by centrifugation at 3000 × *g* at 4 °C for 10 min. Add 1.5 mL of glutathione-sepharose 4B to an Econo-Pac column, wash beads once with 10 mL PBS, close the bottom of the column, add the cleared lysate, tightly close the top of the column, and incubate at 4 °C for 1 h under horizontal rotation.
8. Subsequently, wash beads with 10 mL of column wash buffer and elute the SH2 domain with 2.5 mL GSH elution buffer. Equilibrate a PD10 column with PBS/20% (v/v) glycerol. Apply 2.5 mL of the eluate on the top of the column bed and exchange buffer with 3.5 mL PBS/20% (v/v) glycerol.
9. Determine protein concentration by Bradford assay and store SH2 domains in aliquots at −80 °C until prior use.

3.2 Preparation of Cell Lysates

1. For cells in suspension, transfer cell suspension to a 15 mL Falcon tube and centrifuge for 2 min at 200 × *g*.
2. Wash cells with 10 mL ice-cold PBS and centrifuge for another 2 min at 200 × *g*.
3. Add 1 mL KLB lysis buffer per 5×10^6 cells, carefully resuspend the cell pellet by pipetting up and down, and incubate on ice for 30 min.

4. Transfer the lysate by pipetting to a 1.5 mL autoclaved Eppendorf tube and centrifuge for 10 min at 15,000 × *g* at 4 °C.
5. After centrifugation, transfer the supernatant to a new, precooled, autoclaved Eppendorf tube, keep lysates for immediate use on ice or store lysates at −80 °C; discard the insoluble pellet.
6. For adherent tissue culture cells, wash cell layer with 10 mL ice-cold PBS (sufficient for a 75 cm^2 cell culture flask) and completely remove PBS by vertically tilting the culture flask allowing to drain the buffer to one corner of the flask.
7. Immediately place the flask horizontally on ice and add 1.5 mL KLB lysis buffer per 75 cm^2 of surface area. Scrape cells off the surface with a cell scraper and incubate the flask horizontally on ice for 30 min (*see* **Note 5**).
8. Again tilt the flask vertically on ice, let the lysate completely drain to one corner of the flask and transfer lysate by pipetting to a 1.5 mL autoclaved Eppendorf tube. Proceed with clearing of the lysate by centrifugation and further processing as described for the lysis of cells in suspension.
9. For tissue samples, snap freeze samples in liquid nitrogen, pulverize tissue under liquid nitrogen using mortar and pestle, directly add KLB lysis buffer to the frozen, pulverized tissue (*see* **Note 6**).
10. After thawing, transfer the lysate to an autoclaved Eppendorf tube, incubate on ice for additional 15 min, and proceed with clearing of the extract by centrifugation as described above.

3.3 Bradford Assay

1. Determine protein concentrations of lysates in a 96-well plate using the Bradford assay. Pipette 250 μL of the Bradford reagent in a well of a microtiter plate and add 2–5 μL of prediluted (1:5 in PBS) KLB lysate (*see* **Note 7**).
2. Mix and incubate for 10 min at room temperature. Measure OD at 595 nm on a microplate reader.
3. Perform measurements in triplicate, always include blank controls (Bradford reagent alone) for correction of background absorbance, and use serial dilutions of protein standards in parallel for quantification as recommended by the manufacturer.

3.4 Pull Down

1. Pipette 100 μL of high capacity streptavidin agarose beads in an autoclaved 1.5 mL Eppendorf tube, add 500 μL of PBS, centrifuge beads at 500 × *g* for 2 min, remove supernatant (*see* **Note 8**), and repeat the washing step.
2. Add 50 μg of biotinylated SH2 domain to the washed beads. Adjust to a final volume of 200 μL with PBS and incubate for 1 h at 4 °C by overhead shaking, fixing the Eppendorf tube horizontal to the rotation axis thereby guarantying even dispersion of the beads.

3. Wash beads two times with 500 μL each of PBS at room temperature.
4. Optional: remove GST-tag by PreScission protease digestion (*see* **Note 9**).
5. Add 500 μL of blocking buffer to washed beads and incubate for 30 min at room temperature by overhead shaking.
6. Wash beads three times with 500 μL each of PBS at room temperature.
7. Add 2 mg of protein extract to the washed beads and incubate at 4 °C for at least 2 h or preferably overnight by overhead shaking (*see* **Note 10**).
8. Centrifuge beads and wash beads three times each with 500 μL PBS, three times each with 500 μL TBS-T followed by three washing steps with 500 μL PBS each.
9. To elute bound proteins, remove washing buffer, add 70 μL 0.1 M triethylamine hydrochloride (pH 11.5) to the beads, and incubate for 5 min at room temperature. Centrifuge beads and transfer supernatant to a fresh, autoclaved 1.5 mL Eppendorf tube. Immediately neutralize the eluate by adding 1 μL 3 N HCl, check pH with pH paper and continue with sample preparation for SDS-PAGE, or store the eluate at −20 °C (*see* **Note 11**).
10. For SDS-PAGE, add 4.6 μL 500 mM DTT and 11.5 μL 4× sample buffer to 30 μL of eluate (*see* **Note 12**). Mix sample by short vortexing and spin down sample by centrifugation. Boil sample in a preheated water bath for 5 min at 95 °C. If not directly proceeding with protein separation by SDS-PAGE, store samples at −20 °C.

3.5 SDS-PAGE

1. Place a precast NuPAGE 4–12% Bis–Tris mini gel into the gel chamber. Fix the gel and fill chambers with approximately 750 mL 1× MES buffer. Add 500 μL antioxidant to the inner chamber and remove the comb from the gel.
2. Prepare molecular weight standard by mixing 5 μL of protein ladder with 1 μL 500 mM DTT, 2.5 μL 4× sample buffer, and 1.5 μL H_2O. Load on the gel.
3. Load 30–40 μL of denatured sample per well prepared as described in Subheading 3.4, **step 10**. Leave one lane empty between the sample and the protein standard to prevent contamination. Always use new gel loading tips when loading different samples.
4. Lock the gel apparatus with the lid and connect to the power supply, set voltage to 200 V constant, start separation, and stop the run after 35–45 min.

3.6 Silver Staining

1. Set up solutions for silver staining as recommended by the manufacturer. Use ultrapure, autoclaved water to prepare solutions.
2. Remove the gel carefully from the plastic cover and place the gel in a sterile petri dish (diameter 10 cm) prefilled with fixing solution (50 % ethanol, 10 % acetic acid in ultrapure, autoclaved water) (*see* **Note 13**). Seal the petri dish with parafilm. Incubate overnight at room temperature by horizontally shaking under the fume hood.
3. Perform silver staining according to the instructions of the manufacturer. In brief, wash gel in 20 % (v/v) ethanol/H_2O for 10 min, incubate gel for 1 min in sensitizer solution, wash gel twice in ultrapure, autoclaved water for 1 min each, incubate gel with silver solution for 30 min, wash gel in ultrapure, autoclaved water for 30 s, incubate gel in pre-development solution for 1 min followed by incubation with developer solution (*see* **Note 14**). Depending on the total amount of protein loaded, development will take 10–15 min. Stop development by adding stop solution and incubate for 20 min. Discard stop solution and add ultrapure water.
4. If not immediately proceeding, store the gel at 4 °C in ultrapure, autoclaved water in the petri dish sealed by parafilm to prevent evaporation.
5. For excision of gel bands, rinse the gel twice in the petri dish with ultrapure, autoclaved water, excise bands of interest with a scalpel, and transfer the gel slices using the scalpel in autoclaved, 1.5 mL Eppendorf tubes (*see* **Note 15**). Use a new scalpel for each band. Store gel slices at 4 °C.

3.7 Mass Spectrometry

1. For analysis by MS, we use either our in-house MS facility or commercial services. Investigators are highly recommended to review the conditions of analysis, availability of instrumentation, and data processing of their local facility beforehand. Usually, our samples are in-gel digested by trypsin, peptides are separated on a nanoLC system by reversed phase HPLC using a C18 microparticulate column, ionized by electrospray and analyzed by an Orbitrap mass spectrometer (e.g., Thermo Fisher).
2. Subsequently, MS/MS data are interpreted by the Mascot software. Depending on the amount of data, the interpretation of MS results and identification of candidate proteins can be challenging. To preselect for potential candidate proteins, we use the Mascot score and frequency of individual peptides detected by MS as first benchmark. In case sufficient amounts of protein extract are available, controls such as pure streptavidin beads, biotinylated GST, or phosphotyrosine binding mutants of SH2 domains immobilized to streptavidin beads

should be analyzed in parallel enabling the investigator to eliminate unspecific binders from the candidate list.

3. To further narrow down specificity, candidate proteins may be checked by the CRAPome database (www.crapome.org), a computational tool by which contaminants frequently detected by affinity purification and MS can be evaluated [10].
4. Once a candidate list is set up, the identity and phosphotyrosine status of the candidate proteins should be confirmed by western blot, far-western blot, or immunoprecipitation followed by biological experiments such as overexpression or knockdown of candidates to proof the significance of proteins captured by SH2 domains in cellular signaling. As contaminations present in the sample will substantially affect the quality of MS data and thus the identification of candidate proteins, prevention of contamination of samples during all steps of analysis is essential.

4 Notes

1. To prevent contamination by foreign sources, always use autoclaved tubes, glassware and solutions. Sterile filter (0.2 μm pore size) buffers and solutions that will decompose under heating. To set up buffers and solutions always use fresh reagents. Once opened and for future use, keep containers in a separate place. Label reagents, solutions, and buffers appropriately to prevent erroneous usage and contamination by others. Use sterile, filtered pipette tips during all steps of the protocol. Always wear gloves and appropriate protective clothing such as face and head cover to minimize contamination. Perform if possible all steps of affinity purification and excision of gel bands under a sterile workbench with laminar airflow.
2. Use 37% hydrochloric acid (v/v) for adjusting pH. Pipette hydrochloric acid with great caution under a fume hood. Always wear protective clothing, safety gloves, and safety glasses!
3. Tween 20 is viscous and adheres strongly to the pipette. Place pipette tip filled with Tween 20 in TBS buffer and rinse pipette by carefully pipetting up and down several times to release all of the detergent into the buffer.
4. Adding skim milk stepwise in small portions to TBS-T will facilitate dissolving.
5. To prevent dephosphorylation or protein degradation, keep sample during all steps of cell lysis on ice.
6. Wear protective clothing, safety gloves, and safety glasses when handling liquid nitrogen and pulverizing tissue by mortar and

pestle. Always keep in mind that specimens, especially of human origin, may be infectious. Take appropriate precautions to prevent infection and contamination, perform tissue preparation and cell lysis under a sterile workbench with laminar air flow, and use appropriate disinfectants for disinfection of instruments, containers, and surfaces.

7. To prevent erroneous measurements of protein concentrations due to high concentrations of interfering compounds from the lysis buffer, dilution of KLB lysis buffer is required. Applying larger volumes of buffer will compromise accurate quantification of protein concentrations.
8. Streptavidin agarose is provided as 50% aqueous slurry. Mix beads well before pipetting and use a wide bored pipette tip to remove the bead suspension from the stock. During the washing steps, special care is required not to accidentally aspirate beads by pipetting during removal of the supernatant. To prevent the loss of beads, we do not remove the buffer entirely always leaving a small volume of several microliters of buffer on top of the beads.
9. When removal of the GST tag is preferred, add 60 μL PBS and 7.5 μL Prescission protease (15 U; GE Healthcare) to beads and incubate overnight at 4 °C by overhead shaking. Wash beads three times with 500 μL of PBS at room temperature and proceed with blocking of beads in **step 5**.
10. Using the KLB lysis protocol described above, our protein concentrations are usually in the range of 2–5 mg/mL so that 400 μL to 1 mL of lysate is needed for pull down. Using less than 2 mg input of protein extract per pull down resulted in our hands in the identification of only a few, very likely the most abundant proteins, by MS. When protein concentrations are below 2 mg/mL, protein extracts can be concentrated using ultra centrifugal filters (Merck Millipore, Darmstadt, Germany). However, we do not recommend applying less than 2 mg of protein extract per pull down.
11. After adding of HCl, the pH of the eluate should be in the range of pH 6.5–7.5. Test the pH by applying a few microliters of eluate to a pH paper. In case that the pH exceeds the given range, appropriately adjust the volume of 3 N HCl required for neutralization. If necessary, determine the optimal conditions for neutralization afore by mixing different volumes of 3 N HCl with 70 μL 0.1 M triethylamine hydrochloride and measure the pH using pH paper.
12. The remaining eluate of approximately 40 μL will either serve as backup or for later validation experiments by which, e.g., the efficient enrichment of tyrosine phosphorylated proteins or the pull down of potential binding partners is confirmed by Western blot analysis.

13. To facilitate removal and transport, optionally cover the gel with a clean piece of 3 MM paper of similar size, carefully remove the gel attached to the 3MM paper from the plastic cover, submerge in fixing solution, and remove the 3 MM paper that is floating off.
14. Exchange staining solutions by tilting of the petri dish. Fix gel at the edges with your fingers to prevent slipping of the gel. To prevent contaminations, do not touch gel areas that will be excised later.
15. During excision of bands, do not place your head or upper part of the body directly over the gel to prevent contaminations from hairs and skin. When shifting the gel, only touch the edges of the gel. It is mandatory during the excision of bands to wear protective gloves and clothing. Wearing of a surgical mask and head cover is recommended.

References

1. Liu BA, Jablonowski K, Raina M et al (2006) The human and mouse complement of SH2 domain proteins-establishing the boundaries of phosphotyrosine signaling. Mol Cell 22:851–868
2. Hunter T (2009) Tyrosine phosphorylation: thirty years and counting. Curr Opin Cell Biol 21:140–146
3. Lim WA, Pawson T (2010) Phosphotyrosine signaling: evolving a new cellular communication system. Cell 142:661–667
4. Machida K, Khenkhar M, Nollau P (2012) Deciphering phosphotyrosine-dependent signaling networks in cancer by SH2 profiling. Genes Cancer 3:353–361
5. Machida K, Mayer BJ, Nollau P (2003) Profiling the global tyrosine phosphorylation state. Mol Cell Proteomics 2:215–233
6. Dierck K, Machida K, Voigt A et al (2006) Quantitative multiplexed profiling of cellular signaling networks using phosphotyrosine-specific DNA-tagged SH2 domains. Nat Methods 3:737–744
7. Schweigel H, Geiger J, Beck F et al (2013) Deciphering of ADP-induced, phosphotyrosine-dependent signaling networks in human platelets by Src-homology 2 region (SH2)-profiling. Proteomics 13: 1016–1027
8. Dierck K, Machida K, Mayer BJ, Nollau P (2009) Profiling the tyrosine phosphorylation state using SH2 domains. Methods Mol Biol (Clifton, NJ) 527:131–155. doi:10.1007/978-1-60327-834-8_11, ix
9. Blum H, Beier H, Gross HJ (1987) Improved silver staining of plant proteins, RNA and DNA in polyacrylamide gels. Electrophoresis 8:93–99
10. Mellacheruvu D, Wright Z, Couzens AL et al (2013) The CRAPome: a contaminant repository for affinity purification-mass spectrometry data. Nat Methods 10:730–736

Chapter 24

Analysis of the Global Changes in SH2 Binding Properties Using Mass Spectrometry Supported by Quantitative Stable Isotope Labeling by Amino Acids in Cell Culture (SILAC) Technique

Radoslaw M. Sobota

Abstract

Quantitative mass spectrometry (MS)-based proteomics enables fast and reliable analysis of protein complexes. Its robustness and sensitivity effectively substitute traditional antibody-based approaches. Here, we describe the combination of mass spectrometry and Stable Isotope Labeling by Amino acids in Cell culture (SILAC) in characterization of the SH2 domain binding capacity.

Key words SH2 domains, Quantitative mass spectrometry, SILAC, Peptide pull down

1 Introduction

SH2 domains bind phosphorylated tyrosine residues. The binding specificity of SH2 domains is typically characterized ex vivo with peptide libraries, protein arrays, or amino acid permutation experiments of already known binding sites. With obvious discovery power, peptides library approach has also its limitations—i.e., phosphorylated peptide sequences may not be necessarily reflected in the cellular environment. To overcome the library limits, we propose to use denatured total cell lysates from pervanadate treated cells as a source of natural protein targets. This strategy, together with application of mass spectrometry coupled with SILAC-based quantitation, constitutes a powerful approach for identifying physiological interacting protein partners and determining the in vivo binding specificity of SH2 domains. Our setup can be used in deciphering the specificity among different SH2 domains and possibly distinguishing the specificity bias triggered by disease-related mutations [1].

Kazuya Machida and Bernard A. Liu (eds.), *SH2 Domains: Methods and Protocols*, Methods in Molecular Biology, vol. 1555, DOI 10.1007/978-1-4939-6762-9_24, © Springer Science+Business Media LLC 2017

SILAC is a robust and reliable strategy for relative protein quantification by mass spectrometry [2]. It consists of replacing L-Arginine and L-Lysine in the cell culture medium with variants that contain nonradioactive isotopic ^{13}C and ^{15}N atoms. These amino acids are incorporated into proteins during protein translation. To achieve full labeling, cells need to proliferate, thus, a main requirement for SILAC. To study the binding specificity of multiple SH2 domains, a triple SILAC experiment was set up as shown in Fig. 1. The triple SILAC labeling enables comparison of three different conditions in a single experiment. Briefly, bead-immobilized SH2 domains (which may include mutants) were incubated with hyper-phosphorylated, denatured and subsequently dialyzed—SILAC labeled total cell lysates. Following extensive washes, SH2 domains and bound interacting proteins were combined, and subsequently identified and quantified through mass spectrometry analysis. As controls, competition experiments using tyrosine-phosphorylated peptides were performed to identify SH2-dependent interactions (Fig. 2). In this experimental setup, SH2 domains were incubated with unlabeled short phospho-specific peptides or corresponding nonphosphorylated peptides prior to exposure of the total cell lysates. As a control for phospho-independent competition additional samples without addition of peptide (mock) were included. In the ideal situation, phosphorylated peptides will only compete out proteins from the lysates.

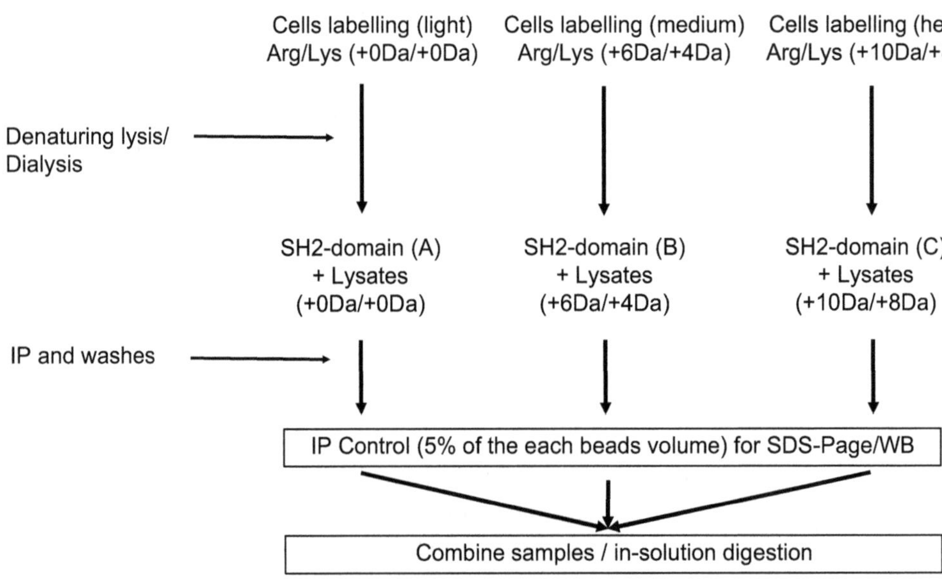

Fig. 1 Workflow for differential SH2 domain binding experiment. Differentially (triple SILAC) labeled cells were harvested and denatured. Following dialysis the equal amount of protein was subjected to pull down experiment with different SH2 domain variants (**a**–**c**). After extensive washing, ~5 % from each sample was taken for western blot analysis to check pull down efficiency. Subsequently remaining beads were mixed and digested prior mass spectrometry analysis

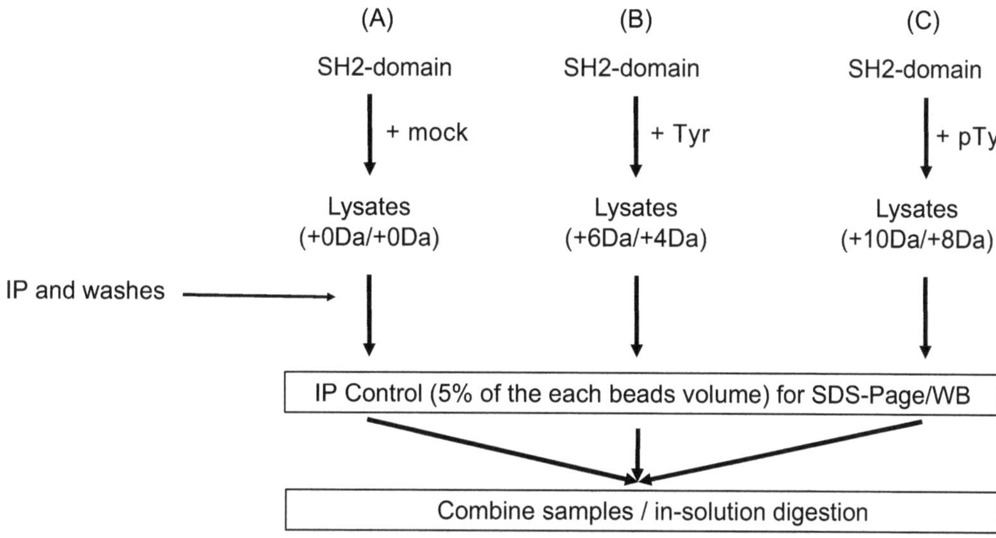

Fig. 2 Workflow for control peptide competition experiment. To assess SH2 domain specificity peptides competition experiment was set up. GST-SH2 domains were exposed to: (**a**) mock conditions (no peptides), (**b**) phosphorylated peptide, (**c**) nonphosphorylated peptide. Following preincubation, pull down experiment was performed with denatured (triple SILAC) labeled cell lysates. Following extensive washing, ~5 % from each sample was taken for western blot analysis to check pull down efficiency, remaining beads were mixed and digested prior mass spectrometry analysis

Finally, both experiments are compared and proteins that overlap constitute high confidence SH2 domain interactors. These strategies combined with SILAC and MS, give rise to the identification of direct SH2 interacting proteins.

2 Materials

2.1 Recombinant Proteins

The source of SH2 domains will depend on the experimental setup. One approach is to express glutathione S-transferase (GST) tagged SH2 domain and mutants in *E. coli*. As a negative control for SH2-specific pull down, we propose to use a "SH2-dead" mutant where the key Arginine residue (βB5 in the secondary structure) for phosphate binding is substituted with Glutamine (R→Q mutant). Substitution effectively abolishes the binding of the SH2 domain to phosphorylated tyrosine residues. This SH2 domain mutant (138R → Q) in the C-SH2 domain of PTPN11, found in melanoma, was used in our studies as a negative control [3, 4] (*see* **Note 1**) (*see also* Chapter 8 and 12).

2.2 Synthetic Peptides

For competition experiment, pairs of phosphorylated and non-phosphorylated synthetic peptides matching to known binding consensus motif for each SH2 domain of interest were prepared. For example, a pair of the synthetic peptides for the SH2 domain of PTPN11 that correspond to the IL-6 receptor Y759 region (gp130) [5] are:

Y759: β-A-TSFTVQYSTVVHSGYRHQVPSV.

pY759: β-A-TSFTVQpYSTVVHSGYRHQVPSV.

2.3 Cells Culture Reagents and Labeling with SILAC

1. Dialyzed Fetal Calf serum (FCS) (10,000 MW cutoff).
2. Dulbecco's Modified Eagle Medium (DMEM) cell culture medium.
3. L-Lysine (Lys + 0 Da) (45 mg/l).
4. L-Arginine (Arg + 0 Da) (30 mg/l).
5. $^{2}H_{4}$-L-Lysine (Lys + 4 Da) (45 mg/l).
6. $^{13}C_{6}$-L-Arginine (Arg + 6 Da) (30 mg/l).
7. $^{13}C_{6}{}^{15}N_{2}$-L-Lysine (Lys + 8 Da) (45 mg/l).
8. $^{13}C_{6}{}^{15}N_{4}$-L-Arginine (Arg + 10 Da) (30 mg/l).
9. L-Proline (200 mg/l).

2.4 Buffers and Other Reagents for Pull Down Assay

1. Denaturing lysis buffer: 8 M Urea in 10 mM Tris/HCl (pH 8.5).
2. Dialysis buffer: 50 mM Tris/HCl (pH 8.0).
3. Dialysis membrane: 10,000 MW cutoff.
4. PD buffer: 150 mM NaCl, 1% NP-40, 50 mM Tris/HCl (pH 7.4).
5. PBS (Phosphate-buffered saline).
6. Tosyl activated magnetic beads (*see* **Note 2**).

2.5 Sample Preparation Reagents—Mass Spectrometry

1. DTT—dithiothreitol 100 mM (!) (*see* **Note 3**).
2. CAA—chloro acetamid 550 mM (!).
3. TFA—trifluoro acetic acid (!).
4. Protease LysC (Wako).
5. Sequencing Grade Trypsin.
6. Sep-Pak C-18 desalting columns.
7. 100 mM ABC—ammonium bicarbonate.
8. Desalting Solvent A (0.5% acetic acid in water).
9. Desalting Solvent B (0.5% acetic acid, 80% acetonitrile (!) in water).

2.6 Mass Spectrometry

Mass spectrometry analysis can be performed with the help of any state-of-the-art instrument platforms like: Orbitrap series or Q-TOF Impact II, 3TOF 5600/6600. All systems should be routinely coupled online with liquid chromatography capable of delivering 100–500 nl/min flow (nanoflow HPLC). Peptide separation is performed on C-18 columns preferable without pre-column.

2.7 Data Analysis Software

Raw mass spectrometry data was processed using MaxQuant software (http://www.maxquant.org) [6, 7] with the following parameters: Human database—Uniprot (http://www.uniprot.org). Fixed modification: Carbamidomethyl cysteine. Variable modifications: Oxidation on methionine, Acetylated N-terminal protein, phospho STY, SILAC amino acids (Arg + 6 Da, Arg + 10 Da, Lys + 4 Da, Lys + 8 Da), two missed cleavages, MS accuracy of 7 ppm, MS/MS accuracy of 0.5 Da. Peptide identification cutoff was set to 1 % False Discovery Rate. Unique and razor peptides were used for protein ratio quantification with a minimum ratio count 1. MaxQuant output was further analyzed using R-based software—G-prox (http://gprox.sourceforge.net) [8] and visualized with Cytoscape software (http://www.cytoscape.org/).

3 Methods

3.1 Cell Labeling

Labeling of HeLa cells with DMEM tissue culture medium:

1. Choose the right Arginine/Lysine free medium.
2. Supplement the missing amino acids with the desired combination:

   for double SILAC (*see* **Notes 4** and **5**).

   - L-Lysine (Lys +0 Da) (45 mg/l).
   - L-Arginine (Arg +0 Da) (30 mg/l).
   - $^{2}H_{4}$-L-Lysine (Lys +4 Da) (45 mg/l).
   - $^{13}C_{6}$-L-Arginine (Arg +6 Da) (30 mg/l).

   for triple SILAC add

   - L-L-Lysine (Lys + 0 Da) (45 mg/l).
   - L-Arginine (Arg + 0 Da) (30 mg/l).
   - $^{2}H_{4}$-L-Lysine (Lys + 4 Da) (45 mg/l).
   - $^{13}C_{6}$-L-Arginine (Arg + 6 Da) (30 mg/l).
   - $^{13}C_{6}{}^{15}N_{2}$-L-Lysine (Lys + 8 Da) (45 mg/l).
   - $^{13}C_{6}{}^{15}N_{4}$-L-Arginine (Arg + 10 Da) (30 mg/l).
3. Supplement with L-Proline (200 mg/l) to decrease possibility of arginine to proline conversion.
4. Supplement with 10 % of dialyzed FCS.
5. Work in a sterile environment.
6. Grow cells for 7 days and take small aliquot for labeling test (*see* **Note 6**).
7. Expand till you reach desired number of cells for experiment.

3.2 Cell Lysis and Dialysis Setup

1. Treat HeLa cells with pervanadate (1 mM) for 15 min prior harvesting.
2. Remove tissue culture medium and rinse cells with PBS.
3. Remove PBS and lyse cells on the plates in the Denaturing lysis buffer.
4. Sonicate the lysate to fragment nucleic acids and decrease viscosity.
5. To remove excess of urea dialyze lysates using a dialysis membrane (10 kDa cutoff) placed into a large container (5 L) of 50 mM Tris/HCl pH 8.0 overnight while stirring.
6. Estimate protein concentration after dialysis using bicinchoninic acid (BCA) assay or Bradford assay.

3.3 Domain Pull Down (PD) (Fig. 1)

1. Take 1–5 mg of protein lysate (depends on MS-setup sensitivity) and transfer into a 15 ml Falcon tube.
2. Dilute the lysate with 150 mM NaCl and 1 % NP-40 in 50 mM Tris/HCl pH 7.4 up to 10 ml volume.
3. Add beads containing immobilized GST-SH2 domains and incubate at 4 °C overnight (*see* **Note 7**).
4. Wash with 10 ml of PD buffer (repeat six times).
5. Wash with 10 ml of PD buffer omitting detergent (repeat six times).
6. Combine beads from the different SILAC conditions (*see* **Note 8**).
7. After the last wash, resuspend beads in denaturing buffer.

3.4 Competitive Domain Pull Down (Fig. 2)

1. To test specificity of the SH2 domain interaction, GST-SH2 domain of interest was tested in competition experiment with synthetic peptides. Pair of phosphorylated and nonphosphorylated synthetic peptides matching to known binding consensus motif were dissolved in methanol (1 mg/ml).
2. For the competition experiment the equal amounts of GST-SH2 domain coupled beads were exposed to: (a) mock conditions (no peptides), (b) phosphorylated peptide, (c) nonphosphorylated for 30 min at 4 °C in PD buffer separately (three tubes) (Fig. 2).
3. Rinse two times with 1 ml of PD buffer to remove excess peptides.
4. Take 1–5 mg of protein lysates (depends on MS-setup sensitivity) in 15 ml Falcon tube.
5. Dilute with 150 mM NaCl and 1% NP-40 in 50 mM Tris/HCl pH 7.4 up to 10 ml volume.
6. Add beads with immobilized GST-SH2 domains and incubate at 4 °C overnight.

7. Wash with 10 ml of PD buffer (repeat six times).
8. Wash with 10 ml of PD buffer omitting detergent (repeat six times).
9. Combine beads from the different SILAC conditions (*see* **Note 8**).
10. After the last wash, resuspend beads in denaturing buffer.

3.5 MS-Sample Preparation

There are a number of different protocols for enzymatic digestion. For the sample handling, we suggest using an in-solution digestion protocol as it gives wider possibilities of the downstream sample processing like direct injection (unfractionated sample) or further sample fractionation with high pH reverse phase chromatography and/or Isoelectric focusing (IEF) on a peptide level.

3.5.1 Enzymatic Digestion

Following extensive washing, collected complexes were subjected to in-solution digestion in denaturing conditions:

1. Resuspend beads in 50 μl (*) of 8 M Urea/50 mM Tris–HCl pH 8.5.
2. Add 20 mM DTT in 25 °C for 20 min.
3. Add 55 mM CAA in room temperature (RT) (protect from light!) for 30 min.
4. Diluted with 100 mM ABC to 600 μl.
5. Add LysC 10 μg (**) per sample for 3–4 h at 25 °C.
6. Diluted with 100 mM ABC to 750 μl.
7. Add Trypsin 10 μg (**) overnight at 25 °C.
8. Acidify with TFA (!) to reach ~1 % TFA in a sample.
9. Centrifuge at 14,000 × *g* for 10 min at RT.
10. Collect the supernatant for desalting step.

*All volumes including dilution steps need to be adjusted to the beads amount!

**Enzymes quantity needs to be adjusted to the expected protein amount (1:50 or 1:100). Final Urea concentration has to be below 1 M!

3.5.2 Desalting

1. Equilibrate the Sep-Pak C-18 column with 2 ml of 100% acetonitrile, followed by 2 ml of Solvent A.
2. Connect the column to a new 5 ml syringe and slowly bind the sample to the column bead.
3. Repeat this step twice.
4. Wash the column once with 5ml Solvent A and expel sharply the residual fluid.
5. Connect the column to a new 1 ml syringe and add 1 ml of Solvent B.

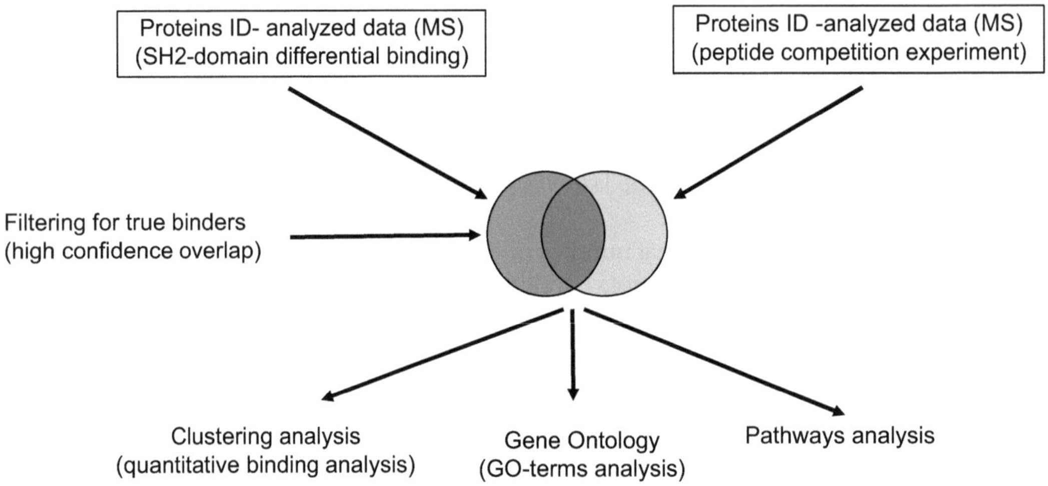

Fig. 3 Proposed data analysis workflow. To obtain final results of SH2 binding properties, differential SH2 domain binding experiment is compared to the reference competition experiment. Overlap between both will constitute high confidence dataset that can be further processed. Downstream data processing includes quantitative and qualitative analysis (clustering), GO-terms analysis, and/or pathway analysis (*see* Note 9)

6. Elute slowly into a low-binding tube.
7. Repeat the previous step twice, and expel any residual fluid with air.
8. Use concentrator to remove residual organic solvent.

3.6 Data Analysis

Data analysis workflow (Fig. 3) will depend on experimental hypothesis. More generally, this protocol focuses on two main issues:

1. Quantitative analysis of SH2 domains differential binding properties (hypothesis driven).
2. Elimination of false-positive binders by parsing main experiments against existing SH2 domains interaction databases, prediction databases, and competition experiment.

Further downstream data analysis may include mutant (domain) dependent quantitative clustering of interacting proteins. Gene ontology (GO terms) analysis or pathway analysis followed by data visualization and independent validation (*see* **Note 9**).

4 Notes

1. Proteins were expressed using BL21 bacteria system and pGEX-4T2, although any expression system can be used.
2. We have chosen magnetic beads due to low background in affinity purifications. If needed, sepharose can be used as well.

3. (!) marks toxic or dangerous chemicals please get familiar with MSDS, wear appropriate PPE, and work under the fume hood.
4. Cell type will determine the culture media to be used. Currently, there is a broad spectrum of SILAC media on the market and in the worst case scenario, custom formulation medium has to be made. Amino acid concentration has to be selected empirically starting from as little as possible (observe cells growth/proliferation rate).
5. Double or Triple SILAC setup depends on number of conditions that have to be compared at once. Please use light amino acids in the same concentration as a comparison.
6. Labeling test is obligatory, since only complete labeling can give reliable quantification results.
7. Amount of GST-SH2 domain per pull down has to be estimated empirically.
8. Before combining beads, remove a small aliquot (5 % of beads) for a pull down test for western blot analysis with anti-phosphotyrosine antibodies.
9. Data analysis and visualization software URLs:

   G-prox (http://gprox.sourceforge.net)—GO-terms annotation, clustering.

   GOrilla: (http://cbl-gorilla.cs.technion.ac.il/)—GO-terms annotation.

   Cytoscape software (http://www.cytoscape.org/)—Visualization.

Acknowledgments

I would like to acknowledge Prof. Fred Schaper (University of Magdeburg, Germany) and Dr. Pia Muller (RWTH-Aachen, Germany) for the great contribution to original work which became foundation for the method described in this chapter. This work was supported by the grants from Lundbeckfonden (Denmark), Kong Christian den Tiendes Fond (Denmark) and Young Investigator Grant 2015—BMRC A*STAR (Singapore) to R.M.S.

References

1. Muller PJ, Rigbolt KT, Paterok D, Piehler J, Vanselow J, Lasonder E, Andersen JS, Schaper F, Sobota RM (2013) Protein tyrosine phosphatase SHP2/PTPN11 mistargeting as a consequence of SH2-domain point mutations associated with Noonan Syndrome and leukemia. J Proteomics 84:132–147. doi:10.1016/j.jprot.2013.04.005
2. Ong SE, Blagoev B, Kratchmarova I, Kristensen DB, Steen H, Pandey A, Mann M (2002) Stable isotope labeling by amino acids in cell culture, SILAC, as a simple and accurate approach to expression proteomics. Mol Cell Proteomics 1(5):376–386
3. Bentires-Alj M, Paez JG, David FS, Keilhack H, Halmos B, Naoki K, Maris JM, Richardson A,

Bardelli A, Sugarbaker DJ, Richards WG, Du J, Girard L, Minna JD, Loh ML, Fisher DE, Velculescu VE, Vogelstein B, Meyerson M, Sellers WR, Neel BG (2004) Activating mutations of the noonan syndrome-associated SHP2/PTPN11 gene in human solid tumors and adult acute myelogenous leukemia. Cancer Res 64(24):8816–8820. doi:10.1158/0008-5472.CAN-04-1923

4. Tartaglia M, Gelb BD (2005) Germ-line and somatic PTPN11 mutations in human disease. Eur J Med Genet 48(2):81–96. doi:10.1016/j.ejmg.2005.03.001
5. Lehmann U, Schmitz J, Weissenbach M, Sobota RM, Hortner M, Friederichs K, Behrmann I, Tsiaris W, Sasaki A, Schneider-Mergener J, Yoshimura A, Neel BG, Heinrich PC, Schaper F (2003) SHP2 and SOCS3 contribute to Tyr-759-dependent attenuation of interleukin-6 signaling through gp130. J Biol Chem 278(1):661–671. doi:10.1074/jbc.M210552200
6. Cox J, Mann M (2008) MaxQuant enables high peptide identification rates, individualized p.p.b.-range mass accuracies and proteome-wide protein quantification. Nat Biotechnol 26(12):1367–1372. doi:10.1038/nbt.1511
7. Cox J, Matic I, Hilger M, Nagaraj N, Selbach M, Olsen JV, Mann M (2009) A practical guide to the MaxQuant computational platform for SILAC-based quantitative proteomics. Nat Protoc 4(5):698–705. doi:10.1038/nprot.2009.36
8. Rigbolt KT, Vanselow JT, Blagoev B (2011) GProX, a user-friendly platform for bioinformatics analysis and visualization of quantitative proteomics data. Mol Cell Proteomics 10(8):O110.007450. doi:10.1074/mcp.O110.007450

Chapter 25

Using Reciprocal Protein-Peptide Array Screening to Unravel Protein Interaction Networks

Huadong Liu, Courtney Voss, and Shawn S.C. Li

Abstract

Protein-protein interactions (PPIs) play a central role in almost all cellular processes. Recent technological advances have enabled the elucidation of an incredibly complex PPI network within the cell. However, protein interactions driven by posttranslational modifications (PTMs) such as phosphorylation, which comprises a significant part of the PPI network, have proven difficult to decipher systematically. Herein, we describe a reciprocal protein-peptide array strategy to uncover PPIs mediated by tyrosine phosphorylation and the Src homology 2 (SH2) domain. This strategy, namely combining peptide and protein domain arrays for PPI mapping, may be applicable for other peptide-binding modules.

Key words Protein array, Peptide array, Protein-protein interaction, SH2 domain, Phosphorylation

1 Introduction

Protein-protein interactions (PPIs) play central roles in nearly all aspects of cell biology, and aberrant PPI networks underlie a host of diseases [1]. To decipher PPI networks in a systematic manner, a number of high-throughput approaches have been developed. These include yeast two-hybrid (Y2H) screens [2], affinity purification followed by mass spectrometry (AP-MS) [3], peptide arrays [4], protein arrays [5], and bioinformatic predictions [6]. Each method has its merits and limitations, and none is ideal for mapping all PPI events. For instance, although systematic Y2H screen of human open-reading frames (ORFs) uncovered numerous PPIs [2], the method is not designed to interrogate interactions mediated by posttranslational modifications (PTMs). Likewise, AP-MS, which is widely used to identify PPIs and is compatible with PTMs, does not provide definitive information on direct physical interactions between the probe and the bait. Peptide or protein array, which involves the immobilization of a large number of peptides or proteins on a solid support, may lead to false positives or false positives.

Kazuya Machida and Bernard A. Liu (eds.), *SH2 Domains: Methods and Protocols*, Methods in Molecular Biology, vol. 1555, DOI 10.1007/978-1-4939-6762-9_25, © Springer Science+Business Media LLC 2017

Because none of the approaches described above is ideal for mapping authentic PPIs, strategies have been developed to achieve a comprehensive and high-confidence PPI network by integrating data obtained from using complementary approaches. For example, Tong et al. developed a strategy that combined computational prediction of interactions based on consensus sequences obtained from phage display with large-scale Y2H screening [7]. Xu et al. integrated computer modeling and peptide microarray technology to identify binding peptides for the ABL1 SH3 domain [8]. Here, we introduce a new strategy that combines binding signals from reciprocal peptide and protein array screens to produce an SH2 domain-ITRM (Immunoreceptor Tyrosine-based Regulatory Motif) interaction network. The combination of phosphopeptide and SH2 domain arrays yielded protein-protein interactions with greater confidence than that might be achieved using either array alone. By reciprocal screening of a protein array containing 78 human SH2 domains and a peptide array comprised of hundreds of phosphotyrosyl peptides, the approach yielded a comprehensive, high-confidence immunoreceptor signaling network mediated by tyrosine phosphorylation [9]. The same approach may be extended to other peptide-binding modules [10] to systematically and accurately identify PPI networks.

2 Materials

2.1 Reagents and Materials

Isopropyl β-D-1-thiogalactopyranoside (IPTG), lysozyme, Triton X-100, and benzonase were purchased from Millipore. SuperAB glass slides, proteinase inhibitors, neutravidin, and streptavidin-Alexa Fluor 647 (S-21374) were purchased from Thermo Fisher. *Escherichia coli* BL21 DE3 cells were purchased from Biolabs. Bicine, NaCl, EDTA, sucrose, Tris, KCl, Tween-20 were purchased from Sigma-Aldrich. The LB media, agar, and PBS were purchased from AMRESCO. The goat anti-GST antibody was purchased from Abcam.

2.2 Buffers

1. Printing buffer: 37 mM Bicine, pH 8.25, 150 mM NaCl, 0.37 mM EDTA, 0.5 M sucrose.
2. Washing buffer: 1× TBST (200 mM Tris-HCl, 150 mM NaCl, and 2.7 mM KCl, 0.1% Tween-20, pH 7.6).
3. Lysis buffer: 1× PBS with 2% Triton X-100, 1 mg/mL lysozyme, 1× proteinase inhibitor, and 0.1 μL/mL benzonase.
4. Elution buffer: 1× PBS with 20 mM glutathione, pH 8.0.
5. Blotting buffer: 3% BSA in TBST.

3 Methods

3.1 Protein Expression

1. Expression plasmids for the GST-SH2 proteins were transformed into *Escherichia coli* BL21 DE3 cells using an established protocol. Briefly, 2 μL plasmid was incubated with 20 μL BL21 cells for 30 min on ice. Following a 45 s heat shock in 45 °C water bath and subsequent 1 min incubation on ice, 80 μL LB medium was added. The cells were recovered for 1 h by shaking in a 37 °C incubator. Cells were then plated onto LB+ agar plates with appropriate antibiotics, and incubated at 37 °C overnight.
2. Two colonies for each SH2 domain were inoculated into 15 mL LB media with antibiotics in a 50 mL conical tube shaking at 37 °C (*see* **Note 1**). The incubator temperature was adjusted to 18 °C once the OD_{600} reached 0.6–1.0. IPTG was added to a final concentration of 1 mM to induce the protein expression and incubation was continued overnight.
3. The bacteria were collected by centrifugation at 3000×*g* for 15 min the following morning. The supernatant was discarded, the pellet was resuspended in 1 mL of 1× PBS and stored at −80 °C.
4. Frozen bacterial pellet was thawed and the following reagents added to the indicated final concentrations: lysozyme (1 mg/mL), Triton X-100 (2%), proteinase inhibitor (1×), and benzonase (0.1 μL/mL). The mixture was incubated for 30 min on ice to allow lysis.
5. The lysate was centrifuged 40,000×*g* for 30 min at 4 °C. The supernatant was mixed with 50 μl glutathione-Sepharose slurry and the resin was then washed three times with 1× PBS and eluted with 100 μL of 50 mM glutathione in PBS (pH 8.0). The protein concentration was determined using Bradford assay following the manufacturer's protocol.
6. For protein array printing, the proteins were diluted to a final concentration of 25 μM in 1× PBS.

3.2 Peptide Complex

1. All phosphopeptides were synthesized with 6-aminohexanoic acid (ahx) at the N-terminus and labeled with biotin.
2. The peptides were synthesized on Tentagel resin on an Intavis-AG MultiPep peptide synthesizer using N-(9-fluorenyl) methoxycarbonyl (Fmoc) chemistry at 1 μmol scale. The dried peptide powder was dissolved in 100 μL water, with the pH adjusted by 100 mM NaOH to help dissolve the peptides. The pH was readjusted to 7.0 with 100 mM HCl upon peptide dissolution. The final peptide concentration was determined by spectrophotometry.

3. To prepare the peptide array, 10 μL peptide (at 5 mM) was added to 50 μL neutravidin solution (at 25 μM in PBS) (*see* **Note 2**). Centricon filters with a 3 kDa MWCO (Amicon) were used to remove excess unbound peptides by buffer exchange (*see* **Note 3**).

3.3 Peptide Array

1. 30 μL of peptide-neutravidin complex (25 μM) was added to the source plate. Neutravidin was used as the negative control.
2. Using a VersArray Chipwriter Pro (Arrayit), the peptide complex was printed onto preactivated SuperAB agarose (*see* **Note 3**) or nitrocellulose slides.
3. The slides were dried at 60% humidity within the Chipwriter chamber, and then stored at 4 °C until future use.
4. The slides were briefly washed three times with 1× TBST, and then incubated in blocking buffer for 1 h at room temperature.
5. Purified GST fusion proteins (GST-SH2) were diluted in 3% BSA/1× TBST at a final concentration of 1 μM and incubated with the slides overnight at 4 °C.
6. The slides were washed 5 × 1 min with 1× TBST and then incubated with goat anti-GST (diluted 1:2000 in the blocking buffer) for 1 h at room temperature.
7. The slides were washed 5 × 1 min in 1× TBST and then incubated with anti-Goat-Alexa Fluor 647 (diluted 1:10,000 in the blocking buffer, 3% BSA in TBST) for 1 h at room temperature.
8. The slides were washed 5 × 1 min again with 1× TBST and then placed inside a 50 mL conical tube and centrifuged at 1000 × *g* for 5 min to remove excess liquid.
9. The slides were imaged in a microarray laser scanner (Tecan) (Fig. 1a).

3.4 Protein Array

1. Aliquots of 30 μL purified GST-SH2 protein (25 μM) was added to the source plate. GST was used as negative control.
2. Using a VersArray Chipwriter Pro, the proteins were printed onto nitrocellulose slides.
3. The slides were dried at 60% humidity and then stored at 4 °C for future use.
4. Slides were washed 3 × 1 min with 1xTBST and then incubated in blocking buffer for 1 h at room temperature.
5. Biotin-labeled peptides were diluted in 3% BSA/1× TBST to a final concentration of 1 μM and incubated with the slides overnight at 4 °C.

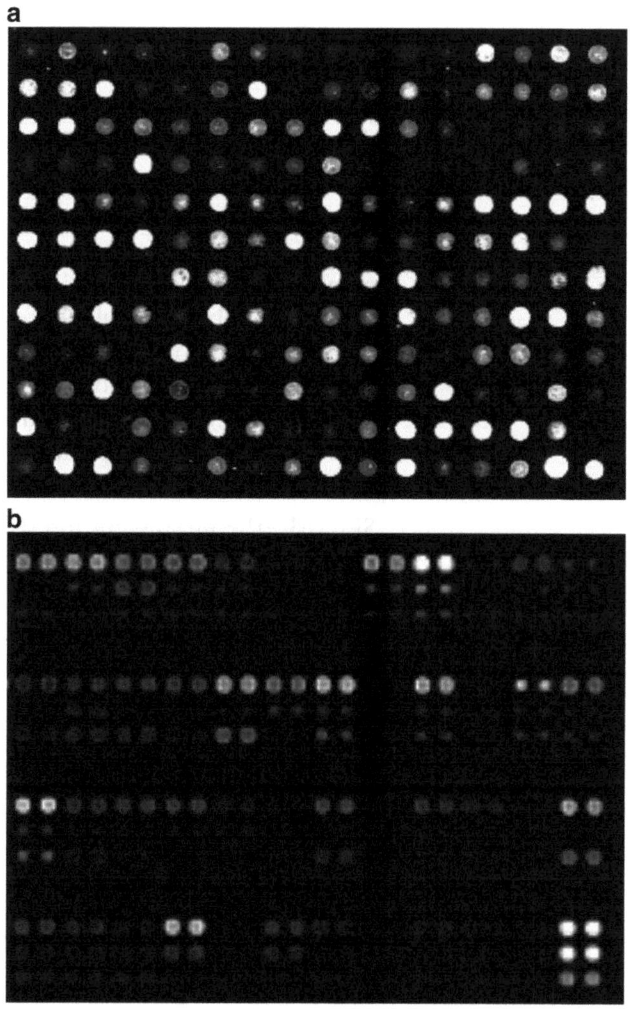

Fig. 1 Representative peptide of domain arrays. (**a**) ApY-containing peptide array was probed with a GST fusion SH2 domain. The peptides were printed in duplication. The bound GST-SH2 protein was visualized by anti-GST-Alexa Fluor 647. (**b**) An SH2 domain array probed by a biotin labeled pY peptide. Proteins were printed in duplication with three different concentrations. The array was visualized by Streptavidin-Alexa Fluor 647

6. The slides were washed 5 × 1 min in 1× TBST and incubated with Streptavidin-Alexa-647 (diluted 1:10,000 in the blocking buffer) for 1 h at room temperature.
7. Slides were washed again 5 × 1 min in 1× TBST, placed inside a 50 mL conical tube, and centrifuged at 1000 × *g* for 5 min to remove excess liquid.
8. The binding signal was measured using a microarray scanner (a representative result is shown in Fig. 1b).

3.5 Data Analysis

1. The signal for each array spot was normalized across the entire protein or peptide array using the following equation:

$$X_{PiDi} = \frac{P_{PiDi} - \mathrm{Min}\left(P_{P1Di}, P_{PnDi}\right)}{\mathrm{Max}\left(P_{P1Di}, P_{PnDi}\right) - \mathrm{Min}\left(P_{P1Di}, P_{PnDi}\right)}$$

where *XPiDi* represents the normalized peptide array binding score for the interaction between protein domain i (*Di*) and peptide i (*Pi*) on the peptide array, and *PPiDi* represents the strength of the original fluorescent signal for protein *Di* to peptide *Pi* on the peptide array. Max denotes the strongest binding signal produced by a protein *Di* while Min represents the weakest signal for a protein on the peptide array that contains "*n*" number of peptides. The value of *XPiDi* ranges from 0 (weakest binding) to 1 (strongest binding).

2. Similarly, the fluorescent intensity of each spot on the protein array was normalized from 0 to 1 according the equation:

$$Y_{PiDi} = \frac{D_{PiDi} - \mathrm{Min}\left(D_{PiD1}, D_{PiDn}\right)}{\mathrm{Max}\left(D_{PiD1}, D_{PiDn}\right) - \mathrm{Min}\left(D_{PiD1}, D_{PiDn}\right)}$$

where *YPiDi* represents the normalized protein array score for the interaction between *Pi* and *Di* from the protein array, and *DPiDi* represents the original fluorescent intensity corresponding to the interaction between *Pi* and *Di* on the protein array.

3. The bidirectional array score, *BPiDi*, for a given peptide-protein pair is defined by the average of the corresponding peptide array score *XPiDi* and protein array score *YPiDi*:

$$B_{PiDi} = \left(X_{PiDi} + Y_{PiDi}\right) / 2$$

4. The consistence score between the peptide and protein array binding signals for protein-peptide interaction is defined as follows:

$$C_{PiDi} = 1 - \left|X_{PiDi} - Y_{PiDi}\right|$$

where *CPiDi* represents the array consistence value of the interaction between peptide *Pi* and protein *Di*.

5. By plotting (Fig. 2) *BPiDi* and *CPiDi*, the data was divided into four squares according to the chosen cutoff (any score between 0 and 1 is acceptable with 0.5 recommended). The interaction pairs above the cutoff in both *BPiDi* and *CPiDi* are considered double-positive interactions, which were used as gold standards. By applying these gold standard interactions back on each probed slide as positive controls, more positive binders were revealed.

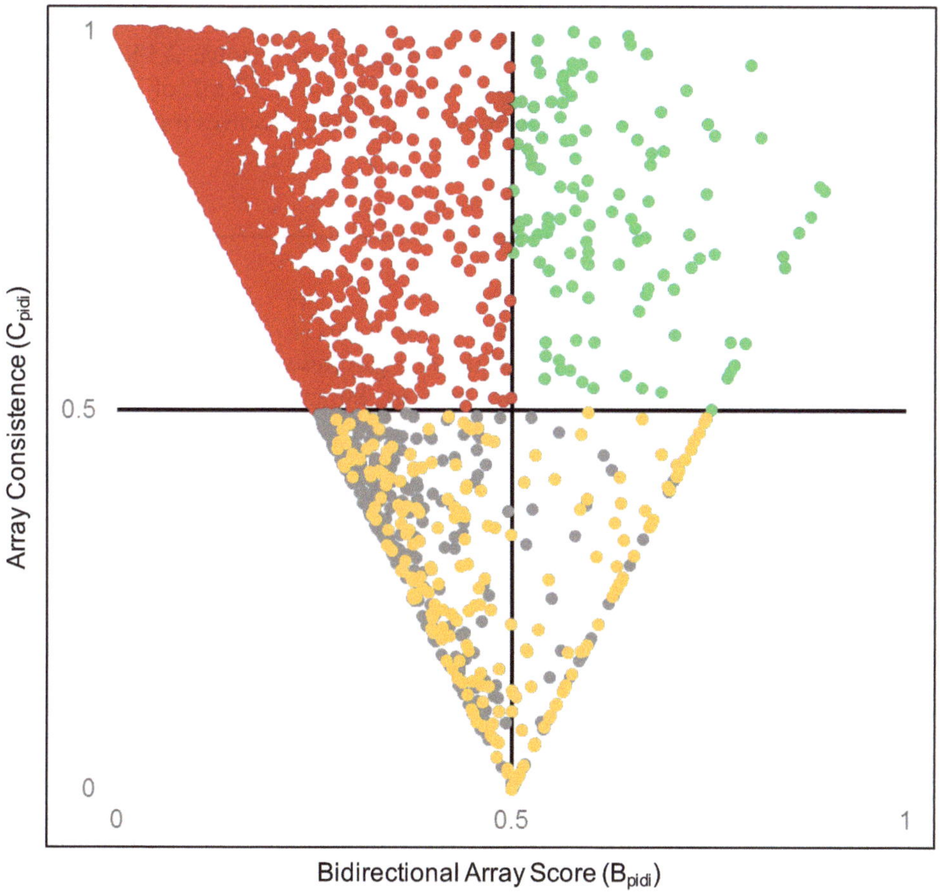

Fig. 2 A representative plot of the reciprocal protein-peptide array binding data. SH2-pY interactions identified in the reciprocal protein-peptide array screen are plotted using the corresponding bidirectional array scores and array consistence scores (*see* main text for definition). The SH2-pY pairs with both scores greater than 0.5 were treated as high-confidence positive interactions (*green dots*). The pairs with high consistence score but low array score were predicted as negatives (in *red*). High-confidence positive pairs were used gold standards to set the cutoff for each array. Signals greater than the gold standards in each array were rescued as positives (in *yellow*) from the low consistence pairs

4 Notes

1. The benefit of using 50 mL conical tubes is that they are easier to assemble in the incubator for high-throughput protein expression. The cap of the tube should be left loose and secured using tape to allow for sufficient aeration.
2. Neutravidin should be dissolved in pure water before an appropriate amount of 10× PBS was added to make the Neutravidin-PBS solution. Neutravidin cannot be dissolved in PBS directly.
3. Agarose-based slides were activated with 50 mM NaIO4 in 0.1 M sodium acetate pH 5.5 for 0.5 h at room temperature,

rinsed twice with water, air-dried and used immediately. To prepare the activation solution, dissolve 2.14 g of $NaIO_4$ in 90 mL deionized water, then add 10 mL of 1 M sodium acetate, pH 5.5. Sodium periodate dissolves more readily in water than in buffer.

Acknowledgments

This work was supported by grants (to S.S.C.L.) from the Canadian Cancer Society and the Ontario Research Fund. S.S.C.L. holds the Canada Research Chair in Functional Genomics and Cellular Proteomics.

References

1. Kathiriya JJ, Pathak RR, Clayman E, Xue B, Uversky VN, Dave V (2014) Presence and utility of intrinsically disordered regions in kinases. Mol Biosyst 10(11):2876–2888. doi:10.1039/c4mb00224e
2. Lamesch P, Li N, Milstein S, Fan C, Hao T, Szabo G, Hu Z, Venkatesan K, Bethel G, Martin P, Rogers J, Lawlor S, McLaren S, Dricot A, Borick H, Cusick ME, Vandenhaute J, Dunham I, Hill DE, Vidal M (2007) hORFeome v3.1: a resource of human open reading frames representing over 10,000 human genes. Genomics 89(3):307–315. doi:10.1016/j.ygeno.2006.11.012
3. Brettner LM, Masel J (2012) Protein stickiness, rather than number of functional protein-protein interactions, predicts expression noise and plasticity in yeast. BMC Syst Biol 6:128. doi:10.1186/1752-0509-6-128
4. Huang H, Li L, Wu C, Schibli D, Colwill K, Ma S, Li C, Roy P, Ho K, Songyang Z, Pawson T, Gao Y, Li SS (2008) Defining the specificity space of the human SRC homology 2 domain. Mol Cell Proteomics 7(4):768–784. doi:10.1074/mcp.M700312-MCP200
5. Melton L (2004) Protein arrays: proteomics in multiplex. Nature 429(6987):101–107. doi:10.1038/429101a
6. Krallinger M, Leitner F, Rodriguez-Penagos C, Valencia A (2008) Overview of the protein-protein interaction annotation extraction task of BioCreative II. Genome Biol 9(Suppl 2):S4. doi:10.1186/gb-2008-9-s2-s4
7. Tong AHY, Drees B, Nardelli G, Bader GD, Brannetti B, Castagnoli L, Evangelista M, Ferracuti S, Nelson B, Paoluzi S, Quondam M, Zucconi A, Hogue CWV, Fields S, Boone C, Cesareni G (2002) A combined experimental and computational strategy to define protein interaction networks for peptide recognition modules. Science 295(5553):321–324
8. Xu Z, Hou T, Li N, Xu Y, Wang W (2012) Proteome-wide detection of Abl1 SH3-binding peptides by integrating computational prediction and peptide microarray. Mol Cell Proteomics 11(1):O111.010389. doi:10.1074/mcp.O111.010389
9. Liu H, Li L, Voss C, Wang F, Liu J, Li SS (2015) A comprehensive immunoreceptor phosphotyrosine-based signaling network revealed by reciprocal protein-peptide array screening. Mol Cell Proteomics 14:1846–1858. doi:10.1074/mcp.M115.047951
10. Biggar KK, Li SSC (2015) Non-histone protein methylation as a regulator of cellular signalling and function. Nat Rev Mol Cell Biol 16(1):13

Chapter 26

Rosette Assay: Highly Customizable Dot-Blot for SH2 Domain Screening

Khong Y. Ng and Kazuya Machida

Abstract

With a growing number of high-throughput studies, structural analyses, and availability of protein-protein interaction databases, it is now possible to apply web-based prediction tools to SH2 domain-interactions. However, in silico prediction is not always reliable and requires experimental validation. Rosette assay is a dot blot-based reverse-phase assay developed for the assessment of binding between SH2 domains and their ligands. It is conveniently customizable, allowing for low- to high-throughput analysis of interactions between various numbers of SH2 domains and their ligands, e.g., short peptides, purified proteins, and cell lysates. The binding assay is performed in a 96-well plate (MBA or MWA apparatus) in which a sample spotted membrane is incubated with up to 96 labeled SH2 domains. Bound domains are detected and quantified using a chemiluminescence or near-infrared fluorescence (IR) imaging system. In this chapter, we describe a practical protocol for rosette assay to assess interactions between synthesized tyrosine phosphorylated peptides and a library of GST-tagged SH2 domains. Since the methodology is not confined to assessment of SH2–pTyr interactions, rosette assay can be broadly utilized for ligand and drug screening using different protein interaction domains or antibodies.

Key words Protein interaction domain, Dot blot, GST fusion protein, SH2 domain, Tyrosine phosphorylation, Reverse-phase assay

1 Introduction

Tyrosine phosphorylation plays a central role in signal transduction controlling many important biological processes. Aberrant activation of tyrosine phosphorylation signaling has been observed in various pathological conditions such as tumors and immunodeficiency disorders [1–3]. Activation of tyrosine kinases leads to tyrosine phosphorylation of signaling proteins, thus providing docking sites for the binding of Src homology 2 (SH2) domains, a modular interaction domain of ~100 amino acids present in many signaling proteins [4]. Based on phosphopeptide library screens and structural studies, binding specificity of SH2 domains is defined by a sequence motif of amino acids, typically three residues C-terminal

Kazuya Machida and Bernard A. Liu (eds.), *SH2 Domains: Methods and Protocols*, Methods in Molecular Biology, vol. 1555,
DOI 10.1007/978-1-4939-6762-9_26, © Springer Science+Business Media LLC 2017

to the phosphotyrosine within the target protein [5]. Because of their modular nature, SH2 domains can be expressed as isolated without losing their specificity in target binding. Thus, the set of 120 SH2 domains currently known to be encoded in the human genome provide an ideal tool for profiling the cellular state of tyrosine phosphorylation [6]. The growing number of SH2 domain-focused high-throughput studies, structural analyses, and availability of protein-protein interaction databases have since progressed to the development of various algorithm-based web prediction tools for SH2 domain-interactions [7–11]. However, these prediction tools can only provide a direction for further focused experimental testing at best but remain inadequate in replacing modular domain-binding assays [12].

We previously developed the rosette assay as an SH2 domain-based profiling tool and were able to use it to detect changes in the global state of tyrosine phosphorylation of cancer cells [13–15]. Rosette assay is a reverse-phase dot blot assay originally developed for assessing interactions between SH2 domains and membrane-immobilized ligands. The assay is customizable for various numbers of SH2 domains and their ligands including short peptides, purified proteins, and cell lysates. Rosette assay is performed in a 96-well chamber plate (MBA or MWA apparatus) in which up to 96 labeled SH2 domains can be individually incubated with membrane-bound ligands. Bound domains are then detected and quantified using a chemiluminescence or near-infrared fluorescence (IR) imaging system. The adaptability, cost of reagents, assay runtime, and instrumental requirements are less demanding for any standard molecular biology laboratories regularly performing immunoblotting experiments.

In this chapter, we describe a practical protocol for rosette assay to quantify interactions between synthesized tyrosine phosphorylated peptides and a library of GST-tagged SH2 domains (*see* **Note 1**). Since the methodology is not confined to the assessment of SH2–pTyr interactions, rosette assay can be readily applied to ligand and drug screening of other protein interaction domains and antibodies.

2 Materials

1. Affinity purified GST-SH2 domains (*see* **Note 2**, Chapter 8).
2. Biotin-labeled synthetic tyrosine phosphorylated peptides (*see* **Note 3**).
3. 2× spotting solution: 100 mM Tris–HCl pH 6.8, 30% glycerol, 2% sodium dodecyl sulfate (SDS) (*see* **Note 4**).
4. Transparency film (US letter or A4 paper size, laser printer-ready).

5. Adobe Illustrator or other vector drawing program.
6. Black & white laser printer with 300 dpi or higher printing resolution.
7. Protran BA79 0.1 μm pore size nitrocellulose membrane 200 × 200 mm size (*see* **Note 5**).
8. Odyssey Pen (LI-COR Biotechnology).
9. White light transilluminator (US letter or A4 paper size).
10. Gel loading tips (~70 mm long, ~0.6 mm orifice diameter).
11. 4 % paraformaldehyde-PBS solution. Prepare fresh by dissolving 4 g of paraformaldehyde in 100 mL of hot PBS (50–60 °C) while stirring. Filter with a Kimwipe and cool down to room temperature. Paraformaldehyde is toxic. Wear suitable protective gloves and clothing to prevent skin, eye, or breathing exposures. Pre-dissolved paraformaldehyde solution is also acceptable.
12. 0.5 % gelatin solution. Prepare fresh by dissolving 0.5 g premium quality type A gelatin (G2500, Sigma) in 100 mL of hot water (50–60 °C) while stirring. Filter with a Kimwipe and keep at 37 °C.
13. Equilibration solution: 10 mM 3-(cyclohexylamino)-1-propane-sulfonic acid (CAPS), pH 11, 20 % (v/v) methanol. Store at 4 °C.
14. Tris-buffered saline with Tween (TBST): 150 mM NaCl, 10 mM Tris–HCl, pH 8.0, 0.05 % Tween 20 (v/v). Filter with a Kimwipe before use.
15. Blocking solution: 10 % (w/v) nonfat dry milk in TBST. Prepare fresh. Filter with a Kimwipe before use.
16. MBA chemotaxis chamber (MBA96, Neuroprobe) or MWA hybridization unit (MWA100-96-1, Gel Company).
17. Membrane support paper pad (MBA only). Cut blotting paper (~0.85 mm thickness) into 125 × 85 mm size and cover with Parafilm. Fix Parafilm to the blotting paper using adhesive.
18. Membrane support plate (MBA only): 96-well plate.
19. Probe reservoir plate: 96-well round-bottom plate.
20. 96-Well plate sealing film.
21. 8- or 12-channel pipette (20–200 μL capacity).
22. Dylight 800-conjugated anti-GST antibody (600-145-200, Rockland Immunochemicals).
23. IRDye 680RD-conjugated streptavidin (926-68079, LI-COR Biotechnology).
24. Odyssey Blocking Buffer (TBS) (LI-COR Biotechnology).

25. Phosphate buffered saline (PBS). Filter with a Kimwipe before use.
26. Odyssey Imager (LI-COR Biotechnology) or other near-infrared fluorescence scanner.
27. Odyssey 3.0 software (LI-COR Biotechnology).
28. ImageStudio (LI-COR Biotechnology) or other densitometry software (*see* **Note 6**).
29. Data analysis programs: Microsoft Excel, GraphPad Prism, Matlab, RStudio, or other computing tools.

3 Methods

In rosette assay, a set of peptide samples are spotted on a gelatin-coated nitrocellulose membrane in register with a 96-well apparatus. After membrane fixation, individual wells are separately incubated with diluted GST-SH2 solutions. The number of samples and SH2 domains to be analyzed is highly customizable: Typically, 4–16 samples and 48 or 96 SH2 probes are used per plate. The whole process of a rosette assay usually takes 3 days: On day 1, lyophilized peptides are dissolved and spotted on two or four membranes (dot blot). On day 2, the dried membrane is fixed, equilibrated, blocked, and incubated with SH2 probes overnight. On day 3, wells are washed individually. Then the entire membrane is incubated with IR-labeled secondary antibodies. And finally, the membrane is washed and bound probes are detected and quantified.

3.1 Preparation of Spotting Template

1. Using a vector drawing PC program such as Adobe Illustrator, design an appropriate spot array template as shown in Fig. 1. Round and square arrays fit the MBA and MWA plates, respectively. As the Protran nitrocellulose with 0.1 μm pore size slightly expands in buffer solution, the array area should be 69.5 × 105 mm as in Fig. 1 to allow for correct alignment with the MBA/MWA plates (*see* **Note 7**).
2. Design a mini array template containing four or eight wells (*see* Fig. 1 top left and **Note 8**).
3. Print the appropriately scaled array image on a transparency film using the laser printer (*see* **Note 9**).
4. Cut nitrocellulose membranes to the size of 75 × 113 mm and soak in 0.5% gelatin solution at 37 °C for 30 min. Remove excessive solution and let the membrane air-dry at room temperature for at least 1 h. Apply the gelatin coating procedure to the small mini array membrane as well. Avoid handling the membranes with bare hand.

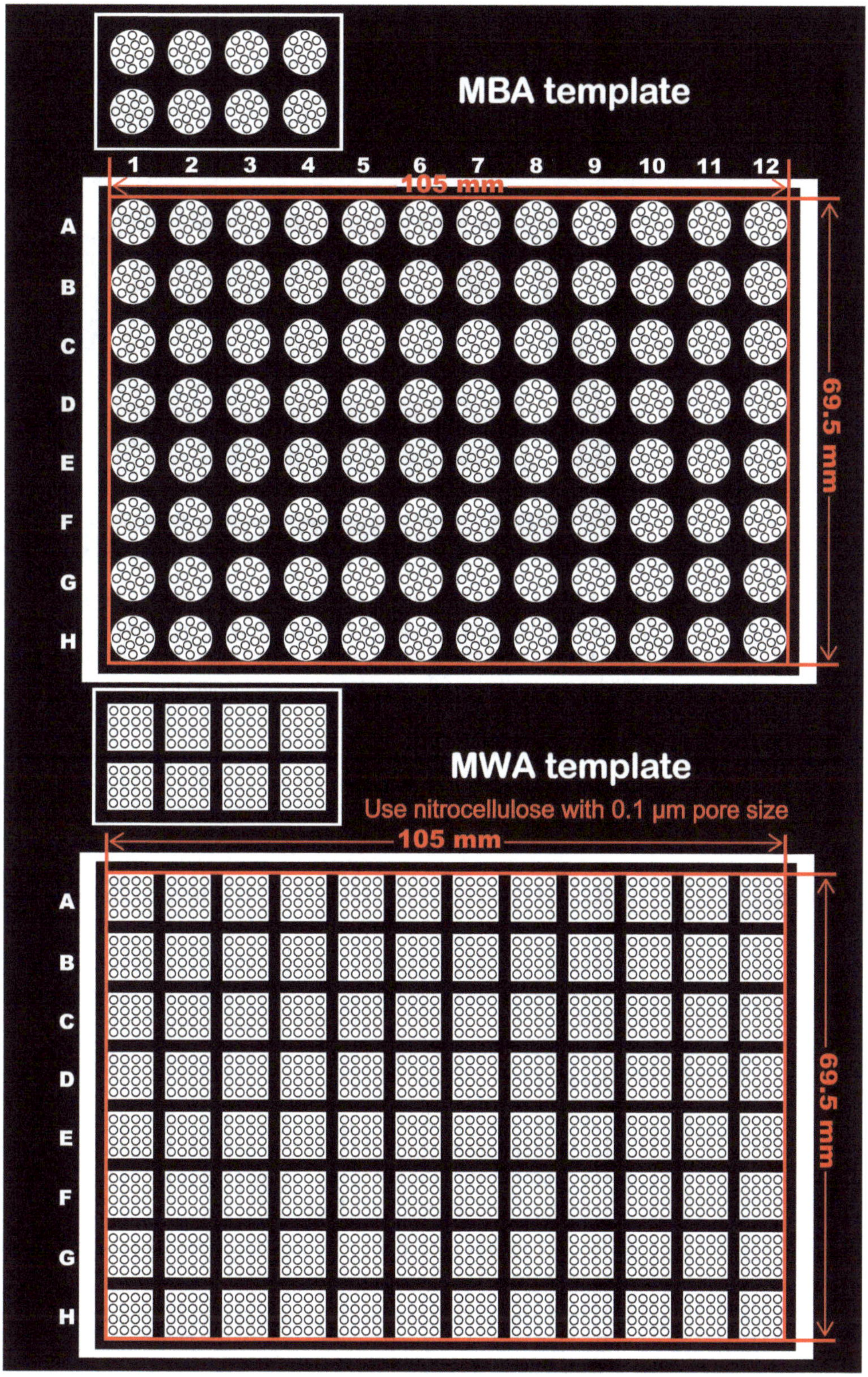

Fig. 1 Templates for the MBA plate (*upper panel*) and MWA plate (*lower panel*) for 12 or 16 peptide arrays, respectively. Due to expansion of nitrocellulose membrane with 0.1 μm pore size in solution, the 96-well array area should be 69.5×105 mm for printing with the laser printer. Print the template image reversely as the membrane should be attached to the nonprinted side of the transparency film

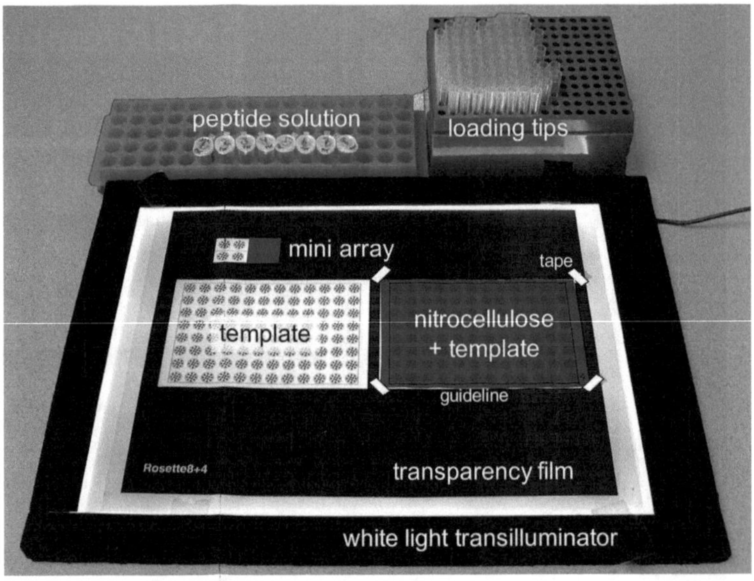

Fig. 2 Experimental setup for spotting. Nitrocellulose membrane is placed on a transparency film printed with the template array for the MBA plate. White light transilluminator allows template spots to be visible through the membrane. Peptides are spotted in 96 replicates using gel-loading tips

5. Position dried membranes on the template and secure the corners with adhesive tape (*see* **Note 10**). Place the template film to the white light illuminator and prepare necessary items for spotting as in Fig. 2.

3.2 Spotting of Peptides

1. Remove lyophilized peptide from the freezer and warm it to room temperature for 10 min.
2. Weigh out approximately 2 mg peptide and dissolve it in a small amount of sterile water or an aqueous solution such as pH 8.0 Tris buffer. Adjust the concentration to ~10 mg/mL (~5 mM) and add the same volume of 2× spotting buffer (*see* **Note 11**).
3. Fill a gel-loading pipette tip with the peptide solution by capillary action, and manually spot the peptide by quickly touching the membrane surface at the spot positions indicated by the template underneath (*see* **Note 12**).
4. Repeat **step 3** with the remaining peptides and/or membranes (*see* **Note 13**).
5. When spotting is done, draw a rectangular guideline around the array area with the Odyssey Pen. This guideline will be used later to align the membrane to the 96-well plate (*see* Figs. 4, 7, and **Note 14**).
6. Allow the spotted membranes to dry at room temperature overnight.

3.3 Preparing Membrane for Probe Binding

1. Fixation: Immerse the membrane in 4% paraformaldehyde-PBS solution and leave it at room temperature for 10 min. Then, gently rinse it three times with TBS-T.
2. Equilibration: Immerse the membrane in equilibration solution with a slow swirling motion (agitation) for 30 min and change the solution every 10 min. Then gently rinse it three times with TBS-T.
3. Blocking: Immerse the membrane in blocking solution with gentle agitation for 1–2 h at room temperature.

3.4 Probe Binding Using MBA Apparatus

Follow the steps in this subheading if you are using the MBA 96-well plate (Figs. 3 and 4). If you are using the MWA 96-well plate, skip to Subheading 3.5.

1. For a typical screen with 48 SH2 domains, add 200 μL of blocking buffer to columns 1, 3, 5, 7, 9, and 11 of the probe reservoir plate at 4 °C (Fig. 5, Plate layout 1). Add 1 μg of respective GST-SH2 probe into the wells above, resulting in a probing concentration of 5 μg/mL or ~150 nM (*see* **Note 15**).

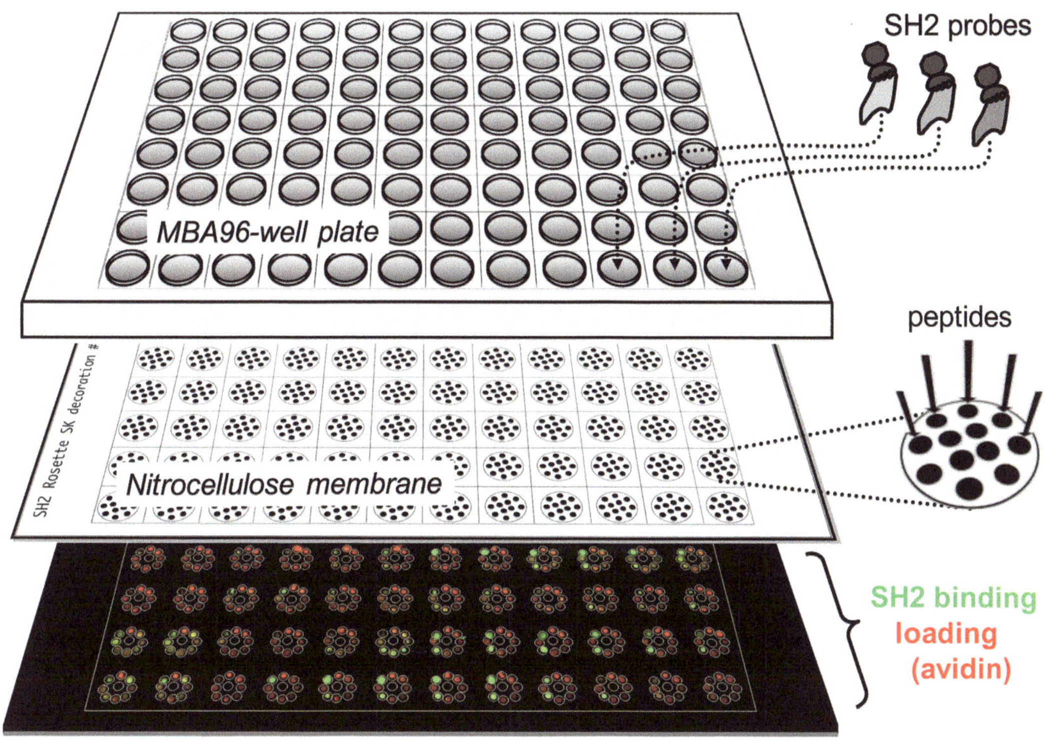

Fig. 3 Overview of SH2 rosette assay using MBA 96-well plate. A set of samples are repeatedly spotted and fixed on a membrane. For example, a membrane may contain 96 replicates of a 12-sample rosette. Each set of samples is incubated with different GST-SH2 probes in separate wells of an MBA 96-well chamber plate. SH2 binding is detected via 800 nm infrared dye-conjugated anti-GST antibody and IR imager. The loaded amount of biotin-tagged peptides can be monitored using 680 nm infrared dye-conjugated avidin

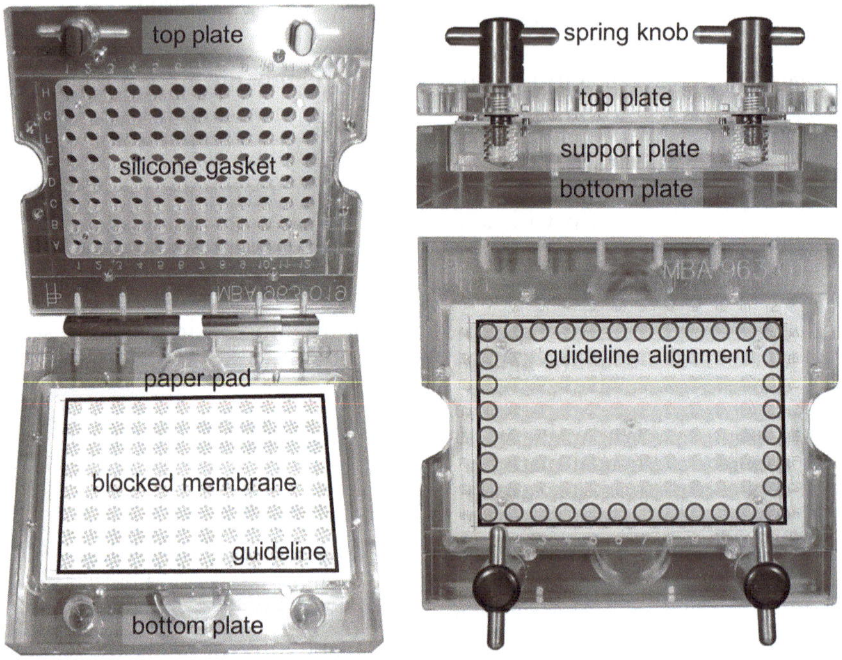

Fig. 4 Setup for MBA 96-well chamber plate. A blocked membrane is placed on Parafilm-covered paper pad above a support plate in register with the MBA 96-well top plate. As peptide spots are colorless, alignment with wells in the top plate is adjusted using the guideline on the membrane. 100 μL of probe solution is then transferred into each well using a multi-channel pipette and the apparatus is incubated overnight at 4 °C

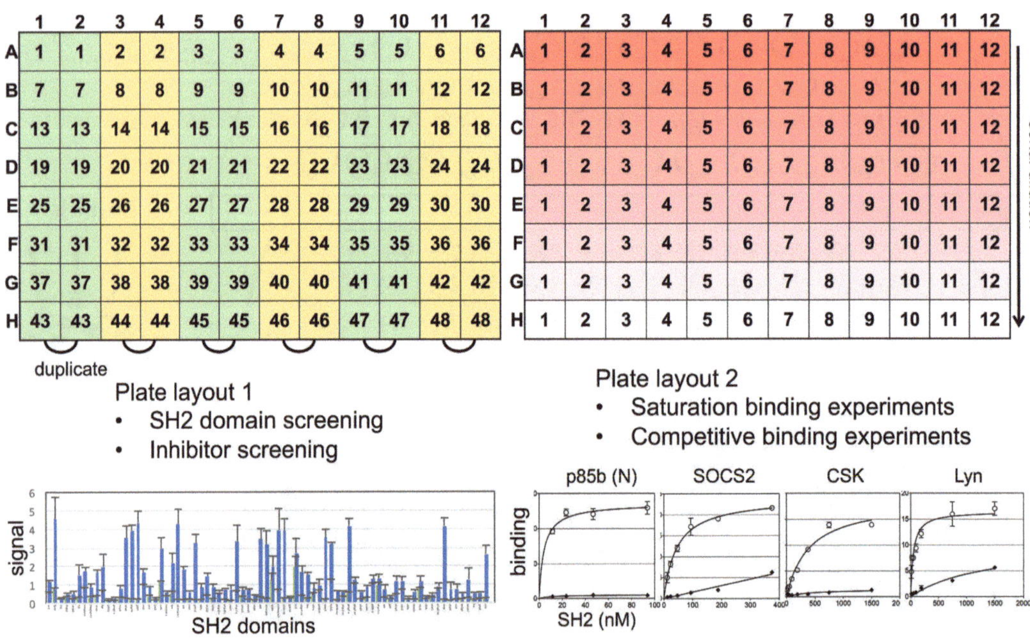

Fig. 5 Rosette assay plate layout. (*Left*) 200 μL of diluted SH2 domain solution is prepared in columns 1, 3, 5, 7, 9, and 11 of a probe reservoir plate and 100 μL is transferred to the MBA/MWA 96-well plate in duplicate as in plate layout 1. This layout is suitable for screening with 48 SH2 domains or inhibitors. (*Right*) The highest concentration of SH2 domain is prepared in row A in a probe reservoir plate and serially diluted from rows B to H. These dilutions are then transferred to an MBA/MWA 96-well plate as in plate layout 2. This layout is suitable for saturation binding or competitive binding experiments

For measurement of the binding affinity or inhibitory concentrations, plates can be set up as shown in Fig. 5 Plate layout 2.

2. Assembling the MBA96 chemotaxis chambers. Flip the top plate of the chamber up and set a gasket to it. Place a membrane supporting plate into the lower plate of the chamber. Place a filter paper pad covered with Parafilm on the top of the membrane supporting plate (Fig. 4). Always wear gloves when handling the plate.
3. Positioning the blocked membrane. Gently lift up the edge of membrane with a forceps and drain off excessive blocking solution and place it on the Parafilm surface with the protein side up. Position the membrane with a forceps to make sure all the spots are properly aligned within the wells of the upper plate with the help of the guidelines (Fig. 4).
4. Close the top plate and press it down firmly before turning the securing knobs clockwise to seal the plates together. Be careful not to tighten too much.
5. Transfer 100 μL of each probe solution from the reservoir plate (**step 1**) into each well using a multi-channel pipette. Cover the chambers with a 96-well plate sealing film and incubate overnight at 4 °C (*see* **Note 16**).

3.5 Probe Binding Using MWA Apparatus

Follow the steps in this subheading if you are using the MWA 96-well plate (Figs. 6 and 7). If you are using the MBA 96-well plate, skip to Subheading 3.6.

1. For a typical screen with 48 SH2 domains, add 200 μL of blocking buffer to columns 1, 3, 5, 7, 9, and 11 of the probe reservoir plate at 4 °C (Fig. 5, Plate layout 1). Add 1 μg of respective GST-SH2 probe into the wells above, resulting in a probing concentration of 5 μg/mL or ~150 nM (*see* **Note 15**). For measurement of the binding affinity or inhibitory concentration, plates can be set up as shown in Fig. 5 Plate layout 2.
2. Loosen the screws and disassemble the MWA 96-well plate. Gently lift up the edge of membrane with a forceps and drain off excessive blocking solution and place it on the Parafilm surface with the protein side up. Position the membrane with a forceps to make sure all the spots are properly aligned within the wells of the upper plate with the help of the guidelines (Fig. 7).
3. Close the top plate and screw down all four corners firmly. Be careful not to dry the membrane from here onward.
4. Transfer 100 μL probe of the probe solutions from the reservoir plate (**step 1**) into each well using a multi-channel pipette. Cover the chambers with a 96-well plate sealing film to avoid evaporation and incubate it overnight at 4 °C (*see* **Note 16**).

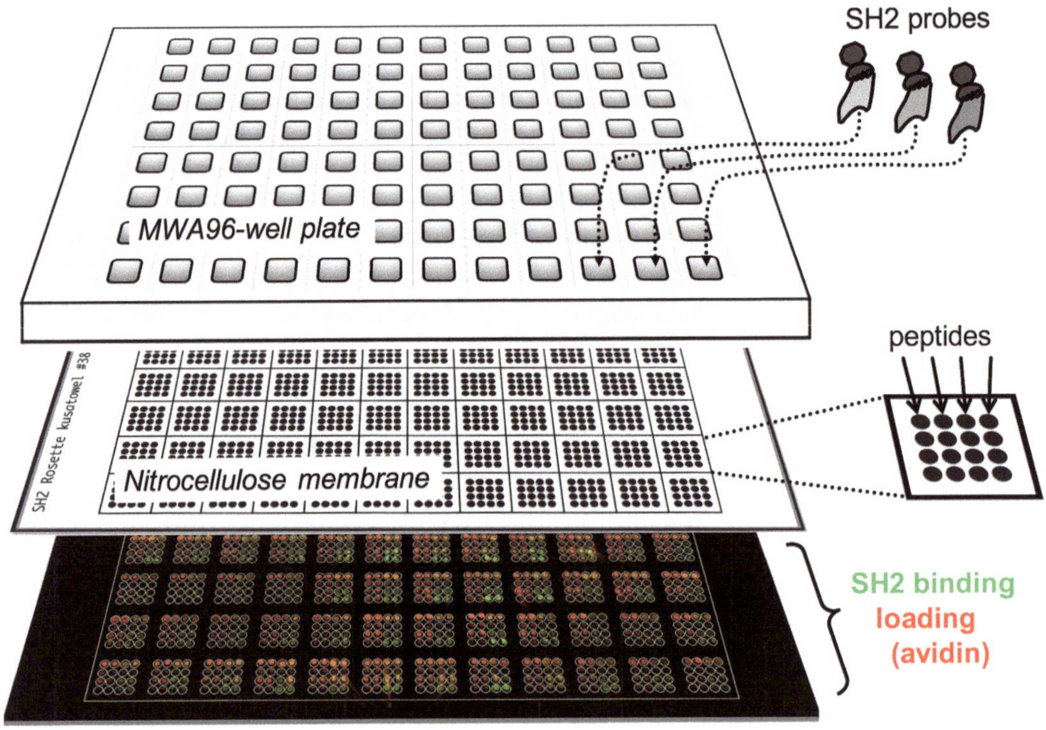

Fig. 6 Overview of SH2 rosette assay using MWA 96-well plate. A set of samples are repeatedly spotted and fixed on a membrane. For example, a membrane above contains 96 replicates of a 16-sample rosette. Each set of samples is incubated with different GST-SH2 probes in separate wells of an MWA 96-well chamber plate. SH2 binding is detected via 800 nm infrared dye-conjugated anti-GST antibody and IR imager. The loaded amount of biotin-tagged peptides can be monitored using 680 nm infrared dye-conjugated avidin

3.6 Secondary Antibody Binding

1. To wash the wells, remove the covering film and use the multi-channel pipette to remove the probe solution from the wells and fill them with 100 μL TBST solution. Repeat the wash with TBST for two additional times.
2. After the final wash, disassemble the MBA/MWA apparatus to remove the membrane and transfer it to a tray containing TBST. Rinse briefly and then wash for 15 min with gentle agitation, replacing the buffer every 5 min (*see* **Note 17**).
3. Incubate the membrane with Odyssey Blocking Buffer-diluted Dylight 800-conjugated anti-GST antibody (1:10,000) and IRDye 680RD-conjugated streptavidin (1:20,000) in a sealed nylon bag (covered with aluminum foil) for 1.5 h at room temperature.

3.7 IR Detection

1. Remove the membrane from the nylon bag and wash with sufficient volume of TBST for 20 min under gentle agitation, replacing the buffer every 5 min (*see* **Note 17**).
2. Rinse the membrane three times with PBS solution.

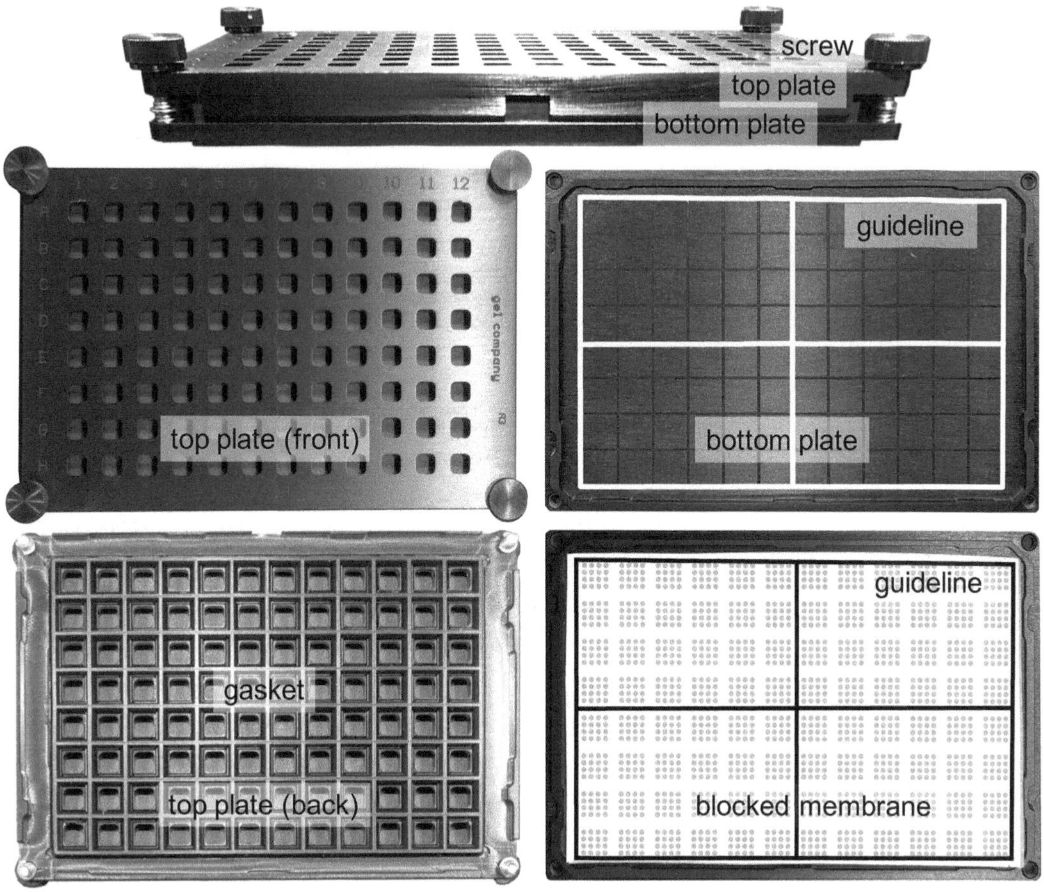

Fig. 7 Setup for MWA 96-well plate. A blocked membrane is placed between the top plate (equipped with gasket) and the bottom plate. Alignment of peptide spots with sectioned squares in the top plate is adjusted using the guidelines on the bottom plate and the membrane. 100 μL of probe solution is transferred to individual squares using a multi-channel pipette and the apparatus is incubated overnight at 4 °C

3. Open the Odyssey Imager and place the membrane facing down on the glass. Cover the membrane with the silicon mat and roll out any air bubbles.
4. Open the Odyssey 3.0 software on the PC and run a preview scan, set appropriate intensity values for the 700 and 800 nm channels, and start scan at the "85 nm" resolution and "high quality" settings (*see* **Note 18**).
5. Once the scan is done, save the image with appropriate file name. Remove the membrane, and clean the scanner glass. Store the membrane wrapped in plastic wrap at −20 °C for future use.

3.8 Quantification

1. Open the ImageStudio program and import the experiment file.
2. On the "Analysis" tab, select "Plate Array," select "96 Well," and insert row and column numbers appropriately (*see* **Note 19**).

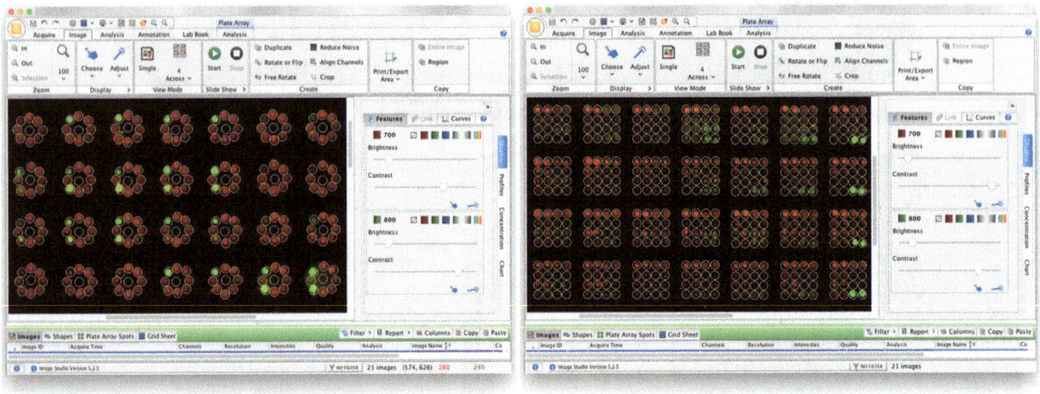

Fig. 8 Quantification of SH2 binding using ImageStudio. (*Left*) Analysis is set to 3 × 3 arrays and spots are rearranged for alignment with the circular MBA plate array. (*Right*) Analysis is set to 4 × 4 arrays and spots are aligned with the square MWA plate array. Both SH2 binding (800 nm channel) and peptide loading (700 nm channel) are quantified at the same time

3. On the "Display" panel, adjust signal intensity of the 700 nm channel (avidin) via "Choose Display" and "Adjust Display" dialogs to visualize all peptide spots. Adjust the position and size of yellow circles to align with individual spots in the array (Fig. 8).
4. Export and analyze quantified data using the data analysis programs (*see* **Note 20**).

4 Notes

1. The protocol described here is optimized for synthesized linear peptides and IR detection system. Lysate samples can be analyzed by using the following protocol modifications: (a) Use nitrocellulose with 0.2 μm pore size; (b) Spotting area of the template should be printed to 69.5 × 106 mm size (instead of 69.5 × 105 mm, *see* Fig. 1); (c) Adjust lysate concentration to 4 μg/μL; (d) Do not perform gelatin coating (Subheading 3.1, **step 4**); (e) Do not perform paraformaldehyde treatment (Subheading 3.3, **step 1**); (f) If samples are whole cell lysates, user may consider counterstaining with a house-keeping gene such as tubulin, actin, GAPDH, etc. as a loading control in the 700 nm channel.
2. We store aliquots of purified GST-SH2 domains at −80 °C (0.1 μg/μL in 10% glycerol-PBS) and thaw them just before use. The diluted probe solution can be kept at 4 °C for up to 2 weeks.
3. We usually design a pair of 12 amino acid peptide, which are either phosphorylated or nonphosphorylated (four and seven residues N- and C-terminal to the pTyr, respectively) with biotinylated N-terminal. In addition, cysteine residues, which are

known to cause nonspecific binding due to intermolecular disulfide bond formation, may be replaced with Ser [16], and the addition of Asp residue to the C-terminal may also improve peptide solubility [17].

4. This solution is colorless as blue dye causes high background with IR detection. 0.03% bromophenol blue can be included if chemiluminescent detection is used. In this case, peptide spots appear bluish and may be helpful for aligning wells with the chamber.

5. Cut four pieces of 75 × 113 mm size membrane from a 200 × 200 mm sheet. Remaining pieces are used for a test array (*see* **Note 8**). Wear protective gloves for membrane handling.

6. For precise quantification, range it is highly recommended to use raw images with a default imaging program for densitometric analysis, rather than using exported/processed images with a third party densitometry measuring program.

7. The template should be customized for each experiment: in our hands, as many as 19 (MBA) and 36 (MWA) spots can be fitted into a single well/grid (MWA has a larger spotting area). With appropriate scaling, the illustration in Fig. 1 can be used as a spotting template (adjust the printed area to 69.5 × 105 mm). For a typical SH2 screening, we print dual or quad templates on a single film and spot all membranes at the same time.

8. Prior to setting up a 96-well scale experiment, assay conditions can be optimized using a test array run (Fig. 1 upper left corner). For example, peptide loading/immobilization and phosphorylation status can be monitored using streptavidin and anti-pTyr antibody, respectively. If the signal appears to be too low with streptavidin and/or anti-pTyr, the sample concentration for spotting should be increased. Minor variation in loading between spots can be normalized using the streptavidin signal.

9. As the nitrocellulose membrane will be placed on the non-printed side of the film, the template image should be inverted. Prepare an inverted image using the drawing program or simply reverse print if possible.

10. Do not tape more than a 2 × 2 mm area to avoid membrane damage upon removal.

11. Determine the solubility of each peptide by starting with a small amount of the peptide. If the peptide does not dissolve in aqueous solutions, try dissolving the peptide with a small amount of DMSO, DMF, Methanol, hydrochloric acid, etc., and dilute with water to ~10 mg/mL. The final concentration of organic solvent, e.g., DMF, should be lower than 20% to avoid damaging the membrane (you may test this on an extra piece of membrane). Once dissolved, the peptide should be spotted as soon as possible. If the samples become too viscous

or turbid at room temperature, warm and keep it on a heat block while spotting (<60 °C).

12. Typically, 0.1–0.2 μL of sample per spot or ~20 μL per 96-well array is required. If a spot is too diffusive/confined, due to suboptimal capillary action, change the concentration of glycerol and/or SDS in the spotting buffer.
13. Complete spotting of four 96-well membranes takes ~4 h. Take occasional breaks to maintain your concentration on precise spotting. Record any unexpected miss-spotting for future reference.
14. Do not use any other pens as common blue and black inks cause unfavorable background in the 700 nm channel of the Odyssey Imager.
15. Optimal concentration (maximum signal and low nonspecific background) should be determined empirically in pilot experiments using the mini array. Empirically, a GST-SH2 probing concentration of 100–200 nM gives a good signal-to-noise ratio for detection.
16. Do not agitate the plate as this may cause the probe solution to leak from one chamber to another and thus, causing cross-well contamination.
17. The speed of agitation and wash time can significantly affect the quality of the result. Do not agitate too rapidly or wash for greater than 30 min.
18. Usually, optimal images can be obtained with intensity levels of 2–3 (700 nm channel) and 5–6 (800 nm channel). Set the parameters similarly when ImageStudio is used for scan.
19. A square array (e.g., Fig. 1, MWA template) fits well with the default ImageStudio array while a circular array (e.g., Fig. 1, MBA template) requires rearrangement of spots for quantification.
20. ImageStudio has different signal and background settings such as "Total," "Signal," and "Signal Trim." Spots with signal saturation are expressed as "infinity," which implies detector saturation. Re-scan the membrane with a lower intensity (*see* **Note 18**).

Acknowledgments

We thank Joshua Jadwin for assistance with editing the chapter, and Bruce Mayer for his continuous encouragement and support. This study was partly supported by grant CA1154966 from the National Institutes of Health and Quest for CURES (QFC) grant from the Leukemia and Lymphoma Society (to K.M.).

References

1. Krause DS, Van Etten RA (2005) Tyrosine kinases as targets for cancer therapy. N Engl J Med 353(2):172–187. doi:10.1056/NEJMra044389
2. O'Shea JJ, Holland SM, Staudt LM (2013) JAKs and STATs in immunity, immunodeficiency, and cancer. N Engl J Med 368(2):161–170. doi:10.1056/NEJMra1202117
3. Young RM, Staudt LM (2013) Targeting pathological B cell receptor signalling in lymphoid malignancies. Nat Rev Drug Discov 12(3):229–243. doi:10.1038/nrd3937
4. Kaneko T, Joshi R, Feller SM, Li SS (2012) Phosphotyrosine recognition domains: the typical, the atypical and the versatile. Cell Commun Signal 10(1):32. doi:10.1186/1478-811x-10-32
5. Huang H, Li L, Wu C, Schibli D, Colwill K, Ma S, Li C, Roy P, Ho K, Songyang Z, Pawson T, Gao Y, Li SS (2008) Defining the specificity space of the human SRC homology 2 domain. Mol Cell Proteomics 7(4):768–784. doi:10.1074/mcp.M700312-MCP200
6. Machida K, Khenkhar M, Nollau P (2012) Deciphering phosphotyrosine-dependent signaling networks in cancer by SH2 profiling. Genes Cancer 3(5–6):353–361. doi:10.1177/1947601912459048
7. Kundu K, Costa F, Huber M, Reth M, Backofen R (2013) Semi-supervised prediction of SH2-peptide interactions from imbalanced high-throughput data. PLoS One 8(5):e62732. doi:10.1371/journal.pone.0062732
8. Tinti M, Kiemer L, Costa S, Miller ML, Sacco F, Olsen JV, Carducci M, Paoluzi S, Langone F, Workman CT, Blom N, Machida K, Thompson CM, Schutkowski M, Brunak S, Mann M, Mayer BJ, Castagnoli L, Cesareni G (2013) The SH2 domain interaction landscape. Cell Rep 3(4):1293–1305. doi:10.1016/j.celrep.2013.03.001
9. Li L, Zhao B, Du J, Zhang K, Ling CX, Li SS (2011) DomPep--a general method for predicting modular domain-mediated protein-protein interactions. PLoS One 6(10):e25528. doi:10.1371/journal.pone.0025528
10. Obenauer JC, Cantley LC, Yaffe MB (2003) Scansite 2.0: Proteome-wide prediction of cell signaling interactions using short sequence motifs. Nucleic Acids Res 31(13):3635–3641
11. Naegle KM, Gymrek M, Joughin BA, Wagner JP, Welsch RE, Yaffe MB, Lauffenburger DA, White FM (2010) PTMScout, a Web resource for analysis of high throughput post-translational proteomics studies. Mol Cell Proteomics 9(11):2558–2570. doi:10.1074/mcp.M110.001206
12. Ehrenberger T, Cantley LC, Yaffe MB (2015) Computational prediction of protein-protein interactions. Methods Mol Biol (Clifton, NJ) 1278:57–75. doi:10.1007/978-1-4939-2425-7_4
13. Machida K, Thompson CM, Dierck K, Jablonowski K, Karkkainen S, Liu B, Zhang H, Nash PD, Newman DK, Nollau P, Pawson T, Renkema GH, Saksela K, Schiller MR, Shin DG, Mayer BJ (2007) High-throughput phosphotyrosine profiling using SH2 domains. Mol Cell 26(6):899–915. doi:10.1016/j.molcel.2007.05.031
14. Machida K, Eschrich S, Li J, Bai Y, Koomen J, Mayer BJ, Haura EB (2010) Characterizing tyrosine phosphorylation signaling in lung cancer using SH2 profiling. PLoS One 5(10):e13470. doi:10.1371/journal.pone.0013470
15. Ng KY, Yin T, Machida K, Wu YI, Mayer BJ (2015) Phosphorylation of Dok1 by Abl family kinases inhibits CrkI transforming activity. Oncogene 34(20):2650–2659. doi:10.1038/onc.2014.210
16. Liu BA, Engelmann BW, Jablonowski K, Higginbotham K, Stergachis AB, Nash PD (2012) SRC homology 2 domain binding sites in insulin, IGF-1 and FGF receptor mediated signaling networks reveal an extensive potential interactome. Cell Commun Signal 10(1):27. doi:10.1186/1478-811x-10-27
17. Kaushansky A, Allen JE, Gordus A, Stiffler MA, Karp ES, Chang BH, MacBeath G (2010) Quantifying protein-protein interactions in high throughput using protein domain microarrays. Nat Protoc 5(4):773–790. doi:10.1038/nprot.2010.36

Chapter 27

Microwestern Arrays for Systems-Level Analysis of SH2 Domain-Containing Proteins

Mark F. Ciaccio and Richard B. Jones

Abstract

The Microwestern Array (MWA) method combines the scalability and miniaturization afforded by the Reverse Phase Lysate Array (RPLA) approach with the electrophoretic separation characteristic of the Western blot. This technology emulates the creation of an array of small Western blots on a single sheet of nitrocellulose allowing for the sensitive and quantitative measurement of hundreds of proteins from hundreds of cell lysates with minimal cost and maximal accuracy, precision, and reproducibility. The MWA is a versatile technology that can be easily configured for purposes such as antibody screening, cell signaling network inference, protein modification/phenotype regression analysis, and genomic/proteomic relationships. Accordingly, configurations for the MWA can be optimized for maximal numbers of proteins analyzed from small numbers of cell lysates, for small numbers of antibodies against large numbers of cell lysates, or for maximal resolution of protein size achieved by increased electrophoretic separation distance. For example, on a single gel, 6 samples can be printed 96 times if a few samples need to be assayed with a large number of antibodies. Alternatively, up to 100 samples can be assayed with four antibodies on a single gel. Intermediate configurations are also discussed.

The efficiency of the MWA is orders of magnitude greater in reagents, labor, and time required per data point relative to the standard Western blotting method and orders of magnitude more sensitive than standard mass spectrometry methods. The MWA is therefore a very attractive approach for capturing global changes in protein abundances and modifications including tyrosine phosphorylation and SH2 domain binding sites.

Key words Microwestern array, Western blot, Reverse phase lysate arrays, Protein quantification, Proteomics, Signal transduction, Phosphorylation, Immunoblots, Antibodies, Systems biology

1 Introduction

Quantitative, accurate, and comprehensive measurement of protein systems is important for many fields of biology but continues to pose a formidable challenge for the systems-level scientists tasked with achieving these goals. The Western blotting method has been used as a "gold standard" methodology by molecular and cellular biologists for measuring proteins levels. While the method is rather inefficient, compared to other multiplexed or array-based methods

Kazuya Machida and Bernard A. Liu (eds.), *SH2 Domains: Methods and Protocols*, Methods in Molecular Biology, vol. 1555, DOI 10.1007/978-1-4939-6762-9_27, © Springer Science+Business Media LLC 2017

such as the Enzyme-Linked ImmunoSorbent Assay (ELISA) or RPLA methods, the electrophoretic separation of proteins provided by the method affords one the ability to discern proteins of different sizes and simultaneously provides the user with an added level of confidence that the proteins observed are indeed the bona fide targets that were intended to be queried [1]. However, Western blots are not a practical solution for many areas of biology that require the simultaneous measurement and comparisons of hundreds to thousands of protein abundances across hundreds of conditions such as current questions in the field of systems biology.

To address the problem of measuring many proteins across many samples, the RPLA has been successfully implemented for observing protein abundances and modifications using minute amounts of sample. RPLAs are performed by arraying cell lysates directly onto nitrocellulose-coated slides and subsequently probing them with antibodies [2] that are then detected through a fluorescent or chemiluminescent reporter. RPLAs however are not applicable for a wide range of antibodies due to convolution of signal from off-target antibody cross-reactivity [3]. Several other antibody-based methods have been developed for quantification of protein modification and abundance including ELISA and flow cytometry bead-based assays [4]. The accuracy of these methods may also be limited by the difficulty in obtaining high quality antibody pairs required for simultaneous capture and detection. Quantitative Mass Spectrometry (qMS) is an orthogonal method to antibody-based methods that can be used to identify a large net of new proteins and modifications [5, 6]. qMS may not be applicable to some scientific questions due to: requirements for large amounts of protein samples; capital costs and expensive maintenance of machinery; the cost of isotopic labeling reagents, e.g., in mouse models of disease; and the difficulty of reproducibly detecting and assaying a defined set of low-abundance protein targets and modifications over a wide range of conditions.

In this protocol, we describe the Microwestern Array (MWA), a recently developed method that combines the scalability and reproducibility of RPLAs with the electrophoretic separation of Western blots [7]. With a sufficient panel of effective antibodies, protein abundances and modifications may be efficiently measured across many samples using the MWA. In addition, antigens containing phosphotyrosines could be assayed on a systems-level with a labeled SH2-domain containing protein. We have shown that the MWA is able to quantify the phosphorylation state of 91 phosphosites modulated after EGF stimulation including many that were previously inaccessible from analogous studies using RPLA or qMS methods [7]. This study provided the first example of how the MWA method could be applied to address scientific questions that were not previously addressable with other protein assays.

The MWA has been used to obtain data sufficient for constructing a Bayesian network-based model of phosphorylation dynamics downstream of the EGFR receptor [7]; study mechanisms of drug synergy in Squamous Cell Carcinoma of the Head and Neck (SCCHN) [8]; to dissect immune-sensing signaling components of dendritic cells [9]; to probe the anticancer activity of bioactive natural products [10]; and to determine the phosphosites that are most predictive of cell phenotype [11]. This technology is likely to be ideal for biological studies that require analysis of many predefined protein targets in a cost-effective, precise, and efficient manner. The MWA may also be used for screening the quality of many antibodies as well be used for clinical experiments where protein samples are limited in supply.

2 Materials

2.1 Equipment and Software

1. LI-COR Odyssey near-infrared imager (LI-COR Odyssey Model).
2. LI-COR Odyssey Image Software (LI-COR Version 3.0 or Image Studio 5.0).
3. High-voltage power supply (Pharmacia EPS600).
4. Nanoplotter with humidifier and GeSiM software (GeSiM 2.1E).
5. Criterion transfer box with plate electrode (Bio-Rad 170-4070).
6. Flat top chamber II electrophoresis unit (The Gel Company 1100-03).
7. Water purification system.
8. 96/48-well sectioning chambers (The Gel Company MWA130-96/48).
9. Ultrasonic cell disrupter such as Bioruptor (Diagenode), or S220 focused ultrasonicator (Covaris).

   If using Covaris ultrasonicator, use the following Materials:

   (a) tissueTUBE™ TT1 (Covaris, 520001).

   (b) tissueTUBE Holder (Covaris, 500095).

   (c) Covaris CP02 Dry Pulverization System.

   (d) 12 mm × 24 mm round-bottom collection tube (Covaris, 520056).

   (e) S-series holder (Covaris, 500199).
10. Paper slicer.
11. Vacuum sealer for plastic bags.
12. Nutator™.

2.2 Materials and Reagents

1. 500 μL Centricons (NMWL: 10000, Millipore UFV5BGC00).
2. Glass casting plates (Moliterno GBS-140-270-2 14×17×2.3 and GBS-140-280-2 14×28×2.3).
3. BlueSlick nonstick glass surface coating (Serva 42500.01).
4. Rubber spacers: 1 mm thick (Serva 42901.01).
5. Binder clips.
6. 30 mL syringes.
7. 19 gauge needles.
8. Methanol.
9. Hydrochloric Acid, HCl.
10. Dithiothreitol, DTT.
11. 384-well black plates (Corning 3676).
12. IR700-Conjugated Protein Marker (LI-COR 928-40000).
13. Kerosene.
14. Filter wicks (Serva 42987.03).
15. Nitrocellulose/filter paper sandwiches.
16. Tween-20.
17. LI-COR Odyssey blocking buffer (LI-COR 927-40000).
18. Goat anti-rabbit Alexa680-conjugated secondary antibody (Invitrogen A21077).
19. Goat anti-mouse IR800-conjugated secondary antibody (LI-COR 926-33210).
20. Tris base.
21. Acetic acid.
22. Sodium dodecyl sulfate, SDS.
23. Glycerol.
24. Ethylenediaminetetraacetic acid, EDTA.
25. Aprotinin.
26. Leupeptin.
27. Pepstatin.
28. β-Glycerophosphate, β-GP.
29. Sodium orthovanadate, Na_3VO_4.
30. Glycine.
31. Sodium chloride, NaCl.
32. Potassium chloride, KCl.
33. Dibasic sodium phosphate, Na_2HPO_4.
34. Sodium phosphate monobasic monohydrate, $NaH_2PO_4{\bullet}H_2O$.
35. Acrylamide.
36. Plastic sealing bags.

2.3 Buffer Compositions

1. 5× Gel Buffer (TAc stock solution): Gel Buffer stock consists of 1.2 M Tris-acetate. Add acetic acid to Tris base to adjust the pH to 6.9 (*see* **Note 1**).
2. Rehydration Buffer: Use 0.240 M Tris-acetate. Prepare by making a 1:5 dilution of the TAc stock solution. Add 0.1% SDS w/v and 20% glycerol v/v. Fill water to appropriate level after adding TAc stock solution and glycerol.
3. Electrode Buffer: Use 0.480 M Tris-acetate (1:2.5 dilution of the TAc stock solution). Add 0.1% SDS w/v.
4. Transfer Buffer (TB): Use 250 mM Tris base with 2 M glycine for the 10× TB stock solution. To make 1 L of 1× buffer measure 100 mL of 10× solution, add 20% methanol (200 mL) and fill to 1 L with ddH_2O. The pH of the solution should be around 8.5 although the pH should it usually unnecessary to alter.
5. Wash Buffer: To make 10× stock, use 24.2 g Tris base and 80 g NaCl. pH to 7.6 with liquid HCl. Dilute 10 to 1 to make 1× TBS. Dilute 10 to 1 and add 1 mL Tween into the 1 L of 1× for TBS with Tween (TBST) (*see* **Note 2**).
6. Acrylamide: Acrylamide with 3.8% crosslink works best for proper hydration and resolution when printing. To make 100 mL of 40% acrylamide (3.8% C), weigh 40.0 g of powdered acrylamide and 1.5 g of bis-acrylamide. Add ddH_2O to 100 mL. Store at 4 °C away from light by keeping in a brown bottle or wrapped in aluminum foil (*see* **Note 3**).

2.4 Preparation of Lysis Buffer

1. Sodium Orthovanadate solution: Prepare a 200 mM solution of sodium orthovanadate in ddH_2O. As the vanadate may be polymerized, additional steps are necessary to activate the solution as a protein tyrosine phosphatase inhibitor. Adjust the pH to 10.0 using either sodium hydroxide or hydrochloric acid as the starting pH of the sodium orthovanadate solution may vary. The color of the solution should be yellow. Boil the solution until it turns colorless which will take approximately 10 min. Cool the solution to room temperature. Next, readjust the pH to 10.0 and repeat boiling and pH adjustment until the solution remains colorless and the pH stabilizes at 10.0. Aliquot the activated sodium orthovanadate at −20 °C for single use.
2. EDTA solution: Weigh out Na_2H_2EDTA and add ddH_2O to make 100 mM stock solution. Add concentrated NaOH and solid NaOH until dissolved, then keep adding NaOH until pH is equal to 8.0. Store at room temperature.
3. DTT, aprotinin, leupeptin, and pepstatin stock solutions: Prepare all solutions in ddH_2O. Prepare stock solutions of aprotinin, leupeptin, and pepstatin at a concentration of 1 mg/mL. Prepare DTT at 1 M. Store all solutions as aliquots at −20 °C.

4. Lysis buffer: To make 10 mL of lysis buffer, prepare 240 mM TAc (2 mL of 1.2 M Stock), 1% SDS w/v (1 mL of 10% stock), 0.5% v/v glycerol (50 μL pure), 5 mM EDTA (100 μL of stock), and add 7 mL of ddH_2O. Add protease and phosphatase inhibitors less than 5 h before lysis. Add 1:1000 dilution of aprotinin, leupeptin, and pepstatin. Add 1:100 dilution of β-GP. Add 1:200 dilution of sodium orthovanadate and 1:20 dilution of DTT (*see* **Note 4**).

3 Methods

3.1 Preparation of Lysates from Cell Culture

This section applies to preparation of protein lysates from adherent cells or cells in suspension. If preparing lysates from tissue, skip to Subheading 3.2.

1. Wash cells twice in PBS to remove any media contamination. Immediately scrape or centrifuge cells into the minimal amount of lysis buffer required to fully dissolve cells. 1.0 mL of Lysis Buffer should be appropriate for a standard 10 cm dish of cells. Keep on ice until the samples are rapidly heated in the following step.
2. Heat at 75–100 °C for 10 min (*see* **Note 5**).
3. Use an ultrasonic cell disruptor to reduce lysate viscosity to near that of water (*see* **Note 6**).
4. Sheer protein samples by passing through a 25-gauge needle at least five times.
5. Heat at 75–100 °C for 2 min and immediately place in a 500 μL Centricon centrifugal concentration device with a 10 kDa nominal molecular weight limit (NMWL).
6. Spin at 12,000×*g* until volume is reduced by 5–10×. Top off with lysis buffer to equalize sample volume across samples (*see* **Note 7**).
7. Freeze at −20 °C until use (*see* **Note 8**).

3.2 Preparation of Lysates from Tissue

This section describes the method for preparing protein samples from tissues. If one is using protein samples from cell culture, skip to Subheading 3.1.

1. If possible, within the confines of the experimental question to be addressed, one should select tissue with minimal fat and blood, both of which will decrease the quality of the blot through reduction in signal, and increase in noise due to smearing of antibody signals and potential fluorescent signals from blood components (*see* **Note 9**).
2. Pulverize tissue to a fine powder after freezing in liquid nitrogen. For small numbers of tissues, this can be accomplished

using a cold mortar and pestle. For larger numbers of samples, we have found the Cryoprep and S2 Focused ultrasonication devices from Covaris to be extremely helpful for maximization of tissue extraction efficiency, and minimization of process variability (*see* **steps 5–25**).

3. If using mortar and pestle method, quickly measure approximately 1 mg of pulverized tissue in a dry environment and place in 1 mL of lysis buffer. Then vortex the mixture for at least 1 min.
4. Heat at 75–100 °C for 10 min (*see* **Note 5**) and skip to **step 26**.
5. If using the Covaris method, proceed as described. First, flash freeze 10 mg of tissue in liquid nitrogen.
6. Place frozen sample in a tissueTUBE™ and submerse tissueTUBE in liquid nitrogen.
7. Assemble the tissueTUBE into the tissueTUBE Holder by tightening washer to fingertight and then loosen by 1/4 of turn.
8. Pulverize with one impact using the Covaris CP02 Dry Pulverization System set to impact setting of 3.
9. Submerse pulverized material and tissueTUBE in liquid nitrogen.
10. Carefully transfer pulverized material to a 12 mm × 24 mm round-bottom collection tube.
11. Resuspend material in 1 mL of SDS lysis buffer (240 mM Tris-Acetate (pH 6.9), 1% (w/v) SDS, 0.5% (v/v) Glycerol, 5 mM EDTA, 50 mM DTT with protease and phosphotase inhibitors).
12. Place collection tube in S-series holder and into the Covaris Sonicator with a degassed and chilled water bath at 7 °C.
13. Disrupt tissue using a duty cycle of 10%, intensity of 4, 200 cycles for 60 s.
14. Remove material to a 1.5 mL eppendorf tube.
15. Needle sheer sample five times with 25-gauge needle.
16. Place sample at 95 °C for 10 min.
17. Cool briefly and then vortex for 1 min.
18. Freeze samples at −80 °C.
19. Thaw samples, place at 95 °C for 1 min.
20. Centrifuge samples at 14,000 × *g* to precipitate any insoluble material.
21. Transfer half of sample to 0.5 mL centricon with 10 K MW cutoff.
22. Spin solution at 12,000 × *g* until concentrated tenfold.
23. Add remaining unconcentrated sample.

24. Spin solution at 12,000 × *g* until concentrated 10× (*see* **Note 7**).
25. Freeze at −80 °C until ready for use (*see* **Note 8**).
26. If using mortar and pestle method, begin here by thawing cell lysate at 37 °C and then centrifuging at full speed to precipitate any insoluble precipitated material.
27. An ultrasonic cell disruptor must then be used to reduce lysate viscosity to near that of water. This is a critical step that differs from the traditional Western blot as samples will be dispensed from a narrow tip during the arraying process (*see* **Note 5**).
28. Needle sheer five times with a 25-gauge needle.
29. Heat at 75–100 °C for 2 min and immediately place 500 μL in a Centricon centrifugal concentration device with a 10 kDa MWCO.
30. Spin at 12,000 × *g* until solution is concentrated 5–10× (*see* **Note 7**).
31. Freeze until use (*see* **Note 8**).

3.3 Pouring the Gel

1. Use nonadhesive glass surface coating (Serva BlueSlick™) on glass and wipe clean. Place rubber spacers on three sides (on the coated side) of one of the glass plates (two long sides and one short side) making sure to leave no gaps between spacer on the corners to prevent leakage. Center one rectangle of Netfix onto one of the glass plates on top of the spacers. Place the second glass plate on top with the coated surface facing the inside, leaving a small lip (about 1 cm) to inject the unpolymerized acrylamide (or use one large and one small plate leaving a gap to pour the gel). Place at least eight clamps along three sides of the sandwiched plates (*see* Fig. 1).
2. We recommend making gels in the laboratory, as purchased, pre-cast gels have not been found to have the correct hydration for the Microwestern Array. To make an acrylamide gel, mix reagents in a 50 mL conical tube. 9.5 % acrylamide works well for proteins with a wide range of molecular weights. However, this concentration may be modified to either 9 % or 10 % depending on whether you wish to obtain better resolution for proteins greater than 150 kDa or smaller than 50 kDa respectively. To make enough gel for one gel sandwich (two gels), add 6 mL 5× Gel Buffer, 9.5 mL 40 % Acrylamide (3.8 % C), 8.0 mL ddH_2O, 6 mL glycerol, and 300 μL 10 % w/v SDS. Mix thoroughly.
3. Place one corner of the sandwiched plates onto an object about 3 in. high so that acrylamide can flow at a diagonal trajectory in two dimensions (or use one small plate and one large plate). Mix 12 μL of TEMED and 150 μL of 10 % APS into acrylamide solution last and mix immediately. Mix well but refrain

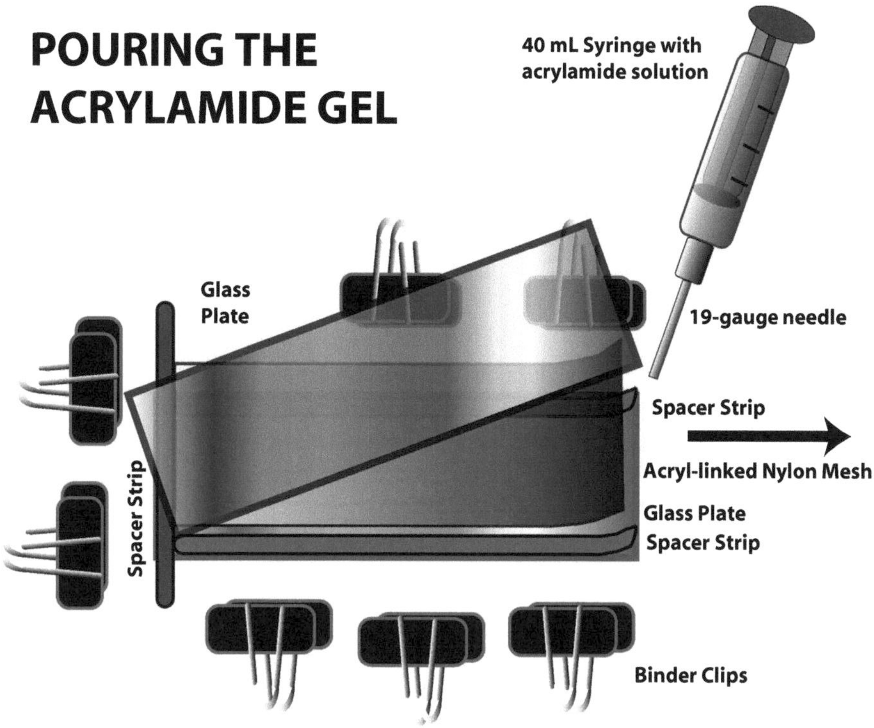

Fig. 1 Procedural schematic for pouring gel

from incorporating air into the mixture which might reduce the efficiency of polymerization. Add 1 mL of isopropanol to the top of the conical tube to eliminate foam. Acrylamide will polymerize quickly so try to complete the pouring steps within 5 min of adding TEMED/APS. Suck acrylamide solution into large syringe with 19-gauge needle. Remove all bubbles from the solution before injecting by tapping syringe against a solid surface. Using the syringe, slowly inject the acrylamide mixture so that it flows down the length of one long side; then runs along the length of the short side on the bottom; and up the second length. One should try to prevent any trapped air bubbles from forming by pouring slowly and tapping the glass if the acrylamide should stop flowing freely in a particular part of the gel plate. The glass plate sandwich will require about 20 mL of acrylamide solution (*see* **Note 10**).

4. Allow the acrylamide gel sandwich to polymerize for at least 1 h. Carefully remove the top plate with a wedge. Gel can then be removed from the bottom glass plate by lifting manually from the plate. The Netfix will provide substantial support for the polymerized gel. Place the gel between two plastic sheets until use. Standard overhead transparency film for copiers

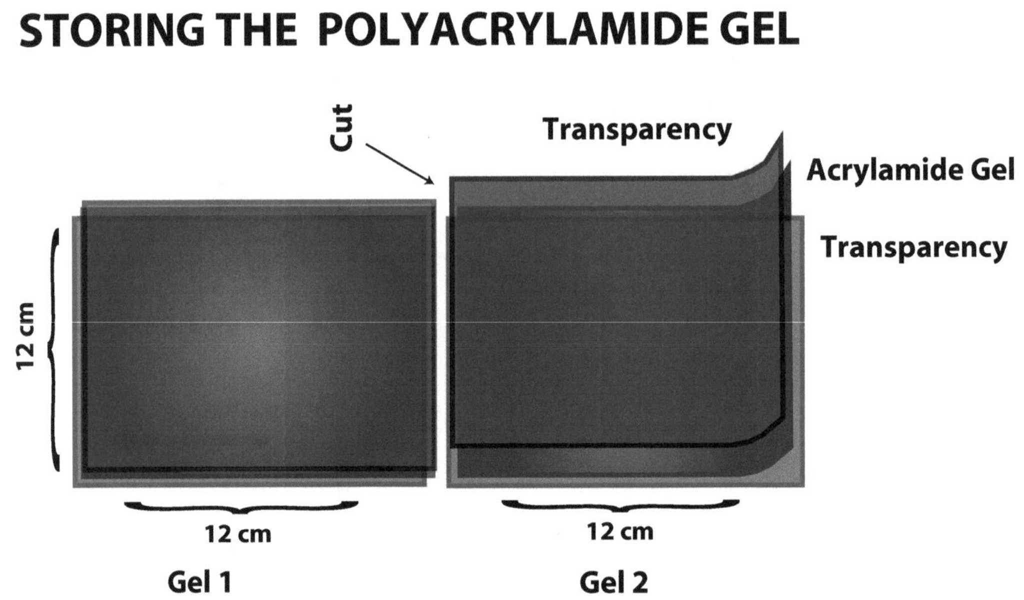

Fig. 2 Procedural schematic for gel storage

works well for this purpose. Press out bubbles from under transparency sheets (*see* Fig. 2).

5. Use a paper slicer to remove the Netfix border. Divide the remaining gel into two equal sections (each must be at least 11.5 cm wide for a 96-well plate). The gels can be vacuum sealed and refrigerated for up to 6 months at 4 °C.

3.4 Print Setup

1. Microdeposition of cell lysates to micro-western array gels will require an appropriate workplate text file format for the default GeSiM software, transfer file, and printing file that each depend on the experiment at hand. These files will be used by the microarrayer to define the layout of number of samples to be printed, the number of gels on which to be printed, and the volume of sample to be printed. One to four gels may be used in conjunction with a four tip microarrayer (*see* Fig. 3). Printing fewer gels will require less time and will result in proteins that are less than 25 kDa to undergo less diffusion in the gel during printing. Increased signal can be obtained from samples of low concentration by arraying more droplets of sample onto the gels. However, the sample spacing must be increased to accommodate the larger spot diameter that accompanies greater printing volume. *See* Table 1 for a summary on suggested configurations.
2. For six or fewer samples, a 96-well format is preferred (*see* Fig. 4). For this format, the number of targets should be set to either one, two, three, or four depending on the number of gels to be printed. Set the block parameters on the GeSiM

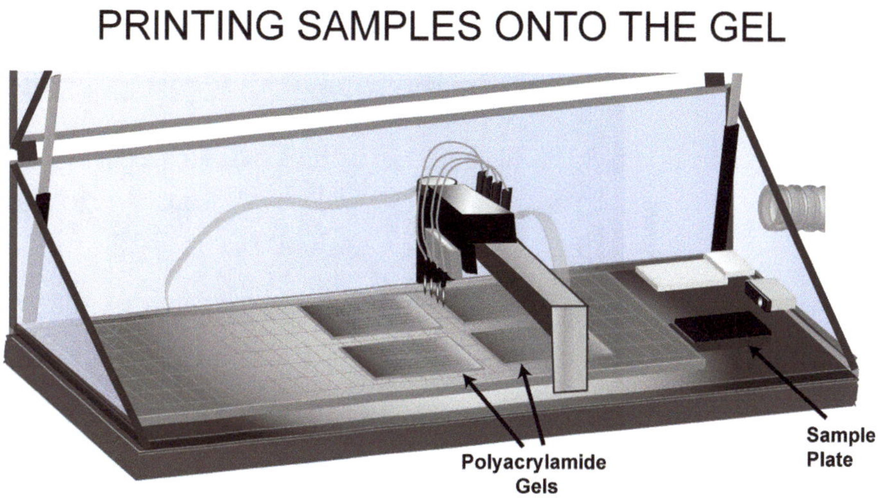

Fig. 3 Procedural schematic for gel printing

Table 1
Suggested array configurations

Configuration	# of samples per gel	# of antibodies per gel	Block size	Block configuration
96-Well	1–6	96 per channel	9 mm × 9 mm	12 × 8 blocks
48-Well	7–17	48 per channel	9 mm × 18 mm	6 × 8 blocks
Quadrant	18–100	4 per channel	36 mm × 55 mm	2 × 2 blocks

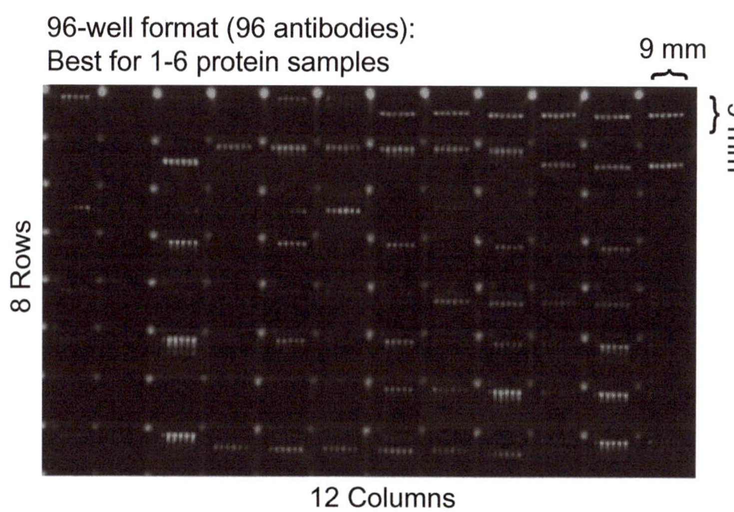

Fig. 4 Example of 96-well Microwestern Array format: Ideal for 1–6 protein samples

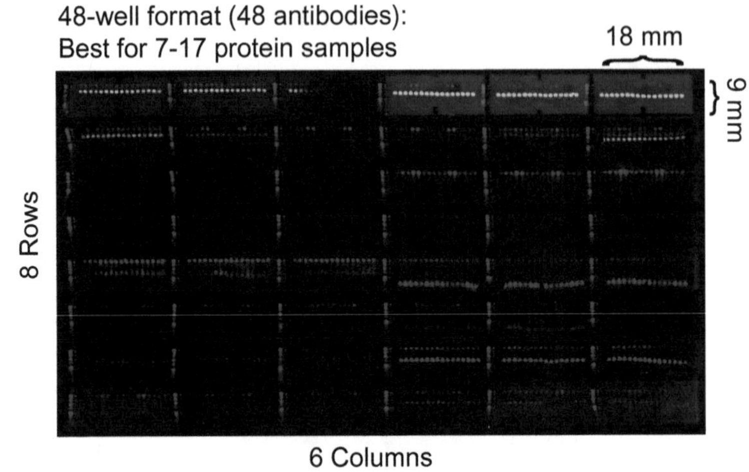

Fig. 5 Example of 48-well Microwestern Array format: Ideal for 7–17 protein samples

Nanoplotter microarrayer to be 9 mm by 9 mm. Define 12 block columns by eight block columns for each of the one to four gel targets. Within each block, eight columns and one row of spots should be defined with a .9 mm pitch between each spot. For each iterative printing pass, approximately 17 droplets should be deposited. For additional information on arraying optimization, refer to the GeSiM NP2 technical manual.

3. For 7–17 samples, a 48-well format (*see* Fig. 5) is optimal. For this format, the block parameters on the GeSiM Nanoplotter microarrayer should be set to 9 mm by 18 mm. Six block columns by eight block rows should be defined for each of the one to four gel targets.
4. For 18–100 samples, a blot format encompassing four equal quadrants of a microtiter format is optimal (*see* Fig. 6). For the quadrant format, the block parameters on the GeSiM Nanoplotter microarrayer should be defined to be 36 mm by 50 mm. Two block columns by two block rows should be defined for each of the one to four gel targets.

3.5 Printing

1. Before printing, heat samples and protein marker at 75–100 °C for 2 min after removing from freezer to redissolve SDS. Centrifuge at top speed for 2 min to collect condensation and remove aggregated molecules.
2. Place LI-COR protein marker in A1 of the 384-well black plate diluted 1 to 1 with lysis buffer. Load subsequent samples horizontally in the A2, A3, A4, … plate location.

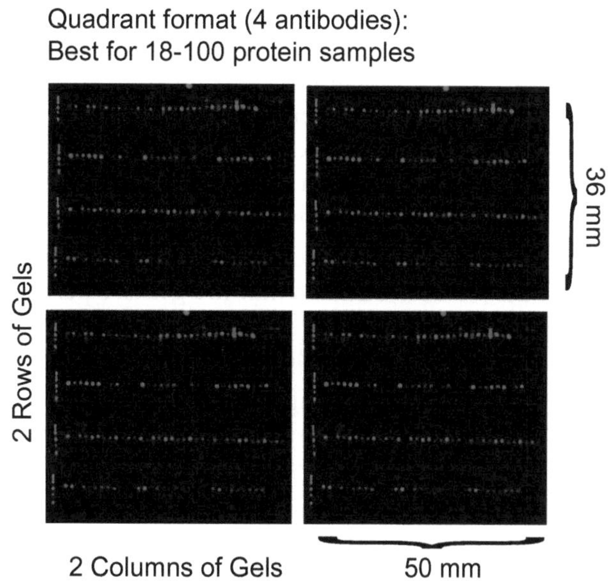

Fig. 6 Example of "Quadrant" Microwestern Array format: Ideal for 18–100 protein samples

3. Place the 384-well plate flat into the plate holder in the GeSiM arrayer.
4. Check tip(s) for proper sample ejection with the onboard stroboscope (*see* **Note 11**).
5. Remove one of the plastic sheets from the gel. For a long print (greater than 4 h), prehydrate gel in rehydration buffer (100 mL) for five minutes with 1 mL of 1 M sodium bisulfate and 1 mL 1 M DTT (save buffer for the following step) (*see* **Note 12**).
6. To keep gel hydrated turn humidifier up to 70%. Input the z-height into the transfer file if any configuration has changed since the last print.
7. Sample intensity will be maximal and not increase when over 150 droplets is deposited. At 150 droplets, the samples should be spaced at least 1.8 mm apart. For 20 droplets the samples should be spaced about 1.0 mm apart. For 17 droplets, spacing should be 0.9 mm apart. The arrayer tips should be defined at 1.5 mm above gel surface while printing and 1.5 mm when traveling between defined gel blocks. When higher concentrations of proteins are desired in the gel, the arraying program should be iteratively repeated after the droplets have been thoroughly absorbed into the gel (5–20 min depending on hydration of gel). Usually, repeating the printing process over the same spot locations ten times results in very concentrated samples in the gels, when the original protein concentration is on the order of 10 mg/mL.

8. Twenty droplets of LI-COR molecular weight protein standard should be printed at a 1:2 dilution for the purpose of alignment of the nitrocellulose membrane under the gasket dividers. In addition, three droplets of ladder at 1:2 dilution should be printed in the column adjacent to the 20 droplet ladder to ensure optimal visualization of protein bands after scanning with the Odyssey near-infrared imaging device.
9. Before starting the microarrayer printing program, one should ensure that the gels are perfectly flat and that the sample plate is not placed at an angle. Any protrusion from the deck may impede the movement of the tips and result in breakage.
10. During printing, one should manually check the stroboscope for the first few samples to make sure that cell lysates are printing properly (*see* **Note 13**).
11. After printing is finished, gels should be left on the deck for 10 min to assure that all droplets are completely absorbed into the polyacrylamide gels.
12. Remove plastic and gently submerge gel with the protein side up into rehydration buffer (100 mL) for five minutes into a dish containing 1 mL of 1 M sodium bisulfite and 1 mL of 1 M DTT. Place the bottom of the gel on a dry paper towel for a few seconds. The gel should be well-hydrated during electrophoresis but uneven pools of buffer on gel should be avoided. If the gel contains excess buffer with evident pools of wetness, the top of the acrylamide may be blotted gently with a Kimwipe or allowed to sit for 10 min for excess buffer to be absorbed into the gel while in the humidified chamber of the arrayer (*see* **Note 14**).
13. Following printing and between print runs, the NP2 standby program should be engaged to keep tips irrigated and hydrated until the next printing activity. Both the humidifier and arrayer water reservoir should be checked intermittently for adequate volumes of fluid. Turn off humidifier but keep computer, arrayer, and diluter on between prints.

3.6 Electrophoresis

1. Prechill the Flat top™ electrophoretic box to 10 °C.
2. Place several drops of kerosene onto the surface of the box and spread it with a Kimwipe to provide a homogenous layer on the surface. Place a clean transparency sheet on the surface of the electrophoresis box. Ensure that no bubbles are present in the kerosene beneath the plastic which would prevent even temperature distribution. Place the hydrated gel on top of the transparency sheet. Orient with protein side on top (proteins will migrate toward red electrode).
3. Cut three filter wicks per electrode and submerge them for several seconds in electrode buffer. Stack three wicks onto the

gel per electrode and run a gloved finger over the length of the wicks to make sure that thorough contact is made between them and the gel. Close the top cover of the electrophoresis box making sure the electrodes fall squarely onto the electrode wicks. Place a small dot of bromophenol blue with a pipette tip on each gel outside the protein range but between the electrodes. Four drops of bromophenol blue should also be placed on the corner of the printed area of the gel to assess the speed of electrophoresis and to aid in properly aligning the printed region of the gel onto the nitrocellulose membrane prior to the stage of electrophoretic transfer (*see* **Note 15**).

4. Operate the power supply at 60 V, 5 W, and unlimited amps for 5 min. Afterward, turn power up to 350 V, 30 W, and unlimited amps. Stop the process when the bromophenol blue has migrated about 8 mm. When running two gels at a time, the proteins may run at different rates. Take the first gel off when it is finished and continue running the slower gel afterward if this occurs.

3.7 Transfer

1. Much of the protein is on the surface of the gel after electrophoresis. Transfer onto a moistened nitrocellulose membrane in transfer buffer but do not submerge. Carefully align when placing so that the gel does not slide laterally relative to the membrane (*see* **Note 16**).
2. Carefully roll out any bubbles in the sandwich without letting the gel slide on the nitrocellulose.
3. The gel can be transferred at 0.8 A for at least 60 min at 4 °C or 0.15 A overnight at 4 °C or over 2 days at 0.1 A and 4 °C (*see* **Note 17**).

3.8 Blotting (Gasket Method)

Follow the steps in this subheading if you are using the 96- or 48-well MWA format. If you are using the quadrant format skip to Subheading 3.9.

1. After transfer, wash for 5 min in TBS (no Tween) to remove methanol.
2. Block in 100% Odyssey Blocking Buffer (LI-COR) for 30 min to 1 h (no Tween).
3. Align gasket on membrane and clamp down using gasket screws. Use a 96/48-well isolation device depending on how the gel was printed.
4. Pipette antibodies appropriately into wells (*see* **Note 18**).
5. Incubate membrane in primary antibody diluted in Odyssey Blocking Buffer (w/ no Tween) for 12–36 h (the exact time will depend on many factors including the antibody performance and protein levels being measured). Use 100 μL per well (*see* **Note 19**).

6. Wash the membrane at least four times for 5 min in TBST while still clamped into the 96-well or 48-well gasketing device using a multichannel pipette and plastic wash boat. 150 µL or more of TBST should be used per well per wash. A Nutator™ should be used to aid sample mixing during antibody wash steps.
7. Incubate in LI-COR near-infrared dye labeled secondary antibody for 1 h (diluted 1:5000–1:10,000) in 20% LI-COR Odyssey Blocking Buffer, 80% TBS (w/ no Tween) while nutating (*see* **Note 20**).
8. Wash membrane three times in TBST for 5 min while still in t96/48-well antibody isolation chamber.
9. Remove membrane from gasketing device and wash one more time in TBST in a container such as a polypropylene pipet tip box top (>5 min).
10. Wash at least one time in TBS without Tween to remove detergent (>5 min).
11. Let the membrane dry completely (*see* **Note 21**).
12. Place the membrane face down on the scanning surface of the LI-COR Odyssey scanner. Place a clean piece of glass over the membrane to keep the nitrocellulose flat.
13. Do a preliminary scan to ensure the gel is properly aligned in the scanner and there are no saturated pixels. If there are saturated pixels, which will appear white, turn down the laser intensity incrementally. After parameters have been optimized, scan with LI-COR scanner at high resolution (42 µm/pixel) and medium quality (*see* **Note 22**).

3.9 Blotting (Quadrant Method)

Follow the steps in this subheading if you are using the Quadrant MWA method. If you are using the 96- or 48-well method, skip to Subheading 3.9.

1. After transfer, wash for 5 min in TBS (no Tween) to remove methanol.
2. Blot full membrane in 100% Odyssey Blocking Buffer for 30 min to 1 h.
3. Place membrane on clean transparency. Cut into quadrants using paper slicer making sure to cut between the protein markers so that each quadrant contains all samples that were arrayed to the membrane.
4. Place each quadrant in a clean black LI-COR blotting box.
5. Incubate in primary antibody diluted in Odyssey Blocking Buffer (w/ no Tween) for 12–36 h. 6 mL of diluted antibodies is sufficient to cover the blot (*see* **Note 23**).
6. Incubate in secondary near-infrared dye labeled secondary antibody for 1 h diluted (1:5000–1:20,000) in 20% Odyssey Blocking Buffer, 80% TBS (w/ no Tween) (*see* **Note 20**).

7. Wash four times in TBST for more than 5 min.
8. Wash at least one additional time in TBS without Tween to remove detergent (>5 min).
9. Let the membrane dry completely (*see* **Note 21**).
10. Place the membrane face down on the scanning surface of the LI-COR Odyssey scanner. Place a clean piece of glass over the membrane to keep the nitrocellulose flat.
11. Do a preliminary scan to ensure the gel is properly aligned in the scanner and that no pixels are saturated at the laser intensity used. Saturated pixels will appear white during the pre-scan, and the laser intensity should be decreased and then pre-scanned again iteratively until no saturated pixels remain. After parameters have been optimized, scan with LI-COR scanner at high resolution (42 μm/pixel) and medium quality setting (*see* **Note 22**).

3.10 Analysis

1. Save scans as both 16-bit (Copy tiff) and 8-bit (Color tiff) files using the LI-COR Odyssey scanning software. 16-bit files will be used for quantitative analysis and 8-bit files for display purposes.
2. Open 16-bit files in the LI-COR Odyssey analysis software. Draw a circle around each band in the appropriate block. Make sure to copy and paste the first object for all subsequent objects so that every object has an identical diameter and area. For two channel scans you can separate the files into a 700- and 800-channel scan before quantification. Name them (highlight + Alt-N) according to the antibody used with an underscore to delineate replicate number, e.g., pSrc_Y416_2. Draw three circles outside the area of printed protein representative of the background fluorescence. Name these with a B-suffix, e.g., pSrc_Y416_2B. Highlight all objects and export the data. Make sure that Trimmed Mean and Minimum Intensity are selected as Data Fields for export.
3. Calculate the net intensity by subtracting the mean of the minimum of the background circles from the trimmed mean of each object. If there is a gradient in background signal across the area of the block, put a row of circles above or below each sample circle outside the printed area and subtract each signal by the corresponding background circle (*see* **Note 24**).
4. Normalize loading variations by dividing by the mean net intensities of Actin, Tubulin, and/or GADPH, or other appropriate control antibodies. Loading variation can also be normalized by dividing the median normalized abundance of all proteins assayed if a sufficient number of total proteins are tested. The appropriate normalization method will depend on the experimental questions to be addressed.

5. Similar to a Western blot, each Microwestern block shows relative not absolute protein abundance. For this reason, in order to compare data related the protein abundance or modification for multiple antibodies across blocks they should be normalized relative to an internal standard within the same block as the data is gathered. For example, if one is examining temporal changes in protein signaling, each intensity can be divided by the starting time point. In this way, fold change in protein abundance can be compared for many antibodies. If several cells lines are being examined, a control cell line can be used for reference. For example, MCF-10a cells may be compared against MCF-7 cells in the same block to get a ratio of protein changes between the two cell lines.
6. Calculate average fold change across technical triplicates as well as the standard error across triplicate measurements.

4 Notes

1. Do not add sodium hydroxide (NaOH) to adjust pH as this will add sodium ions and affect protein migration during electrophoresis.
2. TBST may be unusable after several months due to growth of mold. Make fresh regularly.
3. Make acrylamide fresh at least every 3 months. Use proper ventilation and a face mask when preparing gel solution because unpolymerized acrylamide is a potent neurotoxin.
4. This solution may be stored as frozen aliquots. Do not leave at room temperature for more than 4 h before lysis as the inhibitors will degrade and lose potency.
5. Heat at 75–100 °C soon after lysis (preferably less than 30 min). This causes enzyme denaturation and slows modification of the proteins prior to analysis. You may freeze the samples after this step overnight or longer. As the sample preparation step may take an extended period of time, this is an ideal place to pause protocol.
6. This is a critical step that differs from the traditional Western blot as samples will be dispensed from a narrow tip during the arraying process and high viscosity protein samples will not array. Many cell disruptors may be used and should be optimized so that the samples have a very low viscosity while minimizing the time at which the samples may be exposed to excessive heat. For example, sonicate at medium power on the Sonifier Cell Disruptor (Branson) for 10 s, alternating 1 s on, 1 s off (total 20 s time). Alternatively, use the Biorupter (Diagenode) H-setting alternating 30 s on and 30 s off for

5 min. Make sure all samples are in polypropylene tubes if using the Biorupter. Soft plastic will dampen sonication.

7. Concentrate more for cells lysed at lower density. Concentrate less for samples with high viscosity. Optional: Verify that the final protein sample has more than 5 μg/μL of total protein using a spectrophotometer (NanoDrop). Higher protein concentration will give better signal; however, the sample must maintain a low viscosity free of precipitates in order for the arraying process to be functional.
8. Aliquot lysates if necessary to avoid repeated freeze-thaw cycles. 25 μL aliquots work well for most applications.
9. Tissues with high blood content should be scanned in the 800 channel during the scanning stage to decrease background fluorescence from heme which may cause high background in the 700 channel. Protein from certain tissues may be purified using protocols previously used for the Western blot to improve band resolution; however, the final product should be in the MWA lysis buffer before printing.
10. After gel is poured lay horizontal to prevent leakage. If bubbles form in the gel, bring sandwich to vertical and quickly tap out bubbles and again lay horizontally to avoid leakage. Continue filling.
11. If drops are inconsistent check tip connection and definition on computer. If tip has short electric connectors gently extend them with forceps. If tip is clogged, remove tip and aspirate 200 μL of 50% methanol/50% HCl using a syringe. Make sure to clean off all solution (it will cause corrosion) before touching machine. If the tip still does not work, send in for repair, and replace with new tip. If tip is permanently broken, order a new tip. If arrayer is acting abnormally restart arrayer and computer and reinitialize the program. Call GeSiM or HTS Resources LLC for support if there is an equipment malfunction.
12. Roll out bubbles and excess fluid from gel before printing. For a long print, pre-hydrate gel in rehydration buffer (100 mL) for 5 min with 1 mL 1 M sodium bisulfate and 1 mL 1 M DTT (save buffer for following step). Gently blot off any excess fluid with a cell scraper. Make sure there are no visible puddles of fluid on the gel or smearing will occur during the print. For short prints, prehydration is not necessary.
13. If samples are consistently too thick to print, pause program and dilute samples in the plate with small amount of lysis buffer. You may slightly change the voltage or delay on the stroboscope to optimize detection of spots with stroboscope. Check to make sure that the gel does not curl during the print as this will damage the tips. Adding a bit of water the curled edges should cause it to adhere again to the plastic bottom.

14. Gel hydration is a crucial step in creating well-resolved Microwestern blots. Gels should be well hydrated while running but avoid any visible fluid on the top of the gel. You may leave the gels on the electrophoresis box for 5–10 min to allow rehydration buffer to absorb into the gel. If Microwestern blots appear fuzzy after scanning the gels may be overly hydrated during this step.
15. If the bromophenol blue visibly diffuses after touching the surface of the gel, the gel is overly hydrated. Wait for 10 min for the buffer to absorb into the gel and respot.
16. This step is extremely important. Moving the gel after placing on the nitrocellulose will cause significant loss of data on the final Microwestern by creating a double or smeared image. Precise transfer is more critical than a normal Western blot because much of the protein is on the surface of the gel after electrophoresis and will transfer instantaneously when touching the nitrocellulose.
17. Keep transfer buffer cold for a sharp transfer. Use of a stir bar during transfer will create a more even transfer. Placing rubber bands around the criterion blot sandwiches before placing in box will decrease smears at the edges of the blot.
18. Ensure the membrane does not dry out during this process. Moisten membrane or wells with TBS without Tween during this process if there is danger of the membrane drying.
19. Rotational shakers will cause circular irregularities on the image and should not be used.
20. Use a suitable combination of 680/700 IR-conjugated secondary and 800 IR-conjugated secondary antibody depending on the species of primary antibody used, e.g., 680 anti-rabbit and 800 anti-mouse secondary antibodies. Protect from light from this step on. Use 150 μL per well. Optionally, use 0.01 % SDS in secondary solution to lower background fluorescence. This may also decrease the signal slightly.
21. You may use pressurized air from a building outlet to dry the membrane quickly. Pressurized air should not be used from a can as these may contain fluorescent residue. During this step avoid exposing the membrane to dust or excessive light as these will decrease the quality of the blots.
22. A laser intensity of 1.0 for both 700 and 800 channels and a focus offset 0.0 is usually optimal. *See* the LI-COR Odyssey manual for more information on using this device and software.
23. The quadrant method will have higher signal and higher background due to the antibody being able to diffuse through the membrane. This is usually desired; however, a lower concentration of primary and/or secondary antibody, more washes,

or the addition of 0.01 % SDS to the secondary may be necessary additions to produce a blot with even background.

24. The background may be calculated as the line integral of the circle around the spot if there is very little diffusion of the protein (if there is a quick print or if you are printing directly on nitrocellulose); however, this may cause a nonlinear relationship between signal and protein concentration making your data nonquantitative. Turn your contrast very high on the end image to see if any signal overlaps with your circle. If it does you cannot use this method. Alternatively, you can graph the signal and background values to see if they are correlated. If they are you cannot use this method and should use background calculated from additional spots.

References

1. Burnette W (1981) "Western blotting": electrophoretic transfer of proteins from sodium dodecyl sulfate polyacrylamide gels to unmodified nitrocellulose and radiographic detection with antibody and radioiodinated protein A. Anal Biochem 112:195–203
2. Tibes R, Qiu Y, Lu Y, Hennessy B, Andreeff M, Mills GB, Kornblau SM (2006) Reverse phase protein array: validation of a novel proteomic technology and utility for analysis of primary leukemia specimens and hematopoietic stem cells. Mol Cancer Ther 5:2512–2521
3. Sevecka M, MacBeath G (2006) State-based discovery: a multidimensional screen for small-molecule modulators of EGF signaling. Nat Methods 3:825–831
4. Sachs K, Perez O, Pe'er D, Lauffenburger DA, Nolan GP (2005) Causal protein-signaling networks derived from multiparameter single-cell data. Science 309:1187
5. Hause RJ, Kim HD, Leung KK, Jones RB (2011) Targeted protein-omic methods are bridging the gap between proteomic and hypothesis-driven protein analysis approaches. Expert Rev Proteomics 8:565–575, Review
6. Rikova K, Guo A, Zeng Q, Possemato A, Yu J, Haack H, Nardone J, Lee K, Reeves C, Li Y, Hu Y, Tan Z, Stokes M, Sullivan L, Mitchell J, Wetzel R, Macneill J, Ren JM, Yuan J, Bakalarski CE, Villen J, Kornhauser JM, Smith B, Li D, Zhou X, Gygi SP, Gu TL, Polakiewicz RD, Rush J, Comb MJ (2007) Global survey of phosphotyrosine signaling identifies oncogenic kinases in lung cancer. Cell 131:1190–1203
7. Ciaccio MF, Wagner JP, Chuu CP, Lauffenburger DA, Jones RB (2010) Systems analysis of EGF receptor signaling dynamics with Microwestern arrays. Nat Methods 7:148–155
8. Liu J, Kuo WL, Seiwert TY, Lingen M, Ciaccio MF, Jones RB, Rosner MR, Cohen EE (2011) Effect of complementary pathway blockade on efficacy of combination enzastaurin and rapamycin. Head Neck 33:1774–1782
9. Chevrier N, Mertins P, Artyomov MN, Shalek AK, Iannacone M, Ciaccio MF, Gat-Viks I, Tonti E, DeGrace MM, Clauser KR, Garber M, Eisenhaure TM, Yosef N, Robinson J, Sutton A, Andersen MS, Root DE, von Andrian U, Jones RB, Park H, Carr SA, Regev A, Amit I, Hacohen N (2011) Systematic discovery of TLR signaling components delineates viral-sensing circuits. Cell 147:853–867
10. Chuu CP, Lin HP, Ciaccio MF, Kokontis JM, Hause RJ, Hiipakka RA, Liao S, Jones RB (2012) Caffeic acid phenethyl ester suppresses the proliferation of human prostate cancer cells through inhibition of p70S6K and Akt signaling networks. Cancer Prev Res (Phila) 5:788–797
11. Ciaccio MF, Chen VC, Jones RB, Bagheri N (2015) The DIONESUS algorithm provides scalable and accurate reconstruction of dynamic phosphoproteomic networks to reveal new drug targets. Integr Biol 7:776–791

Part VI

Functional Analysis

Chapter 28

SH2 Binding Site Protection Assay: A Method for Identification of SH2 Domain Interaction Partners by Exploiting SH2 Mediated Phosphosite Protection

Joshua A. Jadwin

Abstract

Over the last two decades there has been a significant effort in the field to characterize the phosphosite binding specificities of SH2 domains with the goal of deciphering the pY signaling code. Although high throughput studies in various formats using most SH2 domains have collectively provided a rich resource of in vitro SH2–pTyr site specificity maps, this data can only be used approximate what is happening in the cell where protein concentrations and localization are not homogenous, as they are for in vitro experiments. Here we describe an in vivo approach, SH2 site protection assay, which can capture the pTyr binding specificity of SH2 domains in the cell. The basis of this approach is SH2–pY site protection, the ability of SH2 domains to prevent the PTP-dependent dephosphorylation of their pY site binding partners. We overexpress a tracer SH2 domain in cells and quantify the change in abundance of tyrosine phosphorylated sites using MS. Since the method is performed in vivo, it has the advantage of identifying SH2–pY interactions as they occur within in the cell.

Key words SH2 domain, pTyr protection, iTRAQ, MS/MS

1 Introduction

There is a significant need for methods that allow for the identification of protein interaction partners as they occur in vivo. This is particularly true for highly dynamic interactions such as phosphotyrosine (pY) dependent Src Homology 2 (SH2) domain binding. Traditionally in vivo interactions have been inferred from data using peptide-tag or antibody-based protein pull downs from cell lysates. While these methods provide a wealth of data about protein binding, they suffer from confounding factors including nonspecific binding and the precipitation of secondary and tertiary interacting partners not directly involved in the binary SH2–pY interaction. More rigorous washings can help to ameliorate these issues, but can also result in the loss of weaker, but potentially important, phospho-dependent interactions. Additionally,

Kazuya Machida and Bernard A. Liu (eds.), *SH2 Domains: Methods and Protocols*, Methods in Molecular Biology, vol. 1555, DOI 10.1007/978-1-4939-6762-9_28, © Springer Science+Business Media LLC 2017

pull-down assays also assess interactions without taking into account specific variations in SH2- and pY-protein concentrations normally occurring within cells. As a result, data from binding assays performed on cell lysate may be biased toward high affinity interactions that do not naturally occur in the cell due to protein compartmentalization.

More recently a number of elegant methods for the identification of protein interactions in vivo have been developed, such as the use of tandem affinity purification [1], synthetic photo-cross-linkable amino acids [2] and in vivo biotinylation of proximal proteins [3]. However, these methods can be limited by the need for multiple washing steps to detect specific interactions, complex SH2 protein engineering, and the labeling of non-direct interaction partners, respectively. In an effort to develop a simple phosphotyrosine specific method for identifying pY-SH2 interactions as they occur in vivo we have developed the SH2 Protection Assay by exploiting the central mechanism of phosphotyrosine signaling.

Phosphotyrosine signaling is dependent on three major components: (1) tyrosine kinases (TK) that initiate signal transduction by phosphorylation of tyrosines residues on nearby and recruited proteins, (2) SH2 domain containing proteins that bind specific phosphorylated tyrosines and recruit downstream signaling proteins thereby translating the phosphotyrosine code to the language of cellular signaling, and (3) protein tyrosine phosphatases (PTPs) that remove phosphate groups from tyrosine residues effectively erasing TK initiated signals (Fig. 1).

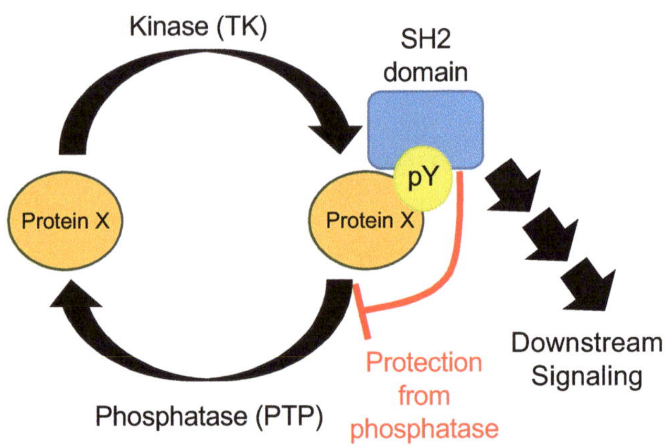

Fig. 1 Phosphotyrosine signaling components and putative SH2 binding site protection. Phosphotyrosine (pY) signaling is dependent on three major components: Tyrosine kinases (TK) that phosphorylate tyrosine residues, Src Homology-2 (SH2) domain containing proteins that bind specific phosphorylated tyrosines, and protein tyrosine phosphatases (PTPs) that remove phosphate from tyrosine residues. pY-bound SH2 domains are thought to protect the phosphorylation of their binding sites from PTPs. This constitutes the basis of the SH2 protection assay

Inherent to the system is the fact that binding of SH2 domains to phosphosites prevents dephosphorylation by PTPs, also known as pY protection [4, 5]. The SH2 Protection assay is based on this principle as well as the high turnover rate of individual phosphosites on tyrosine phosphorylated proteins (half time = 10–50 s) [6], and the relatively low level of tyrosine phosphorylation on a given site [7]. By exploiting these three properties we are able to assay for SH2–pY interactions occurring in vivo simply by monitoring the phosphorylation level of proteins following SH2 domain overexpression. A specific SH2 domain will enhance the phosphorylation of its pY binding sites. This increase in tyrosine phosphorylation can then be identified and quantified for individual pY proteins using anti-pTyr immunoblotting or in a high throughput manner using quantitative phosphospecific mass spectrometry [8] (Fig. 2).

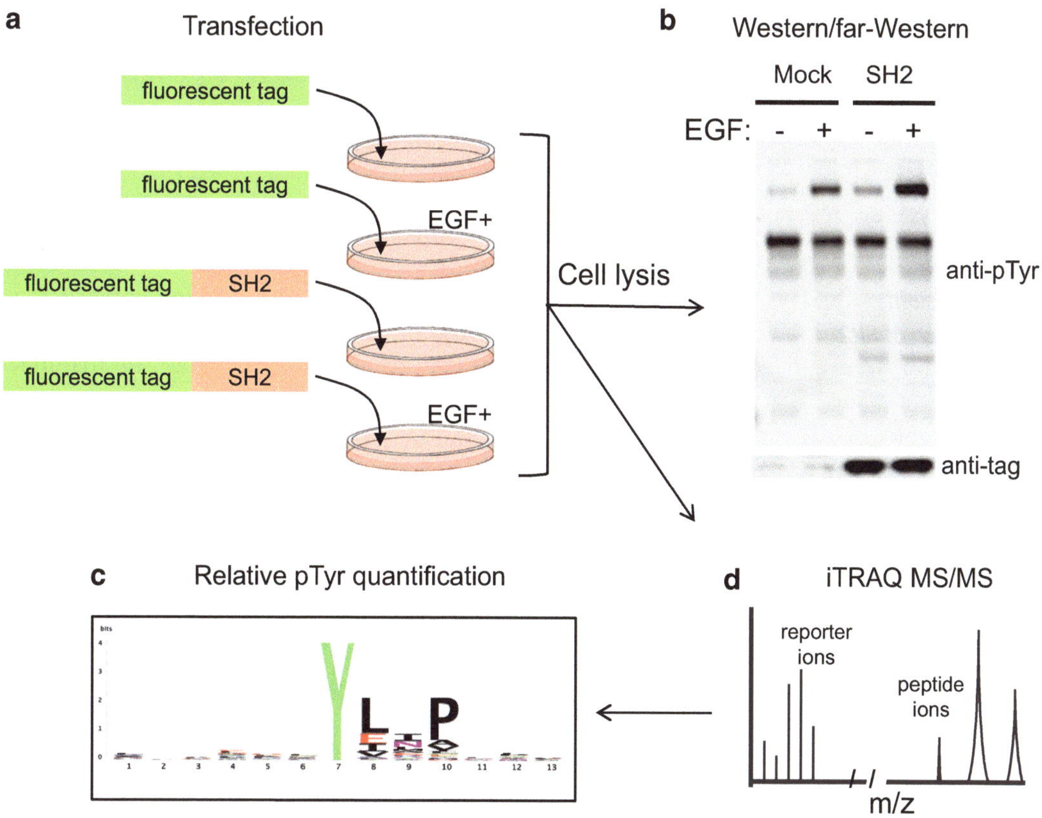

Fig. 2 Experimental outline of SH2 binding site protection assay. (**a**) SH2 domains are cloned as fusion proteins and transfected into cells with or without growth factor stimulation. (**b**) Lysates are run on PAGE for Western/far-Western to visualize SH2 expression dependent increases in tyrosine phosphorylation. (**c**) Lysates can also be analyzed by quantitative pY-specific MS. (**d**) MS data are processed to identify pY sites enriched due to SH2-mediated protection

2 Materials

2.1 SH2 Domain cDNA

1. SH2 domain cDNA can be obtained from Addgene (e.g., Mayer lab clones), DNASU, GE Dharmacon, and the Pawson Lab. Full length cDNA for SH2 domain containing proteins are also available through Addgene (*see* **Note 1**).

2.2 SH2 Domain Expression Constructs

1. SH2 domains should be expressed as fusions with larger fluorescent or functional tags (*see* **Note 2**). Acceptable vectors include, but are not limited to pDEST27 (GST-tagged) and pDEST-tdEOS (tdEOS tagged).
2. Cloning should be performed using standard restriction enzyme or recombination based methods.
3. Plasmid DNA is purified using traditional alkaline lysis-based methods (*see* **Note 3**).

2.3 Cells

COS1 cells or any easily transfectable cells can be used. Protection experiments require that a significant population of cells express your SH2 domain-containing construct of choice (*see* **Note 4**).

1. COS1 cells.
2. COS1 cell growth media: DMEM containing 10% FBS, 1% penicillin/streptomycin.
3. COS1 cell transfection media: DMEM containing 10% FBS.

2.4 Transient Transfection of SH2 Domains

1. Transient expression of SH2 domain containing constructs in COS1 cells is performed using Lipofectamine 2000 (Life Technologies). Transfections are performed per the manufacturer's instructions (*see* **Note 5**).
2. OptiMEM (Life Technologies) is used for mixing DNA and Lipofectamine reagents (*see* **Note 6**).
3. COS1 transfection media: DMEM containing 10% FBS (no antibiotics).

2.5 Growth Factor Stimulation (Optional)

In our experiments, we compare cells stimulated with and without Epidermal Growth Factor (EGF) (*see* **Note 7**).

1. EGF stock solution: 50 μg/mL (*see* **Note 8**).
2. COS1 starvation/stimulation media (0.1% FBS, 1% penicillin/streptomycin).

2.6 Cell Lysis

Cell must be lysed in a manner that preserves their phosphotyrosine state. We recommend combining Kinase Lysis Buffer (KLB) with liquid N_2 snap freezing.

1. Liquid N_2.
2. Kinase lysis buffer: Tris–HCl (25 mM), NaCl (150 mM), EDTA (5 mM), Glycerol (10% v/v), Triton X-100 (10 nM),

Na-pyrophosphate (10 mM), Na-orthovanadate (1 mM), Glycerolphosphate (10 mM), and NaF (1 mM). Add DTT to 1 mM, PMSF to 1 mM, Aprotinin to 15 μg/mL before use.

3. 4× LDS NuPAGE sample buffer (Life Technologies).
4. Bradford assay reagent.

2.7 Phosphotyrosine Immunoblotting

Phosphotyrosine Western blotting of target phosphoproteins with and without overexpression of SH2 domain containing proteins is accomplished using standard Western blotting methods. The materials listed below were used successfully in our experiments but substitute components will likely return similar results.

1. NuPAGE Novex Bis-Tris Midi-Gel Running System (Life Techologies): NuPAGE midi gels, 4× LDS sample buffer, 10× sample reducing agent, 20× NuPAGE running buffer, antioxidant solution.
2. Transfer buffer: CAPS (1 mM, pH 11), methanol (10%, v/v). Store at 4 °C.
3. Nitrocellulose membrane with 0.2 μm pore size.
4. TBS-T: 1 mM Tris–HCl pH 8.0, 150 mM NaCl, and 0.05% Tween 20 (v/v).
5. Nonfat dry milk or LI-COR blocking buffer (LI-COR).
6. pY100 antiphosphotyrosine antibody (Cell Signaling).
7. LICOR Compatible Infrared Dye tagged secondary antibodies (680 and 800 nm).
8. LICOR Odyssey infrared detection station or similar digital Western blot image detection system (*see* **Note 9**).

2.8 Far-Western Blotting

Far-Western blotting can be performed to validate the enhancement of SH2 domain specific pY sites. Refer to [9, 10] for quantitative far-Western blotting method using purified GST-SH2 domains. Various different options for performing SH2 domain far-Western are described in Chapter 17.

1. **Items 1–4** in Subheading 2.7.
2. 5% Milk in TBS-T.
3. GST-tagged recombinant SH2 domain probes. SH2 domain cDNA can be obtained through Addgene (*see* Subheading 2.1).
4. Anti GST–peroxidase conjugate, e.g., Sigma A7340.
5. Western Lightning Plus-ECL Enhanced Chemiluminescence Substrate (PerkinElmer, 103001EA) (*see* **Note 10**).
6. Kodak Image Station 4000 or similar digital luminescence detection system (*see* **Note 11**).

2.9 MS Data Analysis

High-throughput detection and analysis of multiple pY sites requires the use of quantitative phosphotyrosine specific mass spectrometry. The entirety of the material and methods required for phosphospecific mass spectrometry is beyond the scope of this chapter. For our experiments we have utilized iTRAQ MS/MS to quantify phosphotyrosine abundance between up to eight experimental points. A thorough explanation of the methods used can be found in Ref. 11. This method has the advantage of allowing for the relative quantification of phosphotyrosine sites within up to eight difference experimental samples. Other methods including stable isotope labeling of amino acids in cell culture (SILAC) are also likely amenable to this type of analysis.

1. MS lysis buffer (8 M urea, 1 mM sodium orthovanadate (Na_3VO_4)) (*see* **Note 12**).

2.10 LOGO Plot Generation

Differences in MS identified phosphotyrosine peptide amino acid frequency can easily be conveyed using amino acid frequency consensus plots or LOGOs [12, 13]. Multiple online tools are available for this type of analysis. We found that the frequency plot generator available through PhosphoSitePlus provided good results as it allowed us to compare protected sequences against their dataset of previously identified phosphotyrosine peptide sequences [14].

1. Sequence Logo Generators at phosphosite.org or weblogo at berkeley.edu.

3 Methods

3.1 Experimental Design

When designing a protection based interaction SH2–pY assay the following factors should be considered (Fig. 2).

1. SH2 domain construct—SH2 protection experiments can utilize isolated SH2 domains, tandem SH2s, or full-length SH2 domain containing proteins (*see* **Note 13**).
2. SH2 controls—For many SH2 domains missense mutations have been identified that specifically alter the ability of the SH2 to interact with pY. For example, the R86K mutation prevents the interaction GRB2-SH2 with its phosphorylated binding partners (*see* Chapter 12).
3. SH2 expression level—In general, higher expression is better when assessing protection (*see* **Note 14**).
4. Number of samples—This is especially important if the sample will be assayed using both MS and pY-immunoblot. Quantitative MS methods are typically limited to between two and eight samples per run (*see* **Note 15**).

5. Cell treatments—When studying pY signaling multiple reagents, such as growth factors and tyrosine kinase inhibitors are available to modify cellular phosphotyrosine signal levels. In our experiments, the relative effect of protection was greatest as lower levels of protein tyrosine phosphorylation (*see* **Note 16**).

3.2 Expression of SH2 Domain In Vivo

1. Plate COS1 cells in COS1 growth media so that cells will be 70–80% confluent 18–24 h after plating. Typically a 1:4 dilution from a trypsinized confluent dish provides sufficient cell density after 24 h (*see* **Note 17**).
2. Carefully aspirate cell media and replace with transfection media.
3. Set aside two 1.5 mL microcentrifuge tubes for each experimental point. Label them tube A and B.
4. Add the Lipofectamine 2000 reagent and Opti-MEM to tube A and cDNA and Opti-MEM to tube B, as described in Table 1 (*see* **Note 18**).
5. Repeat **steps 1–4** for each experimental point.
6. Gently mix the contents of tubes A and B by pipetting.
7. Allow the mixture to incubate at room temp for 45 min. Mix by inverting the tube 5–10 times every 15 min during the incubation.
8. Add the Lipofectamine/DNA reaction mixture drop wise to the corresponding culture plate in a manner that provides an even distribution and gently swirl the plate to homogenize the mixture.
9. Incubate the cells at 37 °C in 5% CO_2 for 4–6 h.
10. Gently aspirate the media/transfection mixture and replace with complete COS1 growth media (5 mL for 6 cm plates, 10 mL for 10 cm plates).
11. Allow the cells to grow for 18–20 h (*see* **Note 19**).
12. At some point prior to growth factor stimulation make sure that all cell culture plates are labeled both on the lid and dish, as the lids with be discarded during the flash freeze.

Table 1
Suggested transfection condition

	6 cm (4 mL media)		10 cm (8 mL media)	
	Tube A	Tube B	Tube A	Tube B
Lipofectamine 2000	10 μL	–	25 μL	–
Opti-MEM	250 μL	250 μL	500 μL	500 μL
cDNA	–	5–10 μg	–	20–30 μg

13. If you are planning to stimulate cells with EGF, go to Subheading 3.3. If you do not plan to stimulate your cells, pour of the media, aspirate the remainder, and quickly submerge the labeled plate in a container of liquid N_2 and proceed to Subheading 3.4.

3.3 EGF Stimulation (Optional)

1. Gently aspirate cell culture media and replace with COS1 starvation media.
2. Incubate cells at 37 °C in 5 % CO_2 for 12 h.
3. Remove and combine the culture media from all experimental plates. Add 2 mL (6 cm plates) or 4 mL (10 cm plates) back to each dish (*see* **Note 20**).
4. Dilute EGF to twice the desired concentration in the culture media (*see* **Note 21**).
5. Add 2 mL (6 cm plates) or 4 mL (10 cm plates) of stimulation media to the first plate requiring stimulation.
6. Allow cells to incubate in EGF media for your desired length of time (*see* **Notes 22** and **23**).
7. At your desired stimulation time, quickly pour off the media, aspirate the remaining media, and quickly submerge the labeled plate in a container of liquid N_2.

3.4 Cell Lysis

Blot-based assays and MS based assays each require different lysis methods. Follow **steps 1–9** to prepare lysates for Western and far-Western blotting, and follow **steps 10–17** for preparing MS samples.

1. For sample preparation for Western, remove plates from liquid N_2 using tongs or long forceps.
2. Allow cells to thaw on ice for 2–6 min.
3. Add cold KLB buffer. 100–200 μL for 6 cm dishes. 200–400 μL for 10 cm dishes.
4. Tilt plates so that the lysis buffer covers the entire cell culture surface.
5. Allow cells to lyse on ice for 5–10 min.
6. Scrape cells and collect lysates in labeled microcentrifuge tubes.
7. Spin down cell lysates at 16,100 x*g* for 10 min in microcentrifuge at 4 °C.
8. Transfer supernatants into new labeled microcentrifuge tubes on ice.
9. Determine lysate protein concentration using Bradford Assay.
10. For sample preparation for MS, remove plates from liquid N_2 using tongs or long forceps.
11. Allow cells to thaw on ice for 2–6 min.

12. Add 500 μL of MS lysis buffer containing fresh pervanadate to each dish.
13. Tilt plates so that the lysis buffer covers the entire cell culture surface.
14. Allow cells to lyse on ice for 5–10 min.
15. Scrape cells and collect lysates in labeled microcentrifuge tubes.
16. Set aside 100–150 μL of lysate for Western and far-Western analysis.
17. Determine the lysate protein concentration using Bradford Assay (*see* **Note 24**).

3.5 Western and Far-Western

A detailed protocol for Western and far-Western blotting can be found in [9, 15] and Chapter 17.

1. Normalize lysate protein concentrations for all experimental points using the appropriate lysis buffer (i.e., KLB or MS lysis buffer).
2. Run between 10 and 50 μg of lysate on PAGE. Run sufficient replicates for future analyses.
3. Transfer proteins to nitrocellulose membranes using transfer buffer.
4. Blot at least one membrane with anti-phosphotyrosine antibody to assess SH2 associated increases in protein phosphorylation (*see* **Note 25**).

3.6 MS Data Analysis

As noted above the methods for performing quantitative phosphotyrosine specific mass spectrometry are beyond the scope of this chapter. A complete protocol for MS-based phosphosite quantification can be found in Zhang et al. [11]. This section will cover how to compile and quantify iTRAQ MS data obtained from SH2 overexpressing cells. Specifically, this data can be used to identify sites protected by SH2 domains. Phosphosites that are increased in abundance in SH2 expressing cells are considered to be protected and most likely SH2 binding partners (*see* **Note 26**).

1. Identify peptides present in at least two biological replicates and in all experimental points (*see* **Note 27**).
2. Calculate the fold increase of the relative abundance of each pY site ($abundance_{SH2}$/ $abundance_{control}$) (*see* **Note 28**).
3. Set a threshold for significant SH2 domain expression associated increases in tyrosine phosphorylation (*see* **Note 29**).
4. Using the threshold, create the list of sites that are significantly enhanced site with SH2 expression.
5. Sort the list by fold increase in phosphosite abundance (Fig. 3a).

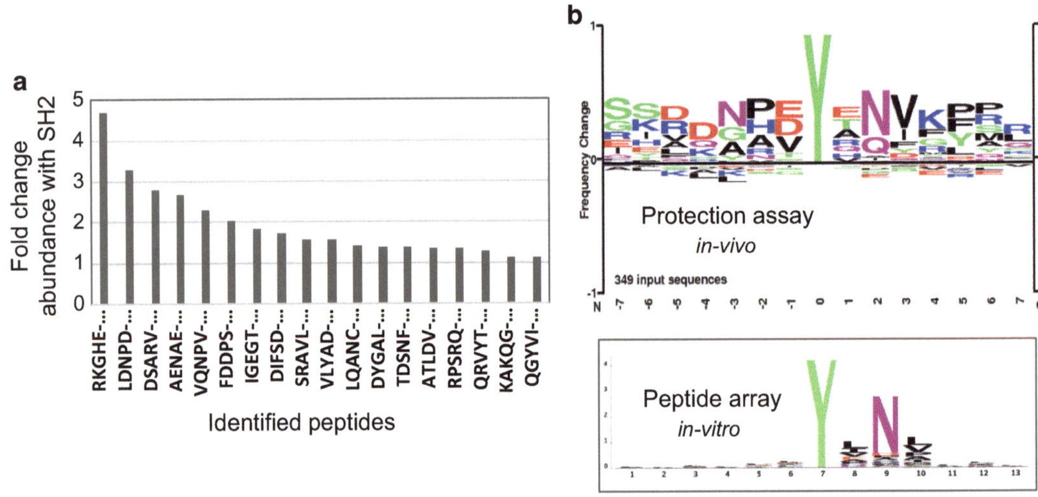

Fig. 3 Phosphopeptides and motifs enhanced by SH2 expression. (**a**) Plot of phosphosites obtained by quantitative MS whose phosphorylation increased by more than two standard deviations in cells expressing the SH2 domain. (**b**) Amino acid consensus sequence logo for enriched phosphosites created by weighting phosphosite abundance data (*upper panel*). As a comparison, the logo plot for the same SH2 using in-vitro peptide array is shown in the *lower panel* (data from the PepSpotDB website) [16]

6. Compare the thresholded list with experimentally identified SH2 interaction sites from the literature (*see* **Note 30**). Sites that have not been previously identified may represent novel SH2 binding sites.
7. Perform orthogonal experiments, i.e., in vivo and in vitro binding assays, to confirm novel SH2 interactions. For example: in vitro phosphopeptide SH2 binding assays [15], SH2 phosphoprotein pull-down assays, SH2 phosphoprotein immunoprecipitation, and in vitro phosphosite protection assay [4].

3.7 LOGO Plot Generation

LOGO plots can be used to visualize the position specific amino acid frequency for peptides whose phosphorylation is enhanced by SH2 expression. The LOGO plot represents the SH2 binding pY motifs that are most likely protected in vivo.

1. Take the list created in Subheading 3.6, **step 4**.
2. Round the fold change of each peptide to nearest tenth.
3. Transform these numbers into whole numbers by multiplying by 10 (*duplication number*).
4. Copy the sequence a *duplication number* of times into a spreadsheet.
5. Compile the duplicate sequences for each pY peptide to create a weighted list of peptides.
6. Copy and paste the entire list into the Sequence Logo Generator at phosphosite.org (*see* **Note 31**).

7. Select "Phospho Tyr" Background and "Frequency Change" algorithm. Press GENERATE.
8. Compared this LOGO plot to in vitro assay derived SH2 binding motifs [16] (Fig. 3b, *see* **Note 32**).

4 Notes

1. Full length proteins have the hypothetical benefit of facilitating physiologic interactions that might stabilize SH2–phosphosite interactions. However, overexpressed non-SH2 protein binding domains within these full length proteins might function as dominant negatives or promote downstream signaling that can feedback (either positively or negatively) on upstream tyrosine phosphorylation. These effects may make results more difficult to interpret.
2. Isolated SH2 domains are poorly expressed in cells. We observed that SH2 domains fused with fluorescent proteins (e.g., tdEOS) or GST tag express well in vivo. These fusions have the additional benefit of being able to be used for imaging or in vitro pull-down assays, respectively.
3. For our experiments, we use polyethylene glycol (PEG)–lithium chloride (LiCl) purification. DNA purification using other standard procedures, e.g., Qiagen Maxi Prep or CeCl, also produces DNA of sufficient purity for transfection.
4. If specific pY signaling pathways are to be examined, their presence in the cell line should be confirmed prior to assessing protection.
5. Lipofectamine 2000 provided ~50 % transfection efficiency and minimal cell toxicity for COS1 cells. Choice of transfection reagent and specific transfection protocol should be made based on these aforementioned factors. SH2 protein expression may also be accomplished via retroviral transduction. However, the effectiveness of this method for use in protection assays has not been verified by our laboratory.
6. Other non-serum media can be used for the transfection mixture.
7. EGF stimulation represents one possible experimental manipulation. Stimulation with other phosphotyrosine signaling modulators such as PDGF, VEGF, and cell adhesion may also provide reliable results in cells types expressing receptors for these ligands.
8. Both experimental and reaction modeling results performed by our lab suggest that the relative increase in phosphorylation

associated with SH2 expression is largest in cells that are not stimulated with EGF or those stimulated with a lower concentration of EGF (less than 5 ng/mL in COS1 cells). However, this feature of the system should be balanced against the sensitivity of pY detection method (e.g., immunoblot or MS) being used, as unstimulated cells typically have much lower levels of tyrosine phosphorylation.

9. We routinely use an infrared detection system for immunoblotting with anti-phosphotyrosine antibodies, and chemiluminescence detection for far-Western blotting with purified SH2 domains. These detection methods can be used interchangeably, although optimal probe concentrations should be determined for each system. In general, chemiluminescence-based detection provides a greater sensitivity and infrared detection has a wider linear dynamic range.
10. Our data suggest that this reagent provides the highest signal to noise ratio in far-Western with purified SH2 domains [15]. However, a number of other ECL products also provide good results.
11. As described in **Note 9**, our results suggest that chemiluminescence-based imaging provides the best signal to noise ratio. We strongly suggest avoiding film as we have found that it provides a much more narrow linear signal range. This may be related to the need to scan images before quantification.
12. Optionally, pervanadate ($H_2O_2 + Na_3VO_4$), a potent inhibitor for tyrosine phosphatases, can be included in the MS lysis buffer. Pervanadate should be prepared fresh for each lysis by adding 32 μL of 30% H_2O_2 to 100 μL of Na-orthovanadate. The solution should be allowed to incubate at room temperature for 0.5–3 h before being added to the 8 M urea solution. Care should be taken to secure the lid of the pervanadate stock solution during incubation as the gas produced by this reaction can cause the lid of a microcentrifuge to pop open.
13. Isolated SH2 domains should be made as fusions with larger proteins, such minimally dimerizing fluorescent proteins [9]. Tandem SH2 domains have been shown to bind more strongly in vitro due to avidity effects [17]. However, we have not assessed the use of tandem SH2 domains for phosphosite protection assays and therefore care should be taken to control for artifacts when using these constructs.
14. We recommend performing titration experiments to assess concentration dependence of the protection effect. For a quantitative analysis, the absolute cellular concentration of SH2 domain can be assessed using recombinant SH2 proteins as a standard for quantitative Western blotting [10].

15. If necessary, the total number of samples assayed by quantitative MS can be increased by duplicating at least one experimental point in the second MS run (e.g., Run1 = A, B, C, D, E, F, G, H and Run 2 = A, I, J, K, L, M, N, O, P) [18].
16. Very low levels of phosphorylation may be difficult to detect by both blot- and MS-based methods. However, we obtained good results detecting protection in SH2 expressing serum starved (unstimulated) cells using MS and to a lesser extent pY-immunoblot.
17. For Western and far-Western-based analyses of SH2 mediated protection we typically use 6 cm culture plates. This provides nearly 1 mg of protein for analysis. For MS analysis we used one 10 cm culture plate for each experimental point.
18. This protocol has worked well for our laboratory when using COS1 cells. The exact amount of transfection reagent, cDNA, and Opti-MEM should be optimized to your cell type, DNA purity, and transfection reagent protocol.
19. Incubations longer than 24 h may result in increased cell death and reduced expression as measured by Western blotting. This is especially true of cells that will be stimulated with EGF, as this step requires an additional 12 h starvation step.
20. There may be some concern about combining media from cells transfected with different constructs. If this is an issue, EGF can be diluted in fresh media containing only 1 % penicillin/streptomycin (0 % FBS). However, this is only feasible if the required stimulation time is less than a few hours as this media tends to be associated with more rapid onset of cell death.
21. We typically use EGF concentrations between 0.5 and 250 ng/mL. The effects of EGF concentrations lower than 0.5 ng/mL are difficult to visualize by anti-phosphotyrosine immunoblotting. Stimulation with EGF concentrations greater than 250 ng/mL does not result in any further increase in EGFR tyrosine phosphorylation.
22. COS1 cells reach their maximum phosphorylation level between 4 and 8 min post-stimulation with EGF. Stimulation time should be optimized for your ligand and cell type of choice.
23. Plates that are incubated with growth factor for more than one minute should be placed back into the cell culture incubator to avoid confounding effects from changes in media pH and temperature.
24. 8 M urea does not interfere with Bradford protein concentration assay.
25. Anti-pY Western blotting only allows for the assessment of protein tyrosine phosphorylation for proteins that are highly

phosphorylated and/or highly expressed. The phosphorylation of other proteins is best assessed using phosphosite specific antibodies or MS.

26. Increases in pY site abundance in response to SH2 expression may be the result of either direct protection or indirect effects on the kinase phosphatase balance (Fig. 1). However, we are unable to deconvolute the contribution of these factors and therefore for the purpose of analysis, we treat all sites as SH2 binding sites.
27. For our MS experiments, we only include pY peptides that are present in at least two of the three biological replicates.
28. iTRAQ analysis provides the normalized average relative abundance of each identified phosphopeptide. By comparing the abundance of different phosphopeptides in the presence or absence of SH2 expression, we can identify the sites that are highly dependent.
29. Threshold values should be determined by the experimenter. However, increases greater than 1.5- to 2-fold or 2 standard deviations are reasonable. These are within the ranges used in both our experiments and multiple other MS studies looking at the effect of specific treatments on protein tyrosine phosphorylation [7, 8, 19].
30. SH2 dependent interactions can be assessed using multiple online databases including PhosphositePlus and the Human Protein Reference Database [14]. Sites with the greatest increases in abundance (i.e., high fold change value) are more likely to be previously identified.
31. Sequence logo generators require that the peptide sequences are the same length, the pY site be at the same position, and they typically do not accept pY for phosphotyrosine. Extreme C or N-terminal peptides may lack the required number of amino acids. This can be rectified by adding "X" as placeholders. For example, input XXXNYPSR for NpYPSR.
32. *See* Refs. 16, 20, 21 for SH2 binding site consensus motifs.

Acknowledgments

I would like to thank Timothy Curran for his work on Mass Spectrometry experiments, Kazuya Machida for assistance with editing, and Bruce Mayer for assistance with concept development. This study was partly supported by grant CA1154966 from the National Institutes of Health.

References

1. Vermeulen M, Hubner NC, Mann M (2008) High confidence determination of specific protein-protein interactions using quantitative mass spectrometry. Curr Opin Biotechnol 19:331–337. doi:10.1016/j.copbio.2008.06.001
2. Hino N, Oyama M, Sato A, Mukai T, Iraha F, Hayashi A, Kozuka-Hata H, Yamamoto T, Yokoyama S, Sakamoto K (2011) Genetic incorporation of a photo-crosslinkable amino acid reveals novel protein complexes with GRB2 in mammalian cells. J Mol Biol 406:343–353. doi:10.1016/j.jmb.2010.12.022
3. Roux KJ, Kim DI, Raida M, Burke B (2012) A promiscuous biotin ligase fusion protein identifies proximal and interacting proteins in mammalian cells. J Cell Biol 196:801–810. doi:10.1083/jcb.201112098
4. Daly RJ, Daum G, Lil N, Fischer EH, Burgess WH, Ullrich A, Schlessinger J (1992) SH2 domains prevent tyrosine dephosphorylation of the EGF receptor: identification of Tyr992 as the high-affinity binding site for SH2 domains of phospholipase CT. EMBO J 11:559–567
5. Wills MKB, Tong J, Tremblay SL, Moran MF, Jones N (2014) The ShcD signaling adaptor facilitates ligand-independent phosphorylation of the EGF receptor. Mol Biol Cell 25:739–752. doi:10.1091/mbc.E13-08-0434
6. Kleiman LB, Maiwald T, Conzelmann H, Lauffenburger DA, Sorger PK (2011) Rapid phospho-turnover by receptor tyrosine kinases impacts downstream signaling and drug binding. Mol Cell 43:723–737. doi:10.1016/j.molcel.2011.07.014
7. Sharma K, D'Souza RCJ, Tyanova S, Schaab C, Wiśniewski JR, Cox J, Mann M (2014) Ultradeep human phosphoproteome reveals a distinct regulatory nature of Tyr and Ser/Thr-based signaling. Cell Rep 8:1583–1594. doi:10.1016/j.celrep.2014.07.036
8. Zhang Y, Wolf-Yadlin A, Ross PL, Pappin DJ, Rush J, Lauffenburger DA, White FM (2005) Time-resolved mass spectrometry of tyrosine phosphorylation sites in the epidermal growth factor receptor signaling network reveals dynamic modules. Mol Cell Proteomics 4:1240–1250. doi:10.1074/mcp.M500089-MCP200
9. Jadwin JA, Oh D, Curran TG, Ogiue-Ikeda M, Jia L, White FM, Machida K, Yu J, Mayer BJ (2016) Time-resolved multimodal analysis of Src Homology 2 (SH2) domain binding in signaling by receptor tyrosine kinases. Elife 5:1–26. doi:10.7554/eLife.11835
10. Machida K, Thompson CM, Dierck K, Jablonowski K, Kärkkäinen S, Liu B, Zhang H, Nash PD, Newman DK, Nollau P, Pawson T, Renkema GH, Saksela K, Schiller MR, Shin D-G, Mayer BJ (2007) High-throughput phosphotyrosine profiling using SH2 domains. Mol Cell 26:899–915. doi:10.1016/j.molcel.2007.05.031
11. Zhang Y, Wolf-Yadlin A, White FM (2007) Quantitative proteomic analysis of phosphotyrosine-mediated cellular signaling networks. Methods Mol Biol (Clifton, NJ) 359:203–212
12. Vacic V, Iakoucheva LM, Radivojac P (2006) Two sample logo: a graphical representation of the differences between two sets of sequence alignments. Bioinformatics 22:1536–1537. doi:10.1093/bioinformatics/btl151
13. Crooks G, Hon G, Chandonia J, Brenner S (2004) WebLogo: a sequence logo generator. Genome Res 14:1188–1190. doi:10.1101/gr.849004.1
14. Hornbeck PV, Kornhauser JM, Tkachev S, Zhang B, Skrzypek E, Murray B, Latham V, Sullivan M (2012) PhosphoSitePlus: a comprehensive resource for investigating the structure and function of experimentally determined post-translational modifications in man and mouse. Nucleic Acids Res 40:D261–D270. doi:10.1093/nar/gkr1122
15. Jadwin JA, Mayer BJ, Kazuya M (2015) Detection and quantification of protein–protein interactions by far-western blotting. In: Kurien B, Scofield H (eds) Western blotting, vol 1312, Methods in molecular biology. Springer, New York, pp 379–398
16. Tinti M, Kiemer L, Costa S, Miller ML, Sacco F, Olsen JV, Carducci M, Paoluzi S, Langone F, Workman CT, Blom N, Machida K, Thompson CM, Schutkowski M, Brunak S, Mann M, Mayer BJ, Castagnoli L, Cesareni G (2013) The SH2 domain interaction landscape. Cell Rep 3:1293–1305. doi:10.1016/j.celrep.2013.03.001
17. Machida K, Mayer BJ, Nollau P (2003) Profiling the global tyrosine phosphorylation state. Mol Cell Proteomics 2:215–233. doi:10.1074/mcp.R300002-MCP200
18. Wolf-Yadlin A, Hautaniemi S, Lauffenburger DA, White FM (2007) Multiple reaction monitoring for robust quantitative proteomic analysis of cellular signaling networks. Proc Natl Acad Sci U S A 104:5860–5865. doi:10.1073/pnas.0608638104
19. Blagoev B, Ong S-E, Kratchmarova I, Mann M (2004) Temporal analysis of phosphotyrosine-dependent signaling networks by quantitative

proteomics. Nat Biotechnol 22:1139–1145. doi:10.1038/nbt1005

20. Liu BA, Engelmann BW, Nash PD (2012) The language of SH2 domain interactions defines phosphotyrosine-mediated signal transduction. FEBS Lett 586:2597–2605. doi:10.1016/j.febslet.2012.04.054

21. Huang H, Li L, Wu C, Schibli D, Colwill K, Ma S, Li C, Roy P, Ho K, Songyang Z, Pawson T, Gao Y, Li SS-C (2008) Defining the specificity space of the human SRC homology 2 domain. Mol Cell Proteomics 7:768–784. doi:10.1074/mcp.M700312-MCP200

Chapter 29

Real-Time Single Molecule Visualization of SH2 Domain Membrane Recruitment in Growth Factor Stimulated Cells

Dongmyung Oh

Abstract

In the last decade, single molecule tracking (SMT) techniques have emerged as a versatile tool for molecular cell biology research. This approach allows researchers to monitor the real-time behavior of individual molecules in living cells with nanometer and millisecond resolution. As a result, it is possible to visualize biological processes as they occur at a molecular level in real time. Here we describe a method for the real-time visualization of SH2 domain membrane recruitment from the cytoplasm to epidermal growth factor (EGF) induced phosphotyrosine sites on the EGF receptor. Further, we describe methods that utilize SMT data to define SH2 domain membrane dynamics parameters such as binding (τ), dissociation (k_d), and diffusion (D) rates. Together these methods may allow us to gain greater understanding of signal transduction dynamics and the molecular basis of disease-related aberrant pathways.

Key words Single molecule, SH2 domain, Membrane trafficking, Total internal reflection fluorescence microscopy (TIRFM), Dissociation rate, Diffusion rate

1 Introduction

Interaction with the cellular environment is critical for cell survival and proliferation. Biochemical signals originating from the extracellular environment are interpreted through molecule–molecule interactions controlled by factors such as specificity, affinity, and duration of binding. These interactions result in the spatial redistribution of downstream molecules in a time-dependent manner. Each molecule involved in a specific signaling pathway will behave differently. Complete understanding of this process requires an appreciation of the highly dynamic nature of molecular interactions that may not be captured by a single biochemical snapshot. For example, binding of epidermal growth factor (EGF) on EGF receptor (EGFR) induces EGFR dimerization which results in autophosphorylation of EGFR tyrosines located on the cytoplasmic C-terminal tail (Fig. 1b) [1]. These phosphotyrosines (pY) serve as docking sites for Src homology 2 (SH2)

Kazuya Machida and Bernard A. Liu (eds.), *SH2 Domains: Methods and Protocols*, Methods in Molecular Biology, vol. 1555, DOI 10.1007/978-1-4939-6762-9_29,

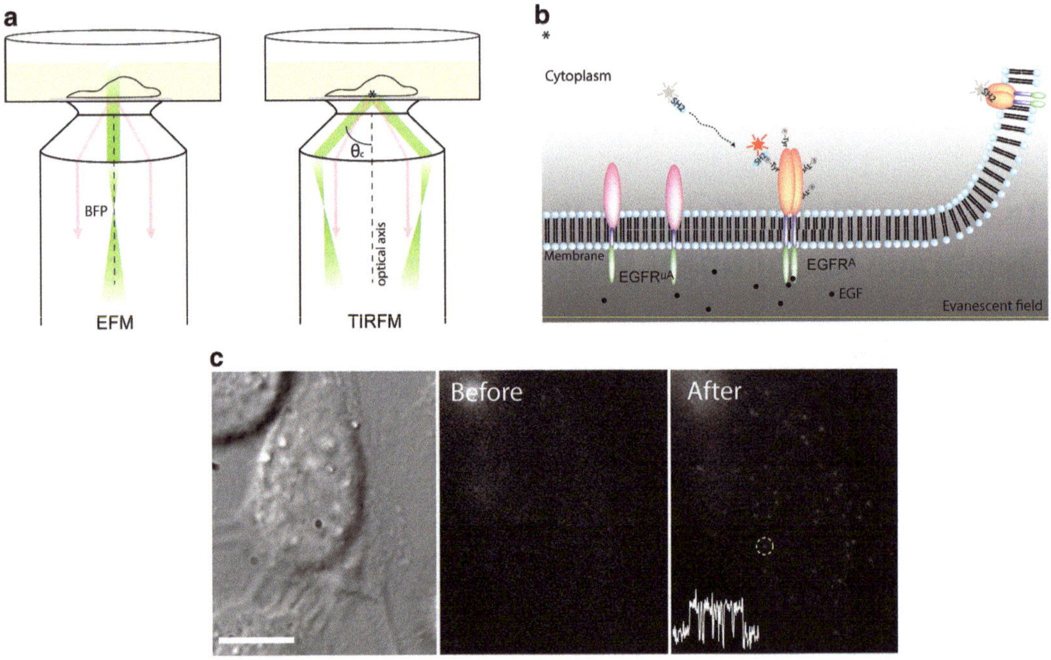

Fig. 1 Single molecule imaging and tracking to study membrane recruitment of SH2-Eos. (**a**) Schematic of epi fluorescence microscopy (EFM) and total internal reflection microscopy (TIRFM) optics geometries. The illumination beam (*green*) passes through the objective along the optical axis (*dashed lines*, EFM) or off the optical axis (TIRFM), focuses on the back focal plane (BFP) of the objective, and illuminates the cell and medium. θ_c is the critical angle to create a total internal reflection of the incident beam and the evanescent field at the glass–medium interface (*asterisk*). (**b**) Illustration of the evanescent field at the glass–medium interface. *Pink ovals* represent unstimulated EGF receptor; *orange ovals* represent EGF receptor activated by EGF-induced dimerization. Several pTyr sites are created on the intracellular domain of the EGF receptor, which serves as docking sites for SH2 domain-containing proteins. In this example, Grb2-SH2 domain was cloned as fusion with Eos (*blue rectangle* and *red star*) and expressed in cells. The evanescent field (*gray color*) illuminates SH2-Eos molecules localized to the bottom membrane region of the cell. Molecules outside of the field (*gray star*) are not detectable. (**c**) Bright field (*left panel*) and fluorescence images (*middle* and *right*) of cells expressing SH2-Eos molecules. Before EGF stimulation, few molecules are detected at the membrane. The number of *bright dots* significantly increases following EGF stimulation. The *inset* shows the single step photoblinking and photobleaching of an individual molecule (*dashed circle*). Fluorescence images were obtained in the photoactivated red channel of Eos with TIRFM. Bar is 10 μm. Some figures were modified from original article [22]

domain-containing proteins, which are recruited in a time and pY-site dependent manner. This recruitment of SH2 domain proteins to the receptor creates a unique signaling complex at the membrane that initiates multiple downstream signaling cascades by recruiting or redistributing other intracellular signaling molecules [2]. Therefore to understand phosphotyrosine signaling pathways, it is essential to first assess the spatial and temporal dynamics of SH2-pY interactions [3].

Recently developed state-of-the-art single molecule imaging techniques can detect real-time molecular interaction events at the single molecule level with up to 1 ms time resolution in living cells. Single molecule tagged fluorophores were first detected in hydrodynamically focused flows of the solution [4], the surface of solids [5] and in live cells [6]. The isolation of green fluorescence protein (GFP) from the jellyfish *Aequorea victoria* [7] and subsequent engineering of this protein has led to the development of fluorescence proteins of multiple colors, higher brightness (quantum yield), shorter maturation time, higher photostability, lower pH sensitivity, and monomerization [8–14]. The high accuracy of single fluorescence molecule imaging has been achieved by significantly increasing the signal to noise by a combination of optical illumination and detection techniques. Single molecule tracking (SMT) has been applied to study fundamental biological processes in real time such as gene expression [13, 15], bacteria cell division [12], DNA recombination mechanisms [11], microtubules and actin dynamics [10, 14], walking of kinesin along microtubules [8, 9], and diffusion and membrane organization [16, 17].

SMT techniques require two essential components, illumination optics and a detection device. By coupling advanced charge-coupled device (CCD) techniques [18] with total internal reflection fluorescence microscopy (TIRFM) optics [19], one can efficiently collect and detect the fluorescence from a single fluorescent molecule. TIRFM can be easily achieved using a conventional epifluorescence microscopy setup by creating a total reflecting beam. In epifluorescence microscopy (EFM), the illumination beam passes through the objective's optical axis and is transmitted through the whole volume of sample (Fig. 1a). The transmitted beam excites all fluorophores within the cell and media so that the detected fluorescence signal includes not only the real fluorescence signal but also nonspecific fluorescence signals such as cellular autofluorescence and media derived background (Fig. 1, EFM). Under this wide field illumination, it is difficult to identify individual fluorescence molecules due to the mixture of nonspecific signals and overwhelming brightness associated with overexpression of the fluorescent-tagged protein. Of note, the issue caused by overexpression could be addressed by integrating inducible gene expression systems [20]. If the illumination beam passes through and off the optical axis (Fig. 1a, TIRFM), it creates reflected and transmitted beams at the surface of glass-cell interface. As the incident beam is shifted away from the optical axis, it will eventually reach the total reflection angle (θ_c) resulting in a transmission angle of 90° against optical axis (Fig. 1a, TIRFM). At critical angle (θ_c) an evanescent wave (marked with asterisk symbol in Fig. 1a and illustrated in gray in Fig. 1b) is generated at the interface. This wave propagates perpendicular to the interface but decays with an exponential constant that is a function of the refraction indices of the glass and medium and reflection angle of the beam [19]. Usually, an evanescent wave propagates into the glass-medium interface at a distance equal to half the wavelength of the illumination beam, i.e., 200–300 nm from the surface of the glass if GFP derivatives are used as a chromophore tag. As such, total internal reflection conditions enable detection of only those molecules within the evanescent field along the bottom membrane (marked with a red star symbol in Fig. 1b). This results in a reduced number of excited fluorescence proteins on the image frame, dramatic suppression of the brightness, and thereby achieving a high signal-to-noise ratio (Fig. 1c).

Here we describe methods for visualizing membrane trafficking and dynamics of individual SH2 molecules as a function of EGF stimulation time using TIRFM based SMT techniques. Further, we describe the process of trajectory analysis to obtain in vivo physical parameters such as membrane recruitment, dissociation, diffusion rate. These in vivo parameters may help better understand the dynamics between SH2 domains and activated receptor tyrosine kinases. In this system, an isolated SH2 domain is expressed as a fusion protein with the photoactivatable fluorescent protein (Eos) (*see* **Note 1**) [21], and imaged under TIRFM. As mentioned above, TIRFM improves signal-to-noise by reducing excitation volume to only the vicinity of the plasma membrane (Fig. 1b). Using Eos, we can further limit the number of excited SH2-Eos by exploiting its propensity for photoactivation. Exposure to 405 nm beam causes a change in the excitation/emission spectrum of a small fraction of Eos molecules. These molecules can then be separately imaged using a different wavelength, corresponding to the photoconverted state of Eos molecule, as isolated single molecules. Tracking these single molecules facilitates measurement of their recruitment from the cytoplasm to EGFR pY sites at the plasma membrane, their binding/dissociation kinetics to/from the plasma membrane, and their diffusion properties at the membrane [22]. The purpose of these experiments is to assess the dynamics of specific SH2-pY interactions in live cells, without interference by other protein-protein interaction mechanisms such as those attributed by SH3 and PDZ domains. A simple example is to detect individual SH2 molecules arriving at the cell membrane after stimulation with EGF [22]. In Fig. 1c, the bright field (left panel) and fluorescence (middle and right panel) images display an A431 cell expressing an SH2 protein (for which the SH2 domain has been known to bind to pY-EGFR). Before EGF stimulation there are few fluorescence dots on the image under TIRFM conditions reflecting the low level of EGFR phosphorylation (middle panel). After the addition of EGF to the sample dish, the SH2 protein is recruited to the membrane indicating an SH2-pY-EGFR binding event (bright dots in right panel). The inset of Fig. 1c (right panel) shows the intensity profile of the circled dot. This intensity trajectory of a single molecule exhibited single step photoblinking and photobleaching, which is characteristic of single fluorescence molecules. A practical protocol for these analyses using SH2-Eos fusion constructs and A431 cells is presented below.

2 Materials

2.1 Preparation of SH2 Entry Vector

1. Grb2 SH2 template cDNA [22]. Sources for human SH2 domain cDNA include Addgene (e.g., Mayer lab clones), DNASU, GE Dharmacon, and the Pawson Lab website.
2. Pfu DNA polymerase (Stratagene, La Jolla, CA, USA).
3. Custom oligonucleotide primers.
4. Entry vector, pENTR/D-TOPO (Invitrogen).
5. pENTR Directional TOPO Cloning Kits (Invitrogen).
6. Chemically competent bacteria (e.g., DH5α).
7. Luria-Bertani (LB) medium and LB agar plate with kanamycin.
8. Miniprep kit.

2.2 Subcloning of SH2-Eos Clone

1. Gateway destination vector: pDEST-mEos3-C or pDEST-tdEos-C [22, 23] (*see* **Note 2**).
2. pENTR-Grb2 SH2 entry vector.
3. Gateway LR reaction kits; Gateway LR Clonase II enzyme, Proteinase K (Invitrogen).
4. TE buffer, pH 8.0.
5. LB medium and LB agar plate with ampicillin.
6. Chemically competent bacteria (DH5α).
7. Miniprep Kit.
8. Tabletop centrifuge.
9. 37 and 43 °C water bath.
10. NanoDrop spectrophotometer.
11. Mini-Sub^R cell GT (Bio-Rad).

2.3 DNA Transfection and Cell Culture

1. Human epithelial carcinoma A431 cell.
2. 0.05 % trypsin–EDTA (1×).
3. 1× Dulbecco's Modification of Eagle's medium (DMEM) with 10 % FBS, 0.1 % streptomycin.
4. Opti-MEM (ThermoFisher Scientific).
5. Cell culture dishes.
6. 35-mm glass bottom dishes (Matek).
7. Ultrasonication Cleaner (e.g., Ultrasonic FS20, Fisher Scientific).
8. UHQ deionized water (DD water).
9. Lipofectamine 2000 (ThermoFisher Scientific).
10. Inverted microscope (Olympus).

2.4 Live Cell Imaging

1. Imaging media: OPTI-MEMR II ((+) L-glutamine, (+) HEPES, (−) phenol red), supplemented with 0.1 % FBS.
2. 25 ng/mL human recombinant EGF.
3. 35 mm glass bottom dishes.
4. Immersion oil (refractive index 1.53).
5. 99 % ethanol (molecular biology grade).
6. Olympus IX81 inverted microscope equipped with 60×, NA 1.49 oil immersion objective lenses.
7. TIRF illumination setup. We used a custom-made system, but commercial lightguide-based setup also applicable.
8. Lasers: 561 nm diode laser (Coherent), 488 nm Argon ion laser (Melles Griot), 405 nm UV laser (Coherent^{R}).
9. Filters set: Recommend a 405/488/561 Laser triple band set (e.g., Chroma TRF69901).
10. TE-cooled EMCCD camera (PhotonMax 512, Princeton Instruments).
11. Bright field Arc lamp U-LH100L-3 (Olympus).
12. ImageJ and Micro-manager plug-in (NIH).

2.5 Data Analysis

1. Image J (NIH).
2. Matlab^{R} (MathWorks, Inc.).
3. Origin^{R} 8.5 (Origin Lab, MA, USA).
4. Single molecule tracking program [8].
5. VbSPT analysis program [24].

3 Methods

3.1 Preparation of SH2 Entry Vector

This section describes preparation of an SH2-EOS construct via Gateway cloning (*see* **Note 3**). If an SH2-Eos expression construct is ready for use in imaging experiment, skip to Subheading 3.4.

1. Design PCR primers for Grb2 SH2 (AA 58–159).
2. Perform PCR and purify PCR product.
3. Mix PCR product, salt solution, and TOPO vector and incubate for 5 min at room temperature.
4. Place the mixture on ice.
5. Add 2 μL mixture into a tube of 50 μL chemically competent bacteria and mix gently.
6. Incubate on ice for 5–30 min.
7. Heat-shock the bacteria for 30 s at 42 °C without shaking.
8. Immediately transfer the tube to ice and cool down.

9. Add 250 μL of room temperature LB without antibiotics.
10. Cap the tube tightly and incubate at 37 °C for 1 h with shaking.
11. Spread 50–150 μL on LB plate with kanamycin and incubate overnight at 37 °C.
12. Pick up several colonies and incubate in 5 mL LB with kanamycin overnight at 37 °C.
13. Purify plasmid DNA using miniprep kit.
14. Analyze the plasmid by PCR and select positive clones.

3.2 LR Reaction

1. Adjust the concentration of both pENTR-Grb2 SH2 entry vector and pDEST-mEos-C destination vector to 100–200 ng/μL.
2. Mix 1 μL of pENTR-Grb2 SH2 entry vector, 1 μL of pDEST-mEos-C destination vector, and 8 μL TE buffer in 1.5 mL tube at room temperature.
3. Thaw Gateway LR clonase II enzyme mix on ice for a few minutes and vortex briefly.
4. Add 2 μL of Gateway LR clonase II enzyme to the vector mixture (**step 2**), vortex briefly and spin down using microcentrifuge.
5. Incubate the mixture for 1–2 h at room temperature.
6. Add 1 μL of Proteinase K solution, vortex briefly, and incubate at 37 °C for 10 min.

3.3 Subcloning of SH2 Expression Clone

1. Thaw 50 μL of DH5α competent bacteria on ice for 5 min.
2. Add 1 μL of mixture (LR recombination reaction prepared in Subheading 3.2) and gently mix. Do not vortex.
3. Keep the tube on the ice for 20 min.
4. Heat-shock the bacteria for 30 s at 42 °C without shaking.
5. Immediately transfer the tube to ice and cool down.
6. Add 250 μL of room temperature LB without antibiotics.
7. Cap the tube tightly and incubate at 37 °C for 1 h with shaking.
8. Spread 50–150 μL of bacteria on LB plate with ampicillin and incubate overnight at 37 °C.
9. Pick up several colonies and incubate in 5 mL LB with ampicillin overnight for at 37 °C.
10. Purify plasmid DNA using miniprep kit.
11. Measure plasmid concentration and verify by sequencing.
12. Store the sequence verified pDEST-SH2-Eos plasmids (−20 °C) and bacteria (−80 °C).

3.4 Clean Glass Bottom Dishes

1. Acid treat the bottom of 35 mm glass bottom dishes using 500 μL of 99.99% sulfuric acid (H_2SO_4) for 5 min.
2. Remove acid by multiple washes with DD water.
3. Place glass bottom dishes in a beaker with 500 mL of DD water and place the beaker in a sonicator water bath. Sonicate for 20 min at maximum power with heat.
4. Rinse dishes with DD water several times.
5. Place dishes into a beaker with 500 mL of 99% ethanol.
6. Remove ethanol by multiple washes with DD water and repeat **steps 3** and **4** (*see* **Note 4**).
7. Dry the glass dishes by blowing the liquid out with a flow of nitrogen stream.

3.5 Plasmid DNA Transfection

1. Culture adherent cells in 35 mm dishes until approximately 70% confluent in complete media for 24–48 h. Optional: If necessary, replace complete media with 500 μL serum and antibiotic free media (*see* **Note 5**).
2. Dilute 2 μL of 100 ng/μL SH2 expression plasmid in 50 μL of Opti-MEM, mix gently and keep at room temperature.
3. Dilute 2 μL of Lipofectamine 2000 in 50 μL of Opti-MEM, mix gently and keep at room temperature for 5 min.
4. Mix diluted DNA (**step 3**) and Lipofectamine (**step 4**) solutions and keep at room temperature for 20 min.
5. Add 100 μL of DNA–Lipofectamine solution (**step 5**) evenly to each cell dish.
6. Incubate cells at 37 °C in a CO_2 incubator for 4–6 h.
7. Remove the transfection media and replace with fresh pre-warmed complete media.
8. Incubate cells for 12–24 h at 37 °C in a CO_2 incubator (*see* **Note 6**).
9. Trypsinize cells and resuspend in complete media.
10. Plate suspended cells on glass bottom dishes at a lower density (<30%) so that each cell is well isolated for imaging.
11. Allow cells to adhere for 10 h at 37 °C in a CO_2 incubator (*see* **Note 7**).
12. Prior to imaging, remove the culture media, wash with PBS, and replace with 1 mL of imaging media.

3.6 TIRF Microscopy

For our experiments we use a objective-based TIRFM system designed and assembled in-house [25]. However, commercial laser illumination TIRFM systems are available from multiple vendors (Olympus, Zeiss) and should work as well for the experiments described here. The commercial TIRFM system typically uses optical fiber to deliver laser excitation beams to the microscope directly and allows adjustment of illumination angle by repositioning the fiber tip (or its conjugate image). These systems allow for users to quickly switch between epi and TIR illumination on the same microscope and are recommended to most biological researchers. For live cell imaging, a heated sample stage for temperature control is critical for maintaining cell viability. We recommend the stable-Z system from Bioptechs, although similar products are available from several other commercial sources. In addition, an objective heater (also available from Bioptechs) is equally important for temperature regulation, because TIRF objectives are oil-immersion type, which will draw a significant amount of thermal energy from the sample if not heated. Finally, a hardware autofocus system is also indispensable for the experiment, without which, small temperature fluctuations during the imaging will inevitably lead to the loss in sharp focusing of the images. Most scientific microscope vendors now offer the option of including an autofocus package in their system.

1. Turn on the microscope 20 min before imaging to warm and stabilize the optical system, the excitation lasers (488 nm ion laser for inactivated Eos molecule illumination, 405 nm diode laser for photoactivation, and 561 nm diode laser for imaging photoactivated Eos molecule) and initiate the cool down process of the EMCCD camera chips.
2. Apply immersion oil on the oil-immersion objective lens and place the glass bottom dishes on the microscope stage. Wait until sample temperature stabilizes.
3. Bring the cells into focus under bright field illumination using a mounted arc lamp.
4. Switch to 488 nm laser illumination (Ex = 480/40 nm, Em = 535/40 nm) to verify fluorescence signal of the cells by epi fluorescence illumination (see Fig. 1a). Select a cell expressing the SH2 proteins based on desired fluorescence brightness.
5. To set up the single molecule imaging of SH2 proteins, start a live acquisition with 561 nm illumination (Ex = 535/10, Em = 575/56 nm). Set the effective illumination power density to be in the range of 100–1000 W/cm^2 (*see* **Note 8**).
6. To create a pool of photoactivated Eos molecule which fused to SH2 protein, turn on 405 nm laser laser while checking fluorescent dots on the camera with 561 nm illumination set up (*see* **Note 9**). Start at the minimal power and gradually increase the power, until the proper number of particle appear on the camera. Turn off 405 nm laser.
7. To avoid spatial overlapping of single-molecule signals due to molecule's diffusion, the spatial density of photo activated Eos molecules should not exceed 0.05–0.1 per μm^2 (*see* **Note 10**). An example of control the number of photoactivated Eos molecule is shown in Fig. 3.

8. Select the appropriate imaging conditions (power, exposuer time and video rate) to maximize spatial and temporal resolution (*see* **Note 11**).
9. To quantify SH2-membrane recruitment as a function of EGF stimulation time, set the micromanager ImageJ plugin to acquire images with a 50 ms exposure time at 10 s intervals spanning up to 1 h. This relatively short exposure and the long interval between image sequence can minimizes the photobleaching effect during measurement. Figure 2a shows

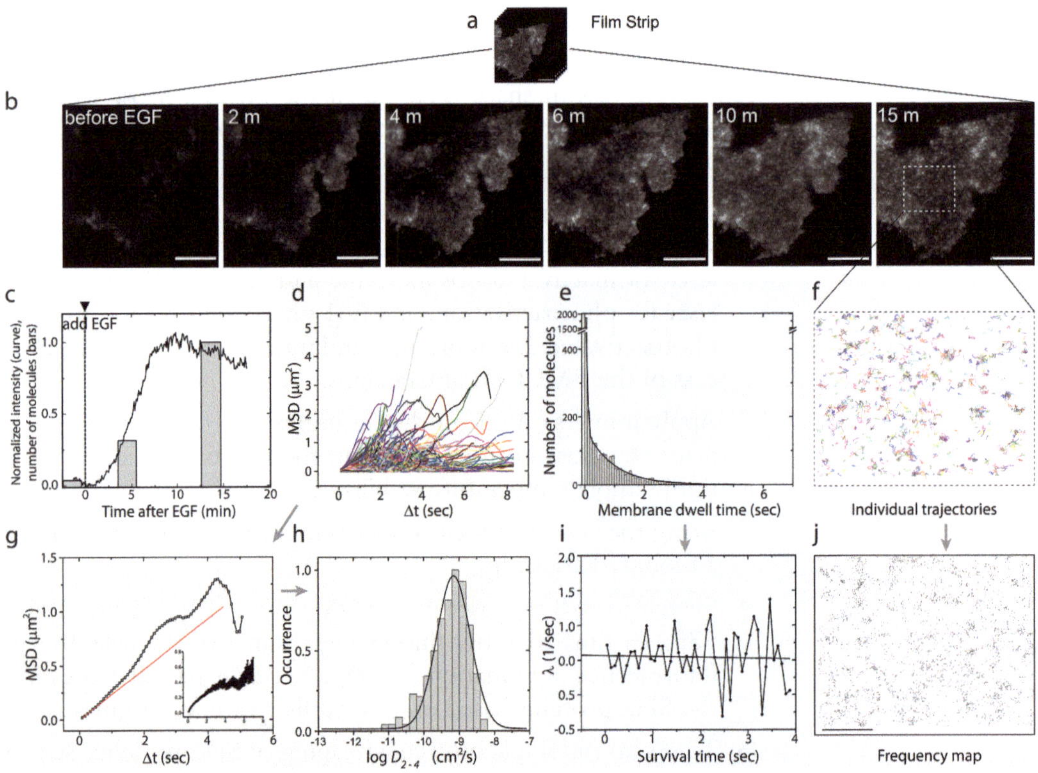

Fig. 2 Single molecule data analysis. (**a**) A Stack of image sequences of representative single molecule experiment. (**b**) Selected images from the stack showing progression of fluorescence intensity following EGF stimulation. (**c**) Normalized intensity profile obtained from image sequences (*black line*), and a normalized number of molecules counted at corresponding time points (*gray bars*). (**d**) Combined MSD vs. Δt curves of all molecules located within a region of interest (ROI, *white dotted box* in **b**). *Each color* represents individual trajectories. (**e**) Distribution of SH2-Eos membrane dwell times. (**f**) Trajectories of SH2-Eos within ROI. (**g**) MSD vs. Δt curve for one trajectory fitted with the normal diffusion equation (*red straight line*) at 2–4 step size. The inset shows the averaged MSD vs. Δt curve for all trajectories shown in panel (**d**). (**h**) Distribution of individual effective diffusion constants (D_{eff}), determined in panel (**g**). (**i**) Hazard rate function $\lambda(t)$ calculated from the dwell time distribution (panel **e**). Here, the hazard rate function (apparent membrane dissociation rate) of SH2-Eos is constant ($\lambda = 0.5\ s^{-1}$). (**j**) Distribution of molecular coordinates (first *x*–*y* coordinate only) obtained from all trajectories in panel (**f**). Scale bar equal to 10 μm. *Gray arrows with background colors* indicate the order of analysis. Some figures were modified from original article [22]

a time-lapse image sequences of SH2-membrane recruitment following EGF stimulation. Figure 2b shows the selected images in Fig. 2a showing the recruitment of SH2 molecules from the cytoplasm to the bottom membrane at specific time points after EGF stimulation.

10. To quantify SH2 dissociation rate on and diffusion rate from the membrane using spt PALM, set the micromanager ImageJ plugin to acquire images with a 50 ms exposure at 80 ms intervals spanning up to 1 min. Repeat this step to measure the changes in the diffusion and dissociation rates as a function of EGF stimulation.
11. Add EGF diluted with 100 μL of imaging media to the sample dish (to final EGF concentration 25 ng/mL) and begin image acquisition as described in **steps 9** and **10**.

3.7 Total Molecule Recruitment Analysis

The following methods are for analysis of total molecule recruitment image sequences obtained using the acquisition settings outlined in Subheading 3.6, **step 9**.

1. Convert raw image sequences by stacking in ImageJ.
2. Open the SH2 membrane recruitment image stack (Fig. 2a) and select the region of interest (ROI) near the cell center (marked with the white dotted box in Fig. 2b).
3. Run "Intensity vs. time" ImageJ plugin. This will return an intensity profile as a function of EGF stimulation time. Figure 2c black curve shows the normalized intensity profile of ROI near the cell center (marked the white dotted box in Fig. 2b).
4. To approximate the number of individual molecules detected, extract individual images from the movie.
5. Use "Particle Counting" plugin to automatically count the number of molecules. Gray bars represent the normalized number of molecules at corresponding image sequences. Alternatively, single particle tracking algorithm can provide the number of the particle within a series of images (*see* **Note 12**).

3.8 Single Molecule Trajectory Analysis

The following methods are for analysis of single molecule trajectory image sequences obtained using the acquisition settings outlined in Subheading 3.6, **step 10**.

1. Identify single molecule trajectories using the standard single particle tracking methods including the automatic single particle tracking algorithm (*see* **Note 12**) [8, 26].
2. Integrate all trajectories obtained from each time-lapse movie into a single image using the Matlab code (*see* **Note 13**). A typical two-dimensional trajectory of SH2-Eos molecules is shown in Fig. 2f (colored lines indicate trajectories of individual molecules that appear in more than five consecutive frames). This

figure provides the graphic visualized of localization and diffusion of membrane-recruited SH2-Eos molecules.

3. To analyze the location and frequency of SH2-Eos binding, plot all the first *x–y* coordinates from all trajectories (black dots in Fig. 2j). Areas with high density are suggestive of clustering sites (*see* **Note 14**). Diffusion and dissociation rate of Grb2 SH2 are different from the cluster to out of the cluster region [23].
4. Repeat **step 3** for the remaining subsequent 1-min long single molecule time-lapse movies to investigate changes in spatial distribution of this first *x–y* coordinates, and binding rates during EGF stimulation [23]. Alternatively, details of SH2-Eos membrane distribution and clustering can be analyzed by live cell spt-PALM image analysis [23].

3.9 Membrane Dissociation Rate Analysis

1. To obtain apparent dissociation rate constant of SH2-Eos molecule from the membrane, first, plot the membrane dwell time distribution (Fig. 2e) (*see* **Note 15**). Dwell time is obtained by multiplying trajectory length (n) and imaging time interval (Δt). Calculate the hazard rate function $\lambda(t)$ as the apparent SH2-Eos dissociation rate at multiple time points. The hazard rate function indicates the probability of finding the SH2-Eos molecule at the time t that dissociates from a binding site, which survived for at least time t (*see* **Note 16**) [22, 23]. Note that this $\lambda(t)$ contains the photobleaching rate of Eos fluorescence tag.
2. To correct for photobleaching, calculate the photobleaching hazard rate function of myristoyl-tagged Eos using single-molecule tracking data (*see* **Note 17**), and subtract this from the measured hazard functions in **step 1**. The result is the corrected hazard rate function of SH2-Eos molecule.
3. After the correction for photobleaching, examine the $\lambda(t)$ to verify that values are constant. If this is the case, the dissociation dynamics can be described by the first-order process with the first-order rate constant λ. If the hazard rate function depends on time, the dissociation dynamics involves multiple steps kinetics and should consider other heterogeneities in molecule's dissociation mechanism.

3.10 Membrane Diffusion Rate Analysis

Mean Square Displacement (MSD) is a measure of the deviation of particle position over the time. There are two analysis methods for determining the effective diffusion constant of SH2 molecule using the MSD calculation: **Method 1**, by fitting the averaged MSD vs. Δt curve over all trajectories to the corresponding diffusion equation, or **Method 2**, using the distribution of individual diffusion rates derived from the MSD vs. Δt curve fit of each trajectory. In our hands, these two methods yielded the similar values: for example, when measuring diffusion constant of Grb2 SH2 molecules, we obtained $D_{240\ ms} = 0.055\ \mu m^2/s$ by **Method 1** and $D_{240\ ms} = 0.063\ \mu m^2/s$ by **Method 2** at a time of EGF stimulation (Fig. 2g inset and h). In general, the **Method 1** has been accepted more widely than the **Method 2** as the reliable method for estimating the diffusion at a particular time scale [27–30]. However, the short trajectory does not contribute to the MSD value at the longer time.

1. Calculate and plot the MSD of each molecule as a function of time lag (Δt) (Fig. 2d, each color represents an individual trajectory). Figure 2g shows an example of MSD vs. Δt curve for an each molecule (*see* **Note 18**).
2. **Method 1**: Average these individual values (± errors) and plot as a function of time lag (Fig. 2g inset). Fit this averaged MSD curve at early time (e.g., 160–320 ms) to the normal diffusion equation: $MSD = 4D(n\Delta t)$ where n is an integer, Δt is the time lag between image sequences (e.g., 80 ms), and D is the effective diffusion constant (*see* **Note 19**). This method is useful determining the average diffusion behavior of SH2-Eos for the particular time scale on the plasma membrane.
3. **Method 2**: Fit each individual MSD vs. Δt curves at the same early time lags to the normal diffusion equation and calculate the diffusion coefficient for each molecule. Figure 2g shows the fit (red line) for a typical trajectory extracted from Fig. 2d.
4. Construct a histogram to plot the distribution of diffusion values obtained in **step 3** and calculate the statistical average (peak value) (Fig. 2h). When the trajectories are sufficiently long, **Method 2** is useful for capturing population heterogeneity if the diffusion modes are mixed, reflecting by a multi-peak in the distribution. Population heterogeneity may suggest that SH2 proteins are binding to several different membrane proteins resulting in various diffusion rates.
5. To investigate EGF-dependent SH2 diffusion dynamics, repeat analyses in Subheadings 3.9 and 3.10 using movies obtained at different EGF stimulation times (e.g., 0, 5, 10 min) (*see* **Note 20**).
6. To further quantify the diffusion heterogeneity, one could analyze the trajectory data with a hidden Markov model (HMM). This analysis is especially useful for the short trajectory data to estimate kinetic rates of interconversion between different subpopulations [24]. We use open source software VbSPT (http://vbspt.sourceforge.net/) .

4 Notes

1. Tandem dimer and monomeric versions of Eos (tdEos and mEos, respectively) have been engineered [21]. We currently use mEos more often.
2. This protocol was generated using C-terminal tdEos-tagged SH2 domain constructs [22, 23]. In our hands, C-terminal and N-terminal fusions gave similar results.
3. For detailed information regarding primer design, TOPO cloning reaction, plasmid analysis and LR reaction, see the manufacturer's pENTR Directional TOPO Cloning protocol (https://goo.gl/XQm03l) and Gateway LR clonase II enzyme mix protocol (https://goo.gl/v94WOy).
4. Additional sonication process after this step can be performed using a 99% Methanol for the purpose of further cleaning. These cleaning processes remove surface contaminants and other potential fluorescence material that might cause background signal. These contaminants can fluoresce along the entire illumination spectra and can be difficult to distinguish from fluorescent protein-tagged SH2 domains.
5. In our hands using A431 cells, sufficient expression of SH2-Eos is observed in the presence of serum and antibiotics during transfection.
6. SH2 expression can be optimized by varying the amount of transfected DNA prior to imaging. In our conditions 200 ng of DNA provided good expression level. As DNA quality and cell condition can affect expression, it is helpful to prepare multiple dishes using different DNA amounts for each experiment. Incubation time following transfection should also be optimized. We observed optimal expression between 12 and 24 h.
7. A431 cells can fully spread in approximately 1–2 h on collagen-coated or 10 h on uncoated glass.
8. To measure illumination power density, close imaging field aperture until the edge of the aperture is visible in your image. Measure the area of the illumination (A) based on the image. Remove objective. Use an optical power media measure laser power (P) at the microscope nose piece. Estimate illumination power density by calculating P/A.
9. Switch back to Epi-illumination for efficient photoactivation by 405-nm laser and switch back to TIRF illumination afterwards to monitor density of single molecules.
10. The photoactivatable fluorescent proteins, such as tdEos and mEos [21], have two chemical forms, native (green, absorption peak at 516 nm) and photoactivated (red, absorption peaks at

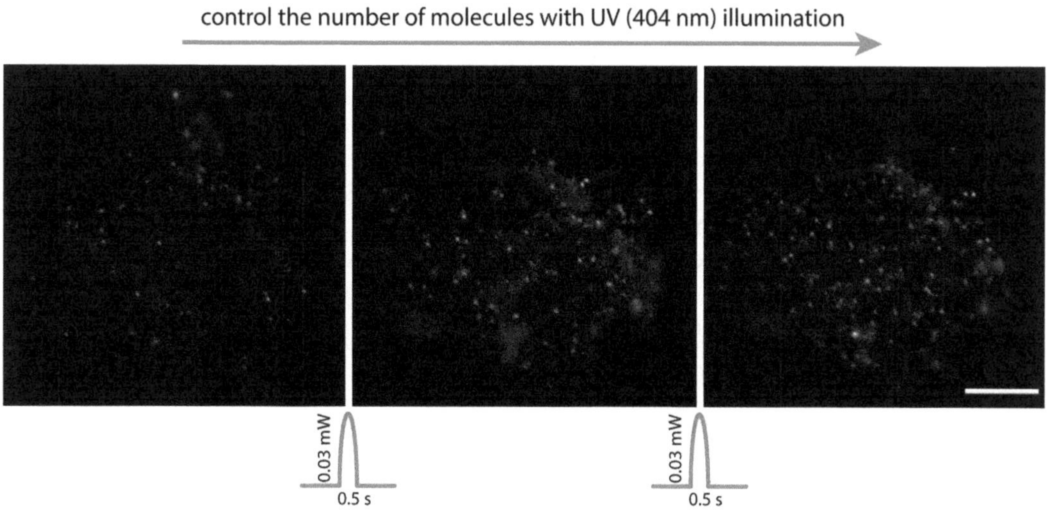

Fig. 3 Photoactivation of SH2-Eos Molecules. *Left panel*, a small fraction of SH2-Eos molecules exist in photoactivated state: ex_{max}: 532 nm, em_{max}: 575 nm, dichroic: 532/10 nm, ex filter: 532/10 nm, em filter: 575/50 nm. *Center* and *right panels*, illumination of cells with a 405 nm beam pulses (power: 0.03 mW, exposure time: 0.5 s) increased the number of fluorescence dots due to photoconversion (ex max: 516 nm). Scale bar equal to 10 μm. Figures were modified from original article [22]

581 nm) [31]. Majority of the molecules are in the native state in left in dark. Illumination with UV light at 405 nm will increase the fraction of photoactivated molecules to a level suitable for single molecule imaging. It is important to optimize the density of photoactivated molecules. A higher number of molecules facilitate more accurate statistical analyses. But with numbers of molecules too high can pose problems for single-molecule detection and tracking because molecular trajectories will spatially overlap. The number of molecules per image can be tuned by differential dosages of UV power for photoactivation (Fig. 3). If needed, photobleaching can be used to reduce the number of photoactivated molecules.

11. For optimal single particle based photoactivatable localization microscopy (sptPALM) experiment, the video frame rate and illumination exposure time should be adjusted thoughtfully. In general, lower illumination power is recommended to reduce the photobleaching effect. However, when imaging highly mobile molecules that require a high video frame rate and short exposure time to maintain temporal resolution (at the expense of spatial resolution), one can increase illumination power to increase fluorescence brightness. This increases the accuracy of molecular coordinates but results in shortened trajectory lengths. On the other hand, when imaging immobile

molecules, using a relatively longer exposure time and lower video rate increases signal to noise. This condition can increase the spatial information.

12. The automatic single particle tracking algorithm based on cross correlation calculation between kernel template and imaged dot. The coordinates of each bright dot on the image are determined with 30 nm resolution [8, 26]. The 10×10 pixel Gaussian kernel is created to approximate the point spread function of a single bright dot with intensity $K(x,y)$. The intensity of each bright dot on the raw image, $I(x,y)$ can be used to calculate cross correlation $C(x,y)$: $C(x,y) = \sum_{i=0}^{n-1}\sum_{j=0}^{k-1} I(x+i, y+j)\left(K(i,j) - s\right)$ where n and k are the dimensions of the image matrix and i and j are integers. The centroid of the peak of the calculated $C(x,y)$ matrix is used as the position of the molecule (x_c, y_c). $(x_c, y_c) = \left(\frac{\sum x(C(x,y))}{\sum(C(x,y))}, \frac{\sum y(C(x,y))}{\sum(C(x,y))}\right)$ [8]. Since the fluorescent protein can be imaged over several frames, the cross-correlation method is applied to all subsequent frames. By centering the scan area on the center of the fluorophore in the previous frame, a trace of the molecule can created [26].

13. The automatic tracking algorithm creates multiple *txt* files containing the *x–y* coordinates of molecules and their corresponding frame numbers. To merge all data into one file, use the simple Matlab code:

    A = dir ('*.txt');, D = [ ] for i = 1:1000;, B = dlmread(A(i).name);, C = [C; B(1:end, 2:3)];, end, >> save ('filename.txt', '-ascii','D');

    Here "filename.txt" contains all trajectory coordinates which can be plotted using graphing software, e.g., Origin pro8 (Fig. 2f).

14. To plot only specific coordinate from each trajectory, first *x–y* coordinates for example, use the following code:

    A = dir ('*.dat');, C = [ ] for i = 1:1000;, b = dlmread(A(i).name);, C = [C; B(1:2:3)];, end >> Save('filename.txt', 'ascii', 'C');,

    Here "filename.txt" contains the first x-y coordinates of all trajectories which can be plotted using graphing software (Fig. 2j).

15. The dissociation rate constant can also be determined by fitting the distribution of dwell times with an exponential function. The exponential function decay constant obtained approximates the dissociation rate constant. However, without proper correction of photobleaching, fitting against a first-order model usually yield poor results (see Fig. 2e).

16. The apparent dissociation rate constant is approximated by the hazard rate function. The hazard rate function is defined as $\lambda(t) = \frac{f'(t)}{P(t)}$, where $P(t)$ is the probability distribution of SH2-Eos membrane recruitment events, and $f'(t)$ is the probability distribution of SH2-Eos membrane dissociation events. Therefore, $f(t) = 1 - P(t)$. $df(t)/dt$ is defined as dissociation rate of SH2 molecule from the membrane. $\lambda(t)$ is the conditional dissociation rate at the specific moment. Here the constant hazard rate suggests that SH2-Eos interacts with its membrane binding sites as a single population with same off rate (i.e., pY-EGFR). A nonlinear hazard rate function suggests the presence of multiple interaction populations, each of their own unique off rate. A simple analogy for the hazard rate is the probability of your car breaking down at any point after purchase: a used car is more likely to break down at an earlier time point than a new car. So the probability of breakdown depends on the condition at the current moment in time after purchase. In this analogy, the breakdown rate is time dependent and as such, the hazard rate would not be constant.
17. We measure photobleaching time of a constitutively membrane-bound form of Eos (myr-EOS). Under the same illumination conditions (same laser power, acquisition time) we measured the hazard rate function of myr-Eos. The true hazard rate function SH2-Eos is determined by subtracting hazard rate of photobleaching from the measured hazard rate: $\lambda(t)_{SH2} = \lambda(t) - \lambda(t)_{photobleaching}$ [22]. To minimize the undesirable photobleaching, I suggest to control the illumination laser power judiciously and obtain all data under the same experimental conditions.
18. The shape of an MSD curve reflects the average diffusion behavior of SH2 molecule on the membrane: confined (concave), normal (straight line) or direct (parabolic) diffusion [28].
19. The accuracy of calculated MSD values decreases as time lag (Δt) increases due to internal averaging (sampling effect) [29] and fluctuation of MSD curves at the longer time lags (Fig. 2d). As a result, only short time regions should be fit to the normal diffusion equation $\text{MSD} = 4D(n\Delta t)$.
20. The dissociation rate and diffusion constant of Grb2 SH2 changes depending on how long the cell had been stimulated by EGF. Both values decrease following EGF stimulation, and reach a constant value after about 20 min [22].

Acknowledgements

I would like to thank Ji Yu, Joshua Jadwin, and Mari Ikeda for assistance with editing the manuscript.

References

1. Schlessinger J (2000) Cell signaling by receptor tyrosine kinases. Cell 103(2):211–225
2. Pawson T (2004) Specificity in signal transduction: from phosphotyrosine-SH2 domain interactions to complex cellular systems. Cell 116(2):191–203
3. Kholodenko BN (2006) Cell-signalling dynamics in time and space. Nat Rev Mol Cell Biol 7(3):165–176. doi:10.1038/nrm1838
4. Nguyen DC, Keller RA, Jett JH, Martin JC (1987) Detection of single molecules of phycoerythrin in hydrodynamically focused flows by laser-induced fluorescence. Anal Chem 59(17):2158–2161. doi:10.1021/ac00144a032
5. Moerner WE, Kador L (1989) Optical detection and spectroscopy of single molecules in a solid. Phys Rev Lett 62(21):2535–2538. doi:10.1103/PhysRevLett.62.2535
6. Sako Y, Yanagida T (2003) Single-molecule visualization in cell biology. Nat Rev Mol Cell Biol Suppl:SS1–SS5
7. Shimomura O, Johnson FH, Saiga Y (1962) Extraction, purification and properties of aequorin, a bioluminescent protein from the luminous hydromedusan, Aequorea. J Cell Comp Physiol 59:223–239
8. Gelles J, Schnapp BJ, Sheetz MP (1988) Tracking kinesin-driven movements with nanometre-scale precision. Nature 331(6155):450–453. doi:10.1038/331450a0
9. Vale RD, Funatsu T, Pierce DW, Romberg L, Harada Y, Yanagida T (1996) Direct observation of single kinesin molecules moving along microtubules. Nature 380(6573):451–453. doi:10.1038/380451a0
10. Watanabe N, Mitchison TJ (2002) Single-molecule speckle analysis of actin filament turnover in lamellipodia. Science 295(5557):1083–1086. doi:10.1126/science.1067470
11. Bustamante C, Bryant Z, Smith SB (2003) Ten years of tension: single-molecule DNA mechanics. Nature 421(6921):423–427. doi:10.1038/nature01405
12. Harry E, Monahan L, Thompson L (2006) Bacterial cell division: the mechanism and its precision. Int Rev Cytol 253:27–94. doi:10.1016/S0074-7696(06)53002-5
13. Yu J, Xiao J, Ren X, Lao K, Xie XS (2006) Probing gene expression in live cells, one protein molecule at a time. Science 311(5767):1600–1603. doi:10.1126/science.1119623
14. Gell C, Bormuth V, Brouhard GJ, Cohen DN, Diez S, Friel CT, Helenius J, Nitzsche B, Petzold H, Ribbe J, Schaffer E, Stear JH, Trushko A, Varga V, Widlund PO, Zanic M, Howard J (2010) Microtubule dynamics reconstituted in vitro and imaged by single-molecule fluorescence microscopy. Methods Cell Biol 95:221–245. doi:10.1016/S0091-679X(10)95013-9
15. Shav-Tal Y, Singer RH, Darzacq X (2004) Imaging gene expression in single living cells. Nat Rev Mol Cell Biol 5(10):855–861. doi:10.1038/nrm1494
16. Schmidt CE, Horwitz AF, Lauffenburger DA, Sheetz MP (1993) Integrin-cytoskeletal interactions in migrating fibroblasts are dynamic, asymmetric, and regulated. J Cell Biol 123(4):977–991
17. Kusumi A, Nakada C, Ritchie K, Murase K, Suzuki K, Murakoshi H, Kasai RS, Kondo J, Fujiwara T (2005) Paradigm shift of the plasma membrane concept from the two-dimensional continuum fluid to the partitioned fluid: high-speed single-molecule tracking of membrane molecules. Annu Rev Biophys Biomol Struct 34:351–378. doi:10.1146/annurev.biophys.34.040204.144637
18. Anderson CM, Georgiou GN, Morrison IE, Stevenson GV, Cherry RJ (1992) Tracking of cell surface receptors by fluorescence digital imaging microscopy using a charge-coupled device camera. Low-density lipoprotein and influenza virus receptor mobility at 4 degrees C. J Cell Sci 101(Pt 2):415–425
19. Axelrod D, Burghardt TP, Thompson NL (1984) Total internal reflection fluorescence. Annu Rev Biophys Bioeng 13:247–268. doi:10.1146/annurev.bb.13.060184.001335
20. Oh D, Yu Y, Lee H, Wanner BL, Ritchie K (2014) Dynamics of the serine chemoreceptor in the Escherichia coli inner membrane: a high-speed single-molecule tracking study. Biophys J 106(1):145–153. doi:10.1016/j.bpj.2013.09.059
21. McKinney SA, Murphy CS, Hazelwood KL, Davidson MW, Looger LL (2009) A bright and photostable photoconvertible fluorescent protein. Nat Methods 6(2):131–133. doi:10.1038/nmeth.1296
22. Oh D, Ogiue-Ikeda M, Jadwin JA, Machida K, Mayer BJ, Yu J (2012) Fast rebinding increases dwell time of Src homology 2 (SH2)-containing proteins near the plasma membrane. Proc Natl Acad Sci U S A 109(35):14024–14029. doi:10.1073/pnas.1203397109
23. Jadwin JA, Oh D, Curran TG, Ogiue-Ikeda M, Jia L, White FM, Machida K, Yu J, Mayer BJ (2016) Time-resolved multimodal analysis of Src Homology 2 (SH2) domain binding in signaling by receptor tyrosine kinases. Elife 5:e11835. doi:10.7554/eLife.11835
24. Persson F, Linden M, Unoson C, Elf J (2013) Extracting intracellular diffusive states and tran-

sition rates from single-molecule tracking data. Nat Methods 10(3):265–269. doi:10.1038/nmeth.2367

25. Niu L, Yu J (2008) Investigating intracellular dynamics of FtsZ cytoskeleton with photoactivation single-molecule tracking. Biophys J 95(4):2009–2016. doi:10.1529/biophysj.108.128751
26. Sako Y, Kusumi A (1994) Compartmentalized structure of the plasma membrane for receptor movements as revealed by a nanometer-level motion analysis. J Cell Biol 125(6):1251–1264
27. Saxton MJ, Jacobson K (1997) Single-particle tracking: applications to membrane dynamics. Annu Rev Biophys Biomol Struct 26:373–399. doi:10.1146/annurev.biophys.26.1.373
28. Saxton MJ (1997) Single-particle tracking: the distribution of diffusion coefficients. Biophys J 72(4):1744–1753. doi:10.1016/s0006-3495(97)78820-9
29. Qian H, Sheetz MP, Elson EL (1991) Single particle tracking. Analysis of diffusion and flow in two-dimensional systems. Biophys J 60(4):910–921. doi:10.1016/s0006-3495(91)82125-7
30. Grebenkov DS (2011) Probability distribution of the time-averaged mean-square displacement of a Gaussian process. Phys Rev E Stat Nonlin Soft Matter Phys 84(3 Pt 1):031124. doi:10.1103/PhysRevE.84.031124
31. Wiedenmann J, Ivanchenko S, Oswald F, Schmitt F, Rocker C, Salih A, Spindler KD, Nienhaus GU (2004) EosFP, a fluorescent marker protein with UV-inducible green-to-red fluorescence conversion. Proc Natl Acad Sci U S A 101(45):15905–15910. doi:10.1073/pnas.0403668101

Chapter 30

SH2 Domain-Based FRET Biosensor for Measuring BCR-ABL Activity in Living CML Cells

Mari Fujioka, Yumi Asano, Shigeyuki Nakada, and Yusuke Ohba

Abstract

Fluorescent proteins (FPs) displaying distinct spectra have shed their light on a wide range of biological functions. Moreover, sophisticated biosensors engineered to contain single or multiple FPs, including Förster resonance energy transfer (FRET)-based biosensors, spatiotemporally reveal the molecular mechanisms underlying a variety of pathophysiological processes. However, their usefulness for applied life sciences has yet to be fully explored. Recently, our research group has begun to expand the potential of FPs from basic biological research to the clinic. Here, we describe a method to evaluate the responsiveness of leukemia cells from patients to tyrosine kinase inhibitors using a biosensor based on FP technology and the principle of FRET. Upon phosphorylation of the tyrosine residue of the biosensor, binding of the SH2 domain to phosphotyrosine induces conformational change of the biosensor and brings the donor and acceptor FPs into close proximity. Therefore, kinase activity and response to kinase inhibitors can be monitored by an increase and a decrease in FRET efficiency, respectively. As in basic research, this biosensor resolves hitherto arduous tasks and may provide innovative technological advances in clinical laboratory examinations. State-of-the-art detection devices that enable such innovation are also introduced.

Key words Förster resonance energy transfer (FRET), Fluorescent protein (FP), Chronic myeloid leukemia (CML), BCR-ABL, Molecular target drug, CrkL, Fluorescence microscopy, Fluorescence cytometry (FCM), Fluorescence lifetime, Src Homology 2 (SH2) domain

1 Introduction

Green FP (GFP) was originally isolated from the luminous organ of the jellyfish *Aequorea victoria* by Dr. Osamu Shimomura [1]. It is indisputable that GFP has continued to shed light on cell biology since its cDNA was isolated in 1992 [2]. The primary reason why GFP is so revolutionary is its ability to be easily incorporated into cells via transfection, which is dependent on a key property of FPs, i.e., its ability to generate an intrinsic fluorophore without cofactors or enzymatic components [3]. It was not long before color variants of GFP were developed and FPs were distributed worldwide. Not only do these tools serve as markers of protein localization, but they are also used for monitoring intracellular

Kazuya Machida and Bernard A. Liu (eds.), *SH2 Domains: Methods and Protocols*, Methods in Molecular Biology, vol. 1555, DOI 10.1007/978-1-4939-6762-9_30, © Springer Science+Business Media LLC 2017

environments under physiological conditions. In addition, based on the principles of the reconstitution of protein fragments or physicochemical energy transfer, protein–protein interactions or conformational changes can be monitored, widening the potential applications of FPs.

Recently, our research group has begun to expand the utility of FPs from basic biological research to the clinic [4]. As in basic research, FP-based biosensors have resolved hitherto arduous tasks and may provide innovative technological advances in clinical laboratory examinations. Specifically, we can now detect a minor population of drug-resistant cells and predict the most effective drug for each individual patient suffering from chronic myeloid leukemia (CML), which had been a utopian ideal for patients and clinicians for many years, using available clinical tests based on conventional concepts.

CML, a hematological malignancy involving the transformation of hematopoietic stem cells in bone marrow, is characterized by the formation of an abnormal chromosome (Philadelphia chromosome, Ph^1) and the expression of its transcript BCR-ABL [5]. BCR-ABL encodes a constitutively active tyrosine kinase that causatively contributes to the malignant transformation of leukemia cells by activating a range of signaling pathways via the tyrosine phosphorylation of its substrates, including SH2 containing proteins CrkL and signal transducer and activator of transcription (STAT) [6, 7]. The emergence of targeted therapy against BCR-ABL has radically changed the treatment of CML, which is now controllable using oral drugs. However, resistance and intolerance remain a concern for a substantial number of patients. Our FRET-based biosensor, referred to as phosphorylation indicator of CrkL *en* substrate (Pickles), may provide a promising future for such patients via the monitoring of drug responsiveness by accurately measuring BCR-ABL activity in living CML cells of from each patient (Fig. 1).

Pickles consists of a variant of yellow FP (YFP), the SH2 containing protein CrkL, and a variant of cyan FP (CFP) from the N-terminus. BCR-ABL phosphorylates CrkL on the tyrosine residue at position 207 (Y207), which induces binding of the SH2 domain and phosphorylated Y207 and thereby brings YFP close to CFP. Because FRET efficiency is inversely proportional to the sixth power of distance between a donor (CFP) and an acceptor (YFP), kinase activity and response to kinase inhibitors can be therefore monitored by an increase and a decrease in FRET efficiency, respectively. FRET efficiency can be determined by calculating the yellow-to-cyan emission ratio, which is conventionally provided by images acquired by fluorescence microscopy (Fig. 2).

Although fluorescence microscopy has been and continues to be the primary tool used to perform FRET experiments, the number of cells that can be observed within a certain period is restricted.

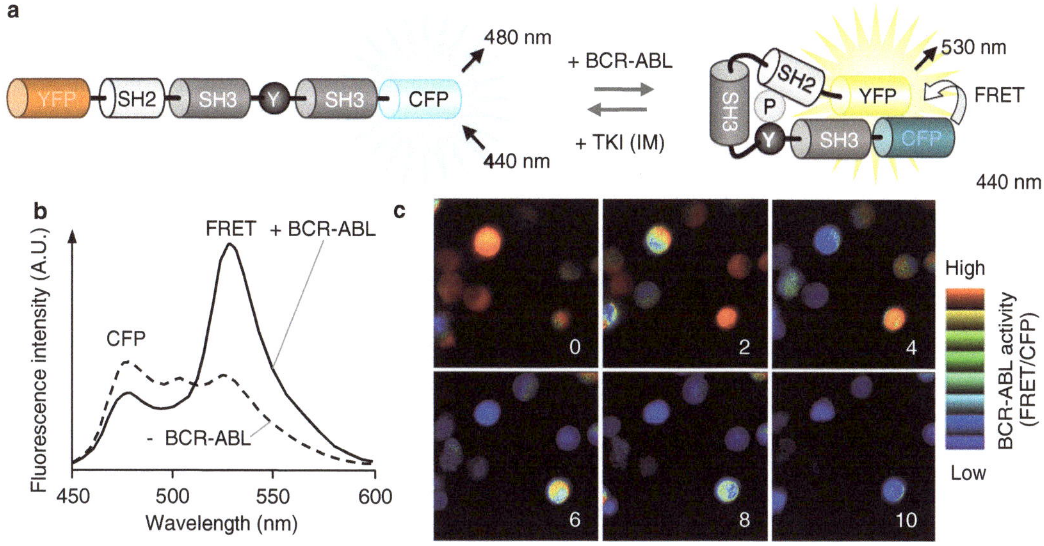

Fig. 1 The FRET-based biosensor Pickles to evaluate BCR-ABL activity in living CML cells. (**a**) Schematic representation of Pickle. The sandwiched region consists of one SH2 and two SH3 domains from human CrkL. P and Y represent a phosphate group and a tyrosine residue corresponding to Y207 of CrkL, respectively. Briefly, Pickles consists of a variant of YFP, CrkL, and a variant of CFP from the N-terminus. Upon Y207 phosphorylation, the SH2 domain binds to this phosphorylated tyrosine, which brings YFP close to CFP and increases the efficiency of FRET from CFP to YFP. (**b**) The fluorescence spectra of Pickles in the presence and absence of BCR-ABL. (**c**) K562 cells were transfected with the expression plasmid for Pickles, transferred to a glass base dish 24 h after transfection, and placed on an incubated stage equipped with a microscope. During time-lapse image acquisition, the cells were treated with IM. The photographs are representative FRET images in intensity-modulated display mode, a sort of pseudocolor mode in which the *red* and *blue colors* represent high and low FRET efficiency (BCR-ABL activity), respectively, at the indicated time point (h) after treatment

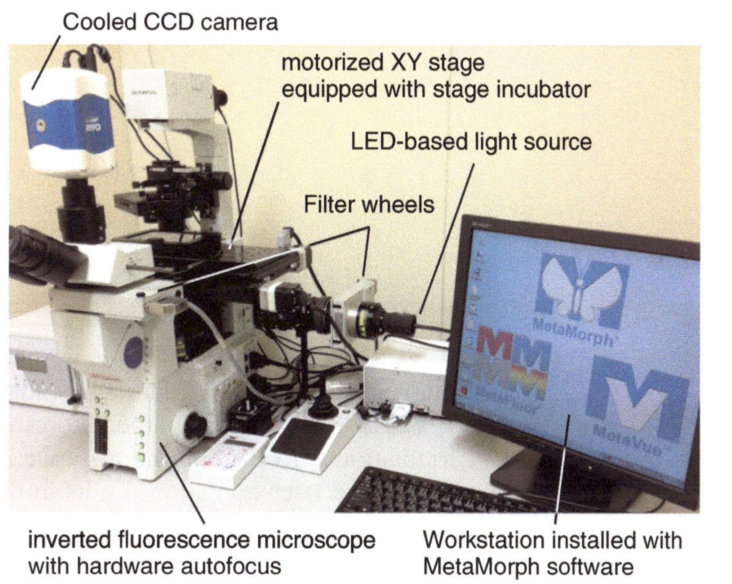

Fig. 2 Overview of the microscope and the peripherals used for data acquisition and analysis in this protocol. *CCD* charge-coupled device, *LED* light-emitting diode

Thus, a high throughput, quantitative FRET measurement technique is necessary to maximize the utility of the biosensor. Fluorescence flow cytometry is a good candidate for handling a vast number of cells in a short period; however, few studies are available regarding the successful application of this technique for the detection of FRET biosensors. The world's first flow cytometer designed to detect both the fluorescence lifetime and the fluorescence intensity [fluorescence lifetime cytometer (FLiCM, Fig. 3a)] might represent a suitable machine to improve the effectiveness and efficiency of present technologies.

The principle underlying the measurement of lifetime using the FLiCM are shown in Fig. 3b–d. The excitation light (laser), the output power of which is sinusoidally modulated at an angular frequency of ω ($2\pi \times 28$ MHz, amplitude modulation), is used to irradiate the sample that passes through the flow cell. The time-dependent component of the modulated laser power $P_E(t)$ is given by Eq. (1), where P_{E0} is the average power of the laser:

$$P_E(t) = P_{E0} sin\omega t. \quad (1)$$

The fluorescence intensity, $P_F(t)$, emanating from a fluorescent molecule that is irradiated by the laser is given by

$$dP_F(t)/dt = -(k_f + k_{nr})P_F(t) + k_f N_0 P_E(t), \quad (2)$$

where k_f and k_{nr} represent rate constants for fluorescence and nonradiative transition, and N_0 is a factor dependent on the molar extinction coefficient and the concentration of the fluorescent molecule. From Eqs. (1) and (2),

$$P_F(t) = k_f N_0 \tau P_{E0} / \sqrt{1+(\tau\omega)^2} \cdot \sin(\omega t - \theta) \quad (3)$$

$$\tau = (k_f + k_{nr})^{-1} \quad (4)$$

$$\theta = tan^{-1}\tau\omega, \quad (5)$$

where τ and θ are the fluorescence lifetime and phase shift, respectively. Furthermore, $k_f N_0 \tau P_{E0} / \sqrt{1+(\tau\omega)^2}$ is the amplitude of the fluorescence intensity that sinusoidally varies over time, and $k_f N_0 \tau P_{E0}$ corresponds to the fluorescence intensity obtained by irradiation with the continuous-wave laser. Therefore, amplitude-modulated laser excitation results in fluorescence emission with a phase shift θ in a manner that is dependent on the substance-specific fluorescence lifetime τ, and fluorescence intensity can be determined based on its amplitude. To obtain the phase shift (θ) and the amplitude ($k_f N_0 \tau P_{E0} / \sqrt{1+(\tau\omega)^2}$), $P_F(t)$ by $P_E(t)$ is

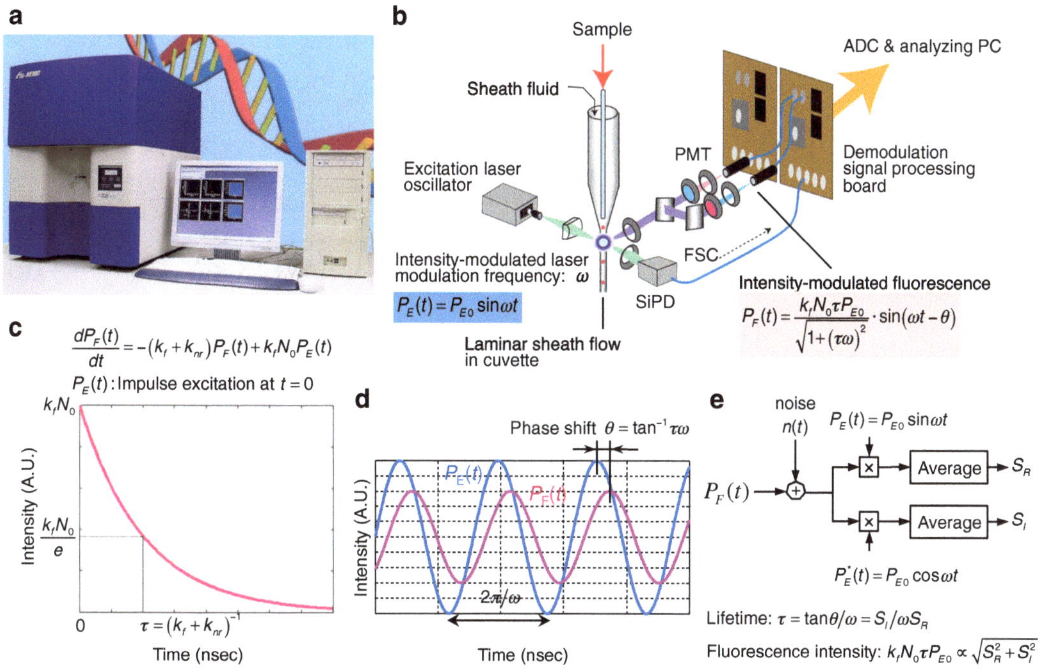

Fig. 3 The FLiCM and the principle of fluorescence lifetime measurement. (**a**) Exterior view of the FLiCM. The FLiCM system consists of the primary flow cytometer unit (*left side*) and a control/analyzing PC (*right side*). The primary FLiCM equipment includes an intensity-modulated laser, an optical system, a flow system, and a signal processing system for modulation and demodulation. FDS software, which is installed on the control/analyzing PC, manages data acquisition, graphical display, and the user interface. (**b**) The interior flow, optical paths, and the electronic circuits of the FLiCM. Samples of cells expressing FPs are carried to a cuvette via laminar sheath flow and intersect the laser beam orthogonally. FSC and SSC or fluorescence emission are detected using a SiPD and a PMT, respectively, and are transformed to electrical signals. The output electrical signals of fluorescence are demodulated in the signal processing circuits and are converted to the fluorescence lifetime or intensity in the control/analyzing PC. (**c**–**e**) Schematic representation of the principle of the measurement of fluorescence lifetime. The decay of the fluorescence intensity of a fluorescent molecule, the lifetime of which consists of a single component, that is excited by an impulse laser at $t=0$. The lifetime τ [s] of the fluorescence of the molecule is defined as the time at which the intensity decays by $1/e$-fold (**c**). Modulated fluorescent emissions (*pink*) in response to modulated laser excitations (*blue*) at an angular frequency of ω. The phase shift θ [rad] in fluorescent emissions is a function of the fluorescence lifetime τ [s] (**d**). Schematic of the signal processing of the FLiCM. The modulated laser intensity $P_E(t)$ and the time-dependent fluorescence intensity $P_F(t)$ are demodulated to the real and imaginary signals S_R and S_I, respectively, from which the lifetime and the intensity of the fluorescence are determined (**e**)

demodulated, i.e., Eqs. (1) and (3) are multiplied, thereby obtaining real and imaginary portions of the demodulation signals, S_R *and* S_I, respectively:

$$S_R = k_f N_0 \tau P_{E0}^2 / 2\sqrt{1+(\tau\omega)^2} \cdot cos\theta \tag{6}$$

$$S_I = k_f N_0 \tau P_{E0}^2 / 2\sqrt{1+(\tau\omega)^2} \cdot sin\theta \tag{7}$$

$$\therefore S_I / S_R = tan\theta. \quad (8)$$

Based on Eqs. (5) and (8), τ is given by

$$\tau = tan\theta / \omega = S_I / \omega S_R. \quad (9)$$

Alternatively, based on Eqs. (6) and (7),

$$\sqrt{S_R^2 + S_I^2} = k_f N_0 \tau P_{E0}^2 / 2\sqrt{1+(\tau\omega)^2} \propto k_f N_0 \tau P_{E0}. \quad (10)$$

Therefore, by calculating the S_I and the S_R, the fluorescence lifetime τ and the fluorescence intensity $k_f N_0 \tau P_{E0}$ can be determined using Eqs. (9) and (10), respectively (Fig. 3b and e). Finally, FRET efficiency (E) is given by

$$E = 1 - \tau_d' / \tau_d \quad (11)$$

where τ_d' and τ_d are the donor lifetimes in the presence and absence of the acceptor, respectively.

Here, we therefore introduce a microscopy-based protocol for evaluating drug responsiveness in leukemia cells from individual patients using FRET biosensors that we have developed and a protocol for the use of a unique instrument, the world's first flow cytometer designed to detect both the fluorescence lifetime and the fluorescence intensity. The fluorescence lifetime cytometer (FLiCM) enables high-throughput quantitative measurements of FRET efficiency in a single living cell, which might represent a suitable machine to improve the effectiveness and efficiency of present technologies.

2 Materials

Prepare all solutions using distilled deionized water (DDW, prepared by autoclaving deionized water to attain a sensitivity of 18 MΩ cm at 25 °C) and analytical grade reagents, unless otherwise specified. Diligently follow all waste disposal regulations when disposing of waste materials. Note that materials that are exposed to blood samples from patients should be treated as infectious waste.

2.1 The Isolation of Mononuclear Cells and Subsequent Cell Culture

1. Phosphate buffered saline (PBS): Dissolve 9.6 g of Dulbecco's PBS (−) in 1 L of DDW and autoclave at 120 °C for 15 min. Store at 4 °C until use.
2. 0.5 M ethylenediamine-*N*,*N*,*N′*,*N′*-tetraacetic acid (EDTA) (pH 8.0) solution: Dissolve 93 g of EDTA, disodium salt, dehydrate [2NA(EDTA•2Na)] and 9 g of NaOH in approximately 400 mL of DDW. Adjust the pH to 8.0 using 10 N

NaOH and add DDW to a total of 500 mL, followed by sterilization using a 0.45 μm filter.

3. PBS buffer: Add 1/250 volume of 0.5 M EDTA (pH 8.0) solution to PBS. Store at 4 °C.
4. Lymphoprep.
5. 100 mM Sodium pyruvate solution (sterile-filtered), representing a 100× concentrated solution.
6. 1 M 4-(2-hydroxyethyl)-1-piperazineethanesulfonic acid (HEPES) solution: Dissolve 11.9 g of HEPES in approximately 40 mL of DDW. Adjust the pH to 7.2 using 10 N NaOH, add DDW to a total of 50 mL, and sterilize using a 0.45 μm filter.
7. RPMI 1640 working medium (complete RPMI-1640): Supplement RPMI-1640 medium with 10% heat-inactivated (56 °C for 30 min) fetal bovine serum (FBS), 100× stabilized penicillin–streptomycin (P/S) solution (f.c. 100 U/mL and 0.1 mg/mL, respectively), and 1 mM sodium pyruvate. Store at 4 °C.
8. RPMI-1640 working medium without phenol red [complete RPMI-1640 phenol red (−)]: Supplement RPMI-1640 medium (no phenol red) with 10% FBS, 1 mM sodium pyruvate, P/S solution, and 15 mM HEPES. Store at 4 °C.
9. Red blood cell (RBC) lysis buffer (10×conc.): Dissolve 82.6 g of NH_4Cl, 11.9 g of $NaHCO_3$, and 378 mg of EDTA·2Na·$2H_2O$ in DDW. Adjust the pH to 7.3 using concentrated HCl (35–37%), add DDW to a total of 1 L, and sterilize using a 0.45 μm filter.

2.2 Transfection

1. Transfection reagent: Nucleofector Kit V (Lonza, Basel, Switzerland). Add the entire Supplement (an adjuvant to transfection) to Nucleofector solution before use. Amaxa cuvettes are included in the kit.
2. Electroporator: Nucleofector II (Lonza).
3. Expression vector pPickles-2.31: *E. coli* JM109 strain was transformed with a plasmid harboring the coding sequence of the FRET-based biosensor Pickles [4] and then inoculated on a Lysogeny/Luria Broth (LB) agar plate containing 100 μg/mL ampicillin. The bacteria were cultured in 100 mL of LB media containing ampicillin, and the plasmids were purified using the JETSTAR 2.0 LFU Plasmid Midi Kit (GENOMED, Löhne, Germany) according to the manufacturer's instructions. At the final step of the preparation, recover the plasmid DNA in endotoxin-free water at a concentration of 3–5 μg/μl and store at −20 °C (*see* **Note 1**).

2.3 Drug Susceptibility Assessment

1. Tyrosine kinase inhibitors for BCR-ABL:

   Imatinib mesylate (IM) 10 mM in DDW.

   Nilotinib (NL): 10 mM in dimethyl sulfoxide (DMSO).

   Dasatinib: 100 μM in DMSO.

   Aliquot the drug solutions into 1.5 mL tubes (app. 50 μl each) and store at −20 °C.
2. Dish coating agent: 0.01 % poly-L-lysine solution.
3. EZVIEW 96-well Glass Bottom LB Culture Plates.

2.4 General Equipment

1. 24-well and 12-well cell culture plates.
2. Centrifuge: RL-101 swinging bucket rotor.
3. CO_2 incubator: CDI-165.

2.5 Equipment for Microscopic Observation (See Fig. 2)

1. Research inverted microscope: IX-81 (Olympus, Tokyo, Japan).
2. Objective lens: UPLSAPO 60XO, NA 1.35 (Olympus).
3. Light source: SORA SE light engine (Lumencor, Beaverton, OR, USA).
4. Cooled charge-coupled device (CCD) camera: CoolSNAP MYO (Photometrics, Tucson, AZ, USA).
5. Filter wheels and shutter control: MAC-5000 (Ludl Electronic Products, Hawthorne, NY, USA).
6. Stage incubator: WP-S-10E Chamlide (Live Cell Instrument, Seoul, Korea).
7. Motorized XY-stage: BIXY FFC (OL) (Chuo Precision Industrial, Tokyo, Japan).
8. Neutral density (ND) filter: 25 % (Olympus).
9. Dichroic mirror: XF2034 (455DRLP) (Omega Optical, Brattleboro, VT, USA).
10. Barrier filters: excitation filters XF1071 (440AF21 for CFP and FRET) and XF1068 (500AF25 for YFP); emission filters XF3075 (480AF30 for CFP) and XF3079 (535AF26 for FRET and YFP) (Omega Optical).
11. Software: MetaMorph Version 7.8 is used for the acquisition and manipulation of microscopic images, and Microsoft Excel Version 14.0 (2010) is for data analyses.

2.6 Flow Cytometry

1. FLiCM (Mitsui Engineering & Shipbuilding; Fig. 3a and Table 1).
2. Falcon™ Tube with Cell Strainer Cap.
3. 8.5 mL polystyrene FACS test tube (75 × 15.7 mm): (*see* **Note 2**).

Table 1
The specifications of the FLiCM

Feature	Specification
Light source	445 nm laser diode (60 mW)
Number of detection channels	
Scattering light	2 ch
Fluorescence intensity	3 ch
Fluorescence lifetime	3 ch
Detector	
Forward scatter	SiPD
Side scatter	PMT
Fluorescence intensity and lifetime	PMT
Detection (fluorescence) wavelength	
SSC	440 ± 5 nm
FL1	482 ± 17.5 nm
FL2	542 ± 13.5 nm
FL3	600 nm long pass
Sample flow rate	6 m/s
Sample flow volume (μl/min)	Lo: 40, Mid: 80, Hi: 160
Maximum acquisition rate	10,000 events/s
Modulation frequency	28 MHz
Size (mm)	500 (W) × 675 (D) × 700 (H)

SiPD silicon photodiode, *PMT* photomultiplier tube

4. Calibration beads: PeakFlow™ Blue or Cell Sorting Set-up beads (UV laser) (Life Technologies, Carlsbad, CA).
5. Data analysis software: FlowJo (Tommy Digital Biology Co., Ltd.).

3 Methods

Perform all procedures at room temperature and on a clean bench unless otherwise specified.

3.1 Mononuclear Cell Isolation from Bone Marrow Cells (BMCs)

1. Dilute BMCs in 1/7 volume of PBS buffer (i.e., BMC: PBS buffer = 7: 1) (*see* **Note 3**).
2. Dampen a cell strainer (100-μm pore size) with PBS buffer.

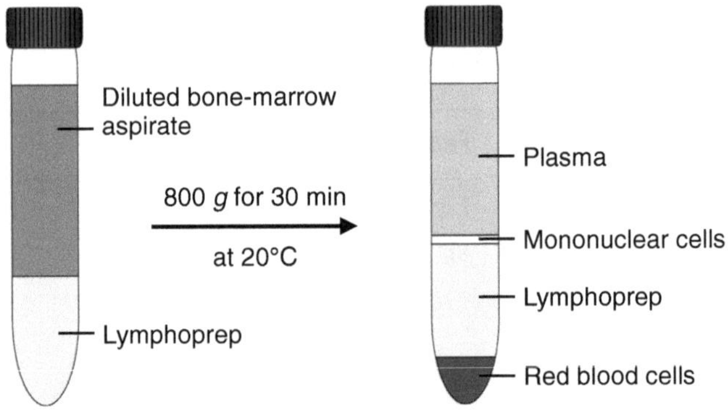

Fig. 4 Mononuclear cell isolation from BMCs. The *left panel* shows a sample tube filled with Lymphoprep and bone-marrow aspirates before centrifugation, and the *right panel* shows the sample after centrifugation

3. Pass the BMCs/PBS buffer through the filter into a fresh 50 mL centrifuge tube to remove the cell clumps and bone fragments.
4. Fill a 15 mL centrifuge tube with 3–4 mL of Lymphoprep (Fig. 4).
5. Carefully layer the filtrated BMCs/PBS buffer (app. 5 mL in each tube) above the Lymphoprep (*see* **Note 4**).
6. Centrifuge at 800 × *g* for 30 min at 20 °C (the brake must be disabled).
7. Recover the mononuclear cell layer fraction using a sterile Pasteur pipette (*see* Fig. 4).
8. Place the recovered aliquot in a 50-mL centrifuge tube; add complete RPMI-1640/PBS to a total volume of 50 mL, and mix by pipetting (*see* **Note 5**).
9. Centrifuge at 300 × *g* for 10 min to remove any remaining Lymphoprep.
10. If the pellet is colored red at this step, perform the following procedure (RBC lysis protocol) to remove the contaminating RBCs (**steps 11–16**) . Otherwise, skip ahead to **step 17**.
11. Dilute 10× conc. RBC lysis buffer to 1× conc. using cold endotoxin-free water (if the number of cells is less than 10^8, prepare 10 mL of 1× conc. RBC lysis buffer).
12. Discard the supernatant (after **step 9**) and suspend the pellet in 1× RBC lysis buffer.
13. Centrifuge at 300 × *g* for 5 min at 20 °C.
14. Discard the supernatant. Add cold PBS to a total of 15 mL and mix by pipetting.
15. Centrifuge at 300 × *g* for 5 min at 20 °C.
16. Repeat **steps 4** and **5**.

17. Discard the supernatant. Add complete RPMI-1640/PBS to a total of 50 mL, and centrifuge at 200 × *g* for 10 min to remove the platelets.
18. Discard the supernatant. Resuspend the cells in complete RPMI 1640, and transfer this suspension to a culture dish (*see* **Note 6**).
19. Place the culture dish in a CO_2 incubator at 37 °C, and culture for 1–4 h.

3.2 Transfection

1. Transfer the culture media containing 1×10^7 cells from a culture dish to a 15 mL centrifuge tube.
2. Centrifuge at 300 × *g* for 5 min. Then, discard all of the supernatant (*see* **Note 7**).
3. Add 100 μl of Solution V and 25 μg of Pickles 2.31 (*see* **Note 8**).
4. Mix gently, and transfer this suspension to an Amaxa cuvette.
5. Electroporate using the Nucleofector (Program T-020).
6. Add ~500 μl of pre-warmed medium using the Amaxa pipette and transfer the cells to a 90-mm dish prefilled with 10 mL of complete RPMI medium.
7. Culture these cells for 12–18 h at 37 °C.

3.3 Drug Treatment

1. Count the cells and transfer them to a 15 mL centrifuge tube.
2. Centrifuge at 300 × *g* for 5 min, and discard the supernatant.
3. Suspend the pellet in complete RPMI-1640 phenol red (−), and aliquot this cell suspension equally into four wells of either a 24-well or 12-well plate (*see* **Note 9**).
4. Add drugs to a final concentration of 2 μM for IM, 4 μM for NL, and 100 nM for DS (*see* **Note 10**).
5. Culture the cells for 24 h at 37 °C in a 5 % CO_2 atmosphere.

3.4 Fluorescence Microscopic Observation

1. Add 75 μl of poly-L-lysine per well to four wells of a 96-well glass bottom culture plate. Incubate for 1 h at 37 °C, and wash twice with 100 μl of PBS.
2. Turn on all devices associated with the microscope and the computer (*see* **Note 11**).
3. Transfer 1.0×10^5 cells to each well of the recoated 96-well plate (*see* **step 1**), and incubate for 10 min in a 5 % CO_2 incubator at 37 °C.
4. Open MetaMorph software.
5. Set the "Configure Illumination" for filter sets as follows (*see* **Note 12**):

   FRET: excitation filter, 440AF21; emission filter, 535AF26;

   CFP: excitation filter, 440AF21; emission filter, 480AF30; and

   YFP: excitation filter, 500AF25; emission filter, 535AF26.

6. Open "Multi Dimensional Acquisition," check "Multiple Wavelengths" in the "Main" tab, and select 4 for the "Number of Wavelengths" in the "Wavelength" tab. Set the exposure times for 4×4 binning to 200 and 50 ms for fluorescence imaging and differential interference contrast (DIC) imaging, respectively, in each configuration tab (*see* **Note 13**).
7. Acquire at least 100 cell images (of FRET, CFP, YFP, and DIC) per well/sample (*see* **Note 14**).

3.5 Microscopic Data Analysis

1. Open MetaMorph software.
2. Stack the images (*see* Fig. 5).

   Open the "Review Multidimensional Data" window. Select all files associated with the same sample, and select "Append Sets" to merge the files. Select the created merge file, and select "View". Check all image and wavelength boxes, and select "Load Images". Save the stacked images.
3. Subtract the background (Fig. 6).

   [Open the stack files of all fluorescence images (FRET, CFP and YFP). Create a background subtraction region (for instance, an area near the corner that is blank; *see* also **Note 15**) on the FRET image. Transfer this region to the CFP and YFP images using the "Transfer Regions" command. Open the

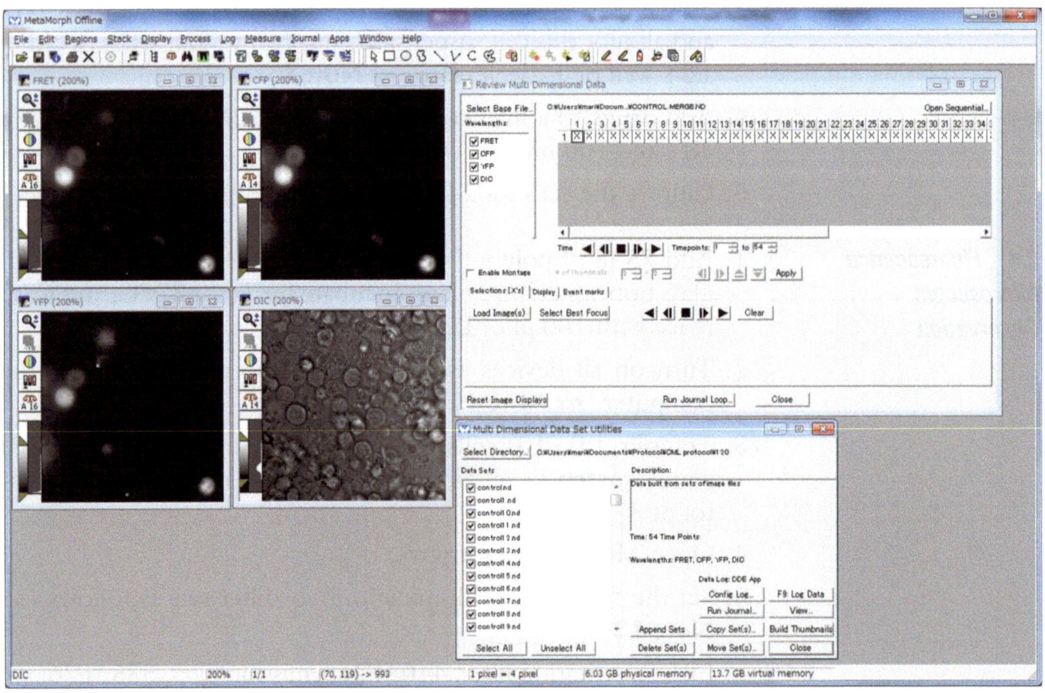

Fig. 5 Example of an application window during construction of a stack from image files. *See* Subheading 3.5 for detailed information

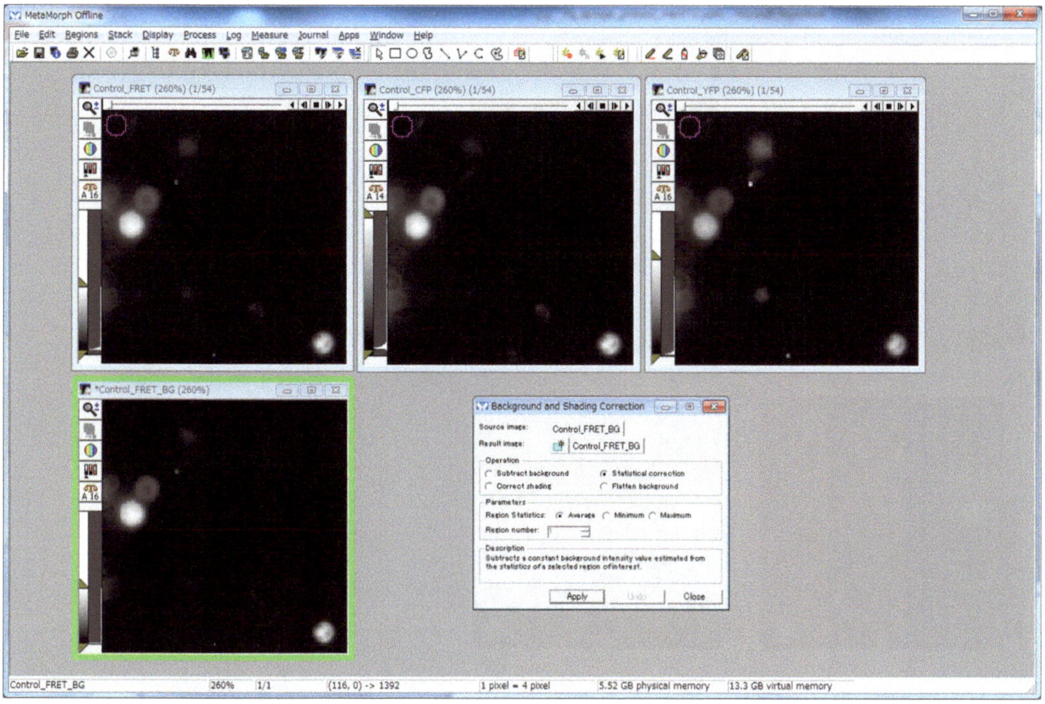

Fig. 6 Example of an application window during background subtraction. *See* Subheading 3.5 for detailed information

"Background and Shading Correction" window. Choose Souse (all planes) and Result Images, and check the "Operation: Statistical Correction" and "Parameters: Average" boxes. Select "Apply".]

Execute steps in parenthesis ([ ]) for both the CFP and YFP stack files. Save the resulting images.

4. Measure the fluorescence intensity (Fig. 7).

   Open the stack files of the background-subtracted fluorescence images (FRET, CFP, and YFP) and that of the DIC images. Create regions around each cell on the FRET image (*see* **Note 15**).

   [Open the "Region Measurement" window. Select "Data Log" to launch Microsoft Excel, and create an Excel file for data export. Choose the current plane of FRET as the source. In the "Configure" tab, confirm that the "Region Label", "Image Name", "Image Plane", "Area", "Average Intensity", "Intensity Standard Dev" boxes are checked. Select "F9: Log Data". Transfer the regions to the CFP and YFP images, and log the data in the same manner.]

   Execute steps in parentheses ([ ]) for all planes (*see* **Note 16**). Save the Excel file.

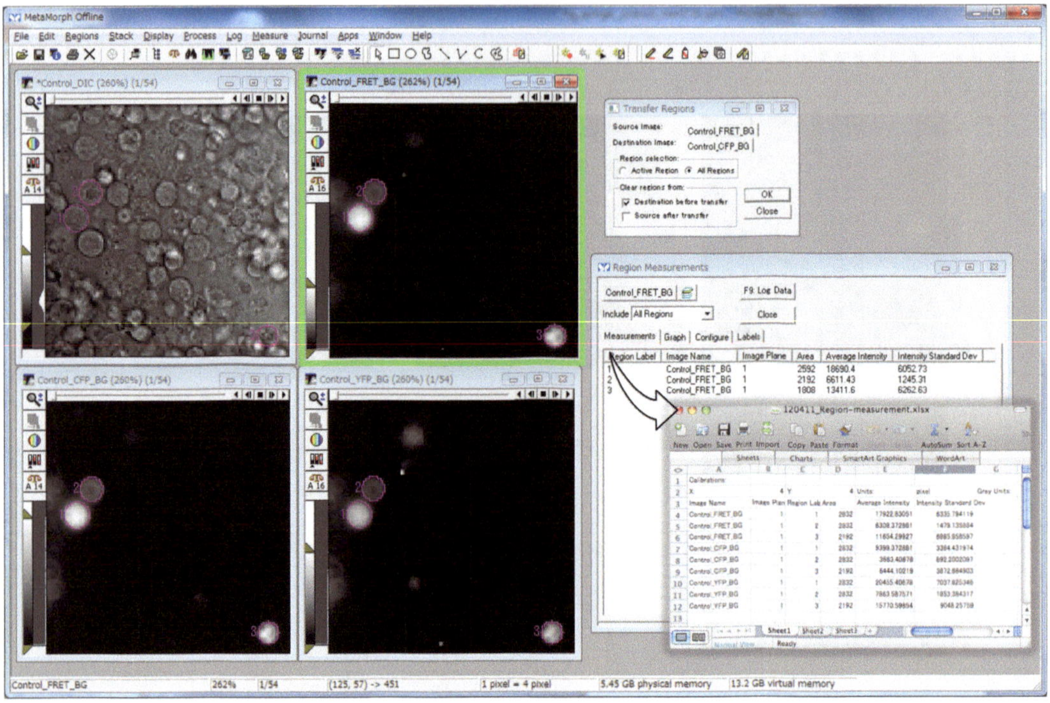

Fig. 7 Example of an application window during the measurement of fluorescence intensity and data logging. The *inset* on the *bottom right corner* shows a window of the Excel file to which log data are being transferred. *See* Subheading 3.5 for detailed information

5. Calculate the FRET efficiency.

   Open the Excel file. Sort the data in ascending order according to the Image Name. Place the Average Intensity and Intensity Standard Dev values for FRET and YFP next to those for CFP. Calculate the following values (*see* **Note 17**) (*see* Fig. 8):

   YFP_estimated = CFP_average intensity × 0.77;

   YFP_min = CFP_average intensity × 1.8665;

   YFP_max = CFP_average intensity × 3.1329;

   Emission ratio = FRET_average intensity/CFP_average intensity; and

   Emission ratio selected = IF (AND (YFP_average intensity > YFP_min, YFP_average intensity < YFP_max, CFP_average intensity > 100), emission _ ratio, "").

   Create a scatter plot using the Emission ratio selected values (Fig. 9). The cells exhibiting an Emission ratio higher than D-FRET (=2.04) are estimated to be drug-resistant cells (*see* **Note 18**).

6. Determine the drug responsiveness.

   The drug responsiveness to each drug is determined based on the presence or absence of a decrease in the median value of

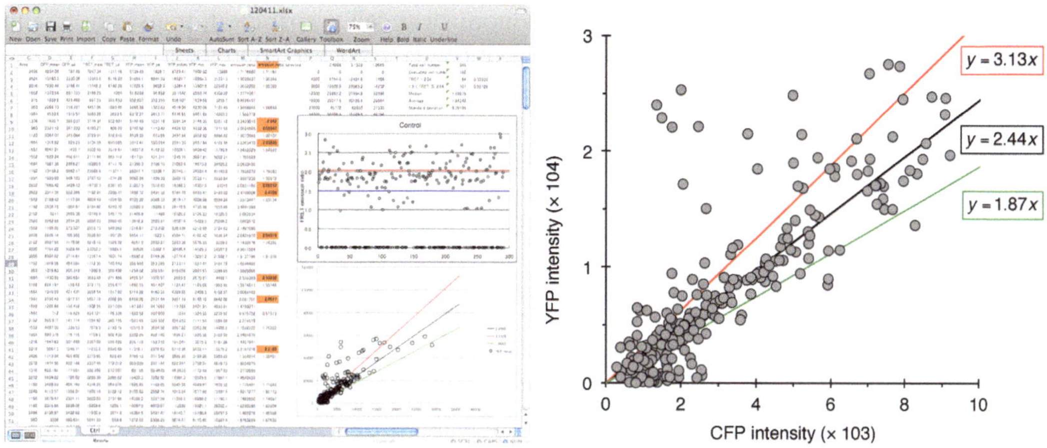

Fig. 8 Example of an application window during the calculation of FRET efficiency using Microsoft Excel. The *right panel* is the enlarged scatter graph within the *left panel* and is an *xy*-plot of the intensities of CFP and YFP in CML cells expressing Pickles. Because Pickles harbors a pair of fluorescent proteins, the intensities of CFP and YFP must display a linear correlation. Therefore, cells displaying intensity ratios of YFP/CFP between 1.87 and 3.13 (i.e., between the *red* and *green lines*) are subjected to analysis. The cells that do not satisfy these criteria (i.e., outside the *red* or *green line*) appear to express incomplete biosensors, which would not be expected to respond appropriately to drug treatment. Therefore, these cells are omitted from analysis

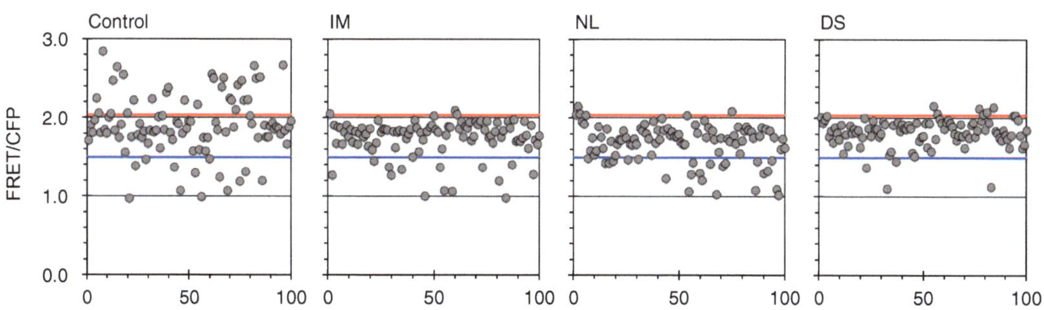

Fig. 9 Plot of the emission ratio (FRET) for all cells analyzed. The *abscissa* indicates the cell number, and the ordinate indicates the FRET efficiency of each cell. When the cells displaying a FRET efficiency greater than 2.04 remain after treatment, the patient is diagnosed as containing drug-resistant cells. Alternatively, the FRET efficiency of the overall cell population provides information about the drug responsiveness of the patient

the FRET efficiency. The presence of drug-resistant cells is ascertained based on the subset of cells displaying a FRET efficiency > 2.04 after drug treatment. In this case, because the FRET values tend to be decreased by all treatments, the patient is expected to respond to all of the drugs to some extent. However, the existence of cells displaying a high FRET efficiency suggests the development drug resistance during his/her therapeutic course.

3.6 Observation via Flow Cytometry

The protocols for mononuclear cell preparation, transfection, and drug treatment are identical to those for fluorescence microscopy. You can prepare these samples together and divide them at **step 3** in Subheading 3.4.

1. Turn on the FLiCM and the control/analyzing computer. Open the FLiCM data station (FDS), the control software for FLiCM, that is installed on this computer.
2. Prepare the flow cytometry system by applying air pressure to the sheath solution tank (*see* **Note 19**).
3. Clean the flow chamber.

   Remove a tube from the sample holder and turn on the 'Cleaning' button in the operation panel. 'Cleaning' mode, which washes the capillary flow channel and removes air bubbles, is completed in 16 s. Execute 'Cleaning' mode three times in a row. Place a test tube filled with FACS Flow in the sample holder.
4. Transfer 1.0×10^5 cells to a 15 mL tube. Centrifuge at $300 \times g$ for 5 min at room temperature, and discard the supernatant.
5. Add 1.5 mL of ice-cold PBS, and resuspend the cells by pipetting.
6. Pass the cells through the 35 μm-nylon filter incorporated into the tube cap of a Falcon™ Tube with Cell Strainer Cap and store on ice until use.
7. After briefly vortexing, transfer the cell suspension to the test tube optimized for FLiCM.
8. Before measurement, select and activate an appropriate format for the graph displayed in FDS. You can select time-waveform, histogram, and scattergram.
9. Place a test tube containing calibration beads diluted in FACS Flow in the sample holder, and push the start button.
10. Select the preview mode in FDS (which allows for measurement without saving). Measure the calibration beads, and verify the proper operation of the FLiCM by confirming that the median values of FSC, SSC, FL1 (fluorescence Ch1), and FL2 (fluorescence Ch2) in the histograms are within the prescribed ranges (FSC, 50–100 ch; SSC, 1000 ch~; FL1, 100 ch~; and FL2, 80 ch~) (*see* **Note 20**).
11. Transfer the cell suspension to a test tube, place the tube in the sample holder, and start the measurement by selecting the preview mode. Tune the voltage of the PMT and the attenuator so that the median values of FSC, SSC, FL1, and FL2 are within suitable ranges (FSC, 30–200 ch; SSC, 50–8000 ch; FL1, 5–2000 ch; and FL2, 10–8000 ch) (*see* **Note 21**).
12. Configure the number of cells to be measured (typically, we prefer to measure 100,000 cells), the folder for data saving,

and the file name in the measurement configuration dialog box in FDS.

13. Select the measure mode in FDS, which enables measurement with saving.
14. When the number of measured cells reaches the set value or the stop button is selected, the measurement is completed, and the acquired data are automatically saved.

3.7 Analysis of the Flow Cytometry Data

1. Open FlowJo software and create a new worksheet.
2. Open the FCS files of all data measured (control, IM, NL, and DS) using "Add Sample".
3. Display a scatter plot of the control sample (by double-clicking on the file name). Change the abscissa and the ordinate to FL1 (donor fluorescence intensity, 482 ± 17.5, CFP) and FL2 (acceptor fluorescence intensity, 542 ± 13.5, YFP), respectively. Create a polygon gate that includes cells expressing sufficient amounts of the biosensor (Fig. 10a, *see* **Note 22**).

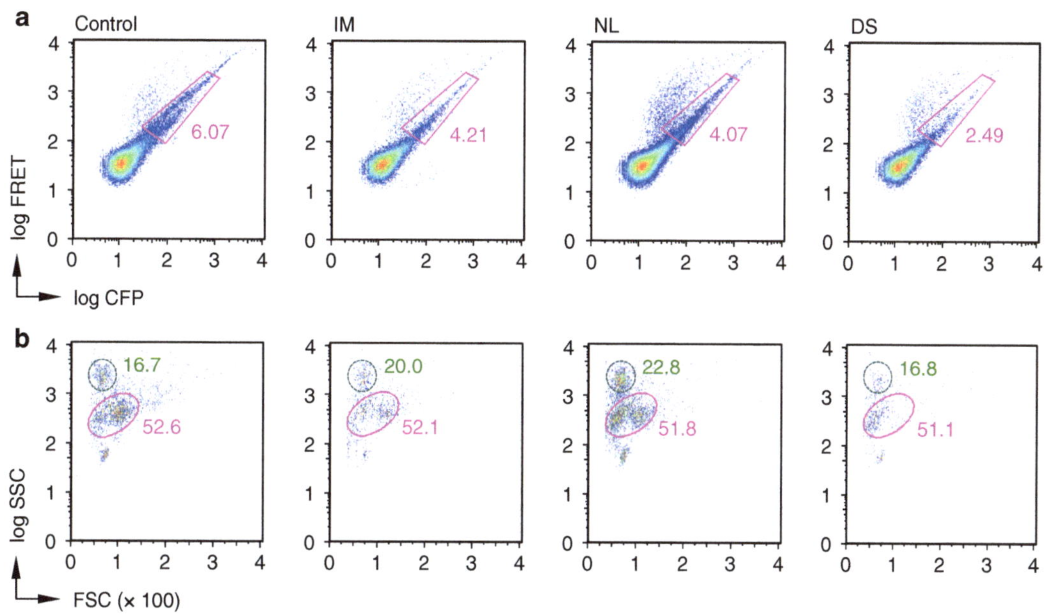

Fig. 10 Selection of cells appropriate for further analysis. (**a**) Selection of cells expressing the appropriate amount of the biosensor. CFP (FL1)-FRET (FL2) scattergrams are shown. All of the examined cells (Control and, IM, NL, or DS treatment) were plotted using an abscissa and an ordinate corresponding to the fluorescence intensity of CFP and FRET, respectively. A gating area was generated to satisfy the following criteria: the fluorescence intensity of both the donor and the acceptor is high; cells in which only either CFP or YFP is expressed are excluded. (**b**) FSC-SSC scattergrams for the cells selected in (**a**). Based on the FSC and SSC values, gating was performed to graphically distinguish between live (*magenta*, *solid circle*) and dead (*green*, *dashed circle*) cells, as shown

4. Select the region created in **step 3**, and open a new gate window displaying a scatter plot consisting of only the cells expressing a sufficient amount of the biosensor.
5. Change the parameters indicated for the abscissa and the ordinate to FSC and SSC, respectively.
6. Create two ellipsoidal gates that contain the live or dead cells (Fig. 10b, *see* **Note 23**).
7. Select either region generated in 6 to open a new window displaying a scatter plot consisting of the live or dead cells exhibiting optimal fluorescence intensities.
8. Select the dead cell subset and right-click on the file name of the workspace. Select "Derived Parameters," and define the emission ratio as FL2/FL1.
9. Change the parameters indicated for the abscissa and the ordinate to Tc1 [fluorescence lifetime of the donor (Tc1)] and Emission ratio. Confirm the location of the dead cells in the FL2/FL1-Tc1 scattergram (Fig. 11a, *see* **Note 24**).
10. Next, select the live cell subset, and create regions that include cells satisfying the following conditions (Fig. 11b) but that

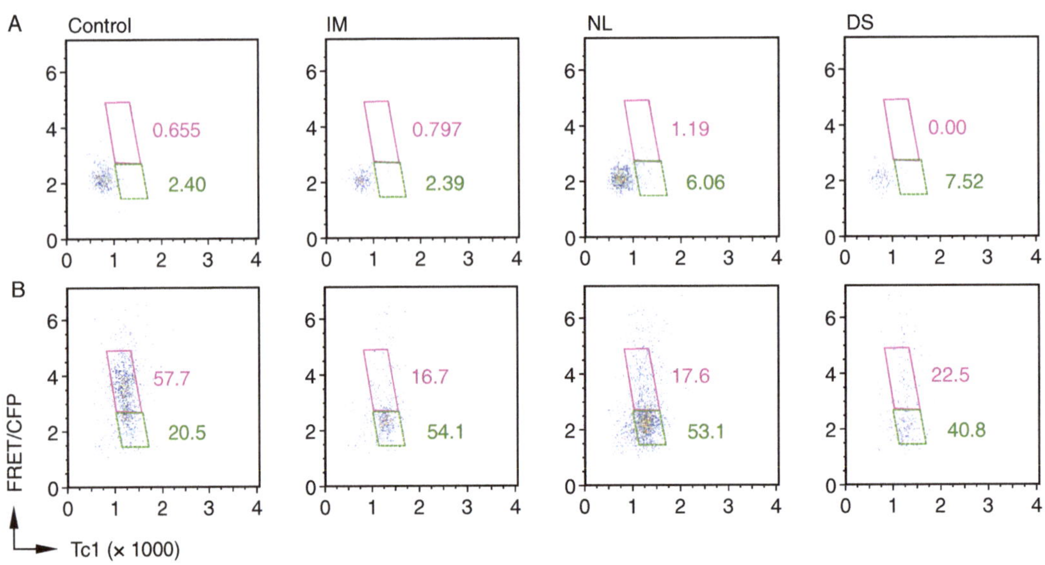

Fig. 11 Tc1-FRET/CFP scattergrams of dead cells (**a**) and live cells (**b**) displaying optimal expression levels of the biosensor. The scattergrams of dead cells are used to determine the gating area for the assessment of drug responsiveness. The regions of interest for the cells displaying FRET^{high} and FRET^{low} are set from the upper left to the lower right of the scatter plot. Because dead cells, which are determined based on their FSC and SSC values, are present in the lower left area, dead cells that cannot be distinguished on the basis of only the FSC-SSC profile might be included in this area, as shown in (**b**). Therefore, the aforementioned criteria are applied to exclude these dead cells

exclude the cells distributed in the lower left region (cells in this region are considered as dead; *see* also Fig. 11a and **Notes 22** and **25**):

(a) 2.8 ≤ Emissi on ratio <5, FRET^{high} population; and

(b) 1.5 < Emission ratio < 2.8, FRET^{low} population.

11. Repeat **steps 3–10** for the files corresponding to the IM, NL, and DS samples.
12. Determination of drug responsiveness.

Any drug that exerts a 20% decrease in the percentage of FRET^{high} cells is considered effective. When there remain cells in the region above Emission ratio = 5, it is possible that the patient will experience relapse and tolerance after treatment with such drug. We are currently accumulating data and evidence to define the diagnostic criteria for drug-resistance. In the present case, all of the drugs evaluated appear to be effective to some extent.

4 Notes

1. Dissolving DNA in TE buffer reduces the transfection efficiency, and contamination of endotoxins decreases the cell viability after electroporation; thus, the use of endotoxin-free water is highly recommended.
2. This tube is required because its inner diameter provides the optimal fit to the diameter of O ring of the sample holder (13 mm) of the FLiCM.
3. Because CML cells are concentrated in the mononuclear cell fraction, but not in the granulocyte fraction, we utilize this protocol in order to isolate leukemia cells.
4. Lymphoprep:diluted BMC should be approximately 3:7.
5. Because the pellet of the leukemia cells tends to agglomerate, we recommend two-step resuspension: e.g., first, add 5 mL of medium (or solution) to the pellet using a 5 mL pipette and mix well by pipetting, and then add the remaining solution and mix again.
6. When the number of cells is less than 10^7, resuspend the cells in 5 mL of RPMI and transfer this suspension to a 60-mm dish. When the cell number exceeds 10^7, use 10 mL of RPMI and a 90-mm dish.
7. Because FBS interferes with electroporation and decreases the transfection efficacy, you should remove the media so that the residual FBS is as little as possible.
8. If the number of cells is less than 1×10^7, proportionally decrease the amount of the plasmid. However, do not change the amount of Solution V.

9. When the number of cells is more than 10^6, resuspend the cells in 8 mL of RPMI and transfer 2 mL to each well of a 12-well plate. When the number of cells is less than 10^6, resuspend cells in 4 mL of RPMI and transfer 1 mL to each well of a 24-well plate.
10. IM and NL should be diluted 1/10 using DDW and then added to the wells. Precipitates will appear if they are directly added to the medium.
11. For reduction of drift induced by temperature changes, we recommend turning on the devices at least 2 h prior to observation.
12. Once you save these configurations, you can skip this step.
13. Although the optimal exposure times may vary depending on the microscope settings, you should keep them proportional to those which have been predetermined so that the data across experiments can be compared. Particular attention should be paid avoiding saturation of the fluorescence intensity. The exposure times of ECFP and FRET should be identical.
14. Leave one corner of the images blank to make it easier to subtract the background.
15. Exclude any dead cells (which can be estimated using the DIC images) and any cells in which the biosensors are eccentrically localized in the image.
16. We made Journals (a sort of macro in MetaMorph software) to semi-automatically perform image data analysis.
17. These coefficients may vary depending on the microscope settings and were determined using 293F cells in our case. These calculations are required to eliminate the cells in which the Pickles biosensors are cleaved. Cleaved biosensors are not expected to display an appropriate response to drug treatment.
18. The D-FRET value is defined as the mean FRET efficiency + 3 × S.D. in 293F cells, which are expected to lack BCR-ABL [4].
19. The reagents required to maintain FLiCM are identical to those used for standard flow cytometers: e.g., Sheath fluid, BD FACS Flow Sheath Fluid (BD Biosciences); Clean liquid, BD FACS Rinse Solution (BD Biosciences); Disinfecting clean liquid, BD FACS Clean Solution (BD Biosciences).
20. When the measurement begins, the sample is extruded using the pressure of the sheath fluid, which generates a laminar flow of cells at the center of the flow cell. The laser beam is focused using two lenses to irradiate the cells passing through the flow cell. The irradiated cells scatter the laser beam and simultaneously emanate fluorescence from the fluorophore expressed in the cells. Forward-scattered light (FSC) is detected using a silicon photodiode (SiPD). The intensity of FSC can be used not

only as an indicator of cell size but also as a trigger generator for the fluorescence lifetime and intensity measurement by regarding the input of the FSC signal as the presence of the cells at the laser spot. Given that the size of the cells varies in a context-dependent manner, the gain and the threshold level for triggering should be appropriately adjusted to obtain optimal values for each cell type. Alternatively, side-scattered light (SSC) and fluorescence are collected using condenser lenses, which are subsequently separated into the desired wavelength regions using dichromatic and band-pass filters and are individually detected using photomultiplier tubes (PMTs).

21. The photons detected by the PMTs are internally transformed into electrical signals. The amplitude of the output signal is proportional to the fluorescence intensity ($k_f N_0 \tau P_{E0}$) but varies by more than one order of magnitude depending on the fluorescent dye used. Therefore, the control voltage of the attenuator should be adjusted to obtain the appropriate output value. The output signal of the PMT undergoes amplification and demodulation in the signal processing circuit, and the resulting data, including scatted light (2 ch), fluorescence intensity (3 ch), and fluorescence lifetime (3 ch), for every event are displayed and stored in the analyzing PC.
22. Ensure that the created gate does NOT include cells expressing only CFP or YFP and cells displaying fluorescence intensity of either the donor or the acceptor is saturated.
23. In general, FSC represents the cell size, whereas SSC represents the complexity of the intracellular architecture (e.g., nuclear and organellar morphology and membrane structure). Therefore, FSC intensity of dead cells is lower than that of living cells because of their reduced size. By contrast, SSC intensity of dead cells becomes greater than that of living cells owing to the membrane collapse of dead cells. According to this information, gates can be created to distinguish live cells from dead cells in most established cell lines, which may be performed at the first step of cytometric data analysis. However, in the case that primary cells are utilized, FSC-SSC scattergrams do not always display a typical pattern, and living cells may not be well demarcated in the scattergram. Dead cells that cannot be distinguished in the scattergram can be discriminated using the Tc1-ratio scattergram, as described below, demonstrating the usefulness of this method.
24. In the presence of FRET, the emission ratio and the fluorescence lifetime are increased and decreased, respectively, and vice versa. Therefore, the populations of dead cells in the lower left region that cannot be distinguished based on their FSC-SSC profile but should be excluded from further analysis

remain in Fig. 11b. Notably, these cells cannot be distinguished based on only their emission ratio or fluorescence lifetime, despite their substantial impact on the final results, demonstrating the benefit of the use of the FLiCM to measure both parameters.

25. Because the FRET efficiency represents BCR-ABL activity, a decrease in the percentage of FRET^{high} cells after drug treatment is implicated in the effectiveness of the drug to suppress BCR-ABL activity in the entire cell population. Alternatively, the presence of cells that remain in the region > FRET^{high} indicate that these cells might display resistance to the relevant drug.

Acknowledgments

This work was supported in part by Grants-in-Aid from the Japan Society for the Promotion of Science (JSPS; #23659196 and #25670184) and the Adaptable and Seamless Technology Transfer Program through Target-driven R&D from Japan Science and Technology Agency (JST, AS2321227F), as well as grants from the Yasuda Medical Foundation, the Japan Leukemia Research Fund, and the Kato Memorial Bioscience Foundation.

References

1. Morise H, Shimomura O, Johnson F et al (1974) Intermolecular energy transfer in the bioluminescent system of Aequorea. Biochemistry 13:2656–2662
2. Prasher D, Eckenrode V, Ward W et al (1992) Primary structure of the Aequorea victoria green-fluorescent protein. Gene 111: 229–233
3. Heim R, Prasher DC, Tsien RY (1994) Wavelength mutations and posttranslational autoxidation of green fluorescent protein. Proc Natl Acad Sci U S A 91:12501–12504
4. Mizutani T, Kondo T, Darmanin S et al (2010) A novel FRET-based biosensor for the measurement of BCR-ABL activity and its response to drugs in living cells. Clin Cancer Res 16:3964–3975
5. Kurzrock R, Estrov Z, Kantarjian H et al (1998) Conversion of interferon-induced, long-term cytogenetic remissions in chronic myelogenous leukemia to polymerase chain reaction negativity. J Clin Oncol 16:1526–1531
6. Groffen J, Stephenson JR, Heisterkamp N et al (1984) Philadelphia chromosomal breakpoints are clustered within a limited region, bcr, on chromosome 22. Cell 36:93–99
7. Heisterkamp N, Groffen J, Stephenson JR et al (1982) Chromosomal localization of human cellular homologues of two viral oncogenes. Nature 299:747–749

Chapter 31

SH2 Domain Histochemistry

Sophia Buhs and Peter Nollau

Abstract

Among posttranslational modifications, the phosphorylation of tyrosine residues is a key modification in cell signaling. Because of its biological importance, characterization of the cellular state of tyrosine phosphorylation is of great interest. Based on the unique properties of endogenously expressed SH2 domains recognizing tyrosine phosphorylated signaling proteins with high specificity we have developed an alternative approach, coined SH2 profiling, enabling us to decipher complex patterns of tyrosine phosphorylation in various normal and cancerous tissues. So far, SH2 profiling has largely been applied for the analysis of protein extracts with the limitation that information on spatial distribution and intensity of tyrosine phosphorylation within a tissue is lost. Here, we describe a novel SH2 domain based strategy for differential characterization of the state of tyrosine phosphorylation in formaldehyde-fixed and paraffin-embedded tissues. This approach demonstrates that SH2 domains may serve as very valuable tools for the analysis of the differential state of tyrosine phosphorylation in primary tissues fixed and processed under conditions frequently applied by routine pathology laboratories.

Key words Signal transduction, Tyrosine phosphorylation, Phosphotyrosine, Src-homology 2, SH2 domain, Cancer, Signal amplification, Histochemistry

1 Introduction

Tyrosine phosphorylation plays a fundamental role in signal transduction controlling many biologically important processes such as cell proliferation, cell death and differentiation [1, 2]. Antibodies are a frequently used tool for the analysis of single, defined phosphotyrosine sites or the overall cellular state of tyrosine phosphorylation [1]. Although the number of available antibodies targeting specific phosphosites is constantly growing, it presently occurs impracticable to comprehensively cover the enormous spectrum of tyrosine phosphorylation sites solely by antibody based approaches. To address this issue, we introduced an alternative strategy based on Src Homology 2 (SH2) domains for surveying the global tyrosine phosphorylation state of signaling proteins [1, 3]. SH2 domains are evolutionary highly conserved, short, modular binding modules of approximately 100 amino acids present in

Kazuya Machida and Bernard A. Liu (eds.), *SH2 Domains: Methods and Protocols*, Methods in Molecular Biology, vol. 1555, DOI 10.1007/978-1-4939-6762-9_31, © Springer Science+Business Media LLC 2017

many signaling proteins. The human genome encodes for 120 different SH2 domains. Dependent on the amino acids flanking the phosphotyrosine residue, SH2 domains selectively recognize phosphotyrosine sites in signaling proteins thus making them ideal analytical tools for scanning of cellular signaling activity mirrored by differential tyrosine phosphorylation [4, 5]. So far, SH2 domains have largely been applied in Far Western blot analysis or reverse phase assays to qualitatively and quantitatively determine the cellular state of tyrosine phosphorylation in protein extracts [6–8]. To gain deeper insights into tyrosine phosphorylation at the cellular level we developed a novel strategy utilizing SH2 domains to detect differential tyrosine phosphorylation in formaldehyde-fixed and paraffin-embedded tissues enabling investigators to study the intensity, distribution and subcellular localization of tyrosine phosphorylation in tissue sections. The detection of tyrosine phosphorylation, especially in primary tissue samples, is challenging as phosphotyrosine levels are in general of relatively low abundance [1–3]. To obtain highly sensitive probes for SH2 histochemistry, we expressed SH2 domains as GST-fusion proteins in bacteria carrying at the C-terminus a monobiotinylation site for in vivo biotinylation and multimerization of probes and a myc-tag for detection (Fig. 1). After bacterial expression, biotinylated SH2 domains were complexed with streptavidin to increase the avidity of the probe. Subsequently, deparaffinized tissue sections were incubated with complexed SH2 domains and bound probes were detected via the myc-tag applying an anti-myc antibody in combination with a commercially available, highly sensitive, tyramide-based amplification system (Fig. 1). In our hands, direct labeling and detection strategies for the analysis of bound SH2 domains on tissue sections using for example anti-GST antibodies or peroxidase-streptavidin conjugates resulted in weak signals or high background staining indicating that SH2 domain staining in combination with a highly sensitive signal amplification system is a prerequisite for the analysis of the cellular state of tyrosine phosphorylation in formaldehyde-fixed and paraffin-embedded tissue samples of primary origin. As an example of our novel staining strategy, representative results of SH2 domain histochemistry applied for the analysis of the phosphotyrosine state of different breast cancer specimens are given in Fig. 2.

2 Materials

Prepare all buffers and solutions with ultrapure water. All reagents need to be of analytical grade. For the handling of hazardous substances and waste disposal follow the regulations of your institution and the recommendations of the manufacturer.

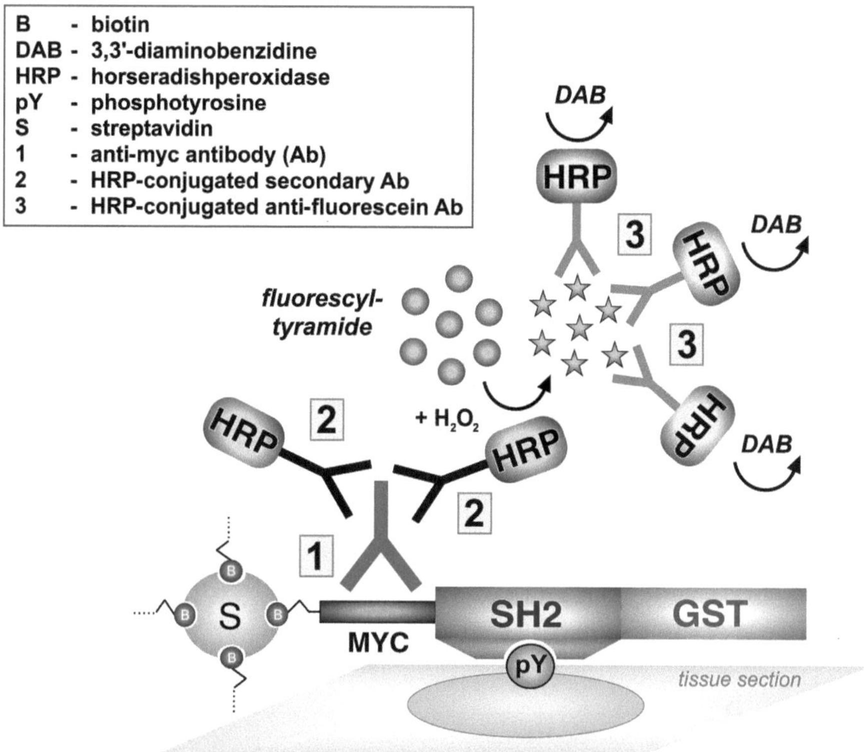

Fig. 1 Schematic illustration of SH2 domain histochemistry and tyramide-based signal amplification. Biotinylated SH2 domains carrying a GST-tag at the N-terminus for purification and a myc-tag and a biotinylation site at the C-terminus are purified from bacteria and complexed with streptavidin in a 4:1 ratio to increase the binding avidity of the SH2 domain. Subsequently, the SH2 domain/streptavidin complex is incubated with a deparaffinized tissue section. For detection, tissue sections are incubated with an anti-myc antibody followed by incubation with a HRP-conjugated secondary antibody. Addition of the amplification reagent results in oxidation of fluorescyl-tyramide and formation of insoluble precipitates on the tissue sample. Precipitates are then detected by a HRP-labeled anti-fluorescein antibody and bound probes are visualized using DAB as a chromogen resulting in brown precipitates at phosphotyrosine sites bound by the SH2 domain

2.1 Staining Components

1. Staining Kit: CSAII Biotin-free Tyramide Signal Amplification System including peroxidase block, protein block (serum free), HRP-labelled anti-mouse immunoglobulins, amplification reagent, anti-fluorescein-HRP, DAB substrate buffer and DAB chromogen. Additionally required: Dako Pen, antibody diluent with background reducing components, hematoxylin and mounting medium (all components from Dako, Glostrup, Denmark).
2. Antibody: c-myc, clone 9E10 (Santa Cruz Biotechnology, CA, USA).
3. SH2 domains: generate SH2-domains as GST fusion proteins carrying a myc-tag followed by a monobiotinylation site at the C-terminus. Express and purify biotinylated SH2 domains from AVB100 E. coli bacteria according to the protocol of Buhs et al. (*see* Chapter 23).

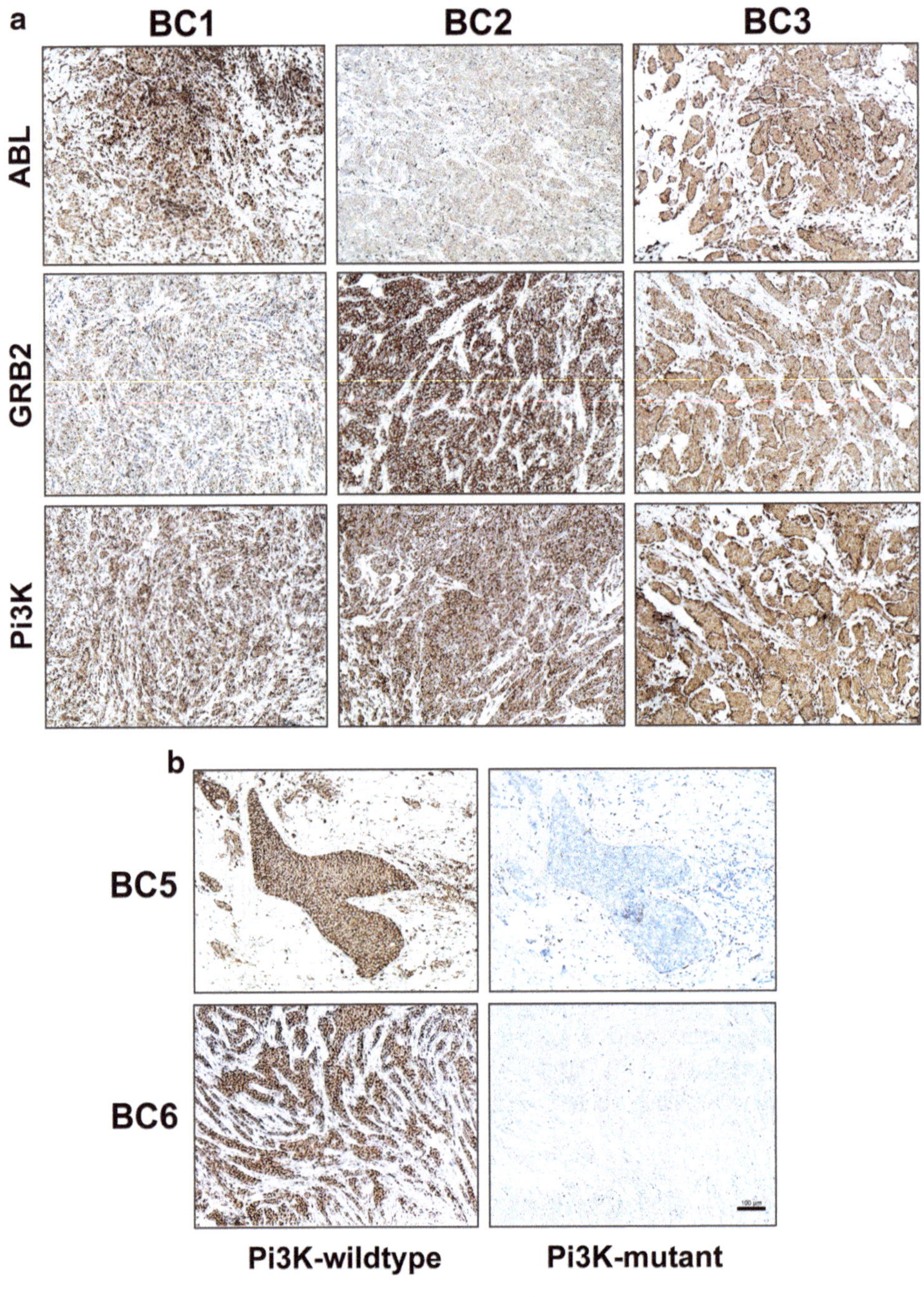

Fig. 2 SH2 domain histochemistry of formalin-fixed and paraffin-embedded breast cancer samples. (**a**) Tissue sections of three breast cancer patients (BC 1–3) were probed with the SH2 domains of ABL2 (Abelson murine leukemia viral oncogene homolog 2), GRB2 (growth factor receptor-bound protein 2) and phosphatidylinositol 3-kinase (Pi3K), respectively. Detection of SH2 domains was performed by the CSAII amplification kit as described. Note the differential staining of cancerous tissue reflecting differences in tyrosine phosphorylation recognized by the different SH2 domains. (**b**) To demonstrate specificity of SH2 domain histochemistry, consecutive sections of two breast cancer samples (BC 4 and 5) were probed with the wildtype and a mutant SH2 domain of Pi3K, respectively. The mutant SH2 domain was generated by substitution of arginine to lysine in the FLVR-motif of the N-terminal SH2 domain of PIK3RI abolishing the binding to phosphotyrosine. Note the lack of SH2 domain binding of the mutant in comparison to the wildtype SH2 domain. Tissue sections were counterstained with hematoxylin; scale bar = 100 μm

4. For complex formation of SH2 domains: Streptavidin (Thermo Fisher, NY, USA).
5. Glass or plastic cuvettes, staining box TX87.1, staining rack TX88.1, stain tray HA51.1 (all from Carl Roth, Karlsruhe, Germany).
6. Solvents and solutions: xylene, absolute ethanol (purity grade ≥ 99.8 %) and ammonia; Triton X-100.
7. Upright light microscope Leica DM IL LED; imaging software LAS V3.8 (both from Leica Microsystems, Wetzlar, Germany).

2.2 Buffers

1. Citrate buffer (10 mM): weigh in 21.01 g citric acid monohydrate, transfer to a graduated 1 L cylinder, add ultrapure water to a final volume of 1 L and stir with a magnetic stir bar until citric acid is dissolved (*see* **Note 1**). Transfer 9 ml of this solution in a 0.5 L graduated cylinder. Prepare 0.1 M sodium citrate solution by dissolving 29.41 g sodium citrate dihydrate in 1 L ultrapure water (*see* **Note 2**). Add 41 ml of this solution to the 9 ml of 0.1 M citric acid and add ultrapure water up to a volume of 500 ml.
2. Tris–HCl (1 M, pH 8.0): weigh in 121.14 g Tris base and transfer into a graduated 1 L cylinder. Add 900 ml ultrapure water and stir with a magnetic stir bar until dissolved. Adjust with HCl (*see* **Note 3**) to pH 8.0 and fill up with ultrapure water to a final volume of 1 L.
3. TBS-T: weigh in 72 g NaCl and transfer into a 1 L beaker glass. Add 80 ml 1 M Tris–HCl (pH 8.0), adjust to 500 ml with ultrapure water and stir until dissolved. Add 4 ml of Tween 20 (*see* **Note 4**). Subsequently, add the solution to a 9 L tank prefilled with 7.5 L ultrapure water and invert carefully several times (final volume 8 L; final concentration of Tween 20: 0.05 %).
4. SH2 blocking buffer: weigh in 10 g skim milk and dissolve in 90 ml TBS-T in a 100 ml cylinder (see **Note 5**). Add 2 ml of 50 mM sodium orthovanadate, 1 ml of 0.5 M EDTA (pH 8.0) and adjust with TBS-T to a final volume of 100 ml.
5. Glassware: cylinder (0.1, 0.5, 1 L), beaker glass (1 L); all from DURAN (Mainz, Germany); square tank (K648.1, Carl Roth, Karlsruhe, Germany).

3 Methods

If not otherwise stated, perform all steps at room temperature. Prevent drying of tissue sections at any time which may rapidly occur and will result in loss of signal or unacceptable high background staining. Always wear protective clothing, safety gloves, and safety glasses.

3.1 Preparing Tissue Sections

1. Cut tissue section of 5 μm thickness from paraffin blocks using a routine microtome and mount sections on glass slides.
2. For deparaffinization of tissue sections, fill a glass cuvette with xylene under a fume hood, place tissue sections in the cuvette, and incubate for 7.5 min. Tissue sections should be fully covered by the organic solvent. Dispose of xylene, refill cuvette with fresh xylene, and incubate for another 7.5 min (*see* **Note 6**). Repeat exchange of solvent additional two times so that tissue sections are incubated in xylene for a total time of 30 min.
3. Use a clean cuvette and fill in absolute ethanol. Transfer tissue sections and incubate for 5 min. Discard ethanol and repeat **step 3** in the same cuvette.
4. Prepare the following solutions with ultrapure water: 96% ethanol (v/v), 80% ethanol (v/v), 70% ethanol (v/v), 50% ethanol (v/v) and 30% ethanol (v/v) (*see* **Note 7**). Fill solutions into separate staining boxes. Prepare another staining box with ultrapure water. Place tissue sections in a staining rack and incubate them in the descending alcohol series for 5 min each followed by a final incubation in ultrapure water for 5 min (*see* **Note 8**).
5. For unmasking of binding sites, discard ultrapure water from the staining box, add 10 mM citrate buffer and place the staining box in a microwave. Heat for 7 min at 700 W and refill evaporated liquid with citrate buffer if required. Continue heating four times for 2 min each and refill citrate buffer between each heating step. To prevent drying, tissue sections must be fully covered with citrate buffer at any time (*see* **Note 9**). Handle the hot liquid with extreme caution, anticipate boiling retardation and always wear protective clothing, safety gloves, and safety glasses.
6. Carefully remove the staining box equipped with the staining rack and the glass slides from the microwave and let the solution cool down to room temperature (*see* **Note 10**).
7. Take out each slide of the staining rack, drip down excessive liquid and encircle the tissue area with the Dako Pen (*see* **Note 11**). Place the slides in a cuvette prefilled with PBS.
8. Wash tissue sections three times with PBS for 5 min each (*see* **Note 12**).
9. Fill PBS/0.2% Triton-X-100 (v/v) in a clean cuvette and incubate tissue sections for 2 min.
10. Wash tissue sections three times with PBS for 5 min each.
11. Place the slides in a cuvette filled with SH2 blocking buffer and incubate at 4 °C for 1 h.

12. During the blocking step, combine 2 μg of SH2 domain and 0.3 μg of streptavidin in a 1.5 ml Eppendorf tube and adjust to a final volume of 20 μl with PBS. Incubate for 1 h and add TBS-T to a final volume of 500 μl (final concentration of SH2 domain 0.1 μM). The amount of complex is sufficient for staining of a single tissue section. When processing several tissue sections in parallel, appropriately scale up amounts and volumes of the components.
13. To create a humidified atmosphere, fill a stain tray with approximately 15 ml of ultrapure water. Remove the tissue sections from the SH2 blocking buffer, drip off excessive liquid and place the glass slides on the stain tray. Entirely cover the encircled area with approximately 500 μl of the SH2 domain/streptavidin complex prepared in step 12 and incubate in the wet chamber overnight at 4 °C.

3.2 Staining of Tissue Sections

For staining of bound SH2 domains the CSAII Biotin-free Tyramide Signal Amplification System is used. The protocol has been modified from the recommendations of the manufacturer. The amounts of reagents given below are sufficient for processing a single tissue section. When a series of tissue sections are stained in parallel, larger volumes of the different reagents are required.

1. After overnight incubation with the SH2 domain/streptavidin complex, rinse tissue section carefully with PBS, tap off excessive liquid and place the glass slide back on the stain tray.
2. Pipette 500 μl ready to use peroxidase blocking solution on the encircled area (*see* **Note 13**) and incubate for 5 min 30 s. Rinse tissue section carefully with TBS-T (*see* **Note 14**) and place the glass slide in a cuvette prefilled with TBS-T.
3. Remove the glass slide from TBS-T and carefully tap off excessive liquid. Again, place the tissue section on the stain tray and pipette 500 μl ready to use protein block serum free solution (component of CSA II Kit) on the encircled area of tissue section and incubate for 5 min 30 s.
4. Drain off liquid and add 500 μl of the c-myc antibody (diluted 1:500 in TBS-T) on the encircled area. Incubate for 30 min. Rinse tissue section carefully with TBS-T. Place tissue section in a cuvette prefilled with fresh TBS-T and incubate for 4 min. Exchange TBS-T and repeat washing additionally two times. Place the tissue section back on the stain tray.
5. Dilute ready to use anti-mouse-immunoglobulins-HRP 1:15 with antibody diluent and add 500 μl on the encircled area of the tissue section. Incubate for 15 min. Wash tissue section as recommended in **step 4** of Subheading 3.2.

6. Add 500 µl of ready to use amplification reagent (component of CSA II Kit) on the encircled area of the tissue section and incubate for 14 min in the dark. Wash tissue section as described in **step 4** of Subheading 3.2.
7. Pipette 500 µl of anti-fluorescein-HRP on the encircled area of tissue section and incubate for 14 min in the dark. Wash tissue section as recommended in **step 4** of Subheading 3.2.
8. Prepare fresh DAB Chromogen by adding one drop of ready to use DAB Chromogen to 1 ml of ready to use DAB substrate buffer. Add 500 µl of substrate solution on the encircled area and incubate for 5–15 min (*see* **Note 15**). Stop staining by rinsing tissue section carefully with ultrapure water. Place tissue section in a cuvette with ultrapure water.
9. Place tissue section in a cuvette filled with hematoxylin ready to use solution for 10 s. Rinse tissue section with ultrapure water to remove staining solution and place tissue section in a cuvette with fresh, ultrapure water. Subsequently, transfer tissue section in a cuvette prefilled with 0.08 % ammonia in water (v/v; 25 % stock solution) for 1 min, remove tissue section, and drip off excessive liquid (*see* **Note 16**).
10. Apply one drop of pre-warmed (40 °C water bath) ready to use mounting medium to the encircled area of the tissue section and place a coverslip on top (*see* **Note 17**). Carefully press down the coverslip and wipe off excessive mounting medium with a paper towel (*see* **Note 18**).

3.3 Data Acquisition and Analysis

Use a standard, upright light microscope for visual inspection of tissue sections and document representative tissue areas with a CCD camera. When tissue microarrays with multiple tissue spots or a large number of tissue sections need to be analyzed, take overview pictures of individual tissue sections or spots, compile individual pictures to a single image using for example Adobe Photoshop and save file in TIFF format. Open the TIFF file in ImageJ (for download of software: imagej.nih.gov/ij/), apply the colour deconvolution plugin (*imagejdocu.tudor.lu*), choose the vector H DAB from the pull down menu and deconvolute the original image into images representing hematoxylin and brownish DAB staining. To get a rapid estimate on overall staining intensity, transform the DAB image to black and white (8-bit image), invert image and measure mean intensity of individual images applying the measurement tool of ImageJ. Intensity values can then be used to group staining intensities of tissue sections in different categories such as negative, weak, moderate or strong staining. As staining by DAB is non-linear, do not apply image analysis for absolute quantification. Image analysis of DAB stained tissues described above should only be applied to provide a framework for categorization of staining intensities of different tissues. Additionally, take into consideration that an average level of signal intensity is determined and important information on differential distribution of staining for example in normal and cancerous tissue is possibly missed. For unambiguous analysis and interpretation of data, tissue sections need to be examined and scored by experts in the field, preferably evaluated independently by at least two experts to minimize the analytical bias.

4 Notes

1. To improve dissolving, fill the cylinder first with approximately 700 ml ultrapure water, add citric acid and stir bar and wait until citric acid is dissolved. Adjust volume to 1 L.
2. Dissolve sodium citrate as recommended in **Note 1**. Stock solutions and citrate buffer can be stored at 4 °C in a 1 L bottle.
3. To adjust the pH use concentrated hydrochloric acid (e.g., 37% hydrochloric acid). Always great caution should be taken when handling acids or bases. Perform pipetting under a fume hood. Wear gloves, protective clothing and safety glasses at any time.
4. As Tween 20 is a viscous liquid, use a wide bored pipette to remove Tween 20 from the stock. Once the appropriate volume of Tween 20 has been drawn into the pipette, place the tip of the pipette in the TBS-buffer, carefully pipet up and down several times while the buffer is stirring to ensure that all of the detergent is transferred to the buffer.
5. Adding skim milk stepwise in small portions and not in bulk to TBS-T improves dissolving.
6. Discard xylene by tilting the cuvette and pouring the solvent in an appropriate waste container. To prevent slipping of slides, fix slides in the cuvette with your fingers. Caution: xylene is a flammable liquid, and vapors may form explosive mixtures with air; xylene may cause acute or chronic health hazards and is hazardous to the aquatic environment.
7. Dilutions of ethanol can be stored in bottles at room temperature under the fume hood and reused for up to three times. Caution: ethanol is highly volatile, highly flammable and vapors may form explosive mixtures with air.
8. To perform deparaffinization in the accurate order, arrange staining boxes close to each other and in the order of the descending alcohol series.
9. Always refill the staining box to a maximum by citrate buffer as substantial amounts of liquid may evaporate during each step of heating. Because of hot steam, special caution should

be taken when opening the door of the microwave. Use microwave safe staining boxes; other glass containers may crack during heating.

10. Cooling down to room temperature may take 30 min or longer. To prevent cracking of the glass container, do not rapidly cool down the staining box with ice.
11. Tissue areas are encircled by the Dako Pen to minimize the amount of probe and staining solutions and to prevent draining off of liquids during incubation. Tissue sections may hardly be visible after deparaffinization. Thus, special care must be taken to inadvertently cover part of the tissue section by the Dako Pen. Always use an appropriate distance to the tissue section when encircling the area of interest.
12. Discard PBS by tilting the cuvette. Fix the slides in the cuvette with your fingers to prevent slipping of the slides.
13. The amount of peroxidase blocking solution required depends on the encircled area. Fully cover the encircled area with blocking solution. Do not add blocking solution in excess as due to low surface tension most of the liquid may drain off.
14. Tissue sections may be flushed by using the TBS-T buffer in a washing flask. Take care as tissue sections may detach when inappropriate, high shear forces are applied during washing.
15. Avoid excessive staining. Time of incubation is variable, dependent on the type of tissue and abundance of tyrosine phosphorylation. Optimal time of incubation needs to be experimentally determined beforehand. When a series of tissue sections is stained in parallel or on different days, keep time of incubation constant to assure comparability of staining intensities.
16. Remove excessive liquid by carefully tapping the edges of the glass slide on a paper towel.
17. To prevent air bubbles, hold bottle downwards, discard the first drop of mounting medium and place the second drop directly to the center of the encircled area of the tissue section.
18. Before hardening of the mounting medium, air bubbles can carefully be squeezed out. Remove excessive mounting medium with a paper towel from the edges of the glass slide. Do not wipe mounting medium on the coverslip. Small traces of mounting medium on top of the coverslip can be removed with cotton swaps soaked in absolute ethanol. In case of inaccurate mounting (uneven mounting, too many air bubbles trapped), remove the coverslip by slight agitation of the glass slide in warm water (40 °C). The coverslip will float off. Do not use forceps for removal or manually slide the coverslip as the tissue section may be damaged or detach. After detachment of the coverslip, repeat the mounting procedure.

References

1. Machida K, Mayer BJ, Nollau P (2003) Profiling the global tyrosine phosphorylation state. Mol Cell Proteomics 2(4):215–233. doi:10.1074/mcp.R300002-MCP200
2. Hunter T (2009) Tyrosine phosphorylation: thirty years and counting. Curr Opin Cell Biol 21(2):140–146. doi:10.1016/j.ceb.2009.01.028
3. Machida K, Khenkhar M, Nollau P (2012) Deciphering phosphotyrosine-dependent signaling networks in cancer by SH2 profiling. Genes Cancer 3(5–6):353–361. doi:10.1177/1947601912459048
4. Liu BA, Jablonowski K, Raina M, Arce M, Pawson T, Nash PD (2006) The human and mouse complement of SH2 domain proteins-establishing the boundaries of phosphotyrosine signaling. Mol Cell 22(6):851–868. doi:10.1016/j.molcel.2006.06.001
5. Tinti M, Kiemer L, Costa S, Miller ML, Sacco F, Olsen JV, Carducci M, Paoluzi S, Langone F, Workman CT, Blom N, Machida K, Thompson CM, Schutkowski M, Brunak S, Mann M, Mayer BJ, Castagnoli L, Cesareni G (2013) The SH2 domain interaction landscape. Cell Rep 3(4):1293–1305. doi:10.1016/j.celrep.2013.03.001
6. Dierck K, Machida K, Voigt A, Thimm J, Horstmann M, Fiedler W, Mayer BJ, Nollau P (2006) Quantitative multiplexed profiling of cellular signaling networks using phosphotyrosine-specific DNA-tagged SH2 domains. Nat Methods 3(9):737–744. doi:10.1038/nmeth917
7. Machida K, Thompson CM, Dierck K, Jablonowski K, Karkkainen S, Liu B, Zhang H, Nash PD, Newman DK, Nollau P, Pawson T, Renkema GH, Saksela K, Schiller MR, Shin DG, Mayer BJ (2007) High-throughput phosphotyrosine profiling using SH2 domains. Mol Cell 26(6):899–915. doi:10.1016/j.molcel.2007.05.031
8. Jadwin JA, Mayer BJ, Machida K (2015) Detection and quantification of protein-protein interactions by far-western blotting. Methods Mol Biol 1312:379–398. doi:10.1007/978-1-4939-2694-7_38

INDEX

Kazuya Machida and Bernard A. Liu (eds.), *SH2 Domains: Methods and Protocols*, Methods in Molecular Biology, vol. 1555, DOI 10.1007/978-1-4939-6762-9,

F

G

H

I

K

L

M

N

O

P

Q

R

T

U

V

MIX
Papier aus verantwortungsvollen Quellen
Paper from responsible sources
FSC® C105338

If you have any concerns about our products,
you can contact us on
ProductSafety@springernature.com

In case Publisher is established outside the EU,
the EU authorized representative is:
Springer Nature Customer Service Center GmbH
Europaplatz 3, 69115 Heidelberg, Germany

Printed by Libri Plureos GmbH
in Hamburg, Germany